SAS® for Forecasting Time Series

second edition

John C. Brocklebank, Ph.D.
David A. Dickey, Ph.D.

The correct bibliographic citation for this manual is as follows: Brocklebank, John C., and David A. Dickey. 2003. *SAS® for Forecasting Time Series, Second Edition*. Cary, NC: SAS Institute Inc.

SAS® for Forecasting Time Series, Second Edition

Jointly co-published by SAS Institute and Wiley 2003.

SAS Institute Inc. ISBN 1-59047-182-2
John Wiley & Sons, Inc. ISBN 0-471-39566-8

SAS Institute Inc., SAS Campus Drive, Cary, North Carolina 27513.

1st printing, May 2003

"Drs. Brocklebank and Dickey have not only done a great job of explaining how to use SAS in forecasting time series, but have also written a good practitioner's text illustrating perils and pitfalls and how to detect them. The authors start at ground zero with illustrated explanations and build to more difficult concepts in a logical progression. For the SAS enthusiast, there is a wealth of SAS code, followed by the SAS output from that code and an abundance of graphs to illustrate what is being seen. If you need a review of time series forecasting or an understanding of how SAS treats time series forecasting, this would be a good book to have on your shelf."

Dr. Alex K. Thompson
Senior Statistician

Contents

Preface

A time series is a set of ordered observations on a quantitative characteristic of a phenomenon at equally spaced time points. The goal of univariate time series analysis is to forecast values of a single historical series. The goal of multivariate time series analysis can be to model the relationships among component series as well as to forecast those components.

Time series analysis can be accomplished most effectively by the SAS procedures ARIMA, STATESPACE, SPECTRA, and VARMAX. To use these procedures properly, you must (1) understand the statistics you need for the analysis and (2) know how to invoke the procedures. *SAS for Forecasting Time Series, Second Edition,* makes it easier for you to apply these procedures to your data analysis problems.

Chapter 1, "Overview of Time Series," reviews the goals and key characteristics of time series. The analysis methods available through SAS/ETS software are presented, beginning with the simpler procedures FORECAST, AUTOREG, and X11 and continuing with the more powerful SPECTRA, ARIMA, and STATESPACE. This chapter shows the interrelationships among the various procedures. It ends with a discussion of linear regression, seasonality in regression, and regression with transformed data.

Chapter 2, "Simple Models: Autoregression," presents the statistical background necessary to model and forecast simple autoregressive (AR) processes. A three-part forecasting strategy is used with PROC ARIMA to identify, estimate, and forecast. The backshift notation is used to write a time series as a weighted sum of past shocks and to compute covariances through the Yule-Walker equations. The chapter ends with an example involving an AR process with regression techniques by overfitting.

Chapter 3, "The General ARIMA Model," extends the class of models to include moving averages and mixed ARMA models. Each model is introduced with its autocovariance function. Estimated autocovariances are used to determine a model to be fit, after which PROC ARIMA is used to fit the model, forecast future values, and provide forecast intervals. A section on time series identification defines the autocorrelation function, partial autocorrelation function, and inverse autocorrelation function. Newer identification techniques are also discussed. A catalog of examples is developed, and properties useful for associating different forms of these functions with the corresponding time series are described. This chapter includes the results of 150 observations generated from each of eight sample series. Stationarity and invertibility, nonstationarity, and differencing are discussed.

Chapter 4, "The ARIMA Model: Introductory Applications," describes the ARIMA model and its introductory applications. Seasonal modeling and model identification are explained, with Box and Jenkins's popular airline data modeled. The chapter combines regression with time series errors to provide a richer class of forecasting models. Three cases are highlighted: Case 1 is a typical regression, case 2 is a simple transfer function, and case 3 is a general transfer function.

New in Chapter 4 for the second edition are several interesting intervention examples involving analyses of

- the effect on calls of charging for directory assistance
- the effect on milk purchases of publicity about tainted milk
- the effect on airline stock volume of the September 11, 2001, terrorist attacks.

Chapter 5, "The ARIMA Model: Special Applications," extends the regression with time series errors class of models to cases where the error variance can change over time—the ARCH and GARCH class. Multivariate models in which individual nonstationary series vary together over time are referred to as "cointegration" or "error correction" models. These are also discussed and illustrated. This chapter presents new developments since the first edition of the book.

Chapter 6, "State Space Modeling," uses the AR model to motivate the construction of the state vector. Next, the equivalence of state space and vector ARMA models is discussed. Examples of multivariate processes and their state space equations are shown. The STATESPACE procedure is outlined, and a section on canonical correlation analysis and Akaike's information criterion is included. The chapter ends with the analysis of a bivariate series exhibiting feedback, a characteristic that cannot be handled with the general ARIMA transfer function approach.

Chapter 7, "Spectral Analysis," describes the SPECTRA procedure and how spectral analysis is used to detect sinusoidal components in time series models. In periodogram analysis, regressions are run on a sequence of values to find hidden periodicities. Spectra for different series, smoothing the periodogram, Fourier coefficients, and white noise tests are covered. The chapter ends with a discussion of cross-spectral analysis. New for the second edition is more in-depth discussion of tests for white noise and the ideas behind spectral analysis.

Chapter 8, "Data Mining and Forecasting," deals with the process of forecasting many time series with little intervention by the user. The goal of the chapter is to illustrate a modern automated interface for a collection of forecasting models, including many that have been discussed thus far. Chapter 8 also examines the SAS/ETS Time Series Forecasting System (TSFS), which provides a menu-driven interface to SAS/ETS and SAS/GRAPH procedures in order to facilitate quick and easy analysis of time series data. The chapter also includes a discussion detailing the use of PROC HPF, an automated high-performance forecasting procedure that is designed to forecast thousands of univariate time series.

Acknowledgments

For the 1986 edition we owe a great debt to students of the SAS Applied Time Series Analysis and Forecasting course, who made valuable suggestions on the subject material and generally stimulated our interest. We are also indebted to Alice T. Allen, Kathryn A. Council, Stephen Ewing, Wayne A. Fuller, Francis G. Giesbrecht, Robert P. Hastings, Herbert Kirk, Stephenie Joyner, Ann A. Lehman, Larry Stewart, and Houston Stokes.

David M. DeLong and Bart Killam enthusiastically reviewed and offered comments on several chapters. For the production of the manuscript, we owe special thanks to Deborah S. Blank for her patience and diligence.

For the second edition we gratefully acknowledge the SAS technical reviewers, Brent Cohen, Evan Anderson, Gul Ege, Bob Lucas, Sanford Gayle, and Youngjin Park. We are also grateful to the two outside reviewers, Houston Stokes of the University of Illinois at Chicago and David Booth of Kent State University.

We would like to acknowledge several people at SAS whose efforts have contributed to the completion of the second edition: Keith Collins, Tom Grant, Julie Platt, John West, Sam Pipkin, Ed Huddleston, Candy Farrell, Patricia Spain, and Patrice Cherry.

Finally, we would like to thank our wives, Vicki H. Brocklebank and Barbara S. Dickey, who were understanding and supportive throughout the writing of this book.

Chapter 1 Overview of Time Series

1.1 Introduction

This book deals with data collected at equally spaced points in time. The discussion begins with a single observation at each point. It continues with k series being observed at each point and then analyzed together in terms of their interrelationships.

One of the main goals of univariate time series analysis is to forecast future values of the series. For multivariate series, relationships among component series, as well as forecasts of these components, may be of interest. Secondary goals are smoothing, interpolating, and modeling of the structure. Three important characteristics of time series are often encountered: *seasonality*, *trend*, and *autocorrelation*.

Seasonality occurs, for example, when data are collected monthly and the value of the series in any given month is closely related to the value of the series in that same month in previous years. Seasonality can be very regular or can change slowly over a period of years.

A trend is a regular, slowly evolving change in the series level. Changes that can be modeled by low-order polynomials or low-frequency sinusoids fit into this category. For example, if a plot of sales over time shows a steady increase of $500 per month, you may fit a linear trend to the sales data. A trend is a long-term movement in the series.

In contrast, autocorrelation is a local phenomenon. When deviations from an overall trend tend to be followed by deviations of a like sign, the deviations are positively autocorrelated. Autocorrelation is the phenomenon that distinguishes time series from other branches of statistical analysis.

For example, consider a manufacturing plant that produces computer parts. Normal production is 100 units per day, although actual production varies from this mean of 100. Variation can be caused by machine failure, absenteeism, or incentives like bonuses or approaching deadlines. A machine may malfunction for several days, resulting in a run of low productivity. Similarly, an approaching deadline may increase production over several days. This is an example of *positive autocorrelation*, with data falling and staying below 100 for a few days, then rising above 100 and staying high for a while, then falling again, and so on.

Another example of positive autocorrelation is the flow rate of a river. Consider variation around the seasonal level: you may see high flow rates for several days following rain and low flow rates for several days during dry periods.

Negative autocorrelation occurs less often than positive autocorrelation. An example is a worker's attempt to control temperature in a furnace. The autocorrelation pattern depends on the worker's habits, but suppose he reads a low value of a furnace temperature and turns up the heat too far and similarly turns it down too far when readings are high. If he reads and adjusts the temperature each minute, you can expect a low temperature reading to be followed by a high reading. As a second example, an athlete may follow a long workout day with a short workout day and vice versa. The time he spends exercising daily displays negative autocorrelation.

1.2 Analysis Methods and SAS/ETS Software

1.2.1 Options

When you perform univariate time series analysis, you observe a single series over time. The goal is to model the historic series and then to use the model to forecast future values of the series. You can use some simple SAS/ETS software procedures to model low-order polynomial trends and autocorrelation. PROC FORECAST automatically fits an overall linear or quadratic trend with autoregressive (AR) error structure when you specify METHOD=STEPAR. As explained later, AR errors are not the most general types of errors that analysts study. For seasonal data you may want to fit a Winters exponentially smoothed trend-seasonal model with METHOD=WINTERS. If the trend is local, you may prefer METHOD=EXPO, which uses exponential smoothing to fit a local linear or quadratic trend. For higher-order trends or for cases where the forecast variable Y_t is related to one or more explanatory variables X_t, PROC AUTOREG estimates this relationship and fits an AR series as an error term.

Polynomials in time and seasonal indicator variables (see **Section 1.3.2**) can be computed as far into the future as desired. If the explanatory variable is a nondeterministic time series, however, actual future values are not available. PROC AUTOREG treats future values of the explanatory variable as known, so user-supplied forecasts of future values with PROC AUTOREG may give incorrect standard errors of forecast estimates. More sophisticated procedures like PROC STATESPACE, PROC VARMAX, or PROC ARIMA, with their transfer function options, are preferable when the explanatory variable's future values are unknown.

One approach to modeling seasonality in time series is the use of seasonal indicator variables in PROC AUTOREG to model a highly regular seasonality. Also, the AR error series from PROC AUTOREG or from PROC FORECAST with METHOD=STEPAR can include some correlation at seasonal lags (that is, it may relate the deviation from trend at time t to the deviation at time $t-12$ in monthly data). The WINTERS method of PROC FORECAST uses updating equations similar to exponential smoothing to fit a seasonal multiplicative model.

Another approach to seasonality is to remove it from the series and to forecast the seasonally adjusted series with other seasonally adjusted series used as inputs, if desired. The U.S. Census Bureau has adjusted thousands of series with its X-11 seasonal adjustment package. This package is the result of years of work by census researchers and is the basis for the seasonally adjusted figures that the federal government reports. You can seasonally adjust your own data using PROC X11, which is the census program set up as a SAS procedure. If you are using seasonally adjusted figures as explanatory variables, this procedure is useful.

An alternative to using X-11 is to model the seasonality as part of an ARIMA model or, if the seasonality is highly regular, to model it with indicator variables or trigonometric functions as explanatory variables. A final introductory point about the PROC X11 program is that it identifies and adjusts for outliers.*

If you are unsure about the presence of seasonality, you can use PROC SPECTRA to check for it; this procedure decomposes a series into cyclical components of various periodicities. Monthly data with highly regular seasonality have a large ordinate at period 12 in the PROC SPECTRA output SAS data set. Other periodicities, like multiyear business cycles, may appear in this analysis. PROC SPECTRA also provides a check on model residuals to see if they exhibit cyclical patterns over time. Often these cyclical patterns are not found by other procedures. Thus, it is good practice to analyze residuals with this procedure. Finally, PROC SPECTRA relates an output time series Y_t to one or more input or explanatory series X_t in terms of cycles. Specifically, cross-spectral analysis estimates the change in amplitude and phase when a cyclical component of an input series is used to predict the corresponding component of an output series. This enables the analyst to separate long-term movements from short-term movements.

Without a doubt, the most powerful and sophisticated methodology for forecasting univariate series is the ARIMA modeling methodology popularized by Box and Jenkins (1976). A flexible class of models is introduced, and one member of the class is fit to the historic data. Then the model is used to forecast the series. Seasonal data can be accommodated, and seasonality can be local; that is, seasonality for month t may be closely related to seasonality for this same month one or two years previously but less closely related to seasonality for this month several years previously. Local trending and even long-term upward or downward drifting in the data can be accommodated in ARIMA models through differencing.

Explanatory time series as inputs to a transfer function model can also be accommodated. Future values of nondeterministic, independent input series can be forecast by PROC ARIMA, which, unlike the previously mentioned procedures, accounts for the fact that these inputs are forecast when you compute prediction error variances and prediction limits for forecasts. A relatively new procedure, PROC VARMAX, models vector processes with possible explanatory variables, the X in VARMAX. As in PROC STATESPACE, this approach assumes that at each time point you observe a vector of responses each entry of which depends on its own lagged values and lags of the other vector entries, but unlike STATESPACE, VARMAX also allows explanatory variables X as well as cointegration among the elements of the response vector. Cointegration is an idea that has become quite popular in recent econometrics. The idea is that each element of the response vector might be a nonstationary process, one that has no tendency to return to a mean or deterministic trend function, and yet one or more linear combinations of the responses are stationary, remaining near some constant. An analogy is two lifeboats adrift in a stormy sea but tied together by a rope. Their location might be expressible mathematically as a random walk with no tendency to return to a particular point. Over time the boats drift arbitrarily far from any particular location. Nevertheless, because they are tied together, the *difference* in their positions would never be too far from 0. Prices of two similar stocks might, over time, vary according to a random walk with no tendency to return to a given mean, and yet if they are indeed similar, their price difference may not get too far from 0.

---

* Recently the Census Bureau has upgraded X-11, including an option to extend the series using ARIMA models prior to applying the centered filters used to deseasonalize the data. The resulting X-12 is incorporated as PROC X12 in SAS software.

1.2.2 How SAS/ETS Software Procedures Interrelate

PROC ARIMA emulates PROC AUTOREG if you choose not to model the inputs. ARIMA can also fit a richer error structure. Specifically, the error structure can be an autoregressive (AR), moving average (MA), or mixed-model structure. PROC ARIMA can emulate PROC FORECAST with METHOD=STEPAR if you use polynomial inputs and AR error specifications. However, unlike FORECAST, ARIMA provides test statistics for the model parameters and checks model adequacy. PROC ARIMA can emulate PROC FORECAST with METHOD=EXPO if you fit a moving average of order d to the dth difference of the data. Instead of arbitrarily choosing a smoothing constant, as necessary in PROC FORECAST METHOD=EXPO, the data tell you what smoothing constant to use when you invoke PROC ARIMA. Furthermore, PROC ARIMA produces more reasonable forecast intervals. In short, PROC ARIMA does everything the simpler procedures do and does it better.

However, to benefit from this additional flexibility and sophistication in software, you must have enough expertise and time to analyze the series. You must be able to identify and specify the form of the time series model using the autocorrelations, partial autocorrelations, inverse autocorrelations, and cross-correlations of the time series. Later chapters explain in detail what these terms mean and how to use them. Once you identify a model, fitting and forecasting are almost automatic.

The identification process is more complicated when you use input series. For proper identification, the ARIMA methodology requires that inputs be independent of each other and that there be no feedback from the output series to the input series. For example, if the temperature T_t in a room at time t is to be explained by current and lagged furnace temperatures F_t, lack of feedback corresponds to there being no thermostat in the room. A thermostat causes the furnace temperature to adjust to recent room temperatures. These ARIMA restrictions may be unrealistic in many examples. You can use PROC STATESPACE and PROC VARMAX to model multiple time series without these restrictions.

Although PROC STATESPACE and PROC VARMAX are sophisticated in theory, they are easy to run in their default mode. The theory allows you to model several time series together, accounting for relationships of individual component series with current and past values of the other series. Feedback and cross-correlated input series are allowed. Unlike PROC ARIMA, PROC STATESPACE uses an information criterion to select a model, thus eliminating the difficult identification process in PROC ARIMA. For example, you can put data on sales, advertising, unemployment rates, and interest rates into the procedure and automatically produce forecasts of these series. It is not necessary to intervene, but you must be certain that you have a property known as stationarity in your series to obtain theoretically valid results. The stationarity concept is discussed in Chapter 3, "The General ARIMA Model," where you will learn how to make nonstationary series stationary.

Although the automatic modeling in PROC STATESPACE sounds appealing, two papers in the *Proceedings of the Ninth Annual SAS Users Group International Conference* (one by Bailey and the other by Chavern) argue that you should use such automated procedures cautiously. Chavern gives an example in which PROC STATESPACE, in its default mode, fails to give as accurate a forecast as a certain vector autoregression. (However, the stationarity of the data is questionable, and stationarity is required to use PROC STATESPACE appropriately.) Bailey shows a PROC STATESPACE forecast considerably better than its competitors in some time intervals but not in others. In *SAS Views: SAS Applied Time Series Analysis and Forecasting*, Brocklebank and Dickey generate data from a simple MA model and feed these data into PROC STATESPACE in the default mode. The dimension of the model is overestimated when 50 observations are used, but the procedure is successful for samples of 100 and 500 observations from this simple series. Thus, it is wise to consider intervening in the modeling procedure through PROC STATESPACE's control options. If a transfer function model is appropriate, PROC ARIMA is a viable alternative.

This chapter introduces some techniques for analyzing and forecasting time series and lists the SAS procedures for the appropriate computations. As you continue reading the rest of the book, you may want to refer back to this chapter to clarify the relationships among the various procedures.

Figure 1.1 shows the interrelationships among the SAS/ETS software procedures mentioned. Table 1.1 lists some common questions and answers concerning the procedures.

Figure 1.1 How SAS/ETS Software Procedures Interrelate

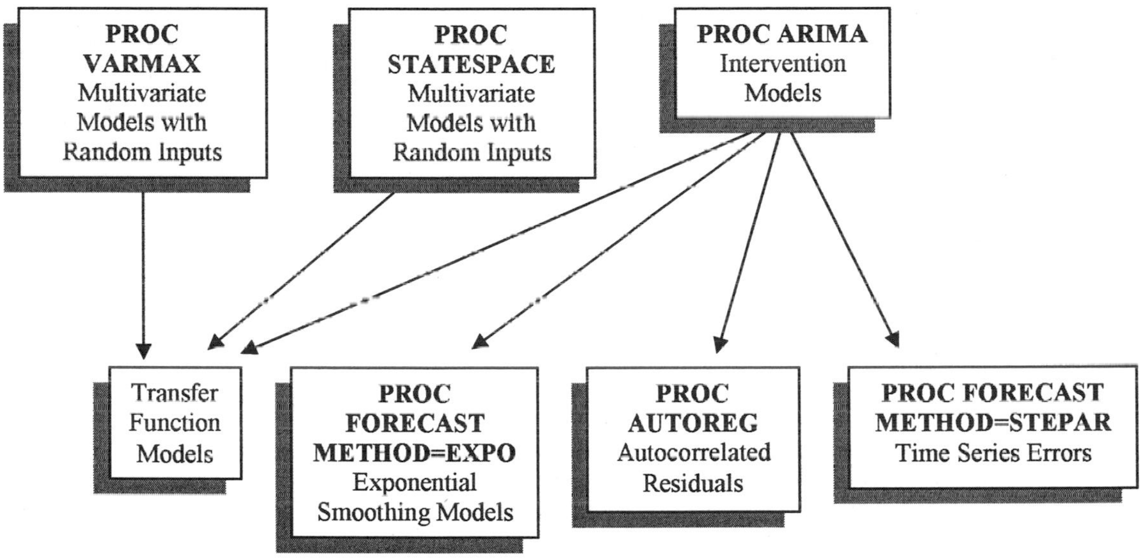

Table 1.1 Selected Questions and Answers Concerning SAS/ETS Software Procedures

Questions

1. Is a frequency domain analysis (F) or time domain analysis (T) conducted?
2. Are forecasts automatically generated?
3. Do predicted values have 95% confidence limits?
4. Can you supply leading indicator variables or explanatory variables?
5. Does the procedure run with little user intervention?
6. Is minimal time series background required for implementation?
7. Does the procedure handle series with embedded missing values?

Answers

SAS/ETS Procedures	1	2	3	4	5	6	7
FORECAST	T	Y	Y	N′	Y	Y	Y
AUTOREG	T	Y*	Y	Y	Y	Y	Y
X11	T	Y*	N	N	Y	Y	N
X12	T	Y*	Y	Y	Y	N	Y
SPECTRA	F	N	N	N	Y	N	N
ARIMA	T	Y*	Y	Y	N	N	N
STATESPACE	T	Y	Y*	Y	Y	N	N
VARMAX	T	Y	Y	Y	Y	N	N
MODEL	T	Y*	Y	Y	Y	N	Y
Time Series Forecasting System	T	Y	Y	Y	Y	Y	Y

* = requires user intervention
′ = supplied by the program
F = frequency domain analysis

N = no
T = time domain analysis
Y = yes

1.3 Simple Models: Regression

1.3.1 Linear Regression

This section introduces linear regression, an elementary but common method of mathematical modeling. Suppose that at time *t* you observe Y_t. You also observe explanatory variables X_{1t}, X_{2t}, and so on. For example, Y_t could be sales in month *t*, X_{1t} could be advertising expenditure in month *t*, and X_{2t} could be competitors' sales in month *t*. **Output 1.1** shows a simple plot of monthly sales versus date.

Output 1.1
Producing a Simple Plot of Monthly Data

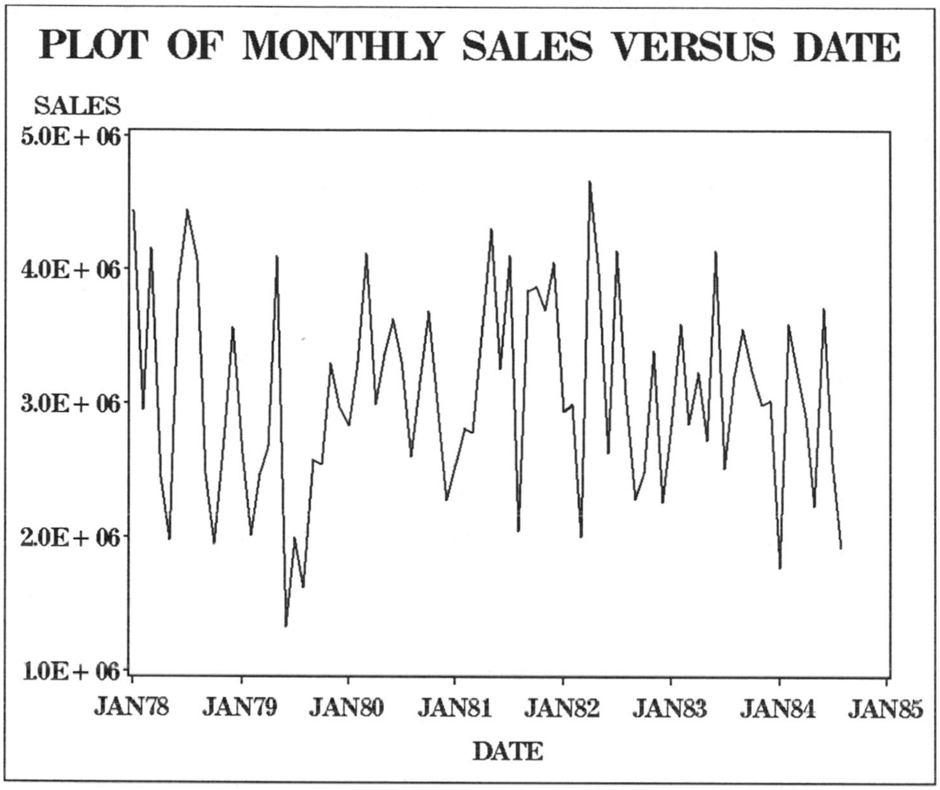

A multiple linear regression model relating the variables is

$$Y_t = \beta_0 + \beta_1 X_{1t} + \beta_2 X_{2t} + \varepsilon_t$$

For this model, assume that the errors ε_t

- have the same variance at all times t
- are uncorrelated with each other (ε_t and ε_s are uncorrelated for t different from s)
- have a normal distribution.

These assumptions allow you to use standard regression methodology, such as PROC REG or PROC GLM. For example, suppose you have 80 observations and you issue the following statements:

```
TITLE "PREDICTING SALES USING ADVERTISING";
TITLE2 "EXPENDITURES AND COMPETITORS' SALES";
PROC REG DATA=SALES;
   MODEL SALES=ADV COMP / DW;
   OUTPUT OUT=OUT1 P=P R=R;
RUN;
```

Output 1.2 shows the estimates of β_0, β_1, and β_2 ❶. The standard errors ❷ are incorrect if the assumptions on ε_t are not satisfied. You have created an output data set called OUT1 and have called for the Durbin-Watson option to check on these error assumptions.

Output 1.2 Performing a Multiple Regression

```
                 PREDICTING SALES USING ADVERTISING
                EXPENDITURES AND COMPETITORS' SALES

                          The REG Procedure
                            Model: MODEL1
                      Dependent Variable: SALES

                        Analysis of Variance

                           Sum of           Mean
Source          DF        Squares         Square      F Value       Prob>F

Model            2  2.5261822E13   1.2630911E13        51.140       0.0001
Error           77  1.9018159E13   246989077881
C Total         79    4.427998E13

           Root MSE    496979.95722     R-square        0.5705
           Dep Mean  3064722.70871      Adj R-sq        0.5593
              C.V.         16.21615

                        Parameter Estimates

                   ❶ Parameter   ❷ Standard    T for H0:
  Variable  DF       Estimate          Error   Parameter=0     Prob > |T|

  INTERCEP   1        2700165   373957.39855         7.221         0.0001
  ADV        1      10.179675     1.91704684         5.310         0.0001
  COMP       1      -0.605607     0.08465433        -7.154         0.0001

                   Durbin-Watson D               1.394 ❸
                   (For Number of Obs.)            80
                   1st Order Autocorrelation     0.283 ❹
```

The test statistics produced by PROC REG are designed specifically to detect departures from the null hypothesis (H_0: ε_t uncorrelated) of the form

$$H_1: \varepsilon_t = \rho\varepsilon_{t-1} + e_t$$

where $|\rho| < 1$ and e_t is an uncorrelated series. This type of error term, in which ε_t is related to ε_{t-1}, is called an AR (autoregressive) error of the first order.

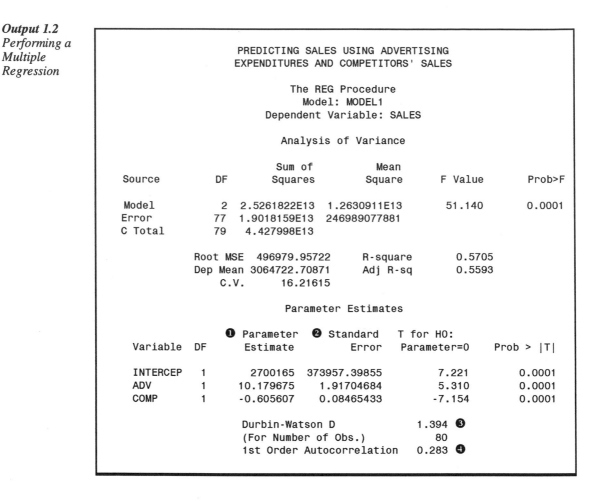

The Durbin-Watson option in the MODEL statement produces the Durbin-Watson test statistic ❸

$$d = \Sigma_{t=2}^{n}\left(\hat{\varepsilon}_t - \hat{\varepsilon}_{t-1}\right)^2 \;/\; \Sigma_{t=1}^{n}\hat{\varepsilon}_t^2$$

where

$$\hat{\varepsilon}_t = Y_t - \hat{\beta}_0 - \hat{\beta}_1 X_{1t} - \hat{\beta}_2 X_{2t}$$

If the actual errors ε_t are uncorrelated, the numerator of *d* has an expected value of about $2(n-1)\sigma^2$ and the denominator has an expected value of approximately $n\sigma^2$. Thus, if the errors ε_t are uncorrelated, the ratio *d* should be approximately 2.

Positive autocorrelation means that ε_t is closer to ε_{t-1} than in the independent case, so $|\varepsilon_t - \varepsilon_{t-1}|$ should be smaller. It follows that *d* should also be smaller. The smallest possible value for *d* is 0. If *d* is significantly less than 2, positive autocorrelation is present.

When is a Durbin-Watson statistic significant? The answer depends on the number of coefficients in the regression and on the number of observations. In this case, you have *k*=3 coefficients (β_0, β_1, and β_2 for the intercept, ADV, and COMP) and n=80 observations. In general, if you want to test for positive autocorrelation at the 5% significance level, you must compare *d*=1.349 to a critical value. Even with *k* and *n* fixed, the critical value can vary depending on actual values of the independent variables. The results of Durbin and Watson imply that if *k*=3 and *n*=80, the critical value must be between d_L=1.59 and d_U=1.69. Since *d* is less than d_L, you would reject the null hypotheses of uncorrelated errors in favor of the alternative: positive autocorrelation. If *d*>2, which is evidence of negative autocorrelation, compute *d'*=4–*d* and compare the results to d_L and d_U. Specifically, if *d'* (1.954) were greater than 1.69, you would be unable to reject the null hypothesis of uncorrelated errors. If *d'* were less than 1.59 you would reject the null hypothesis of uncorrelated errors in favor of the alternative: negative autocorrelation. Note that if

$$1.59 < d < 1.69$$

you cannot be sure whether *d* is to the left or right of the actual critical value *c* because you know only that

$$1.59 < c < 1.69$$

Durbin and Watson have constructed tables of bounds for the critical values. Most tables use *k'*=*k*–1, which equals the number of explanatory variables, excluding the intercept and *n* (number of observations) to obtain the bounds d_L and d_U for any given regression (Draper and Smith 1998).*

Three warnings apply to the Durbin-Watson test. First, it is designed to detect first-order AR errors. Although this type of autocorrelation is only one possibility, it seems to be the most common. The test has some power against other types of autocorrelation. Second, the Durbin-Watson bounds do not hold when lagged values of the dependent variable appear on the right side of the regression. Thus, if the example had used last month's sales to help explain this month's sales, you would not know correct bounds for the critical value. Third, if you incorrectly specify the model, the Durbin-Watson statistic often lies in the critical region even though no real autocorrelation is present. Suppose an important variable, such as X_{3t}=product availability, had been omitted in the sales example. This omission could produce a significant *d*. Some practitioners use *d* as a lack-of-fit statistic, which is justified only if you assume a priori that a correctly specified model cannot have autocorrelated errors and, thus, that significance of *d* must be due to lack of fit.

---

* Exact p-values for *d* are now available in PROC AUTOREG as will be seen in **Output 1.2A** later in this section.

The output also produced a first-order autocorrelation, ❹ denoted as

$$\hat{\rho} = 0.283$$

When n is large and the errors are uncorrelated,

$$n^{1/2}\hat{\rho}/\left(1-\hat{\rho}^2\right)^{1/2}$$

is approximately distributed as a standard normal variate. Thus, a value

$$n^{1/2}\hat{\rho}/\left(1-\hat{\rho}^2\right)^{1/2}$$

exceeding 1.645 is significant evidence of positive autocorrelation at the 5% significance level. This is especially helpful when the number of observations exceeds the largest in the Durbin-Watson table—for example,

$$\sqrt{80}\ (.283)/\sqrt{1-0.283^2} = 2.639$$

You should use this test only for large n values. It is subject to the three warnings given for the Durbin-Watson test. Because of the approximate nature of the $n^{1/2}\hat{\rho}/\left(1-\hat{\rho}^2\right)^{1/2}$ test, the Durbin-Watson test is preferable. In general, d is approximately $2(1-\hat{\rho})$.

This is easily seen by noting that

$$\hat{\rho} = \sum \hat{\varepsilon}_t \hat{\varepsilon}_{t-1} / \sum \hat{\varepsilon}_t^{\,2}$$

and

$$d = \sum (\hat{\varepsilon}_t - \hat{\varepsilon}_{t-1})^2 / \sum \hat{\varepsilon}_t^{\,2}$$

Durbin and Watson also gave a computer-intensive way to compute exact p-values for their test statistic d. This has been incorporated in PROC AUTOREG. For the sales data, you issue this code to fit a model for sales as a function of this-period and last-period advertising.

```
PROC AUTOREG DATA=NCSALES;
   MODEL SALES=ADV ADV1 / DWPROB;
RUN;
```

The resulting **Output 1.2A** shows a significant d=.5427 (p-value .0001 < .05). Could this be because of an omitted variable? Try the model with competitor's sales included.

```
PROC AUTOREG DATA=NCSALES;
   MODEL SALES=ADV ADV1 COMP / DWPROB;
RUN;
```

Now, in **Output 1.2B**, d =1.8728 is insignificant (p-value .2239 > .05). Note also the increase in R-square (the proportion of variation explained by the model) from 39% to 82%. What is the effect of an increase of $1 in advertising expenditure? It gives a sales increase estimated at $6.04 this period but a decrease of $5.18 next period. You wonder if the true coefficients on ADV and ADV1 are the same with opposite signs; that is, you wonder if these coefficients add to 0. If they do, then the increase we get this period from advertising is followed by a decrease of equal magnitude next period. This means our advertising dollar simply shifts the timing of sales rather than increasing the level of sales. Having no autocorrelation evident, you fit the model in PROC REG asking for a test that the coefficients of ADV and ADV1 add to 0.

```
PROC REG DATA = SALES;
   MODEL SALES = ADV ADV1 COMP;
   TEMPR: TEST ADV+ADV1=0;
RUN;
```

Output 1.2C gives the results. Notice that the regression is exactly that given by PROC AUTOREG with no NLAG= specified. The p-value (.077>.05) is not small enough to reject the hypothesis that the coefficients are of equal magnitude, and thus it is possible that advertising just shifts the timing, a temporary effect. Note the label TEMPR on the test.

Note also that, although we may have information on our company's plans to advertise, we would likely not know what our competitor's sales will be in future months, so at best we would have to substitute estimates of these future values in forecasting our sales. It appears that an increase of \$1.00 in our competitor's sales is associated with a \$0.56 decrease in our sales.

From **Output 1.2C** the forecasting equation is seen to be

$$\text{PREDICTED SALES} = 35967 - 0.563227\text{COMP} + 6.038203\text{ADV} - 5.188384\text{ADV1}$$

Output 1.2A *Predicting Sales from Advertising*

```
                       AUTOREG Procedure

                  Dependent Variable = SALES

               Ordinary Least Squares Estimates

         SSE              5.1646E9    DFE              77
         MSE              67072080    Root MSE   8189.755
         SBC              1678.821    AIC        1671.675
         Reg Rsq            0.3866    Total Rsq    0.3866
         Durbin-Watson      0.5427    PROB<DW      0.0001


Variable      DF        B Value   Std Error   t Ratio Approx Prob

Intercept      1          14466      8532.1     1.695      0.0940
ADV            1       6.560093      0.9641     6.804      0.0001
ADV1           1      -5.015231      0.9606    -5.221      0.0001
```

Output 1.2B
Predicting Sales from Advertising and Competitor's Sales

```
                PREDICTING SALES USING ADVERTISING
              EXPENDITURES AND COMPETITOR'S SALES

                         AUTOREG Procedure

                    Dependent Variable = SALES

                  Ordinary Least Squares Estimates

          SSE            1.4877E9    DFE              76
          MSE            19575255    Root MSE   4424.393
          SBC            1583.637    AIC        1574.109
          Reg Rsq          0.8233    Total Rsq    0.8233
          Durbin-Watson  1.8728      PROB<DW      0.2239

 Variable      DF       B Value   Std Error   t Ratio Approx Prob

 Intercept      1         35967      4869.0     7.387      0.0001
 COMP           1     -0.563227      0.0411   -13.705      0.0001
 ADV            1      6.038203      0.5222    11.562      0.0001
 ADV1           1     -5.188384      0.5191    -9.994      0.0001
```

Output 1.2C
Predicting Sales from Advertising and Competitor's Sales

```
                PREDICTING SALES USING ADVERTISING
              EXPENDITURES AND COMPETITOR'S SALES

                   Dependent Variable: SALES

                       Analysis of Variance

Sum of         Mean
Source          DF      Squares        Square       F Value       Prob>F

Model            3 6931264991.2 2310421663.7        118.028       0.0001
Error           76 1487719368.2 19575254.845
C Total         79 8418984359.4

Root MSE    4424.39316      R-square       0.8233
Dep Mean   29630.21250      Adj R-sq       0.8163
C.V.          14.93203
                        Parameter Estimates

                 Parameter      Standard    T for H0:
 Variable  DF     Estimate         Error    Parameter=0    Prob > |T|

 INTERCEP  1         35967  4869.0048678          7.387        0.0001
 COMP      1     -0.563227    0.04109605        -13.705        0.0001
 ADV       1      6.038203    0.52224284         11.562        0.0001
 ADV1      1     -5.188384    0.51912574         -9.994        0.0001

Durbin-Watson D              1.873
(For Number of Obs.)            80
1st Order Autocorrelation    0.044

                PREDICTING SALES USING ADVERTISING
              EXPENDITURES AND COMPETITOR'S SALES

Dependent Variable: SALES
Test: TEMPR     Numerator:63103883.867  DF:    1   F value:    3.2237
Denominator:  19575255  DF:    76   Prob>F:    0.0766
```

1.3.2 Highly Regular Seasonality

Occasionally, a very regular seasonality occurs in a series, such as an average monthly temperature at a given location. In this case, you can model seasonality by computing means. Specifically, the mean of all the January observations estimates the seasonal level for January. Similar means are used for other months throughout the year. An alternative to computing the twelve means is to run a regression on monthly indicator variables. An indicator variable takes on values of 0 or 1. For the January indicator, the 1s occur only for observations made in January. You can compute an indicator variable for each month and regress Y_t on the twelve indicators with no intercept. You can also regress Y_t on a column of 1s and eleven of the indicator variables. The intercept now estimates the level for the month associated with the omitted indicator, and the coefficient of any indicator column is added to the intercept to compute the seasonal level for that month.

For further illustration, **Output 1.3** shows a series of quarterly increases in North Carolina retail sales; that is, each point is the sales for that quarter minus the sales for the previous quarter. **Output 1.4** shows a plot of the monthly sales through time. Quarterly sales were computed as averages of three consecutive months and are used here to make the presentation brief. A model for the monthly data will be shown in Chapter 4. Note that there is a strong seasonal pattern here and perhaps a mild trend over time. The change data are plotted in **Output 1.6**. To model the seasonality, use S1, S2, and S3, and for the trend, use time, T1, and its square T2. The S variables are often referred to as indicator variables, being indicators of the season, or dummy variables. The first CHANGE value is missing because the sales data start in quarter 1 of 1983 so no increase can be computed for that quarter.

Output 1.3 *Displaying North Carolina Retail Sales Data Set*

OBS	DATE	CHANGE	S1	S2	S3	T1	T2
1	83Q1	.	1	0	0	1	1
2	83Q2	1678.41	0	1	0	2	4
3	83Q3	633.24	0	0	1	3	9
4	83Q4	662.35	0	0	0	4	16
5	84Q1	-1283.59	1	0	0	5	25
(More Output Lines)							
47	94Q3	543.61	0	0	1	47	2209
48	94Q4	1526.95	0	0	0	48	2304

Output 1.4
Plotting North Carolina Monthly Sales

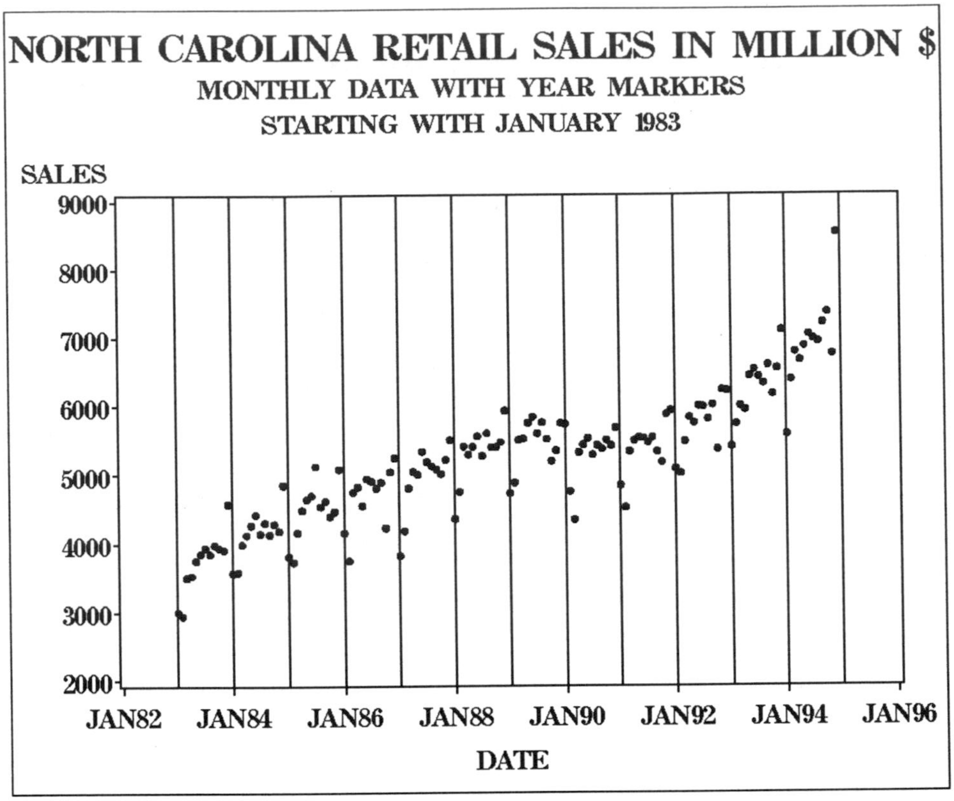

Now issue these commands:

```
PROC AUTOREG DATA=ALL;
   MODEL CHANGE = T1 T2  S1  S2  S3 / DWPROB;
RUN;
```

This gives **Output 1.5**.

Output 1.5
Using PROC AUTOREG to Get the Durbin-Watson Test Statistic

```
                         AUTOREG Procedure

Dependent Variable = CHANGE

                    Ordinary Least Squares Estimates

               SSE              5290128    DFE              41
               MSE             129027.5    Root MSE    359.204
               SBC             703.1478    AIC        692.0469
               Reg Rsq           0.9221    Total Rsq    0.9221
               Durbin-Watson     2.3770    PROB<DW      0.8608


      Variable     DF        B Value   Std Error   t Ratio Approx Prob
      Intercept     1     679.427278       200.1     3.395      0.0015
      T1            1     -44.992888     16.4428    -2.736      0.0091
      T2            1       0.991520      0.3196     3.102      0.0035
      S1            1   -1725.832501       150.3   -11.480      0.0001
      S2            1    1503.717849       146.8    10.240      0.0001
      S3            1    -221.287056       146.7    -1.508      0.1391
```

PROC AUTOREG is intended for regression models with autoregressive errors. An example of a model with autoregressive errors is

$$Y_t = \beta_0 + \beta_1 X_{1t} + \beta_2 X_{2t} + Z_t$$

where

$$Z_t = \rho Z_{t-1} + \varepsilon_t$$

Note how the error term Z_t is related to a lagged value of itself in an equation that resembles a regression equation; hence the term "autoregressive." The term ε_t represents the portion of Z_t that could not have been predicted from previous Z values and is often called an unanticipated "shock" or "white noise." It is assumed that the e series is independent and identically distributed. This one lag error model is fit using the /NAG=1 option in the MODEL statement. Alternatively, the options /NLAG=5 BACKSTEP can be used to try 5 lags of Z, automatically deleting those deemed statistically insignificant.

Our retail sales change data require no autocorrelation adjustment. The Durbin-Watson test has a p-value 0.8608>0.05; so there is no evidence of autocorrelation in the errors. The fitting of the model is the same as in PROC REG because no NLAG specification was issued in the MODEL statement. The parameter estimates are interpreted just as they would be in PROC REG; that is, the predicted change PC in quarter 4 (where S1=S2=S3=0) is given by

$$PC = 679.4 - 44.99\,t + 0.99\,t^2$$

and in quarter 1 (where S1=1, S2=S3=0) is given by

$$PC = 679.4 - 1725.83 - 44.99\,t + 0.99\,t^2$$

etc. Thus the coefficients of S1, S2, and S3 represent shifts in the quadratic polynomial associated with the first through third quarters and the remaining coefficients calibrate the quadratic function to the fourth quarter level. In **Output 1.6** the data are dots, and the fourth quarter quadratic predicting function is the smooth curve. Vertical lines extend from the quadratic, indicating the seasonal shifts required for the other three quarters. The broken line gives the predictions. The last data point for 1994Q4 is indicated with an extended vertical line. Notice that the shift for any quarter is the same every year. This is a property of the dummy variable model and may not be reasonable for some data; for example, sometimes seasonality is slowly changing over a period of years.

Output 1.6
Plotting Quarterly Sales Increase with Quadratic Predicting Function

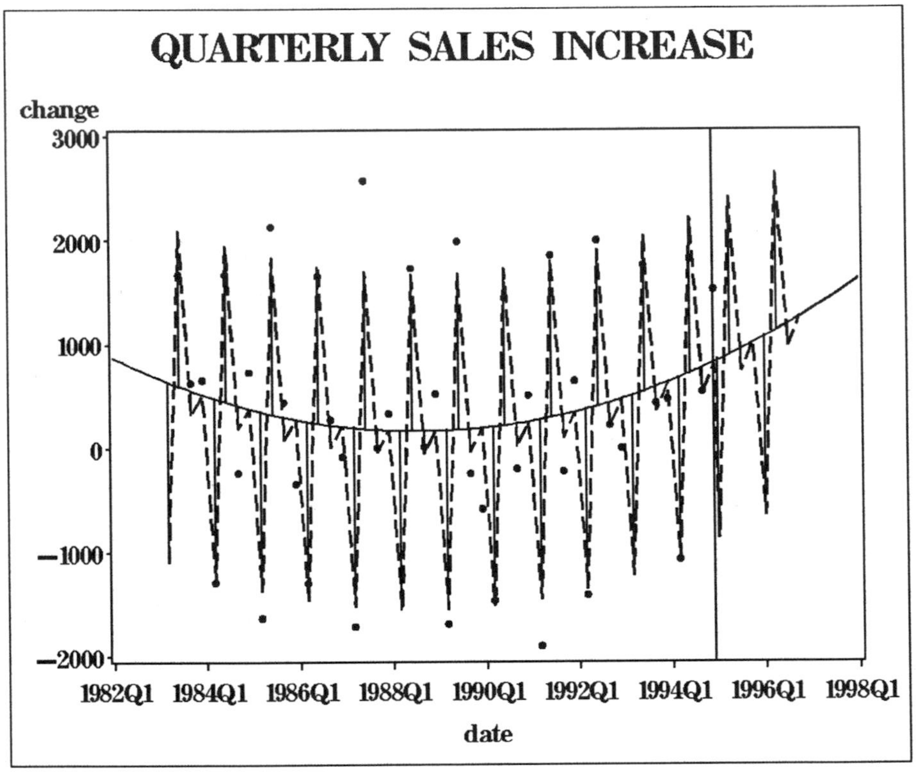

To forecast into the future, extrapolate the linear and quadratic terms and the seasonal dummy variables the requisite number of periods. The data set extra listed in **Output 1.7** contains such values. Notice that there is no question about the future values of these, unlike the case of competitor's sales that was considered in an earlier example. The PROC AUTOREG technology assumes perfectly known future values of the explanatory variables. Set the response variable, CHANGE, to missing.

Output 1.7
Data Appended for Forecasting

OBS	DATE	CHANGE	S1	S2	S3	T1	T2
1	95Q1	.	1	0	0	49	2401
2	95Q2	.	0	1	0	50	2500
3	95Q3	.	0	0	1	51	2601
4	95Q4	.	0	0	0	52	2704
5	96Q1	.	1	0	0	53	2809
6	96Q2	.	0	1	0	54	2916
7	96Q3	.	0	0	1	55	3025
8	96Q4	.	0	0	0	56	3136

Combine the original data set—call it NCSALES—with the data set EXTRA as follows:

```
DATA ALL;
   SET NCSALES EXTRA;
RUN;
```

Now run PROC AUTOREG on the combined data, noting that the extra data cannot contribute to the estimation of the model parameters since CHANGE is missing. The extra data have full information on the explanatory variables and so predicted values (forecasts) will be produced. The predicted values P are output into a data set OUT1 using this statement in PROC AUTOREG:

```
OUTPUT OUT=OUT1 PM=P;
```

Using PM= requests that the predicted values be computed only from the regression function without forecasting the error term Z. If NLAG= is specified, a model is fit to the regression residuals and this model can be used to forecast residuals into the future. Replacing PM= with P= adds forecasts of future Z values to the forecast of the regression function. The two types of forecast, with and without forecasting the residuals, point out the fact that part of the predictability comes from the explanatory variables, and part comes from the autocorrelation—that is, from the momentum of the series. Thus, as seen in **Output 1.5**, there is a total R-square and a regression R-square, the latter measuring the predictability associated with the explanatory variables apart from contributions due to autocorrelation. Of course in the current example, with no autoregressive lags specified, these are the same and P= and PM= create the same variable. The predicted values from PROC AUTOREG using data set ALL are displayed in **Output 1.8**.

Output 1.8
Plotting Quarterly Sales Increase with Prediction

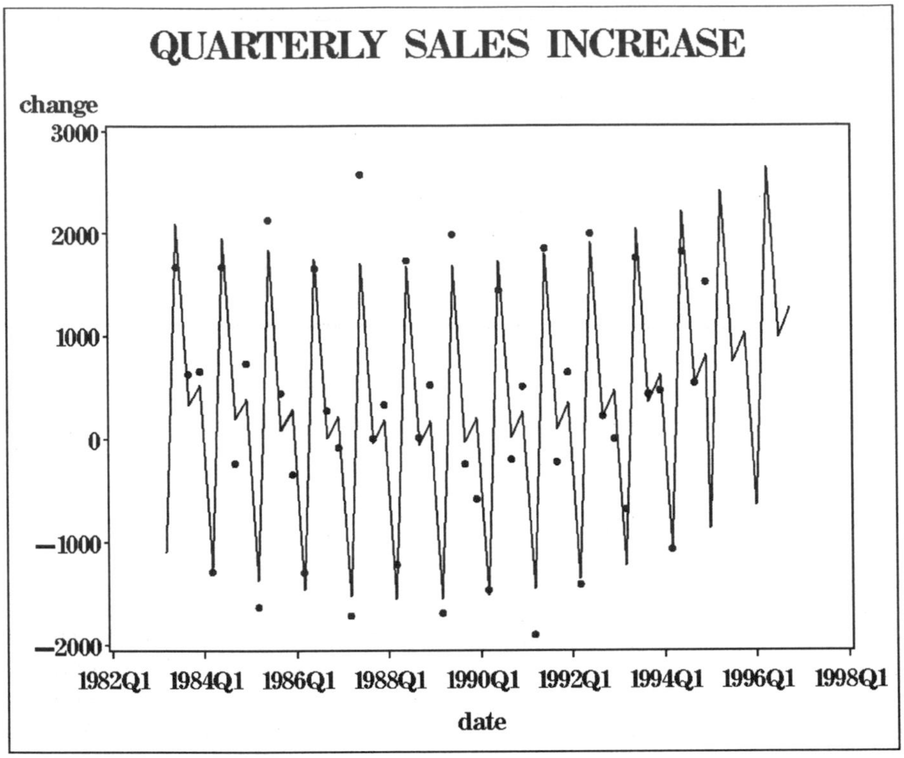

Because this example shows no residual autocorrelation, analysis in PROC REG would be appropriate. Using the data set with the extended explanatory variables, add P and CLI to produce predicted values and associated prediction intervals.

```
PROC REG;
   MODEL CHANGE = T  T2  S1  S2  S3  /  P  CLI;
   TITLE "QUARTERLY SALES INCREASE";
RUN;
```

Output 1.9
Producing Forecasts and Prediction Intervals with the P and CLI Options in the Model Statement

QUARTERLY SALES INCREASE

Dependent Variable: CHANGE

Analysis of Variance

Source	DF	Sum of Squares	Mean Square	F Value	Prob>F
Model	5	62618900.984	12523780.197	97.063	0.0001
Error	41	5290127.6025	129027.5025		
C Total	46	67909028.586			

Root MSE	359.20398	R-square	0.9221
Dep Mean	280.25532	Adj R-sq	0.9126
C.V.	128.17026		

Parameter Estimates

Variable	DF	Parameter Estimate	Standard Error	T for H0: Parameter=0	Prob > |T|
INTERCEP	1	679.427278	200.12467417	3.395	0.0015
T1	1	-44.992888	16.44278429	-2.736	0.0091
T2	1	0.991520	0.31962710	3.102	0.0035
S1	1	-1725.832501	150.33120614	-11.480	0.0001
S2	1	1503.717849	146.84832151	10.240	0.0001
S3	1	-221.287056	146.69576462	-1.508	0.1391

Quarterly Sales Increase

Obs	Dep Var CHANGE	Predict Value	Std Err Predict	Lower95% Predict	Upper95% Predict	Residual
1	.	-1090.4	195.006	-1915.8	-265.0	.
2	1678.4	2097.1	172.102	1292.7	2901.5	-418.7
3	633.2	332.1	160.650	465.1	1129.3	301.2
4	662.4	515.3	156.028	-275.6	1306.2	147.0
5	-1283.6	-1246.6	153.619	-2035.6	-457.6	-37.0083

(more output lines)

Obs	Dep Var CHANGE	Predict Value	Std Err Predict	Lower95% Predict	Upper95% Predict	Residual
49	.	-870.4	195.006	-1695.9	-44.9848	.
50	.	2412.3	200.125	1581.9	3242.7	.
51	.	742.4	211.967	-99.8696	1584.8	.
52	.	1020.9	224.417	165.5	1876.2	.
53	.	-645.8	251.473	-1531.4	239.7	.
54	.	2644.8	259.408	1750.0	3539.6	.
55	.	982.9	274.992	69.2774	1896.5	.
56	.	1269.2	291.006	335.6	2202.8	.

Sum of Residuals	0
Sum of Squared Residuals	5290127.6025
Predicted Resid SS (Press)	7067795.5909

For observation 49 an increase in sales of –870.4 (i.e., a decrease) is predicted for the next quarter with confidence interval extending from –1695.9 to –44.98. This is the typical after-Christmas sales slump.

What does this sales change model say about the level of sales, and why were the levels of sales not used in the analysis? First, notice that a cubic term in time, bt^3, when differenced becomes a quadratic term: $bt^3 - b(t-1)^3 = b(3t^2 - 3t + 1)$. Thus a quadratic plus seasonal model in the differences is associated with a cubic plus seasonal model in the levels. However if the error term in the differences satisfies the usual regression assumptions, which it seems to do for these data, then the error term in the original levels can't possibly satisfy them—the levels appear to have a nonstationary error term. Ordinary regression statistics are invalid on the original level series. If you ignore this, the usual (incorrect here) regression statistics indicate that a degree 8 polynomial is required to get a good fit. A plot of sales and the forecasts from polynomials of varying degree is shown in **Output 1.10**. The first thing to note is that the degree 8 polynomial, arrived at by inappropriate use of ordinary regression, gives a ridiculous forecast that extends vertically beyond the range of our graph just a few quarters into the future. The degree 3 polynomial seems to give a reasonable increase while the intermediate degree 6 polynomial actually forecasts a decrease. It is dangerous to forecast too far into the future using polynomials, especially those of high degree. Time series models specifically designed for nonstationary data will be discussed later. In summary, the differenced data seem to satisfy assumptions needed to justify regression.

Output 1.10 *Plotting Sales and Forecasts of Polynomials of Varying Degree*

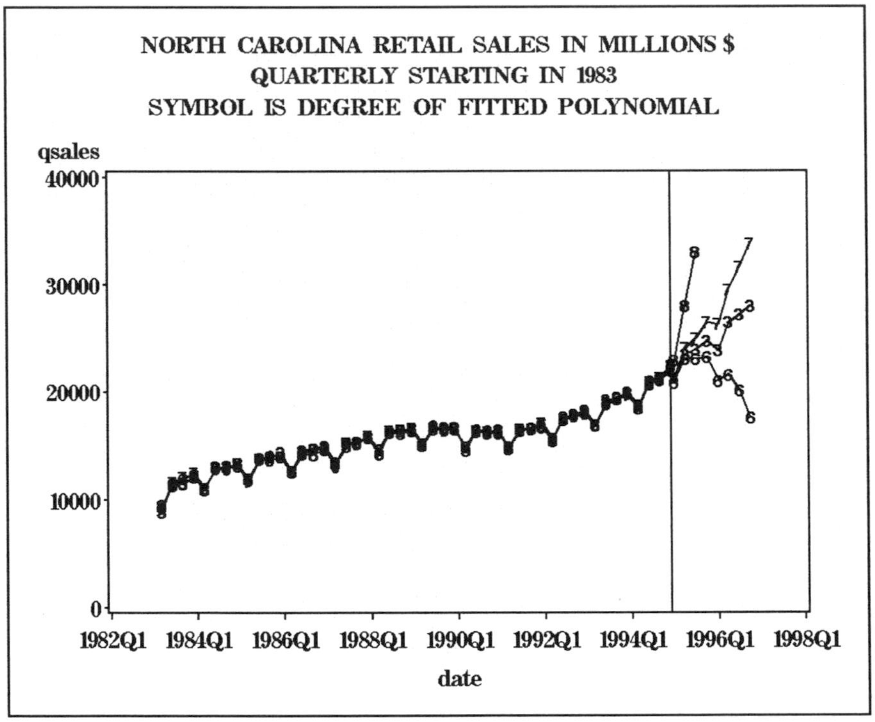

1.3.3 Regression with Transformed Data

Often, you analyze some transformed version of the data rather than the original data. The logarithmic transformation is probably the most common and is the only transformation discussed in this book. Box and Cox (1964) suggest a family of transformations and a method of using the data to select one of them. This is discussed in the time series context in Box and Jenkins (1976, 1994).

Consider the following model:

$$Y_t = \beta_0 \left(\beta_1^{X_t}\right) \varepsilon_t$$

Taking logarithms on both sides, you obtain

$$\log(Y_t) = \log(\beta_0) + \log(\beta_1)X_t + \log(\varepsilon_t)$$

Now if

$$\eta_t = \log(\varepsilon_t)$$

and if η_t satisfies the standard regression assumptions, the regression of $\log(Y_t)$ on 1 and X_t produces the best estimates of $\log(\beta_0)$ and $\log(\beta_1)$.

As before, if the data consist of (X_1, Y_1), (X_2, Y_2), ..., (X_n, Y_n), you can append future known values X_{n+1}, X_{n+2}, ..., X_{n+s} to the data if they are available. Set Y_{n+1} through Y_{n+s} to missing values (.). Now use the MODEL statement in PROC REG:

```
MODEL LY=X / P CLI;
```

where

```
LY=LOG(Y);
```

is specified in the DATA step. This produces predictions of future LY values and prediction limits for them. If, for example, you obtain an interval

$$-1.13 < \log(Y_{n+s}) < 2.7$$

you can compute

$$\exp(-1.13) = .323$$

and

$$\exp(2.7) = 14.88$$

to conclude

$$.323 < Y_{n+s} < 14.88$$

Note that the original prediction interval had to be computed on the log scale, the only scale on which you can justify a *t* distribution or normal distribution.

When should you use logarithms? A quick check is to plot Y against X. When

$$Y_t = \beta_0 \left(\beta_1^{X_t}\right) \varepsilon_t$$

the overall shape of the plot resembles that of

$$Y = \beta_0 \left(\beta_1^{X}\right)$$

See **Output 1.11** for several examples of this type of plot. Note that the curvature in the plot becomes more dramatic as β_1 moves away from 1 in either direction; the actual points are scattered around the appropriate curve. Because the error term ε is multiplied by $\beta_0\left(\beta_1^X\right)$, the variation around the curve is greater at the higher points and lesser at the lower points on the curve.

Output 1.11
Plotting Exponential Curves

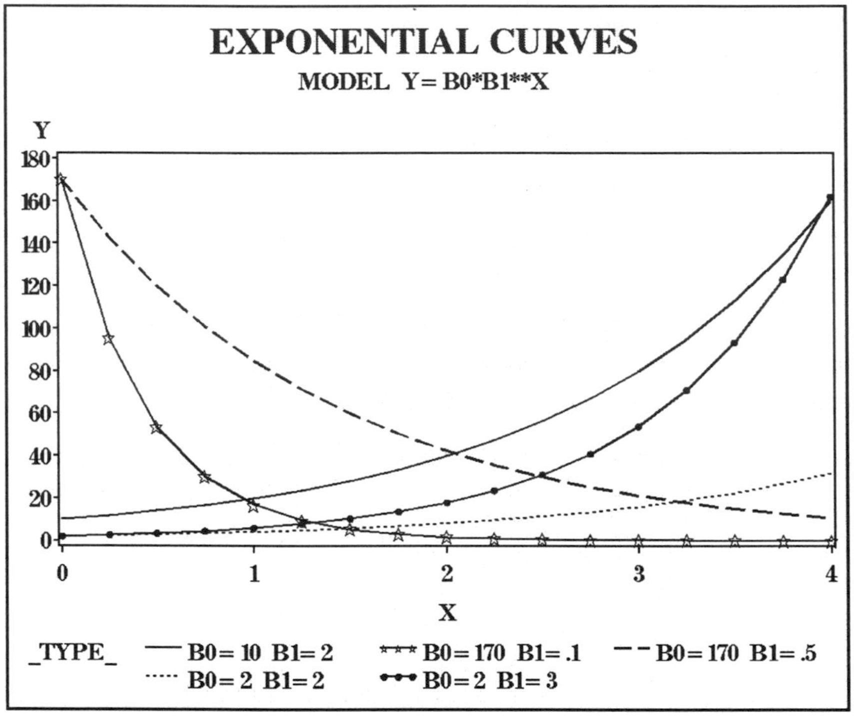

Output 1.12 shows a plot of U.S. Treasury bill rates against time. The curvature and especially the variability displayed are similar to those just described. In this case, you simply have $X_t=t$. A plot of the logarithm of the rates appears in **Output 1.13**. Because this plot is straighter with more uniform variability, you decide to analyze the logarithms.

Output 1.12
Plotting Ninety-Day Treasury Bill Rates

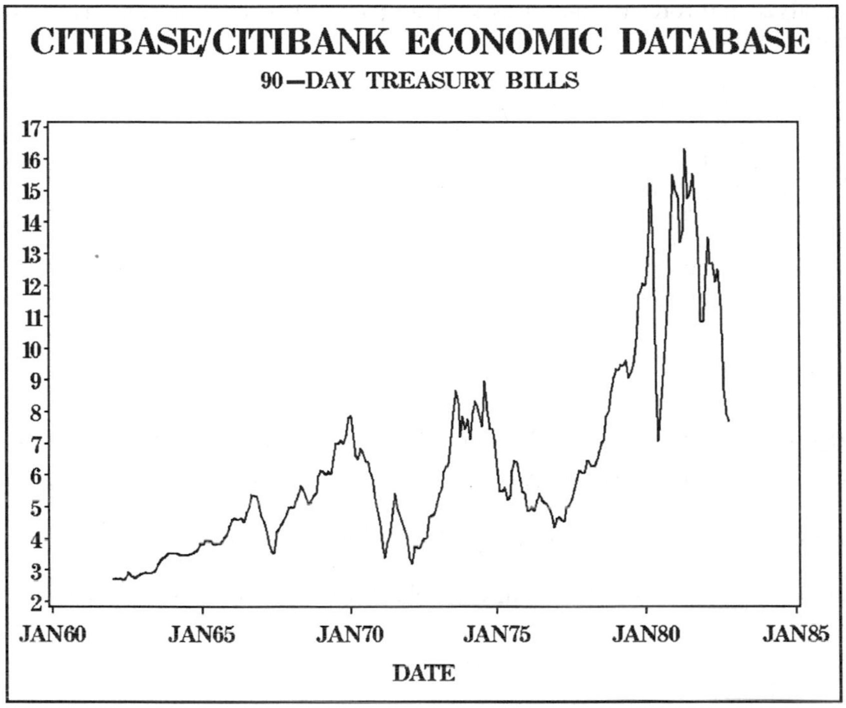

Output 1.13
Plotting Ninety-Day Logged Treasury Bill Rates

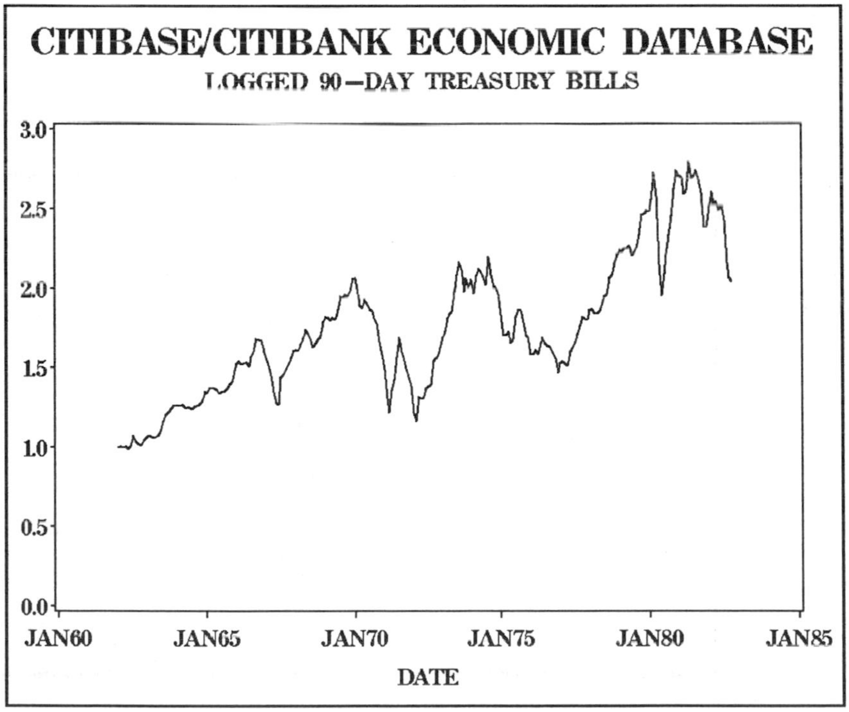

To analyze and forecast the series with simple regression, you first create a data set with future values of time:

```
DATA TBILLS2;
   SET TBILLS END=EOF;
   TIME+1;
   OUTPUT;
   IF EOF THEN DO I=1 TO 24;
      LFYGM3=.;
      TIME+1;
      DATE=INTNX('MONTH',DATE,1);
      OUTPUT;
   END;
   DROP I;
RUN;
```

Output 1.14 shows the last 24 observations of the data set TBILLS2. You then regress the log T-bill rate, LFYGM3, on TIME to estimate $\log(\beta_0)$ and $\log(\beta_1)$ in the following model:

$$\text{LFYGM3} = \log(\beta_0) + \log(\beta_1)*\text{TIME} + \log\ (\varepsilon_t)$$

You also produce predicted values and check for autocorrelation by using these SAS statements:

```
PROC REG DATA=TBILLS2;
   MODEL LFYGM3=TIME / DW P CLI;
   ID DATE;
   TITLE 'CITIBASE/CITIBANK ECONOMIC DATABASE';
   TITLE2 'REGRESSION WITH TRANSFORMED DATA';
RUN;
```

The result is shown in **Output 1.15**.

Output 1.14 *Displaying Future Date Values for U.S. Treasury Bill Data*

```
           CITIBASE/CITIBANK ECONOMIC DATABASE

           OBS      DATE     LFYGM3     TIME

             1      NOV82       .        251
             2      DEC82       .        252
             3      JAN83       .        253
             4      FEB83       .        254
             5      MAR83       .        255

                  (More Output Lines)

            20      JUN84       .        270
            21      JUL84       .        271
            22      AUG84       .        272
            23      SEP84       .        273
            24      OCT84       .        274
```

Output 1.15 *Producing Predicted Values and Checking Autocorrelation with the P, CLI, and DW Options in the MODEL Statement*

```
                  CITIBASE/CITIBANK ECONOMIC DATABASE
                     REGRESSION WITH TRANSFORMED DATA

                      Dependent Variable: LFYGM3

                         Analysis of Variance

                             Sum of         Mean
    Source          DF      Squares       Square      F Value       Prob>F
    Model            1     32.68570     32.68570      540.633       0.0001
    Error          248     14.99365      0.06046
    C Total        249     47.67935

            Root MSE       0.24588     R-square       0.6855
            Dep Mean       1.74783     Adj R-sq       0.6843
            C.V.          14.06788

                          Parameter Estimates

                      Parameter      Standard    T for H0:
    Variable  DF       Estimate         Error   Parameter=0    Prob > |T|

    INTERCEP   1       1.119038    0.03119550        35.872        0.0001
    TIME       1       0.005010    0.00021548        23.252        0.0001

                     REGRESSION WITH TRANSFORMED DATA
                  Dep Var   Predict    Std Err  Lower95%  Upper95%
  Obs    DATE     LFYGM3      Value    Predict   Predict   Predict  Residual

   1     JAN62     1.0006     1.1240     0.031    0.6359    1.6122   -0.1234
   2     FEB62     1.0043     1.1291     0.031    0.6410    1.6171   -0.1248
   3     MAR62     1.0006     1.1341     0.031    0.6460    1.6221   -0.1334
   4     APR62     1.0043     1.1391     0.030    0.6511    1.0271   -0.1040
   5     MAY62     0.9858     1.1441     0.030    0.6562    1.6320   -0.1583

(More Output Lines)

251      NOV82          .     2.3766     0.031    1.8885    2.8648         .

(More Output Lines)


270      JUN84          .     2.4718     0.035    1.9827    2.9609         .
271      JUL84          .     2.4768     0.035    1.9877    2.9660         .
272      AUG84          .     2.4818     0.035    1.9926    2.9711         .
273      SEP84          .     2.4868     0.035    1.9976    2.9761         .
274      OCT84          .     2.4919     0.036    2.0025    2.9812         .

Sum of Residuals                     0
Sum of Squared Residuals       14.9936
Predicted Resid SS (Press)     15.2134

DURBIN-WATSON D              0.090 ❶
(FOR NUMBER OF OBS.)           250 ❷
1ST ORDER AUTOCORRELATION    0.951 ❸
```

Now, for example, you compute:

$$1.119-(1.96)(0.0312) < \log\ (\beta_0) < 1.119+(1.96)(0.0312)$$

Thus,

$$2.880 < \beta_0 < 3.255$$

is a 95% confidence interval for β_0. Similarly, you obtain

$$1.0046 < \beta_1 < 1.0054$$

which is a 95% confidence interval for β_1. The growth rate of Treasury bills is estimated from this model to be between 0.46% and 0.54% per time period. Your forecast for November 1982 can be obtained from

$$1.888 < 2.377 < 2.865$$

so that

$$6.61 < \text{FYGM3}_{251} < 17.55$$

is a 95% prediction interval for the November 1982 yield and

$$\exp(2.377) = 10.77$$

is the predicted value. Because the distribution on the original levels is highly skewed, the prediction 10.77 does not lie midway between 6.61 and 17.55, nor would you want it to do so.

Note that the Durbin-Watson statistic ❶ is d=0.090. However, because n=250 ❷ is beyond the range of the Durbin-Watson tables, you use $\hat{\rho} = 0.951$ ❸ to compute

$$n^{1/2}\hat{\rho}/\left(1-\hat{\rho}^2\right)^{1/2} = 48.63$$

which is greater than 1.645. At the 5% level, you can conclude that positive autocorrelation is present (or that your model is misspecified in some other way). This is also evident in the plot, in **Output 1.13**, in which the data fluctuate around the overall trend in a clearly dependent fashion. Therefore, you should recompute your forecasts and confidence intervals using some of the methods in this book that consider autocorrelation.

Suppose X=log(y) and X is normal with mean M_x and variance σ_x^2. Then y = exp(x) and y has median exp(M_x) and mean exp(M_x+ ½σ_x^2) For this reason, some authors suggest adding half the error variances to a log scale forecast prior to exponentiation. We prefer to simply exponentiate and think of the result, for example, exp(2.377) = 10.77, as an estimate of the median, reasoning that this is a more credible central estimate for such a highly skewed distribution.

Chapter 2 Simple Models: Autoregression

2.1 Introduction

2.1.1 Terminology and Notation

Often you can forecast a series Y_t simply based on past values Y_{t-1}, Y_{t-2}, ... For example, suppose Y_t satisfies

$$Y_t - \mu = \rho\left(Y_{t-1} - \mu\right) + e_t \tag{2.1}$$

where e_t is a sequence of uncorrelated $N\left(0,\sigma^2\right)$ variables. The term for such an e_t sequence is white noise.

Assuming equation 2.1 holds at all times t, you can write, for example,

$$Y_{t-1} - \mu = \rho\left(Y_{t-2} - \mu\right) + e_{t-1}$$

and when you substitute in equation 2.1, you obtain

$$Y_t - \mu = e_t + \rho e_{t-1} + \rho^2\left(Y_{t-2} - \mu\right)$$

When you continue like this, you obtain

$$Y_t - \mu = e_t + \rho e_{t-1} + \rho^2 e_{t-2} + \ldots + \rho^{t-1} e_1 + \rho^t\left(Y_0 - \mu\right) \tag{2.2}$$

If you assume $|\rho| < 1$, the effect of the series values before you started collecting data (Y_0, for example) is minimal. Furthermore, you see that the mean (expected value) of Y_t is μ.

Suppose the variance of Y_{t-1} is $\sigma^2/\left(1-\rho^2\right)$. Then the variance of

$$\rho\left(Y_{t-1} - \mu\right) + e_t$$

is

$$\rho^2\sigma^2 / \left(1-\rho^2\right)+\sigma^2 = \sigma^2 / \left(1-\rho^2\right)$$

which shows that the variance of Y_t is also $\sigma^2 / \left(1-\rho^2\right)$.

2.1.2 Statistical Background

You can define Y_t as an accumulation of past shocks e_t to the system by writing the mathematical model shown in equation 2.1, model 1, as

$$Y_t = \mu + \Sigma_{j=0}^{\infty} \rho^j e_{t-j} \qquad (2.3)$$

that is, by extending equation 2.2 back into the infinite past. This again shows that if $|\rho| < 1$ the effect of shocks in the past is minimal. Equation 2.3, in which the series is expressed in terms of a mean and past shocks, is often called the "Wold representation" of the series. You can also compute a covariance between Y_t and Y_{t-j} from equation 2.3. Calling this covariance

$$\gamma(j) = \mathrm{cov}\left(Y_t, Y_{t-j}\right)$$

you have

$$\gamma(j) = \rho^{|j|}\sigma^2 / \left(1-\rho^2\right) = \rho^{|j|}\mathrm{var}\left(Y_t\right)$$

An interesting feature is that $\gamma(j)$ does not depend on *t*. In other words, the covariance between Y_t and Y_s depends only on the time distance | t–s | between these observations and not on the values *t* and *s*.

Why emphasize variances and covariances? They determine which model is appropriate for your data. One way to determine when model 1 is appropriate is to compute estimates of the covariances of your data and determine if they are of the given form—that is, if they decline exponentially at rate ρ as lag *j* increases. Suppose you observe this $\gamma(j)$ sequence:

$$\gamma(0) = 243,\ \gamma(1) = 162,\ \gamma(2) = 108,\ \gamma(3) = 72,\ \gamma(4) = 48,\ \gamma(5) = 32,\ \gamma(6) = 21.3, \ldots$$

You know the variance of your process, which is

$$\mathrm{var}\left(Y_t\right) = \gamma(0) = 243$$

and you note that

$$\gamma(1) / \gamma(0) = 2/3$$

Also,

$$\gamma(2) / \gamma(1) = 2/3$$

and, in fact,

$$\gamma(j) / \gamma(j-1) = 2/3$$

all the way through the sequence. Thus, you decide that model 1 is appropriate and that $\rho = 2/3$. Because

$$\gamma(0) = \sigma^2 / \left(1-\rho^2\right)$$

you also know that

$$\sigma^2 = \left(1 - (2/3)^2\right)(243) = 135$$

$\sigma^2 = (1-\rho^2)\cdot\gamma(0)$

2.2 Forecasting

How does your knowledge of ρ help you forecast? Suppose you know $\mu = 100$ (in practice, you use an estimate like the mean, $\overline{Y}$, of your observations). If you have data up to time n, you know that in the discussion above

$$Y_{n+1} - 100 = (2/3)(Y_n - 100) + e_{n+1}$$

At time n, e_{n+1} has not occurred and is not correlated with anything that has occurred up to time n. You forecast e_{n+1} by its unconditional mean 0. Because Y_n is available, it is easy to compute the forecast of Y_{n+1} as

$$\hat{Y}_{n+1} = 100 + (2/3)(Y_n - 100)$$

and the forecast error as

$$Y_{n+1} - \hat{Y}_{n+1} = e_{n+1}$$

Similarly,

$$\begin{aligned} Y_{n+2} - 100 &= (2/3)\left(Y_{n+1} - 100\right) + e_{n+2} \\ &= (2/3)\left[(2/3)\left(Y_n - 100\right) + e_{n+1}\right] + e_{n+2} \end{aligned}$$

and you forecast Y_{n+2} as

$$100 + (2/3)^2 (Y_n - 100)$$

with forecast error

$$e_{n+2} + (2/3)e_{n+1}$$

Similarly, for a general μ and ρ, a forecast L steps into the future is

$$\mu + \rho^{L}(Y_n - \mu)$$

with error

$$e_{n+L} + \rho e_{n+L-1} + \ldots + \rho^{L-1} e_{n+1}$$

A forecasting strategy now becomes clear. You do the following:

1. Examine estimates of the autocovariances $\gamma(j)$ to see if they decrease exponentially.
2. If so, assume model 1 holds and estimate μ and ρ.
3. Calculate the prediction

$$\hat{Y}_{n+L} = \mu + \rho^{L}(Y_n - \mu)$$

and the forecast error variance

$$\sigma^2\left(1+\rho^2+\rho^4+\ldots+\rho^{2L-2}\right)$$

You must substitute estimates, like $\hat{\rho}$, for your parameters.

For example, if $\mu=100$, $\rho=2/3$, and $Y_n=127$, the forecasts become 118, 112, 108, 105.3, 103.6, 102.4, … . The forecast error variances, based on

$$\text{var}(e)=\sigma^2=135$$

become 135, 195, 221.7, 233.5, 238.8, and 241.1. The forecasts decrease exponentially at rate $\rho=2/3$ to the series mean $\mu=100$. The forecast error variance converges to the series variance

$$\sigma^2/\left(1-\rho^2\right)=\left\{135/\left[1-(2/3)^2\right]\right\}=243=\gamma(0)$$

This shows that an equation like equation 2.1 helps you forecast in the short run, but you may as well use the series mean to forecast a stationary series far into the future.

In this section $Y_t-\mu=\rho\,(Y_{t-1}-\mu)+e_t$ was expanded as an infinite sum of past shocks e_t showing how past shocks accumulate to determine the current deviation of Y from the mean. At time n + L this expansion was

$$Y_{n+L}-\mu=\left\{e_{n+L}+\rho\, e_{n+L-1}+\cdots+\rho^{L-1}e_{n+1}\right\}+\rho^L\left\{e_n+\rho\, e_{n-1}+\cdots\right\}$$

which, substituting $Y_n-\mu=e_n+\rho\, e_{n-1}+\cdots$, shows that

1. The best (minimum prediction error variance) prediction of Y_{n+L} is $\mu+\rho^L(Y_n-\mu)$
2. The error in that prediction is $\left\{e_{n+L}+\rho\, e_{n+L-1}+\cdots+\rho^{L-1}e_{n+1}\right\}$ so the prediction error variance is $\sigma^2\left[1+\rho^2+\rho^4+\cdots+\rho^{2L-2}\right]$
3. The effect of shocks that happened a long time ago has little effect on the present Y if $|\rho|<1$.

The future shocks (e's) in item 2 have not yet occurred, but from the historic residuals an estimate of σ^2 can be obtained so that the error variance can be estimated and a prediction interval calculated. It will be shown that for a whole class of models called ARMA models, such a decomposition of Y_{n+L} into a prediction that is a function of current and past Y's, plus a prediction error that is a linear combination of future shocks (e's) is possible. The coefficients in these expressions are functions of the model parameters, like ρ, which can be estimated.

2.2.1 Forecasting with PROC ARIMA

As an example, 200 data values Y_1, Y_2, …, Y_{200} with mean $\overline{Y}=90.091$ and last observation Y_{200} = 140.246 are analyzed with these statements:

```
PROC ARIMA DATA=EXAMPLE;
   IDENTIFY VAR=Y CENTER;
   ESTIMATE P=1 NOCONSTANT;
   FORECAST LEAD=5;
RUN;
```

Output 2.1 shows the results when you use PROC ARIMA to identify, estimate, and forecast. The CENTER option tells PROC ARIMA to use the series mean, $\bar{Y}$, to estimate μ. The estimates of $\gamma(j)$ are called covariances ❶ with *j* labeled LAG ❷ on the printout. The covariances 1199, 956, 709, 524, 402, 309, . . . decrease at a rate of about .8. Dividing each covariance by the variance 1198.54 (covariance at lag 0) gives the estimated sequence of correlations. ❸ (The correlation at lag 0 is always $\rho(0)=1$ and in general $\rho(j)=\gamma(j)/\gamma(0)$.) The correlation plot ❹ shows roughly exponential decay. The ESTIMATE statement produces an estimate ❺ $\hat{\rho}=0.80575$ that you can test for significance with the *t* ratio. ❻ Since *t*=18.91 exceeds the 5% critical value, $\hat{\rho}$ is significant. If ρ were 0, this *t* would have approximately a standard normal distribution in large samples. Thus a *t* exceeding 1.96 in magnitude would be considered significant at about the 5% level. Also, you have an estimate of σ^2, 430.7275 ❼. You forecast Y_{201} by

$$90.091 + .80575(140.246 - 90.091) = 130.503$$

with forecast standard error

$$(430.73)^{.5} = 20.754 \text{ ❽}$$

Next, you forecast Y_{202} by

$$90.091 + (.80575)^2(140.246 - 90.091) = 122.653$$

with forecast standard error

$$(430.73(1 + .80575^2))^{.5} = 26.6528 \text{ ❽}$$

Output 2.1 Using PROC ARIMA to Identify, Estimate, and Forecast

```
                                The ARIMA Procedure

                         Name of Variable                   Y
                         Mean of Working Series             0
                         Standard Deviation          34.61987
                         Number of Observations           200

                                      Autocorrelations
 ❷         ❶                 ❸                                ❹
 Lag    Covariance    Correlation    -1 9 8 7 6 5 4 3 2 1 0 1 2 3 4 5 6 7 8 9 1      Std Error
   0      1198.535        1.00000    |                    |********************|             0
   1       955.584        0.79729    |                  . |****************    |      0.070711
   2       708.551        0.59118    |                 .  |************        |      0.106568
   3       524.036        0.43723    |                .   |*********           |      0.121868
   4       402.374        0.33572    |                .   |*******             |      0.129474
   5       308.942        0.25777    |                .   |*****               |      0.133755

 (More Output Lines)

  19    -47.371450        -.03952    |                .  *|   .                |      0.147668
  20    -82.867591        -.06914    |                .  *|   .                |      0.147720
  21      -140.527        -.11725    |                . **|   .                |      0.147882
  22      -113.545        -.09474    |                . **|   .                |      0.148346
  23    -88.683505        -.07399    |                .  *|   .                |      0.148648
  24    -50.803423        -.04239    |                .  *|   .                |      0.148832

                             "." marks two standard errors
```

Output 2.1 *Using PROC ARIMA to Identify, Estimate, and Forecast (continued)*

```
                       Inverse Autocorrelations

 Lag    Correlation    -1 9 8 7 6 5 4 3 2 1 0 1 2 3 4 5 6 7 8 9 1
   1       -0.57652    |        ************|  .                 |
   2        0.09622    |                 .  |**.                 |
   3        0.02752    |                 .  |* .                 |
   4       -0.07210    |                 . *|  .                 |
   5        0.04054    |                 .  |* .                 |

  (More Output Lines)

  19        0.11277    |                 .  |**.                 |
  20       -0.18424    |                ****|  .                 |
  21        0.20996    |                 .  |****                |
  22       -0.12659    |                 ***|  .                 |
  23        0.04992    |                 .  |* .                 |
  24       -0.01756    |                 .  |  .                 |


                       Partial Autocorrelations

 Lag    Correlation    -1 9 8 7 6 5 4 3 2 1 0 1 2 3 4 5 6 7 8 9 1
   1        0.79729    |                 .  |****************    |
   2       -0.12213    |                 .**|  .                 |
   3        0.01587    |                 .  |  .                 |
   4        0.03245    |                 .  |* .                 |
   5       -0.00962    |                 .  |  .                 |

  (More Output Lines)

  19       -0.06557    |                 . *|  .                 |
  20        0.01500    |                 .  |  .                 |
  21       -0.10473    |                 .**|  .                 |
  22        0.14816    |                 .  |***                 |
  23       -0.03625    |                 . *|  .                 |
  24        0.03510    |                 .  |* .                 |

                  Autocorrelation Check for White Noise

 To        Chi-             Pr >
Lag      Square     DF     ChiSq    --------------------Autocorrelations-------------------
  6      287.25      6    <.0001     0.797     0.591     0.437     0.336     0.258     0.227
 12      342.46     12    <.0001     0.240     0.261     0.243     0.198     0.157     0.111
 18      345.17     18    <.0001     0.071     0.054     0.042     0.037     0.036     0.008
 24      353.39     24    <.0001    -0.040    -0.069    -0.117    -0.095    -0.074    -0.042
```

Output 2.1 *Using PROC ARIMA to Identify, Estimate, and Forecast (continued)*

```
                          Conditional Least Squares Estimation

                              ❺          Standard          ❻         Approx
        Parameter          Estimate         Error     t Value    Pr > |t|       Lag
        AR1,1               0.80575       0.04261       18.91      <.0001         1

                          Variance Estimate          430.7275 ❼
                          Std Error Estimate         20.75397
                          AIC                        1781.668
                          SBC                        1784.966
                          Number of Residuals             200
                     * AIC and SBC do not include log determinant.

                             Autocorrelation Check of Residuals

  To        Chi-               Pr >
 Lag      Square      DF      ChiSq     -------------------Autocorrelations-------------------
   6        5.46       5     0.3623      0.103     -0.051     -0.074     -0.020     -0.063     -0.060
  12        9.46      11     0.5791      0.014      0.110      0.074      0.007      0.034     -0.002
  18       11.30      17     0.0400     -0.048      0.002      0.007     -0.001      0.065      0.042
  24       20.10      23     0.6359     -0.042      0.043     -0.185     -0.006     -0.032     -0.005
  30       24.49      29     0.7043      0.064     -0.007      0.033      0.028     -0.098     -0.056
  36       27.06      35     0.8290      0.029      0.029      0.074      0.002     -0.036     -0.046


                                   The ARIMA Procedure

                                 Model for variable Y
                   Data have been centered by subtracting the value      00.00064
                               No mean term in this model.
                                 Autoregressive Factors

                              Factor 1:  1 - 0.80575 B**(1)

                                Forecasts for variable Y

          Obs        Forecast      Std Error          95% Confidence Limits
          201        130.5036        20.7540 ❽         89.8265         171.1806
          202        122.6533        26.6528           70.4149         174.8918
          203        116.3280        29.8651           57.7936         174.8625
          204        111.2314        31.7772           48.9492         173.5136
          205        107.1248        32.9593           42.5257         171.7239
```

In the manner previously illustrated, PROC ARIMA produced the forecasts and standard errors. ❽ The coefficients are estimated through the least squares (LS) method. This means that

$$.80575 = \Sigma\left(Y_t - \overline{Y}\right)\left(Y_{t-1} - \overline{Y}\right) \;/\; \Sigma\left(Y_{t-1} - \overline{Y}\right)^2$$

where $\overline{Y}$ is the mean of the data set and the sums run from 2 to 200. One alternative estimation scheme is the maximum-likelihood (ML) method and another is unconditional least squares (ULS). A discussion of these methods in the context of the autoregressive order 1 model follows. The likelihood function for a set of observations is simply their joint probability density viewed as a function of the parameters. The first observation Y_1 is normal with mean μ and variance $\sigma^2/(1-\rho^2)$. Its probability density function is

$$\frac{\sqrt{1-\rho^2}}{\sqrt{2\pi\sigma^2}}\exp\left(-\frac{(Y_1-\mu)^2(1-\rho^2)}{2\sigma^2}\right)$$

For the rest of the observations, t=2,3,4,…, it is most convenient to note that $e_t = Y_t - \rho Y_{t-1}$ has a normal distribution with mean $\mu - \rho\mu = (1-\rho)\mu$ and variance σ^2.

Each of these probability densities is thus given by

$$\frac{1}{\sqrt{2\pi\sigma^2}}\exp\left(-\frac{[(Y_t-\mu)-\rho(Y_{t-1}-\mu)]^2}{2\sigma^2}\right)$$

Because $Y_1, e_2, e_3, \ldots, e_n$ are independent, the joint likelihood is the product of these n probability density functions, namely

$$\frac{\sqrt{1-\rho^2}}{\left(\sqrt{2\pi\sigma^2}\right)^n}\exp\left(-\frac{(1-\rho^2)(Y_1-\mu)^2+[(Y_2-\mu)-\rho(Y_1-\mu)]^2+\ldots+[(Y_n-\mu)-\rho(Y_{n-1}-\mu)]^2}{2\sigma^2}\right)$$

Now substituting the observations for Y in the expression above produces an expression involving μ, ρ, and σ^2. Viewed in this way, the expression above is called the likelihood function for the data and clearly depends on assumptions about the model form. Using calculus, it can be shown that the estimate of σ^2 that maximizes the likelihood is USS/n, where USS represents the unconditional sum of squares:

$$\text{USS} = (1-\rho^2)(Y_1-\mu)^2+[(Y_2-\mu)-\rho(Y_1-\mu)]^2+\ldots+[(Y_n-\mu)-\rho(Y_{n-1}-\mu)]^2$$

The estimates that minimize USS are the unconditional least squares (ULS) estimates—that is, USS is the objective function to be minimized by the ULS method. The minimization can be modified as in the current example by inserting $\bar{Y}$ in place of μ, leaving only ρ to be estimated.

The conditional least squares (CLS) method results from assuming that Y_0 and all other Ys that occurred before we started observing the series are equal to the mean. Thus it minimizes a slightly different objective function,

$$[Y_1-\mu]^2+[(Y_2-\mu)-\rho(Y_1-\mu)]^2+\ldots+[(Y_n-\mu)-\rho(Y_{n-1}-\mu)]^2$$

and, as with the other methods, it can be modified by inserting $\bar{Y}$ in place of μ, leaving only ρ to be estimated. The first term cannot be changed by manipulating ρ, so the CLS method with $\bar{Y}$ inserted also minimizes

$$[(Y_2-\bar{Y})-\rho(Y_1-\bar{Y})]^2+\ldots+[(Y_n-\bar{Y})-\rho(Y_{n-1}-\bar{Y})]^2$$

In other words the CLS estimate of ρ could be obtained by regressing deviations from the sample mean on their lags with no intercept in this simple centered case.

If full maximum likelihood estimation is desired, the expression USS/*n* is substituted for σ^2 in the likelihood function and the resulting expression, called a concentrated likelihood, is maximized. The log of the likelihood is

$$-(n/2)\log(2\pi/n)-(n/2)-(n/2)\log(\text{USS})+(1/2)\log(1-\rho^2)$$

The ML method can be run on centered data by inserting $Y_t-\bar{Y}$ in USS in place of $Y_t-\mu$.

For the series 14 15 14 10 12 10 5 6 6 8 the sample average is 10. The three rows in **Output 2.2** display the objective functions just discussed for conditional least squares, unconditional least squares, and maximum likelihood for an autoregressive order 1 model fit to these data. The negative of the likelihood is shown so that a minimum is sought in each case. The right panel in each row plots the function to be minimized over a floor of (ρ,μ) pairs, with each function truncated by a convenient ceiling plane. Crosshairs in the plot floors indicate the minimizing values, and it is seen that these estimates can vary somewhat from method to method when the sample size is very small. Each plot also shows a vertical slicing plane at $\mu=10$, corresponding to the sample mean. The left plots show the cross section from the slicing planes. These then are the objective functions to be minimized when the sample mean, 10, is used as an estimate of the population mean. The slicing plane does not meet the floor at the crosshair mark, so the sample mean differs somewhat from the estimate that minimizes the objective function. Likewise the ρ that minimizes the cross section plot is not the same as the one minimizing the surface plot, although this difference is quite minor for ULS and ML in this small example.

Output 2.2 *Objective Functions*

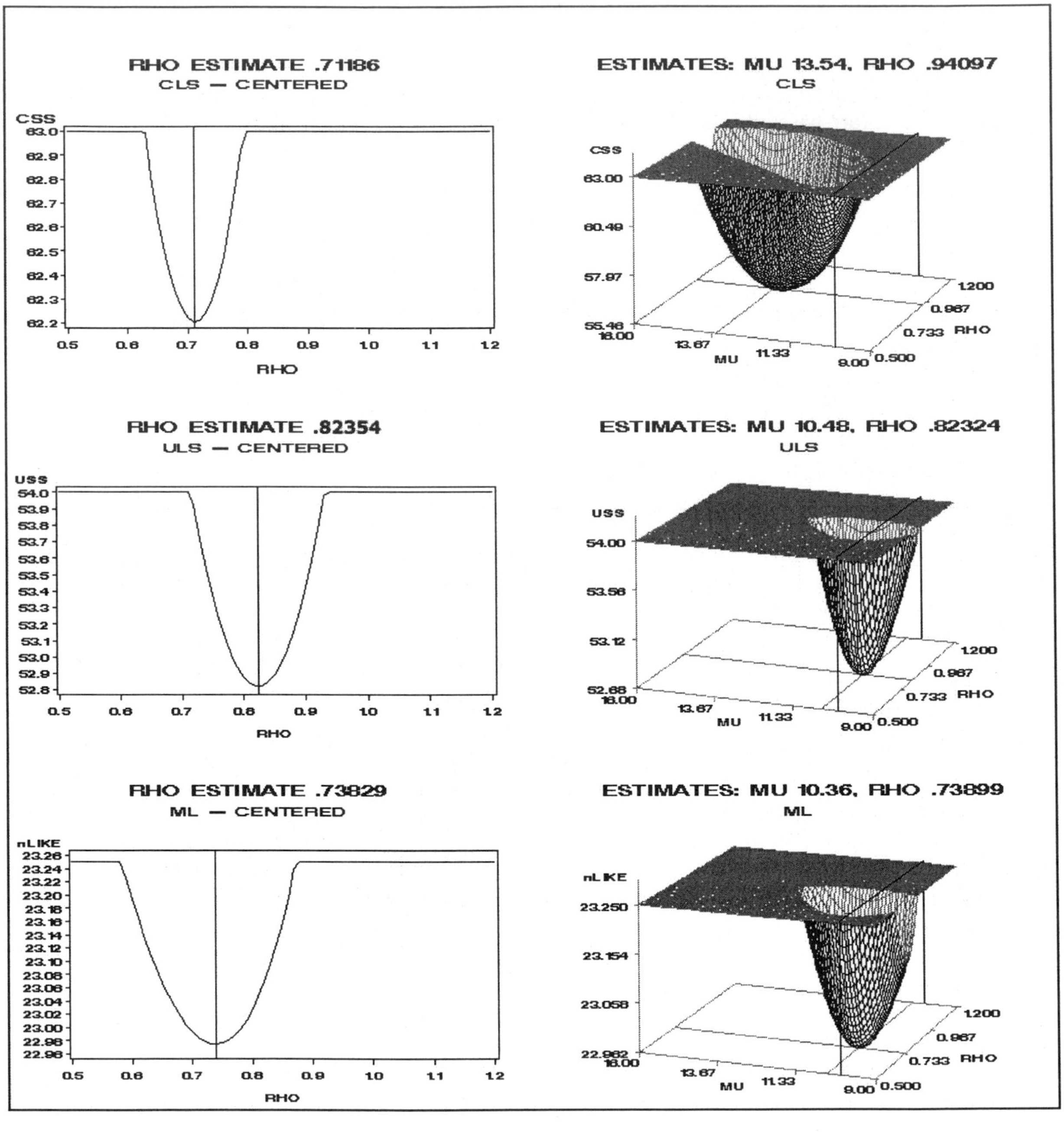

The minimizing values for the right-side ULS plot are obtained from the code

```
PROC ARIMA DATA=ESTIMATE;
   IDENTIFY VAR=Y NOPRINT;
   ESTIMATE P=1 METHOD = ULS OUTEST=OUTULS PRINTALL;
RUN;
```

with METHOD=ML for maximum likelihood and no method specification for CLS. The OUTEST data set holds the estimates and related information and PRINTALL shows the iterative steps used to search for the minima. The use of the CENTER option in the IDENTIFY statement along with the NOCONSTANT option in the ESTIMATE statement will produce the ρ estimate that minimizes the objective function computed with the sample mean (10). A partial output showing the iterations for our small series is shown in **Output 2.3**. The second column in each segment is the objective function that is being minimized and should end with the height of the lowest point in each plot. The estimates correspond to the coordinate(s) on the horizontal axis (or the floor) corresponding to the minimum.

Output 2.3 *Using PROC ARIMA to Get Iterations for Parameter Estimates*

Conditional Least Squares Estimation

Iteration	SSE	MU	AR1,1	Constant	Lambda	R Crit
0	62.26767	10.0000	0.6885	3.114754	0.00001	1
1	58.98312	11.2357	0.7053	3.310709	1E-6	0.216536
2	57.57647	11.4867	0.7957	2.347267	1E-7	0.1318
3	56.72048	12.1199	0.8211	2.168455	1E-8	0.0988
4	56.17147	12.4242	0.8643	1.686007	1E-9	0.074987
5	55.81877	12.8141	0.8858	1.463231	1E-10	0.060663
6	55.62370	13.0412	0.9073	1.209559	1E-11	0.045096
7	55.52790	13.2380	0.9191	1.070524	1E-12	0.03213
8	55.48638	13.3505	0.9282	0.959077	1E-12	0.021304
9	55.46978	13.4302	0.9332	0.897121	1E-12	0.013585
10	55.46351	13.4751	0.9366	0.854795	1E-12	0.008377
11	55.46123	13.5040	0.9385	0.831106	1E-12	0.005073
12	55.46041	13.5205	0.9396	0.816055	1E-12	0.003033
13	55.46013	13.5300	0.9400	0.807508	1E-12	0.001801
14	55.46003	13.5364	0.9407	0.802304	1E-12	0.001065
15	55.45999	13.5399	0.9410	0.79931	1E-12	0.000628

Conditional Least Squares Estimation

Iteration	SSE	MU	AR1,1	Constant	Lambda	R Crit
0	62.26767	10.0000	0.6885	3.114754	0.00001	1
1	58.98312	11.2357	0.7053	3.310709	1E-6	0.216536
2	57.57647	11.4867	0.7957	2.347267	1E-7	0.1318
3	56.72048	12.1199	0.8211	2.168455	1E-8	0.0988
4	56.17147	12.4242	0.8643	1.686007	1E-9	0.074987

Output 2.3
Using PROC ARIMA to Get Iterations for Parameter Estimates (continued)

Unconditional Least Squares Estimation

Iteration	SSE	MU	AR1,1	Constant	Lambda	R Crit
0	54.31645	12.4242	0.8643	1.686007	0.00001	1
1	52.98938	10.6007	0.8771	1.302965	1E-6	0.165164
2	52.77450	10.8691	0.8370	1.771771	1E-7	0.065079
3	52.70017	10.5107	0.8357	1.726955	1E-8	0.036672
4	52.68643	10.5479	0.8262	1.833486	1E-9	0.01526
5	52.68382	10.4905	0.8254	1.831752	1E-10	0.006528
6	52.68330	10.4928	0.8237	1.849434	1E-11	0.00267
7	52.68321	10.4841	0.8235	1.850351	1E-12	0.001104
8	52.68319	10.4838	0.8232	1.853158	1E-12	0.000458

Conditional Least Squares Estimation

Iteration	SSE	MU	AR1,1	Constant	Lambda	R Crit
0	62.26767	10.0000	0.6885	3.114754	0.00001	1
1	58.98312	11.2357	0.7053	3.310709	1E-6	0.216536
2	57.57647	11.4867	0.7957	2.347267	1E-7	0.1318
3	56.72048	12.1199	0.8211	2.168455	1E-8	0.0988
4	56.17147	12.4242	0.8643	1.686007	1E-9	0.074987

Maximum Likelihood Estimation

Iter	Loglike	MU	AR1,1	Constant	Lambda	R Crit
0	-23.33779	12.4242	0.8643	1.686007	0.00001	1
1	-22.97496	10.3212	0.7696	2.378179	1E-6	0.233964
2	-22.96465	10.5093	0.7328	2.808362	1E-7	0.058455
3	-22.96211	10.3352	0.7438	2.647467	1E-8	0.028078
4	-22.96176	10.3827	0.7374	2.726623	1E-9	0.010932
5	-22.96169	10.3548	0.7397	2.69579	1E-10	0.004795
6	-22.96168	10.3648	0.7385	2.709918	1E-11	0.002018
7	-22.96168	10.3600	0.7390	2.70409	1E-12	0.00087

Conditional Least Squares Estimation

Iteration	SSE	AR1,1	Lambda	R Crit
0	62.26767	0.6885	0.00001	1
1	62.20339	0.7119	1E-6	0.03213
2	62.20339	0.7119	1E-7	3.213E-7

Conditional Least Squares Estimation

Iteration	SSE	AR1,1	Lambda	R Crit
0	62.26767	0.6885	0.00001	1
1	62.20339	0.7119	1E-6	0.03213
2	62.20339	0.7119	1E-7	3.213E-7

Output 2.3 *Using PROC ARIMA to Get Iterations for Parameter Estimates (continued)*

```
              Unconditional Least Squares Estimation

Iteration          SSE     AR1,1     Lambda     R Crit
        0     54.09537    0.7119    0.00001          1
        1     52.89708    0.7967       1E-6   0.133727
        2     52.82994    0.8156       1E-7   0.031428
        3     52.82410    0.8212       1E-8   0.009357
        4     52.82357    0.8229       1E-9   0.002985
        5     52.82353    0.8235      1E-10   0.000973

               Conditional Least Squares Estimation

Iteration          SSE     AR1,1     Lambda     R Crit
        0     62.26767    0.6885    0.00001          1
        1     62.20339    0.7119       1E-6    0.03213
        2     62.20339    0.7119       1E-7   3.213E-7

                  Maximum Likelihood Estimation

     Iter      Loglike     AR1,1     Lambda     R Crit
        0    -22.98357    0.7119    0.00001          1
        1    -22.97472    0.7389       1E-6   0.042347
        2    -22.97471    0.7383       1E-7   0.001059
        3    -22.97471    0.7383       1E-8   0.000027
```

Notice that each method begins with conditional least squares starting with the sample mean and an estimate, .6885, of the autoregressive coefficient. The CLS estimates, after a few iterations, are substituted in the ULS or ML objective function when one of those methods is specified. In more complex models, the likelihood function is more involved, as are the other objective functions. Nevertheless the basic ideas presented here generalize nicely to all models handled by PROC ARIMA.

You have no reason to believe that dependence of Y_t on past values should be limited to the previous observation Y_{t-1}. For example, you may have

$$Y_t - \mu = \alpha_1(Y_{t-1} - \mu) + \alpha_2(Y_{t-2} - \mu) + e_t \tag{2.4}$$

which is a second-order autoregressive (AR) process. One way to determine if you have this process is to examine the autocorrelation plot by using the following SAS statements:

```
PROC ARIMA DATA=ESTIMATE;
   IDENTIFY VAR=Y;
RUN;
```

You thus need to study the form of autocorrelations for such AR processes, which is facilitated by writing the models in backshift notation.

2.2.2 Backshift Notation B for Time Series

A convenient notation for time series is the backshift notation B where

$$B(Y_t) = Y_{t-1}$$

That is, B indicates a shifting back of the time subscript. Similarly,

$$B^2(Y_t) = B(Y_{t-1}) = Y_{t-2}$$

and

$$B^5(Y_t) = Y_{t-5}$$

Now consider the process

$$Y_t = .8Y_{t-1} + e_t$$

In backshift notation this becomes

$$(1 - .8B)Y_t = e_t$$

You can write

$$Y_t = (1 - .8B)^{-1} e_t$$

and, recalling that

$$(1 - X)^{-1} = 1 + X + X^2 + X^3 + \ldots$$

for $|X| < 1$, you obtain

$$Y_t = (1 + .8B + .8^2B^2 + .8^3B^3 + \ldots)e_t$$

or

$$Y_t = e_t + .8e_{t-1} + .64e_{t-2} + \ldots$$

It becomes apparent that the backshift allows you to execute the computations, linking equations 2.1 and 2.3 in a simplified manner. This technique extends to higher-order processes. For example, let

$$Y_t = 1.70Y_{t-1} - .72Y_{t-2} + e_t \qquad (2.5)$$

Comparing equations 2.5 and 2.4 results in $\mu = 0$, $\alpha_1 = 1.70$, and $\alpha_2 = -.72$. You can rewrite equation 2.5 as

$$(1 - 1.70B + .72B^2)Y_t = e_t$$

or as

$$Y_t = (1 - 1.70B + .72B^2)^{-1} e_t \qquad (2.6)$$

Algebraic combination shows that

$$9 / (1 - .9B) - 8 / (1 - .8B) = 1 / (1 - 1.70B + .72B^2)$$

Thus, you can write Y_t as

$$Y_t = \Sigma_{j=0}^{\infty} W_j e_{t-j}$$

where

$$W_j = 9(.9^j) - 8(.8^j)$$

You can see that the influence of early shocks e_{t-j} is minimal because .9 and .8 are less than 1. Equation 2.6 allows you to write Y_t as

$$\begin{aligned} Y_t &= e_t + 1.7e_{t-1} + 2.17e_{t-2} + 2.47e_{t-3} + 2.63e_{t-4} \\ &\quad + 2.69e_{t-5} + 2.69e_{t-6} + 2.63e_{t-7} + 2.53e_{t-8} + \ldots \end{aligned} \tag{2.7}$$

which you can also accomplish by repeated back substitution as in equation 2.3. Note that the weights W_j initially increase before tapering off toward 0.

2.2.3 Yule-Walker Equations for Covariances

You have learned how to use backshift notation to write a time series as a weighted sum of past shocks (as in equation 2.7); you are now ready to compute covariances $\gamma(j)$. You accomplish this by using the Yule-Walker equations. These equations result from multiplying the time series equation, such as equation 2.5, by Y_{t-j} and computing expected values.

For equation 2.5, when you use j=0, you obtain

$$E(Y_t^2) = 1.70E(Y_t Y_{t-1}) - .72E(Y_t Y_{t-2}) + E(Y_t e_t)$$

$E(y_t^2) = E(\alpha_1 y_t y_{t-1} + \alpha_2 y_t y_{t-2} + y_t a_t)$
$= \alpha_1 E(y_t y_{t-1}) + \alpha_2 E(y_t y_{t-2}) + E(y_t a_t)$
$= \alpha_1 Cov(y_t, y_{t-1}) + \alpha_2 Cov(y_t, y_{t-2}) + Cov(y_t, a_t)$
$= \alpha_1 \gamma(1) + \alpha_2 \gamma(2) + \sigma_a^2$ since $\mu = 0$.

or

$$\gamma(0) = 1.70\gamma(1) - .72\gamma(2) + \sigma^2$$

where E stands for expected value. Using equation 2.7 with all subscripts lagged by 1, you see that Y_{t-1} involves only $e_{t-1}, e_{t-2}, \ldots$. Thus,

$$E(Y_{t-1}e_t) = 0$$

When you use j=1, you obtain

$$E(Y_t Y_{t-1}) = 1.70E(Y_{t-1}^{\ 2}) - .72E(Y_{t-1}Y_{t-2}) + E(Y_{t-1}e_t)$$

Furthermore,

$$E(Y_{t-1}Y_{t-2}) = \gamma(1)$$

because the difference in subscripts is

$$(t-1) - (t-2) = 1$$

Also recall that

$$\gamma(1) = \gamma(-1)$$

Using these ideas, write your second Yule-Walker equation as

$$\gamma(1) = 1.70\gamma(0) - .72\gamma(1)$$

In the same manner, for all j>0,

$$\gamma(j) = 1.70\gamma(j-1) - .72\gamma(j-2) \tag{2.8}$$

If you assume a value for σ^2 (for example, $\sigma^2 = 10$), you can use the Yule-Walker equations to compute autocovariances $\gamma(j)$ and autocorrelations

$$\rho(j) = \gamma(j)/\gamma(0)$$

The autocorrelations do not depend on σ^2. The Yule-Walker equations for j=0, j=1, and j=2 are three equations in three unknowns: $\gamma(0)$, $\gamma(1)$, and $\gamma(2)$. Solving these (using $\sigma^2 = 10$), you get $\gamma(0) = 898.1$, $\gamma(1) = 887.6$, and $\gamma(2) = 862.4$. Using equation 2.8, you then compute

$$\gamma(3) = 1.7(862.4) \ - \ .72(887.6) = 827.0$$

and

$$\gamma(4) = 1.7(827.0) \ - \ .72(862.4) = 785$$

and so forth.

Thus, the Yule-Walker equations for a second-order AR process (see equation 2.4) are

$$\gamma(0) = \alpha_1\gamma(1) \ + \ \alpha_2\gamma(2) \ + \ \sigma^2$$

and

$$\gamma(j) = \alpha_1\gamma(j-1) + \alpha_2\gamma(j-2), \ \ j > 0$$

You have also seen that PROC ARIMA gives estimates of $\gamma(j)$. With that in mind, suppose you have a time series with mean 100 and the covariance sequence as follows:

j	0	1	2	3	4	5	6	7	8	9	10	11	12	13	14
$\gamma(j)$	390	360	277.5	157.5	19.9	−113.8	−223.7	−294.5	−317.6	−292.2	−223.9	−125.5	−13.2	95.5	184.4

The last two observations are 130 and 132, and you want to predict five steps ahead. How do you do it? First, you need a model for the data. You can eliminate the first-order AR model 1 based on the failure of $\gamma(j)$ to damp out at a constant exponential rate. For example,

$$\gamma(1) / \gamma(0) = .92$$

but

$$\gamma(2) / \gamma(1) = .77$$

If the model is a second-order autoregression like equation 2.4, you have the Yule-Walker equations

$\gamma(0) = \alpha_1\gamma(1) + \alpha_2\gamma(2) + \sigma^2$ $\quad 390 = \alpha_1(360) + \alpha_2(277.5) + \sigma^2 \quad$ $Y_t = \alpha_1 Y_{t-1} + \alpha_2 Y_{t-2} + e_t$ AR(2)

$\gamma(1) = \alpha_1\gamma(0) + \alpha_2\gamma(1)$ $\quad 360 = \alpha_1(390) + \alpha_2(360)$

and

$\gamma(2) = \alpha_1\gamma(1) + \alpha_2\gamma(0)$ $\quad 277.5 = \alpha_1(360) + \alpha_2(390)$

These can be solved with $\alpha_1 = 1.80$, $\alpha_2 = -.95$, and $\sigma^2 = 5.625$. Thus, in general, if you know or can estimate the $\gamma(j)$s, you can find or estimate the coefficients from the Yule-Walker equations. You can confirm this diagnosis by checking to see that

$$\gamma(j) = 1.80\gamma(j - 1) - .95\gamma(j - 2)$$

for *j*=3, 4, ..., 14.

To predict, you first write your equation

$$Y_t - 100 = 1.80(Y_{t-1} - 100) - .95(Y_{t-2} - 100) + e_t \qquad (2.9)$$

Assuming your last observation is Y_n, you now write

$$Y_{n+1} = 100 + 1.80(Y_n - 100) - .95(Y_{n-1} - 100) + e_{n+1}$$

and

$$\hat{Y}_{n+1} = 100 + 1.80(132 - 100) - .95(130 - 100) = 129.1$$

where you recall that 130 and 132 were the last two observations. The prediction error is

$$Y_{n+1} - \hat{Y}_{n+1} = e_{n+1}$$

with variance $\sigma^2 = 5.625$, and you compute the one-step-ahead prediction interval from

$$129.1 - 1.96(5.625)^{.5}$$

to

$$129.1 + 1.96(5.625)^{.5}$$

The prediction of Y_{n+2} arises from

$$Y_{n+2} = 100 + 1.80(Y_{n+1} - 100) - .95(Y_n - 100) + e_{n+2}$$

and is given by

$$\hat{Y}_{n+2} = 100 + 1.80(\hat{Y}_{n+1} - 100) - .95(Y_n - 100) = 122$$

The prediction error is

$$1.8(Y_{n+1} - \hat{Y}_{n+1}) + e_{n+2} = 1.8e_{n+1} + e_{n+2}$$

with variance

$$\sigma^2(1 + 1.8^2) = 23.85$$

Using equation 2.9, you compute predictions, replacing unknown Y_{n+j} with predictions and e_{n+j} with 0 for *j*>0. You also can monitor prediction error variances. If you express Y_t in the form of equation 2.7, you get

$$Y_t - 100 = e_t + 1.8e_{t-1} + 2.29e_{t-2} + 2.41e_{t-3} + \ldots \qquad (2.10)$$

The prediction error variances for one, two, three, and four steps ahead are then σ^2, $\sigma^2(1+1.8^2)$, $\sigma^2(1+1.8^2+2.29^2)$, and $\sigma^2(1+1.8^2+2.29^2+2.41^2)$.

Surprisingly, the weights on e_{t-j} seem to increase as you move further into the past. However, if you continue to write out the expression for Y_t in terms of e_t, you see that the weights eventually taper off toward 0, just as in equation 2.7. You obtained equation 2.10 by writing the model

$$\left(1-1.80B+.95B^2\right)\left(Y_t-\mu\right)=e_t$$

as

$$\begin{aligned}\left(Y_t-\mu\right)&=\left(1-1.80B+.95B^2\right)^{-1}e_t\\&=\left(1+1.80B+2.29B^2+2.41B^3+...\right)e_t\end{aligned}$$

Now replace B with an algebraic variable M. The key to tapering off the weights involves the characteristic equation

$$1-1.80M+.95M^2=0$$

If all values of M (roots) that solve this equation are larger than 1 in magnitude, the weights taper off. In this case, the roots are $M=.95\pm.39i$, which is a complex pair of numbers with magnitude 1.03. In equation 2.5, the roots are 1.11 and 1.25. The condition of roots having a magnitude greater than 1 is called *stationarity* and ensures that shocks e_{t-j} in the distant past have little influence on the current observation Y_t.

A general review of the discussion so far indicates that an AR model of order *p*, written as

$$\begin{aligned}\left(Y_t-\mu\right)&=\alpha_1\left(Y_{t-1}-\mu\right)+\alpha_2\left(Y_{t-2}-\mu\right)+...\\&\quad+\alpha_p\left(Y_{t-p}-\mu\right)+e_t\end{aligned}\qquad(2.11)$$

can be written in backshift form as

$$\left(1-\alpha_1B-\alpha_2B^2-...-\alpha_pB^p\right)\left(Y_t-\mu\right)=e_t$$

and can be written as an infinite weighted sum of current and past shocks e_t with

$$\begin{aligned}\left(Y_t-\mu\right)&=\left(1-\alpha_1B-\alpha_2B^2-...-\alpha_pB^p\right)^{-1}e_t\\&=\left(1+W_1B+W_2B^2+W_3B^3+...\right)e_t\end{aligned}$$

where you can find the W_js. The W_js taper off toward 0 if all Ms satisfying

$$1-\alpha_1M-\alpha_2M^2-...-\alpha_pM^p=0$$

are such that $|M|>1$.

You have also learned how to compute the system of Yule-Walker equations by multiplying equation 2.11 on both sides by $\left(Y_{t-j}-\mu\right)$ for *j*=0, *j*=1, *j*=2, . . . and by computing expected values. You can use these Yule-Walker equations to estimate coefficients α_j when you know or can estimate values of the covariances $\gamma(j)$. You have also used covariance patterns to distinguish the second-order AR model from the first-order model.

2.3 Fitting an AR Model in PROC REG

Chapter 3, "The General ARIMA Model," shows that associating autocovariance patterns with models is crucial for determining an appropriate model for a data set. As you expand your set of models, remember that the primary way to distinguish among them is through their covariance functions. Thus, it is crucial to build a catalog of their covariance functions as you expand your repertoire of models. The covariance functions are like fingerprints, helping you identify the model form appropriate for your data.

Output 2.4 shows a plot of the stocks of silver at the New York Commodity Exchange in 1000 troy ounces from December 1976 through May 1981 (Fairchild Publications 1981). If you deal only with AR processes, you can fit the models by ordinary regression techniques like PROC REG or PROC GLM. You also can simplify the choice of the model's order, as illustrated in **Output 2.5**, and thus simplify your analysis.

Assuming a fourth-order model is adequate, you regress Y_t on Y_{t-1}, Y_{t-2}, Y_{t-3}, and Y_{t-4} using these SAS statements:

```
DATA SILVER;
   TITLE 'MONTH END STOCKS OF SILVER';
   INPUT SILVER @@;
   T=_N_;
   RETAIN DATE '01DEC76'D LSILVER1-LSILVER4;
   DATE=INTNX('MONTH',DATE,1);
   FORMAT DATE MONYY.;
   OUTPUT;
   LSILVER4=LSILVER3;
   LSILVER3=LSILVER2;
   LSILVER2=LSILVER1;
   LSILVER1=SILVER;
CARDS;
 846 827 799 768 719 652 580 546 500 493 530 548 565
 573 632 645 674 693 706 661 648 604 647 684 700 723
 741 734 708 728 737 729 678 651 627 582 521 519 496
 501 555 541 485 476 515 606 694 788 761 794 836 846
;
RUN;

PROC PRINT DATA=SILVER;
RUN;

PROC REG DATA=SILVER;
   MODEL SILVER=LSILVER1 LSILVER2
                LSILVER3 LSILVER4 / SS1;
RUN;

PROC REG DATA=SILVER;
   MODEL SILVER=LSILVER1 LSILVER2;
RUN;
```

Output 2.4 *Plotting Monthly Stock Values*

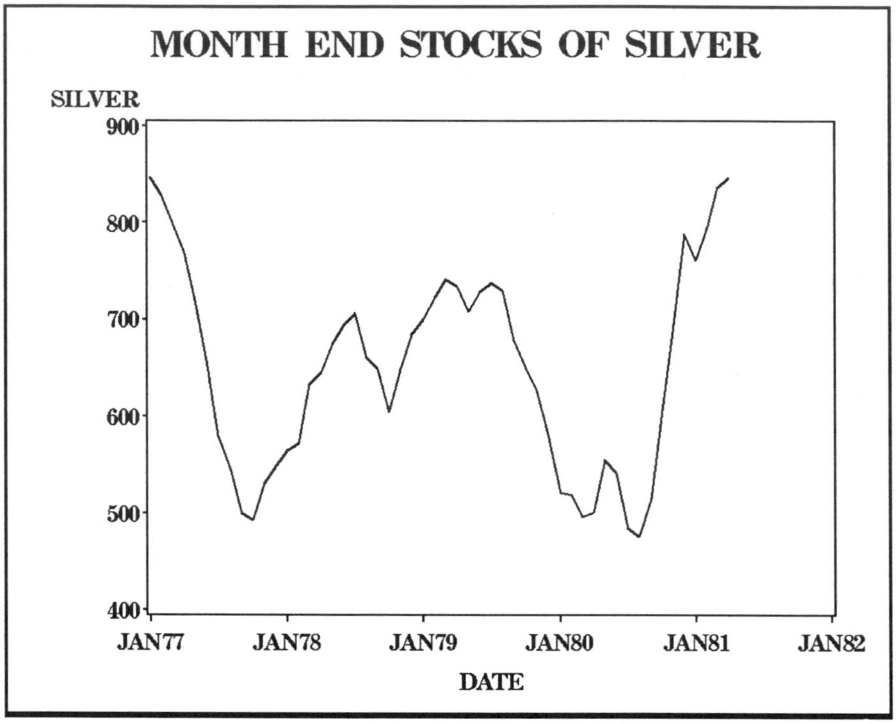

Output 2.5 *Using PROC PRINT to List the Data and PROC REG to Fit an AR Process*

MONTH END STOCKS OF SILVER

Obs	SILVER	T	DATE	LSILVER1	LSILVER2	LSILVER3	LSILVER4
1	846	1	JAN77	.	.	.	.
2	827	2	FEB77	846	.	.	.
3	799	3	MAR77	827	846	.	.
4	768	4	APR77	799	827	846	.
5	719	5	MAY77	768	799	827	846
(More Output Lines)							
48	788	48	DEC80	694	606	515	476
49	761	49	JAN81	788	694	606	515
50	794	50	FEB81	761	788	694	606
51	836	51	MAR81	794	761	788	694
52	846	52	APR81	836	794	761	788

Output 2.5 *Using PROC PRINT to List the Data and PROC REG to Fit an AR Process (continued)*

MONTH END STOCKS OF SILVER

The REG Procedure
Model: MODEL1
Dependent Variable: SILVER

Analysis of Variance

Source	DF	Sum of Squares	Mean Square	F Value	Pr > F
Model	4	417429	104357	95.30	<.0001
Error	43	47085	1095.00765 ❷		
Corrected Total	47	464514			

Root MSE	33.09090	R-Square	0.8986
Dependent Mean	636.89583	Adj R-Sq	0.8892
Coeff Var	5.19565		

Parameter Estimates

Variable	DF	Parameter Estimate	Standard Error	t Value	Pr > |t|	Type I SS
Intercept	1	102.84126	37.85904	2.72	0.0095	19470543
LSILVER1	1	1.38589	0.15156	9.14	<.0001	387295
LSILVER2	1	-0.44231	0.26078	-1.70	0.0971	28472
LSILVER3	1	0.00921	0.26137	0.04	0.9720	1061.93530
LSILVER4	1	-0.11236	0.15185	-0.74	0.4633	599.56290 ❶

MONTH END STOCKS OF SILVER
The REG Procedure
Model: MODEL1
Dependent Variable: SILVER

Analysis of Variance

Source	DF	Sum of Squares	Mean Square	F Value	Pr > F
Model	2	457454	228727	220.26	<.0001
Error	47	48808	1038.45850		
Corrected Total	49	506261			

Root MSE	32.22512	R-Square	0.9036
Dependent Mean	642.76000	Adj R-Sq	0.8995
Coeff Var	5.01355		

Parameter Estimates

Variable	DF	Parameter ❸ Estimate	Standard Error	t Value ❹	Pr > |t|
Intercept	1	77.95372	30.21038	2.58	0.0131
LSILVER1	1	1.49087	0.11589	12.86	<.0001
LSILVER2	1	-0.61144	0.11543	-5.30	<.00

Output 2.5 shows that lags 3 and 4 may not be needed because the overall *F* statistic for these two lags is computed ❶❷ as

((1062 + 600) / 2) / 1095 = .76

This is insignificant compared to the F^2_{43} distribution. Alternatively, a TEST statement could be used to produce F.

You have identified the model through overfitting, and now the final estimated model ❸ is

$$Y_t = 77.9537 + 1.4909Y_{t-1} - .6114Y_{t-2} + e_t$$

which becomes

$$Y_t - 647 = 1.4909(Y_{t-1} - 647) - .6114(Y_{t-2} - 647) + e_t$$

All parameters are significant according to their *t* statistics. ❹

The fact that M=1 almost solves the characteristic equation

$$1 - 1.49M + .61M^2$$

suggests that this series may be nonstationary.

In Chapter 3, you extend your class of models to include moving averages and mixed ARMA models. These models require more sophisticated fitting and identification techniques than the simple regression with overfitting used in the silver example.

Chapter 3 The General ARIMA Model

3.1 Introduction

3.1.1 Statistical Background

The general class of autoregressive moving average (ARMA) models is developed in this chapter. As each new model is introduced, its autocovariance function $\gamma(j)$ is given. This helps you use the estimated autocovariances C(j) that PROC ARIMA produces to select an appropriate model for the data. Using estimated autocovariances to determine a model to be fit is called model identification. Once you select the model, you can use PROC ARIMA to fit the model, forecast future values, and provide forecast intervals.

3.1.2 Terminology and Notation

The moving average of order 1 is given by

$$Y_t = \mu + e_t - \beta e_{t-1} \tag{3.1}$$

where e_t is a white noise (uncorrelated) sequence with mean 0 and variance σ^2. Clearly,

$$\text{var}\left(Y_t\right) = \gamma(0) = \sigma^2\left(1 + \beta^2\right)$$

$$\text{cov}\left(Y_t, Y_{t-1}\right) = \gamma(1) = E\left(\left(e_t - \beta e_{t-1}\right)\left(e_{t-1} - \beta e_{t-2}\right)\right) = -\beta\sigma^2$$

and

$$\text{cov}\left(Y_t, Y_{t-j}\right) = 0$$

for $j>1$.

If you observe the autocovariance sequence $\gamma(0) = 100,\ \gamma(1) = 40,\ \gamma(2) = 0,\ \gamma(3) = 0,\ \ldots$, you are dealing with an MA process of order 1 because $\gamma(j) = 0$ for $j>1$. Also, you know that $-\beta\sigma^2 = 40$ and $\left(1+\beta^2\right)\sigma^2 = 100$, so $\beta = -.5$ and $\sigma^2 = 80$. The model is

$$Y_t = \mu + e_t + .5e_{t-1}$$

If each autocovariance $\gamma(j)$ is divided by $\gamma(0)$, the resulting sequence of autocorrelations is $\rho(j)$. For a moving average like equation 3.1,

$$\rho(0) = 1$$

$$\rho(1) = -\beta/\left(1+\beta^2\right)$$

and

$$\rho(j) = 0$$

for $j>1$. Note that

$$-1/2 \le -\beta / \left(1 + \beta^2\right) \le 1/2$$

regardless of the value β. In the example, the autocorrelations for lags 0 through 4 are 1, .4, 0, 0, and 0.

The general moving average of order q is written as

$$Y_t = \mu + e_t - \beta_1 e_{t-1} - \ldots - \beta_q e_{t-q}$$

and is characterized by the fact that $\gamma(j)$ and $\rho(j)$ are 0 for $j>q$. In backshift notation you write

$$Y_t = \mu + \left(1 - \beta_1 B - \beta_2 B^2 - \ldots - \beta_q B^q\right)e_t$$

Similarly, you write the mixed autoregressive moving average model ARMA(p,q) as

$$\left(Y_t - \mu\right) - \alpha_1\left(Y_{t-1} - \mu\right) - \ldots - \alpha_p\left(Y_{t-p} - \mu\right)$$
$$= e_t - \beta_1 e_{t-1} - \ldots - \beta_q e_{t-q}$$

or in backshift notation as

$$\left(1-\alpha_1 B-\ldots-\alpha_p B^p\right)\left(Y_t-\mu\right)=\left(1-\beta_1 B-\ldots-\beta_q B^q\right)e_t$$

For example, the model

$$(1-.6B)Y_t=(1+.4B)e_t$$

is an ARMA(1,1) with mean $\mu = 0$. In practice, parameters are estimated and then used to estimate prediction error variances for several periods ahead. PROC ARIMA provides these computations.

3.2 Prediction

3.2.1 One-Step-Ahead Predictions

You can further clarify the example above by predicting sequentially one step at a time. Let n denote the number of available observations. The next (n+1) observation in the sequence satisfies

$$Y_{n+1} = .6Y_n + e_{n+1} + .4e_n$$

First, predict Y_{n+1} by

$$\hat{Y}_{n+1} = .6Y_n + .4e_n$$

with error variance σ^2. Next,

$$\begin{aligned} Y_{n+2} &= .6Y_{n+1} + e_{n+2} + .4e_{n+1} \\ &= .6\left(.6Y_n + e_{n+1} + .4e_n\right) + e_{n+2} + .4e_{n+1} \end{aligned}$$

so predict Y_{n+2} by removing "future *e*s" (subscripts greater than n):

$$.36Y_n + .24e_n = .6\hat{Y}_{n+1}$$

The prediction error is $e_{n+1} + e_{n+2}$, which has variance $2\sigma^2$. Finally,

$$Y_{n+3} = .6Y_{n+2} + e_{n+3} + .4e_{n+2}$$

and

$$\hat{Y}_{n+3} = .6\hat{Y}_{n+2} + 0$$

so the prediction error is

$$\begin{aligned} Y_{n+3} - \hat{Y}_{n+3} &= .6\left(Y_{n+2} - \hat{Y}_{n+2}\right) + e_{n+3} + .4e_{n+2} \\ &= .6\left(e_{n+1} + e_{n+2}\right) + e_{n+3} + .4e_{n+2} \end{aligned}$$

and the prediction error variance is $2.36\sigma^2$. This example shows that you can readily compute predictions and associated error variances after model parameters or their estimates are available.

The predictions for the model

$$Y_t = .6Y_{t-1} + e_t + .4e_{t-1}$$

can be computed recursively as follows:

Observation	10	5	−3	−8	1	6	—	—	—
Prediction	(0)	10	1	−3.4	−6.64	3.656	4.538	2.723	1.634
Residual	10	−5	−4	−4.6	7.64	2.344	—	—	—

Start by assuming the mean (0) as a prediction of Y_1 with implied error $e_1 = 10$. Predict Y_2 by $0.6Y_1 + 0.4e_1 = 10$, using the assumed $e_1 = 10$. The residual is $r_2 = 5 - 10 = -5$. Using r_2 as an estimate of e_2, predict Y_3 by

$$0.6Y_2 + 0.4r_2 = 0.6(5) + 0.4(-5) = 1$$

The residual is $r_3 = Y_3 - 1 = -4$. Then predict Y_4 by

$$0.6Y_3 + 0.4r_3 = 0.6(-3) + 0.4(-4) = -3.4$$

and Y_5 by −6.64 and Y_6 by 3.656. These are one-step-ahead predictions for the historic data. For example, you use only the data up through $t = 3$ (and the assumed $e_1 = 10$) to predict Y_4. The sum of squares of these residuals, $100 + 25 + \cdots + 2.344^2 = 226.024$, is called the conditional sum of squares associated with the parameters 0.6 and 0.4. If you search over AR and MA parameters to find those that minimize this conditional sum of squares, you are performing conditional least squares estimation, the default in PROC ARIMA. An estimate of the white noise variance is given by dividing the conditional sum of squares by n minus the number of estimated parameters; that is, $n - 2 = 6 - 2 = 4$ for this ARMA(1,1) with mean 0.

3.2.2 Future Predictions

Predictions into the future are of real interest, while one-step-ahead computations are used to start the process. Continuing the process as shown, estimate e_ts as 0 for t beyond n ($n = 6$ observations in the example); that is, estimate future Ys by their predictions.

The next three predictions are as follows:

$\hat{Y}_7$ is 0.6(6) + 0.4(2.344) = 4.538 with error e_7

$\hat{Y}_8$ is 0.6(4.538) + 0.4(0) = 2.723 with error $e_8 + e_7$

$\hat{Y}_9$ is 0.6(2.723) + 0.4(0) = 1.634 with error $e_9 + e_8 + 0.6e_7$.

PROC ARIMA provides these computations for you. The illustration simply shows what PROC ARIMA is computing.

Note that the prediction of Y_{7+j} is just $(.6)^j \hat{Y}_7^j$ and thus declines exponentially to the series mean (0 in the example). The prediction error variance increases from var(e_t) to var(Y_t). In a practical application, the form

$$Y_t - \alpha Y_{t-1} = e_t - \beta e_{t-1}$$

and parameter values

$$Y_t - .6Y_{t-1} = e_t + .4e_{t-1}$$

are not known. They can be determined through PROC ARIMA.

In practice, estimated parameters are used to compute predictions and standard errors. This procedure requires sample sizes much larger than those in the example above.

Although they would not have to be, the forecasting methods used in PROC ARIMA are tied to the method of estimation. If you use conditional least squares, the forecast is based on the expression of $Y_t - \mu$ as an infinite autoregression. For example, suppose $Y_t = \mu + e_t - \beta e_{t-1}$, a simple MA(1). Note that $e_t = Y_t - \mu + \beta e_{t-1}$, so at time $t-1$ you have $e_{t-1} = Y_{t-1} - \mu + \beta e_{t-2}$; substituting this second expression into the first, you have $e_t = (Y_t - \mu) + \beta(Y_{t-1} - \mu) + \beta^2 e_{t-2}$. Continuing in this fashion, and assuming that $|\beta| < 1$ so that $\beta^j e_{t-j}$ converges to 0 as j gets large, you find $e_t = \sum_{j=0}^{\infty} \beta^j (Y_{t-j} - \mu)$, which can alternatively be expressed as $(Y_t - \mu) = -\sum_{j=1}^{\infty} \beta^j (Y_{t-j} - \mu) + e_t$. Thus the forecast of Y_t given data up to time $t-1$ is $\hat{Y}_t = \mu - \sum_{j=1}^{\infty} \beta^j (Y_{t-j} - \mu)$. The expression $\sum_{j=1}^{\infty} \beta^j (Y_{t-j} - \mu)$ depends on Y values prior to time 1, the "infinite past." PROC ARIMA assumes Y values before time 1 are just equal to μ and, of course, the parameters are replaced by estimates. The truncated sum $\sum_{j=1}^{t-1} \beta^j (Y_{t-j} - \mu)$ is not necessarily the best linear combination of lagged Y values for predicting $Y_t - \mu$.

When ML or ULS estimation is used, optimal linear forecasts based on the finite past are computed. Suppose you want to minimize $E\{[(Y_t - \mu) - \phi_1(Y_{t-1} - \mu) - \phi_2(Y_{t-2} - \mu)]^2\}$ by finding ϕ_1 and ϕ_2; that is, you want the minimum variance forecast of Y_t based on its two predecessors. Note that here ϕ_1 and ϕ_2 are just coefficients; they do not represent autoregressive or moving average parameters. Using calculus you find

$$\begin{pmatrix} \phi_1 \\ \phi_2 \end{pmatrix} = \begin{pmatrix} \gamma(0) & \gamma(1) \\ \gamma(1) & \gamma(0) \end{pmatrix}^{-1} \begin{pmatrix} \gamma(1) \\ \gamma(2) \end{pmatrix}$$

This would give the best forecast of Y_3 based on a linear combination of Y_1 and Y_2:

$$\hat{Y}_3 = \mu + \phi_1(Y_2 - \mu) + \phi_2(Y_1 - \mu)$$

Likewise, to forecast Y_5 using Y_1, Y_2, Y_3, Y_4, the four ϕ_js are computed as

$$\begin{pmatrix} \phi_1 \\ \phi_2 \\ \phi_3 \\ \phi_4 \end{pmatrix} = \begin{pmatrix} \gamma(0) & \gamma(1) & \gamma(2) & \gamma(3) \\ \gamma(1) & \gamma(0) & \gamma(1) & \gamma(2) \\ \gamma(2) & \gamma(1) & \gamma(0) & \gamma(1) \\ \gamma(3) & \gamma(2) & \gamma(1) & \gamma(0) \end{pmatrix}^{-1} \begin{pmatrix} \gamma(1) \\ \gamma(2) \\ \gamma(3) \\ \gamma(4) \end{pmatrix}$$

Here $\gamma(h)$ is the autocovariance at lag h and the equations for the ϕ_js can be set up for any ARMA structure and any number of lags. For the MA(1) example with parameter β, to predict the fifth Y you have

$$\begin{pmatrix}\phi_1\\ \phi_2\\ \phi_3\\ \phi_4\end{pmatrix} = \begin{pmatrix}1+\beta^2 & -\beta & 0 & 0\\ -\beta & 1+\beta^2 & -\beta & 0\\ 0 & -\beta & 1+\beta^2 & -\beta\\ 0 & 0 & -\beta & 1+\beta^2\end{pmatrix}^{-1} \begin{pmatrix}-\beta\\ 0\\ 0\\ 0\end{pmatrix}$$

For reasonably long time series whose parameters are well inside the stationarity and invertibility regions, the best linear combination forecast used when ML or ULS is specified does not differ by much from the truncated sum used when CLS is specified. (See **Section 3.3.1.**)

For an MA(1) process with lag 1 parameter $\beta = 0.8$, the weights on past Y, used in forecasting 1 step ahead, are listed below. The top row shows the first 14 weights assuming infinite past $(-(0.8)^j)$, and the next two rows show finite past weights for $n = 9$ and $n = 14$ past observations.

lag	Y_{t-1}	Y_{t-2}	Y_{t-3}	Y_{t-4}	Y_{t-5}	Y_{t-6}	Y_{t-7}	...	Y_{t-13}	Y_{t-14}
Infinite past	−.80	−.64	−.51	−.41	−.33	−.26	−.21		−.05	−.04
$n = 9$, finite past	−.79	−.61	−.47	−.35	−.25	−.16	−.08		—	—
$n = 14$, finite past	−.80	−.64	−.51	−.41	−.32	−.26	−.21		−.02	−.02

Despite the fairly large β and small n values, the weights are quite similar. Increasing n to 25 produces weights indistinguishable, out to 2 decimal places, from those for the infinite past.

If $\beta = 1$, the series $Y_t = e_t - 1e_{t-1}$ is said to be "noninvertible," indicating that you cannot get a nice, convergent series representation of e_t as a function of current and lagged Y values. Not only does this negate the discussion above, but since a reasonable estimate of e_t cannot be extracted from the data, it eliminates any sensible model-based forecasting. In the moving average of order q, $Y_t = e_t - \beta_1 e_{t-1} - \cdots - \beta_q e_{t-q}$ has an associated polynomial equation in the algebraic variable M, $1 - \beta_1 M - \cdots - \beta_q M^q = 0$, whose roots must satisfy $|M| > 1$ in order for the series to be invertible. Note the analogy with the characteristic equation computed from the autoregressive coefficients.

Fortunately in practice it is rare to encounter a naturally measured series that appears to be noninvertible. However, when differences are taken, noninvertibility can be artificially induced. For example, the time series $Y_t = \alpha_0 + \alpha_1 t + e_t$ is a simple linear trend plus white noise. Some practitioners have the false impression that any sort of trend in a time series should be removed by taking differences. If that is done, one sees that

$$Y_t - Y_{t-1} = (\alpha_0 + \alpha_1 t + e_t) - (\alpha_0 + \alpha_1 (t-1) + e_{t-1}) = \alpha_1 + e_t - e_{t-1}$$

so that in the process of reducing the trend $\alpha_0 + \alpha_1 t$ to a constant α_1, a noninvertible moving average has been produced. Note that the parameters of $Y_t = \alpha_0 + \alpha_1 t + e_t$ are best estimated by the ordinary least squares regression of Y_t on t, this being a fundamental result of basic statistical theory. The practitioner perhaps was confused by thinking in a very narrow time series way.

3.3 Model Identification

3.3.1 Stationarity and Invertibility

Consider the ARMA model

$$(1-\alpha_1 B-\alpha_2 B^2-\ldots-\alpha_p B^p)(Y_t-\mu)$$
$$=(1-\beta_1 B-\beta_2 B^2-\ldots-\beta_q B^q)e_t$$

The model is stationary if all values of M such that

$$1-\alpha_1 M-\alpha_2 M^2-\ldots-\alpha_p M^p=0$$

are larger than 1 in absolute value. Stationarity ensures that early values of *e* have little influence on the current value of Y. It also ensures that setting a few values of *e* to 0 at the beginning of a series does not affect the predictions very much, provided the series is moderately long. In the ARMA(1,1) example, the prediction of Y_6 with 0 as an estimate of e_1 differs from the prediction using the true e_1 by the quantity .01 e_1. Any MA process is stationary. One AR example is

$$(1-1.3B+.3B^2)Y_t=e_t$$

which is not stationary (the roots of $1-1.3M+.3M^2=0$ are M=1 and M=10/3). Another example is

$$(1-1.3B+.42B^2)Y_t=e_t$$

which is stationary (the roots of $1-1.3M+.42M^2=0$ are M=10/7 and M=10/6).

A series satisfies the invertibility condition if all Ms for which

$$1-\beta_1 M-\beta_2 M^2-\ldots-\beta_q M^q=0$$

are such that $|M|>1$. The invertibility condition ensures that Y_t can be expressed in terms of e_t and an infinite weighted sum of previous Ys. In the example,

$$e_t=(1+.4B)^{-1}(1-.6B)Y_t$$

and

$$e_t=Y_t-Y_{t-1}+.4Y_{t-2}-.16Y_{t-3}+.064Y_{t-4}-\ldots$$

so

$$Y_t=e_t+Y_{t-1}-.4Y_{t-2}+.16Y_{t-3}-.064Y_{t-4}+\ldots$$

The decreasing weights on lagged values of Y allow you to estimate e_t from recent values of Y. Note that in **Section 3.2.1** the forecast of Y_{n+1} was $.6Y_n+.4e_n$, so the ability to estimate e_n from the data was crucial.

3.3.2 Time Series Identification

You need to identify the form of the model. You can do this in PROC ARIMA by inspecting data-derived estimates of three functions:

- autocorrelation function (ACF)
- inverse autocorrelation function (IACF)
- partial autocorrelation function (PACF).

These functions are defined below. A short catalog of examples is developed, and properties useful for associating different forms of these functions with the corresponding time series forms are summarized.

In PROC ARIMA, an IDENTIFY statement produces estimates of all these functions. For example, the following SAS statements produce lists and plots of all three of these functions for the variable Y in the data set SERIES:

```
PROC ARIMA DATA=SERIES;
   IDENTIFY VAR=Y;
RUN;
```

3.3.2.1 Autocovariance Function $\gamma(j)$

Recall that $\gamma(j)$ is the covariance between Y_t and Y_{t-j}, which is assumed to be the same for every t (stationarity). See the listing below of autocovariance functions for Series 1–8 (in these examples, e_t is white noise with variance $\sigma^2 = 1$).

Series	Model
1	$Y_t = .8Y_{t-1} + e_t$, AR(1), $\gamma(1) > 0$
2	$Y_t = -.8Y_{t-1} + e_t$, AR(1), $\gamma(1) < 0$
3	$Y_t = .3Y_{t-1} + .4Y_{t-2} + e_t$, AR(2)
4	$Y_t = .7Y_{t-1} + .49Y_{t-2} + e_t$, AR(2)
5	$Y_t = e_t + .8e_{t-1}$, MA(1)
6	$Y_t = e_t - .3e_{t-1} + .4e_{t-2}$, MA(2)
7	$Y_t = e_t$, (white noise)
8	$Y_t = .6Y_{t-1} + e_t + .4e_{t-1}$, ARMA(1,1)

For an AR(1) series $Y_t - \rho Y_{t-1} = e_t$ (such as Series 1 and 2), the covariance sequence is

$$\gamma(j) = \rho^{|j|}\sigma^2/(1-\rho^2)$$

For an AR(2) series $Y_t - \alpha_1 Y_{t-1} - \alpha_2 Y_{t-2} = e_t$ (such as Series 3 and 4), the covariance sequence begins with values $\gamma(0)$ and $\gamma(1)$, followed by $\gamma(j)$ that satisfy

$$\gamma(j) - \alpha_1\gamma(j-1) - \alpha_2\gamma(j-2) = 0$$

The **covariances** may oscillate with a period depending on α_1 and α_2 (such as Series 4). Beginning values are determined from the Yule-Walker equations.

For a general AR(p) series

$$Y_t - \alpha_1 Y_{t-1} - \alpha_2 Y_{t-2} - \ldots - \alpha_p Y_{t-p} = e_t$$

beginning values are $\gamma(0), \ldots, \gamma(p-1)$, from which $\gamma(j)$ satisfies

$$\gamma(j) - \alpha_1\gamma(j-1) - \alpha_2\gamma(j-2) - \ldots - \alpha_p\gamma(j-p) = 0$$

for $j>p$. The fact that $\gamma(j)$ satisfies the same difference equation as the series ensures that $|\gamma(j)| < H\lambda^j$, where $0 < \lambda < 1$ and H is some finite constant. In other words, $\gamma(j)$ may oscillate, but it is bounded by a function that decreases exponentially to zero.

For MA(1),

$$Y_t - \mu = e_t - \beta e_{t-1}$$

$$\gamma(0) = (1 + \beta^2)\sigma^2$$

$$\gamma(1) = \gamma(-1) = -\beta\sigma^2$$

and $\gamma(j) = 0$ for $|j| > 1$.

For a general MA(q)

$$Y_t - \mu = e_t - \beta_1 e_{t-1} - \beta_2 e_{t-2} - \ldots - \beta_q e_{t-q}$$

the q beginning values are $\gamma(0), \gamma(1), \ldots, \gamma(q)$. Then $\gamma(j)=0$ for $|j|>q$.

For an ARMA(1,1) process

$$(Y_t - \mu) - \alpha(Y_{t-1} - \mu) = e_t - \beta e_{t-1}$$

there is a dropoff from $\gamma(0)$ to $\gamma(1)$ determined by α and β. For $j>1$, the pattern

$$\gamma(j) = \alpha\gamma(j-1)$$

occurs. Thus, an apparently arbitrary drop followed by exponential decay characterizes the ARMA(1,1) covariance function.

For the ARMA(p,q) process

$$(Y_t - \mu) - \alpha_1(Y_{t-1} - \mu) - \ldots - \alpha_p(Y_{t-p} - \mu) = e_t - \beta_1 e_{t-1} - \ldots - \beta_q e_{t-q}$$

there are

$$r = \max(p-1, q)$$

beginning values followed by behavior characteristic of an AR(p); that is,

$$\gamma(j) - \alpha_1\gamma(j-1) - \ldots - \alpha_p\gamma(j-p) = 0$$

for $j>r$. For a white noise sequence, $\gamma(j) = 0$ if $j \neq 0$.

3.3.2.2 ACF

Note that the pattern, rather than the magnitude of the sequence $\gamma(j)$, is associated with the model form. Normalize the autocovariance sequence $\gamma(j)$ by computing autocorrelations

$$\rho(j) = \gamma(j) \;/\; \gamma(0)$$

Note that

$$\rho(0) = 1$$

for all series and that

$$\rho(j) = \rho(-j)$$

The ACFs for the eight series previously listed are listed below.

Series	Model, ACF
1	$Y_t = .8Y_{t-1} + e_t,\ \rho(j) = .8^{\lvert j \rvert}$
2	$Y_t = -.8Y_{t-1} + e_t,\ \rho(j) = (-.8)^{\lvert j \rvert}$
3	$Y_t = .3Y_{t-1} + .4Y_{t-2} + e_t,\ \rho(1) = .5000,$ $\rho(j) = .3\rho(j-1) + .4\rho(j-2)$ for $j > 1$
4	$Y_t = .7Y_{t-1} - .49Y_{t-2} + e_t,\ \rho(1) = .4698,$ $\rho(j) = .7\rho(j-1) - .49\rho(j-2)$ for $j > 1$
5	$Y_t = e_t + .8e_{t-1},\ \rho(1) = .4878,\ \rho(j) = 0$ for $j > 1$
6	$Y_t = e_t - .3e_{t-1} - .4e_{t-2},\ \rho(1) = -.144,$ $\rho(2) = -.32,\ \rho(j) = 0$ for $j > 2$
7	$Y_t = e_t,\ \rho(0) = 1,\ \rho(j) = 0$ for $j > 0$
8	$Y_t - .6Y_{t-1} = e_t + .4e_{t-1},\ \rho(0) = 1,\ \rho(1) = .7561,$ $\rho(j) = .6\rho(j-1)$ for $j > 1$

The ACFs are plotted in **Output 3.1**.

Output 3.1
Plotting Actual Autocorrelations for Series 1–8

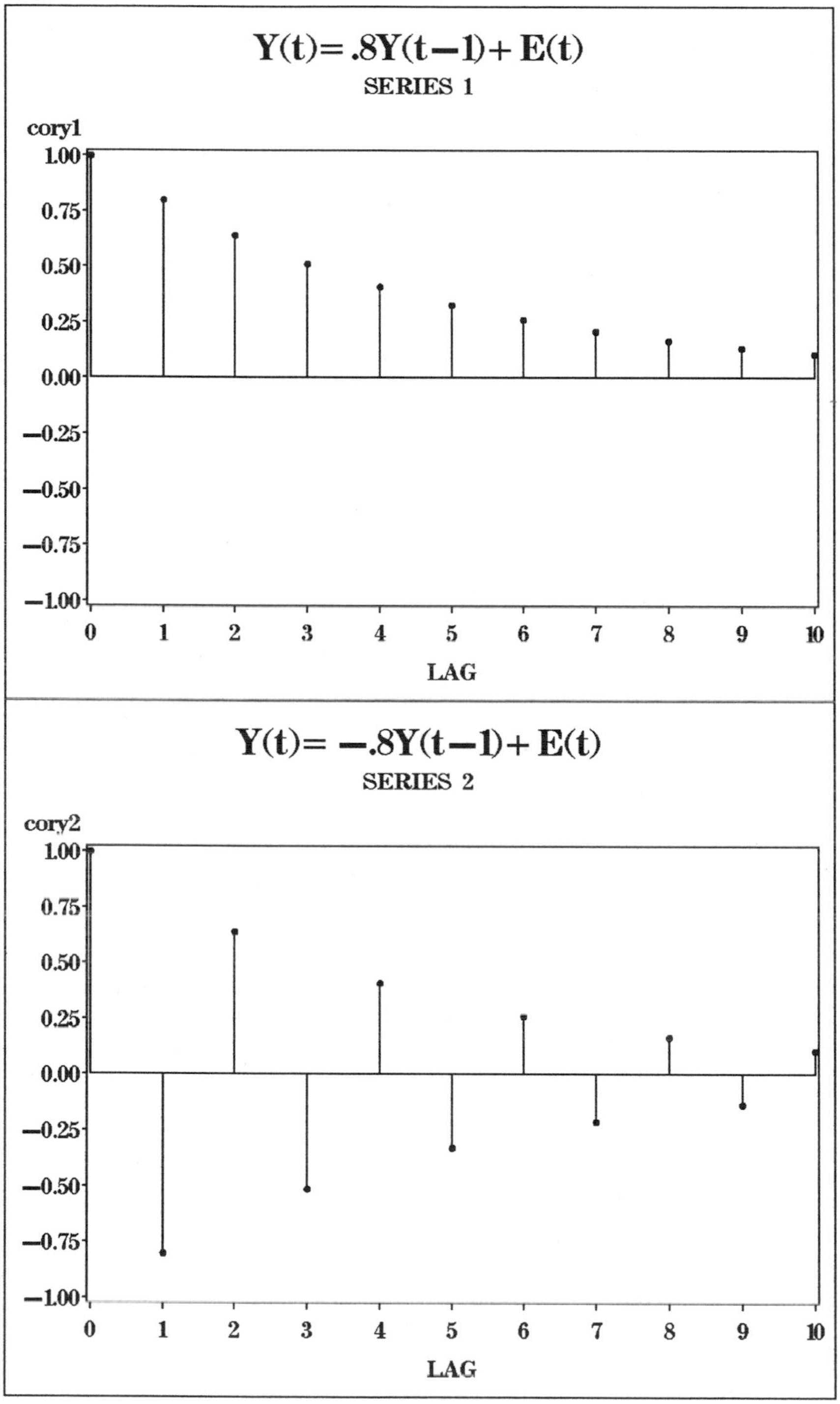

Output 3.1
Plotting Actual Autocorrelations for Series 1–8 (continued)

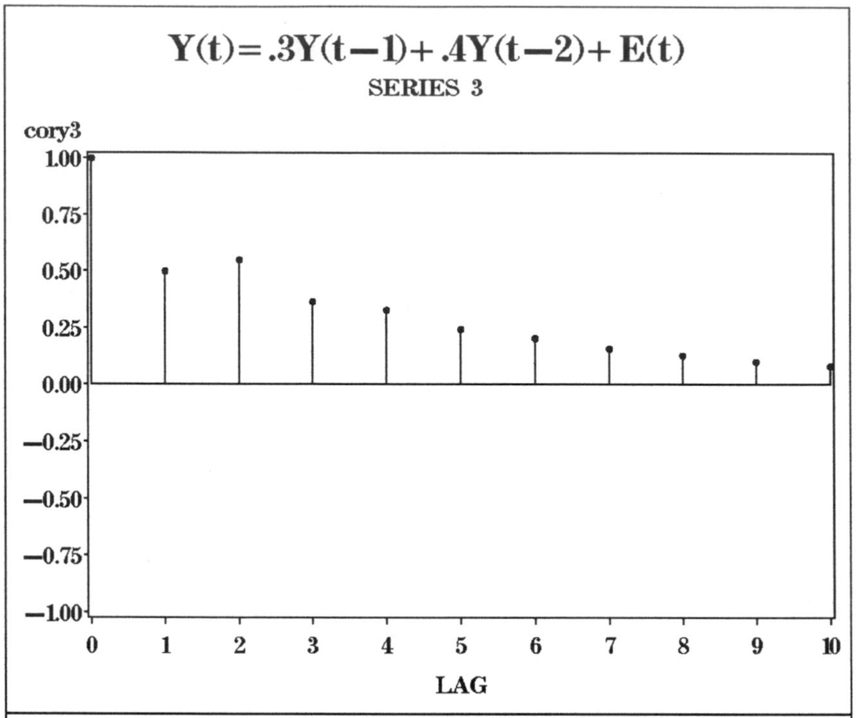

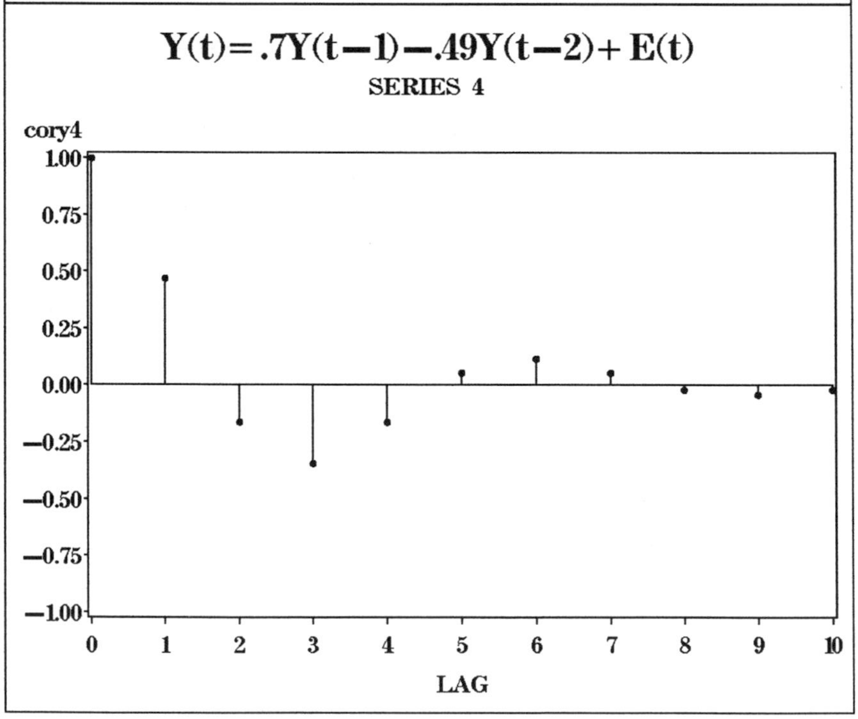

Output 3.1
Plotting Actual Autocorrelations for Series 1–8 (continued)

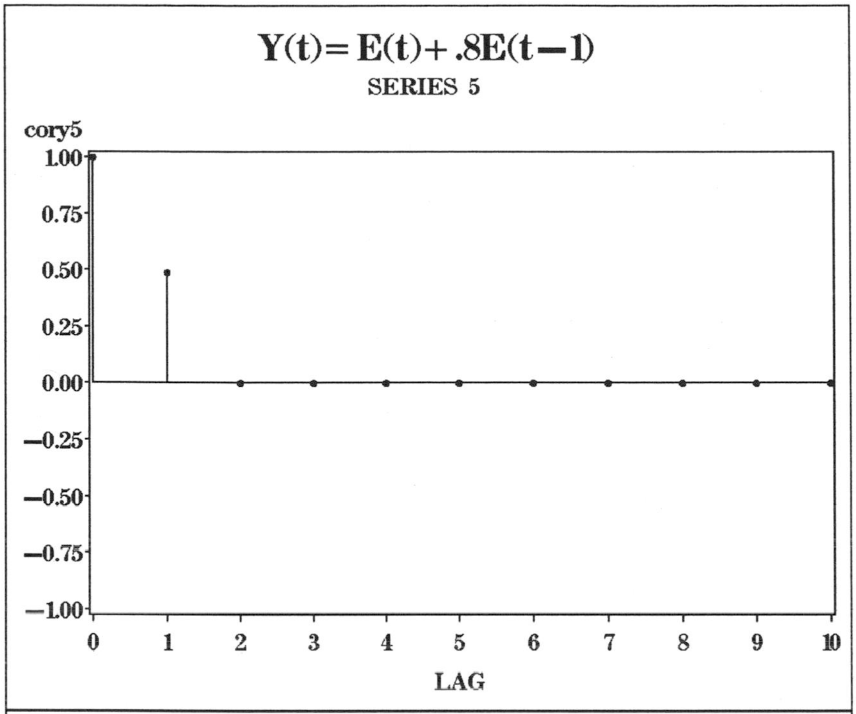

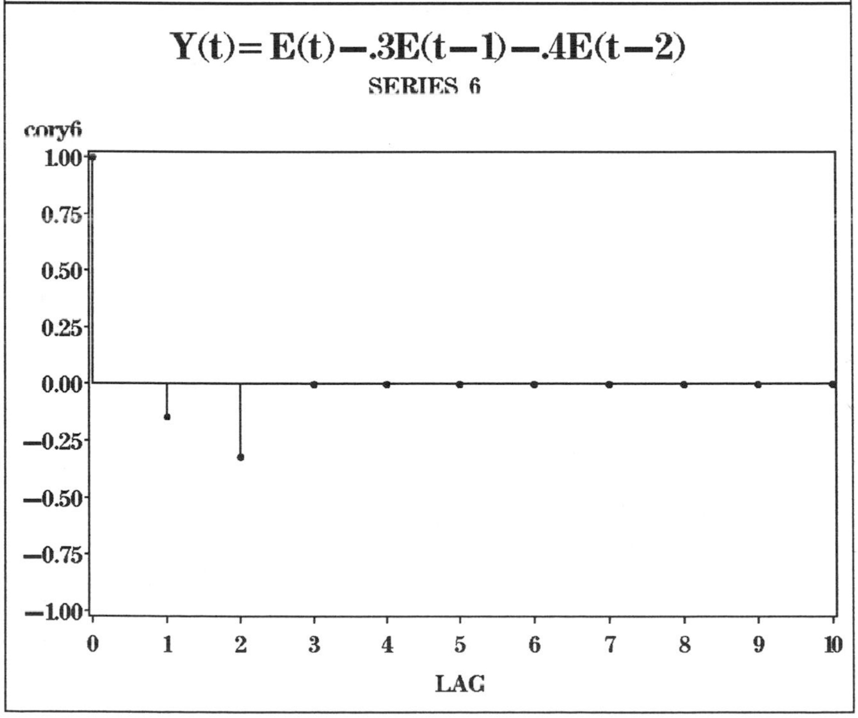

Output 3.1
Plotting Actual Autocorrelations for Series 1–8 (continued)

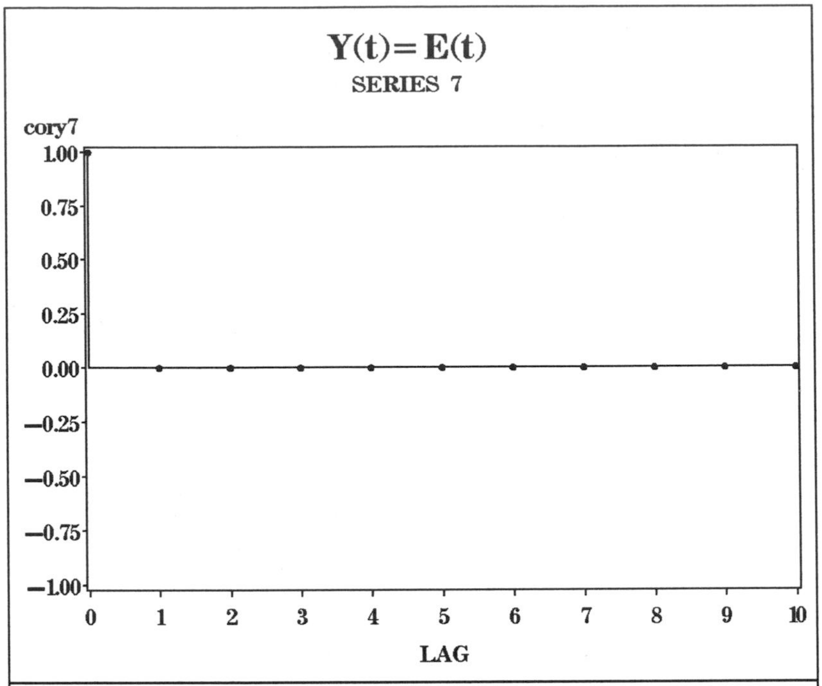

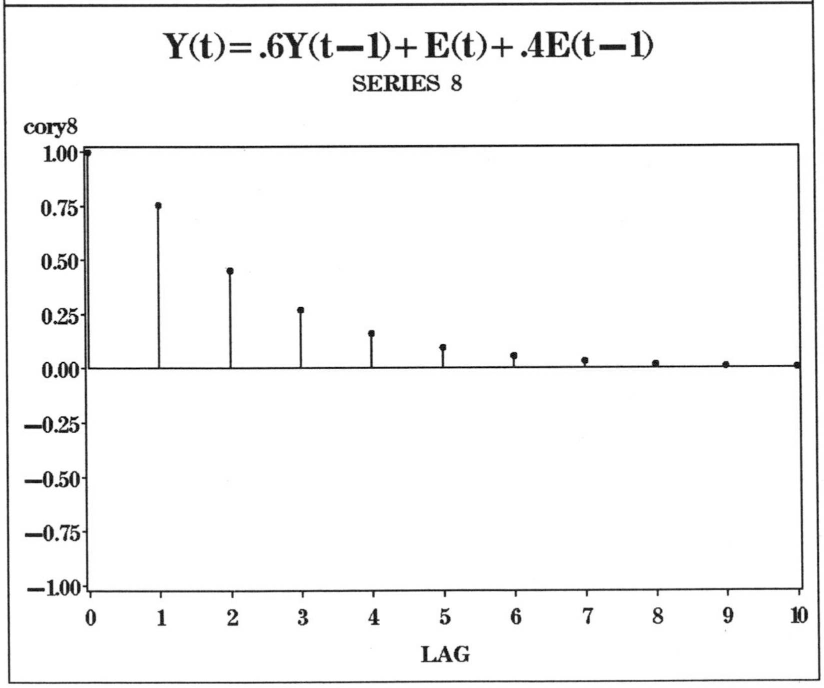

3.3.2.3 PACF

The PACF is motivated by the regression approach to the silver example in Chapter 2, "Simple Models: Autoregression." First, regress Y_t on Y_{t-1} and call the coefficient on Y_{t-1} $\hat{\pi}_1$. Next, regress Y_t on Y_{t-1}, Y_{t-2} and call the coefficient on Y_{t-2} $\hat{\pi}_2$. Continue in this manner, regressing Y_t on Y_{t-1}, Y_{t-2},. . . , Y_{t-j} and calling the last coefficient $\hat{\pi}_j$. The $\hat{\pi}_j$ values are the estimated partial autocorrelations.

In an autoregression of order *p*, the coefficients $\hat{\pi}_j$ estimate 0s for all *j*>p. The theoretical partial autocorrelations π_j estimated by the $\hat{\pi}_j$ are obtained by solving equations similar to the regression normal equations.

$$\begin{bmatrix} \gamma(0) & \gamma(1) & \cdots & \gamma(j-1) \\ \gamma(1) & \gamma(0) & \cdots & \gamma(j-2) \\ \cdot & \cdot & & \cdot \\ \cdot & \cdot & & \cdot \\ \cdot & \cdot & & \cdot \\ \gamma(j-1) & \gamma(j-2) & \cdots & \gamma(0) \end{bmatrix} \begin{bmatrix} b_1 \\ b_2 \\ \cdot \\ \cdot \\ \cdot \\ b_j \end{bmatrix} = \begin{bmatrix} \gamma(1) \\ \gamma(2) \\ \cdot \\ \cdot \\ \cdot \\ \gamma(j) \end{bmatrix}$$

For each *j*, let $\pi_j = b_j$. (A new set of equations is needed for each *j*.) As with autocorrelations, the π_j sequence is useful for identifying the form of a time series model. The PACF is most useful for identifying AR processes because, for an AR(*p*), the PACF is 0 beyond lag *p*. For MA or mixed (ARMA) processes, the theoretical PACF does not become 0 after a fixed number of lags.

You can solve the previous set of equations for the catalog of series. When you observe an estimated PACF $\hat{\pi}_j$, compare its behavior to the behavior shown next to choose a model. The following is a list of actual partial autocorrelations for Series 1–8:

		Lag				
Series	Model	1	2	3	4	5
1	$Y_t = .8Y_{t-1} + e_t$	0.8	0	0	0	0
2	$Y_t = -.8\ Y_t + e_t$	−0.8	0	0	0	0
3	$Y_t = .3Y_{t-1} + .4Y_{t-2} + e_t$	0.5	0.4	0	0	0
4	$Y_t = .7Y_{t-1} - .49Y_{t-2} + e_t$	0.4698	−0.4900	0	0	0
5	$Y_t = e_t + .8\ e_{t-1}$	0.4878	−0.3123	0.2215	−0.1652	0.1267
6	$Y_t = e_t - .3\ e_{t-1} - .4e_{t-2}$	−0.144	−0.3480	−0.1304	−0.1634	−0.0944
7	$Y_t = e_t$	0	0	0	0	0
8	$Y_t = .6Y_{t-1} + e_t + .4e_{t-1}$	0.7561	−0.2756	0.1087	−0.0434	0.0173

Plots of these values against lag number, with A used as a plot symbol for the ACF and P for the PACF, are given in **Output 3.2**. A list of actual autocorrelations for Series 1–8 follows:

		Lag				
Series	Model	1	2	3	4	5
1	$Y_t = .8Y_{t-1} + e_t$	0.8	0.64	0.512	0.410	0.328
2	$Y_t = -.8\ Y_t + e_t$	−0.8	0.64	−0.512	0.410	−0.328
3	$Y_t = .3Y_{t-1} + .4Y_{t-2} + e_t$	0.500	0.550	0.365	0.330	0.245
4	$Y_t = .\ 7Y_{t-1} - .49Y_{t-2} + e_t$	0.470	−0.161	−0.343	−0.161	0.055
5	$Y_t = e_t + .8\ e_{t-1}$	0.488	0	0	0	0
6	$Y_t = e_t - .3\ e_{t-1} - .4e_{t-2}$	−0.144	−0.32	0	0	0
7	$Y_t = e_t$	0	0	0	0	0
8	$Y_t = .6Y_{t-1} + e_t + .4e_{t-1}$	0.756	0.454	0.272	0.163	0.098

Output 3.2 shows the plots.

Output 3.2
Plotting Actual Autocorrelations and Actual Partial Autocorrelations for Series 1–8

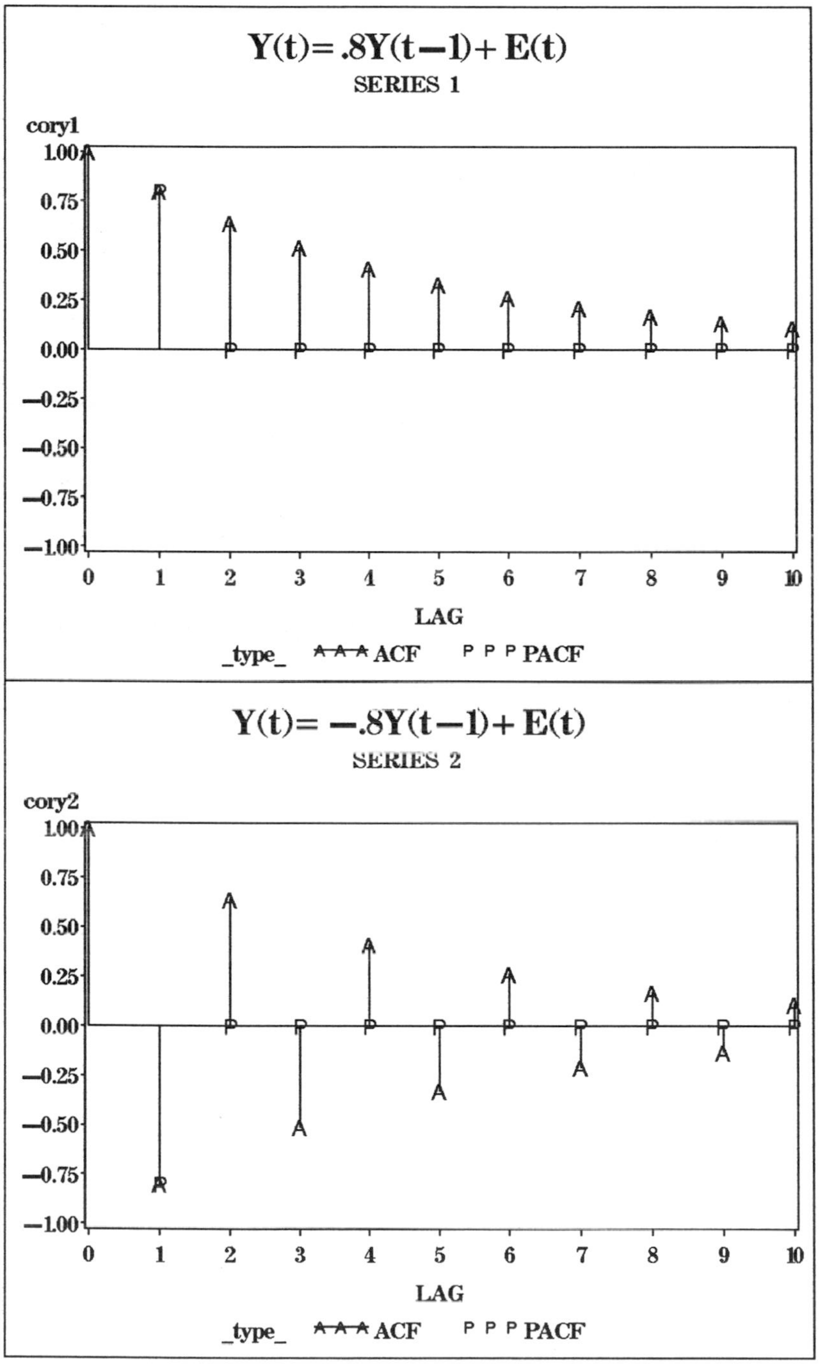

Output 3.2
Plotting Actual Autocorrelations and Actual Partial Autocorrelations for Series 1–8 (continued)

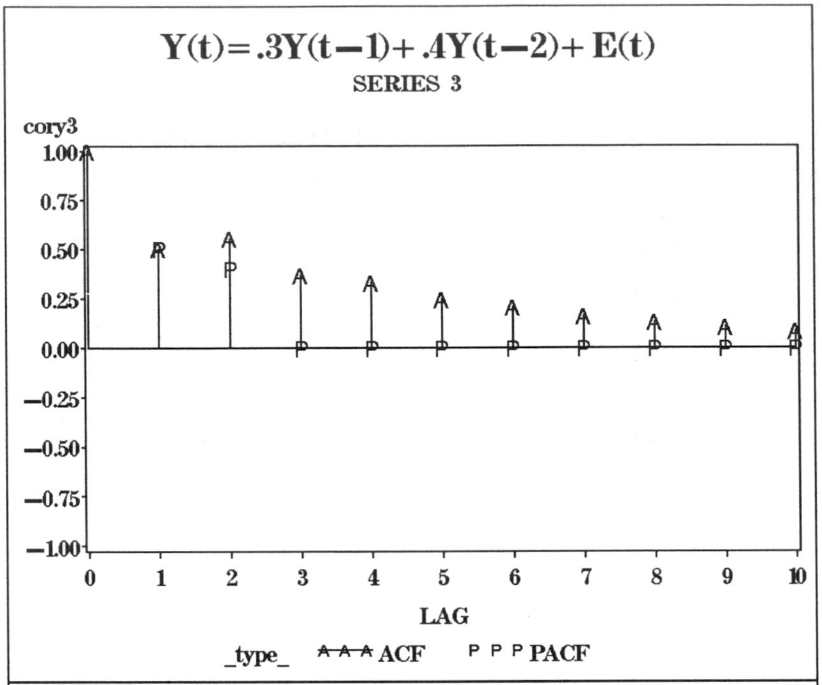

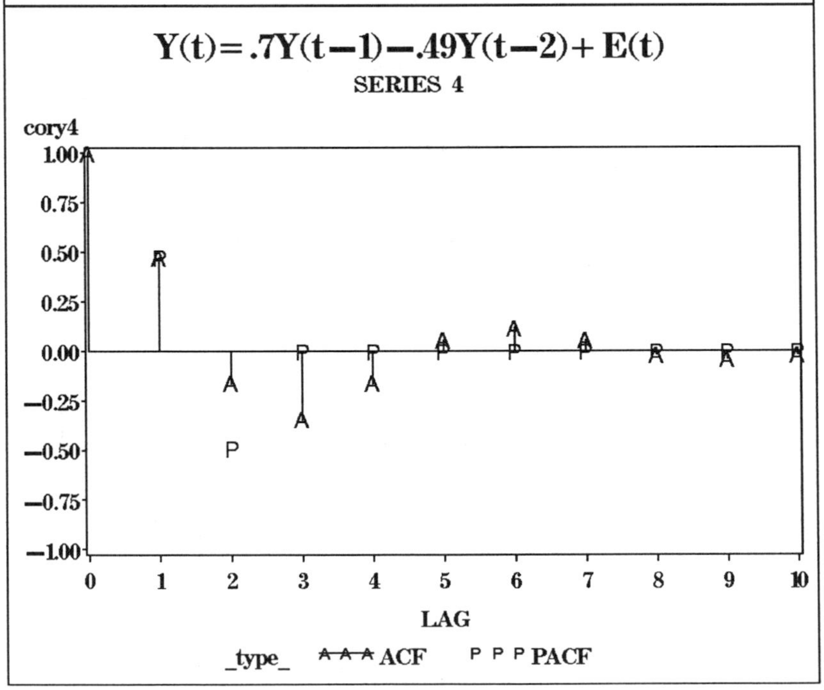

Output 3.2 *Plotting Actual Autocorrelations and Actual Partial Autocorrelations for Series 1–8 (continued)*

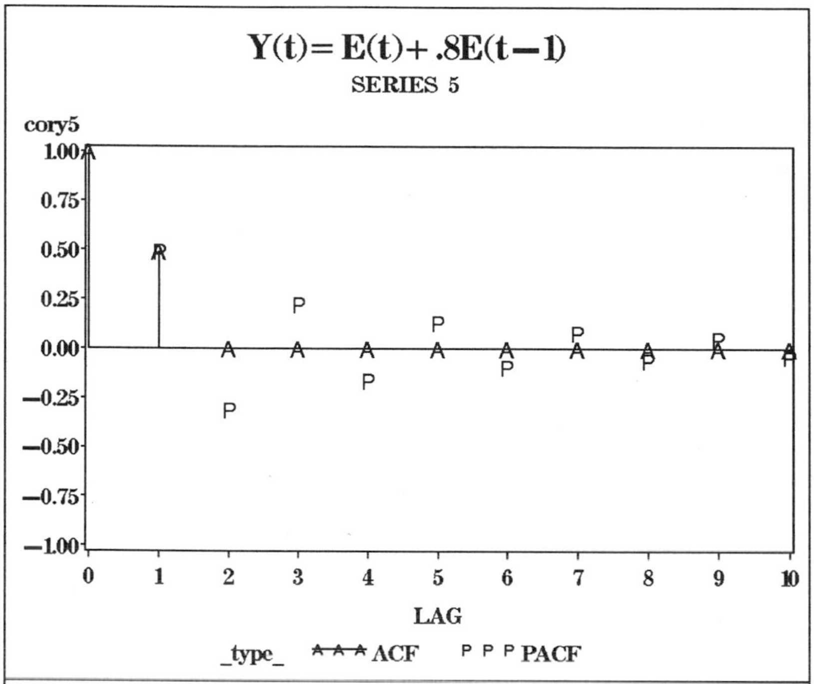

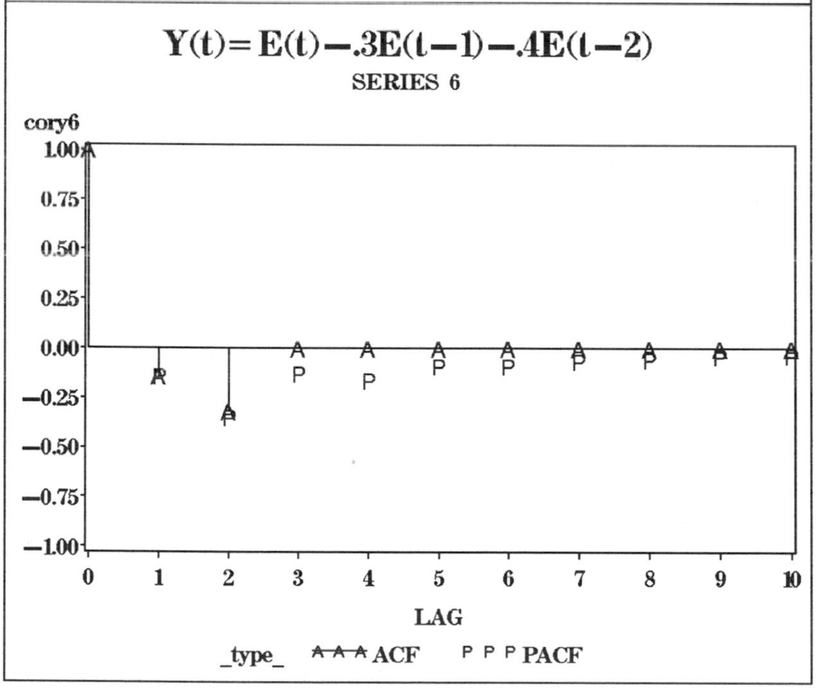

Output 3.2
Plotting Actual Autocorrelations and Actual Partial Autocorrelations for Series 1–8 (continued)

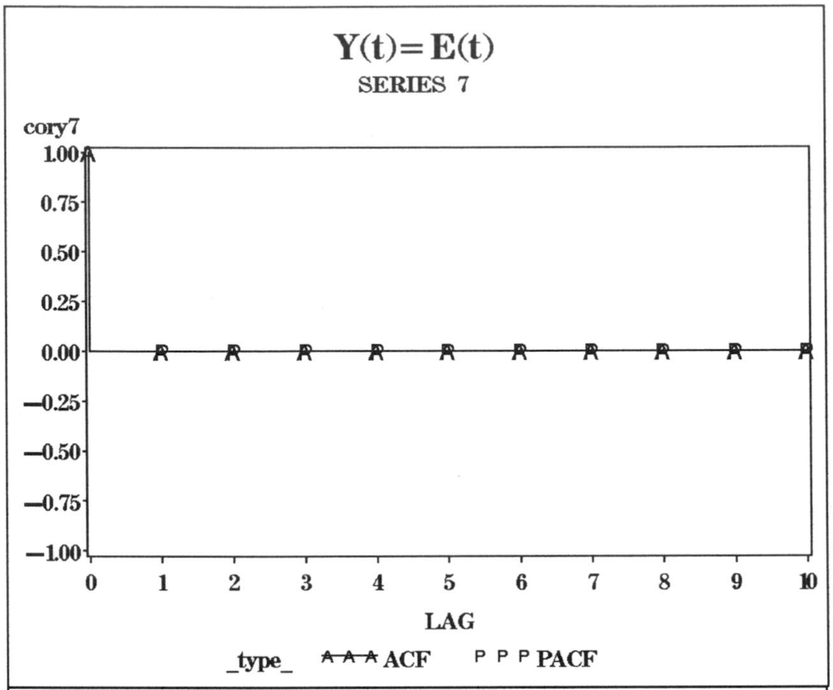

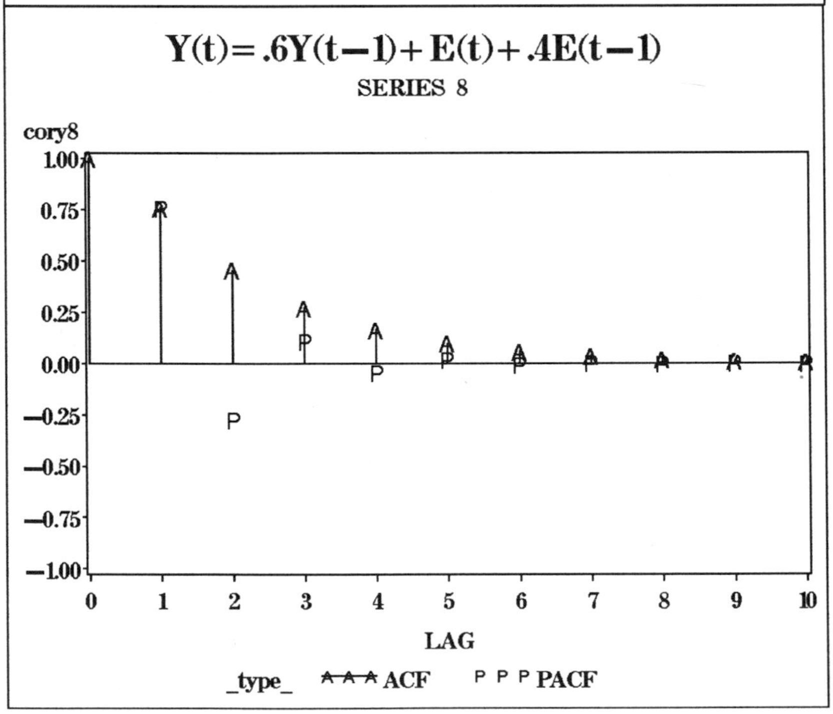

3.3.2.4 Estimated ACF

Begin the PROC ARIMA analysis by estimating the three functions defined above. Use these estimates to identify the form of the model. Define the estimated autocovariance C(j) as

$$C(j) = \Sigma\left(Y_t - \overline{Y}\right)\left(Y_{t+j} - \overline{Y}\right) / n$$

where the summation is from 1 to $n-j$ and $\overline{Y}$ is the mean of the entire series. Define the estimated autocorrelation by

$$r(j) = C(j) / C(0)$$

Compute standard errors for autocorrelations in PROC ARIMA as follows:

- For autocorrelation $r(j)$, assign a variance $\left(\Sigma r^2(i)\right)/n$ where the summation runs from $-j+1$ to $j-1$.
- The standard error is the square root of this variance.
- This is the appropriate variance under the hypothesis that $\gamma(i) = 0$ for $i \geq j$ while $\gamma(i) \neq 0$ for $i < j$.

The group of plots in **Output 3.3** illustrates the actual (A) and estimated (E) ACFs for the series. Each data series contains 150 observations. The purpose of the plots is to indicate the amount of sampling error in the estimates.

Output 3.3 *Plotting Actual and Estimated Autocorrelations for Series 1–8*

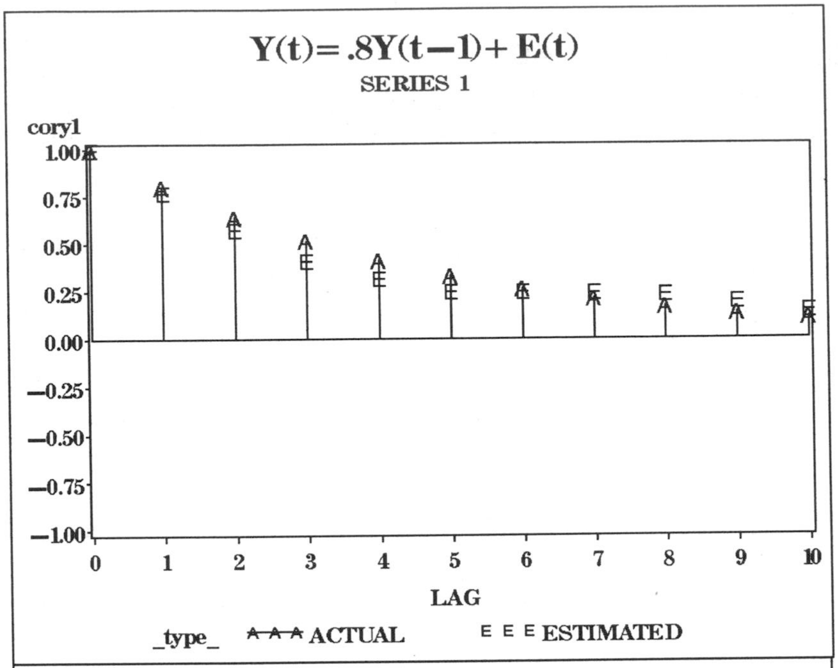

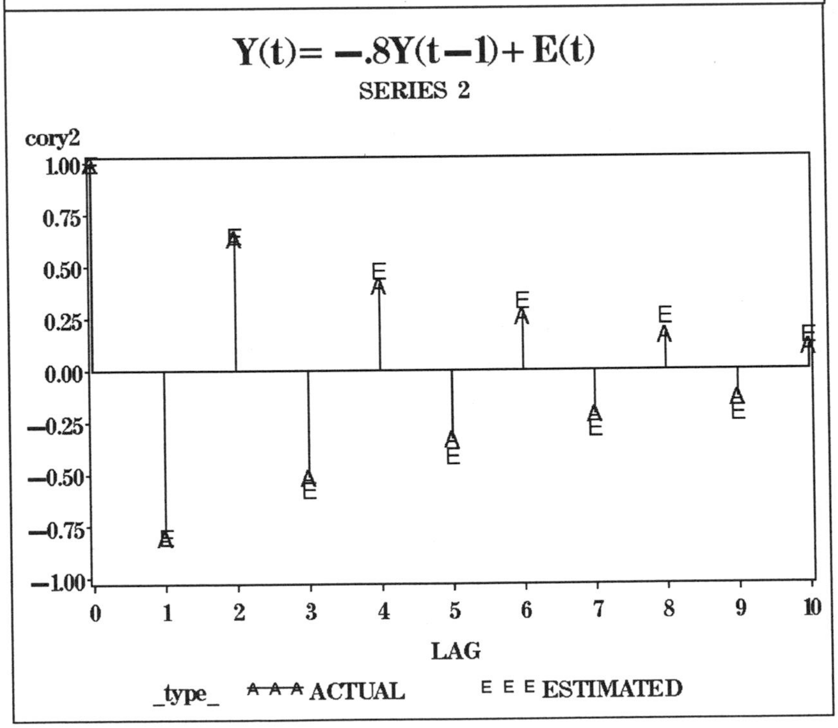

Output 3.3
Plotting Actual and Estimated Autocorrelations for Series 1–8 (continued)

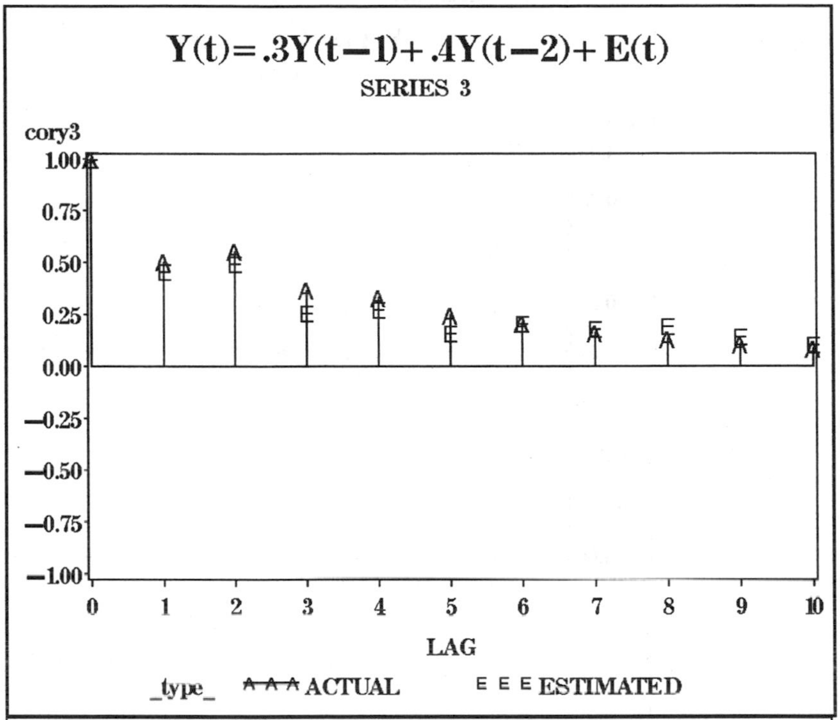

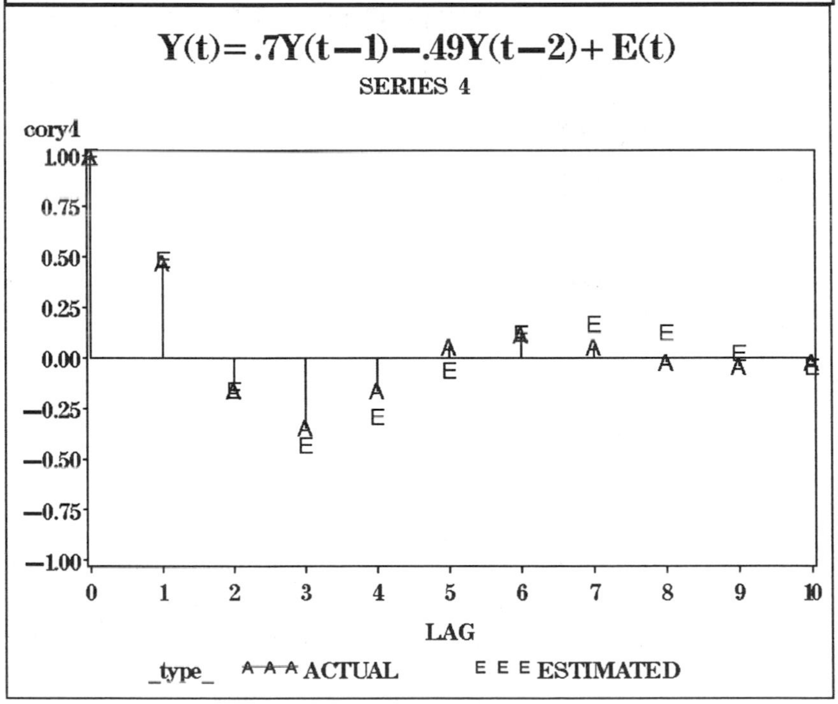

Output 3.3
Plotting Actual and Estimated Autocorrelations for Series 1–8 (continued)

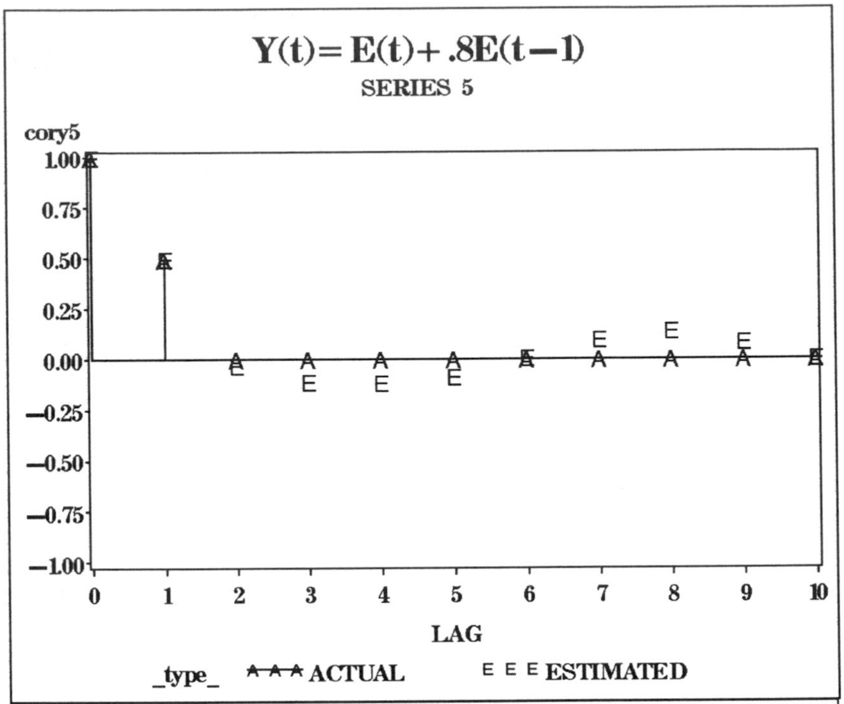

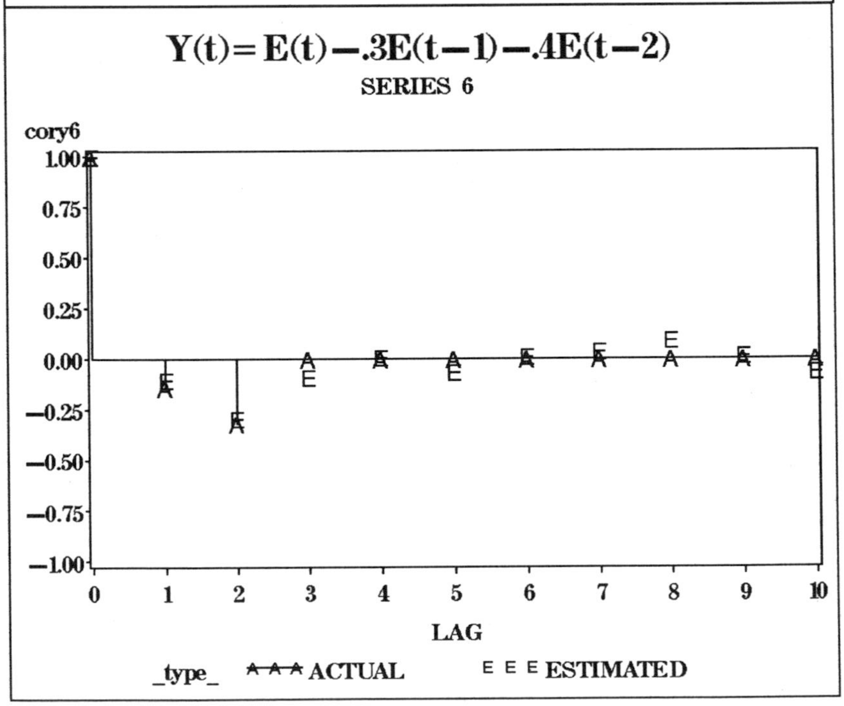

Output 3.3
Plotting Actual and Estimated Autocorrelations for Series 1–8 (continued)

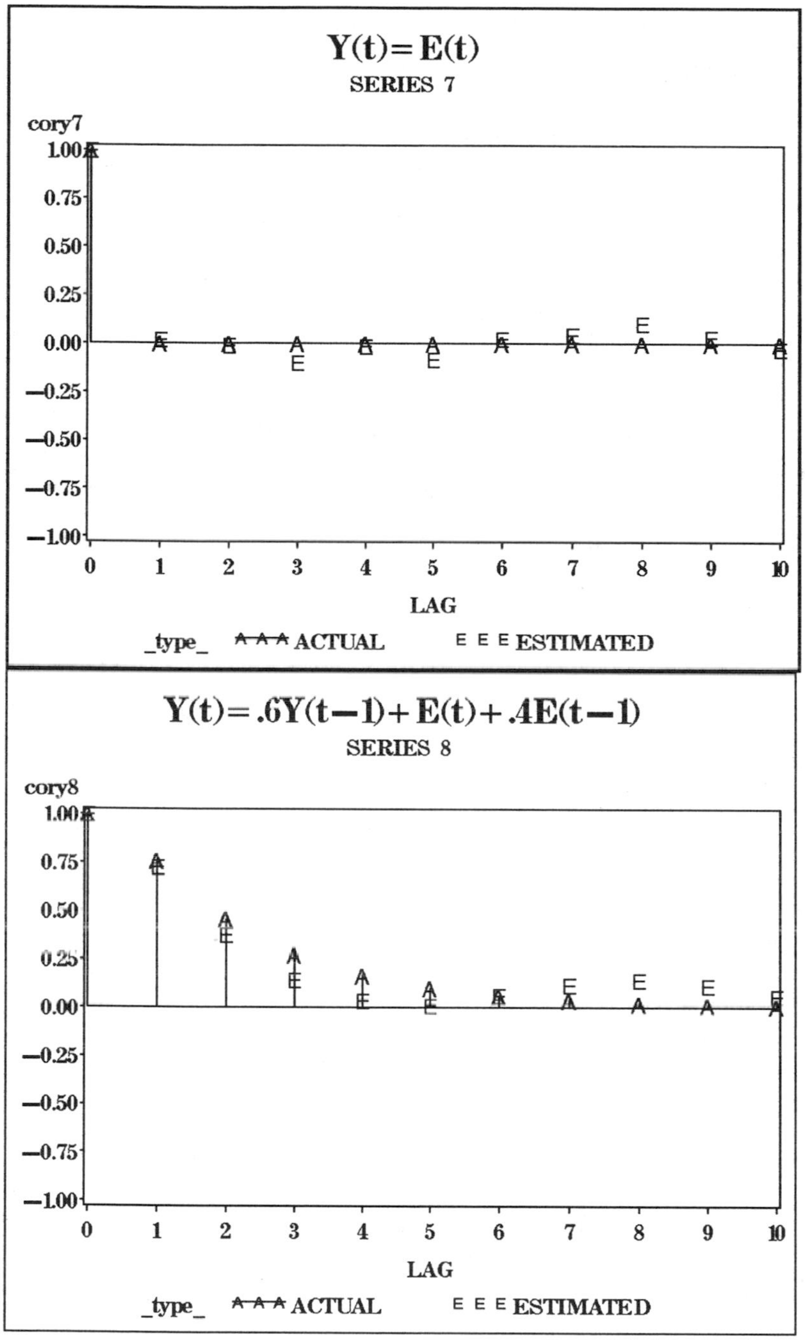

3.3.2.5 Estimated PACF

The partial autocorrelations are defined in **Section 3.3.2.3** as solutions to equations involving the covariances $\gamma(j)$. To estimate these partial autocorrelations, substitute estimated covariances C(j) for the actual covariances and solve. For j large enough that the actual partial autocorrelation π_j is 0 or nearly 0, an approximate standard error for the estimated partial autocorrelation is $n^{-1/2}$.

The next group of plots, in **Output 3.4,** illustrate the actual (A) and estimated (E) PACFs for the series.

Output 3.4
Plotting Actual and Estimated Partial Autocorrelations for Series 1–8

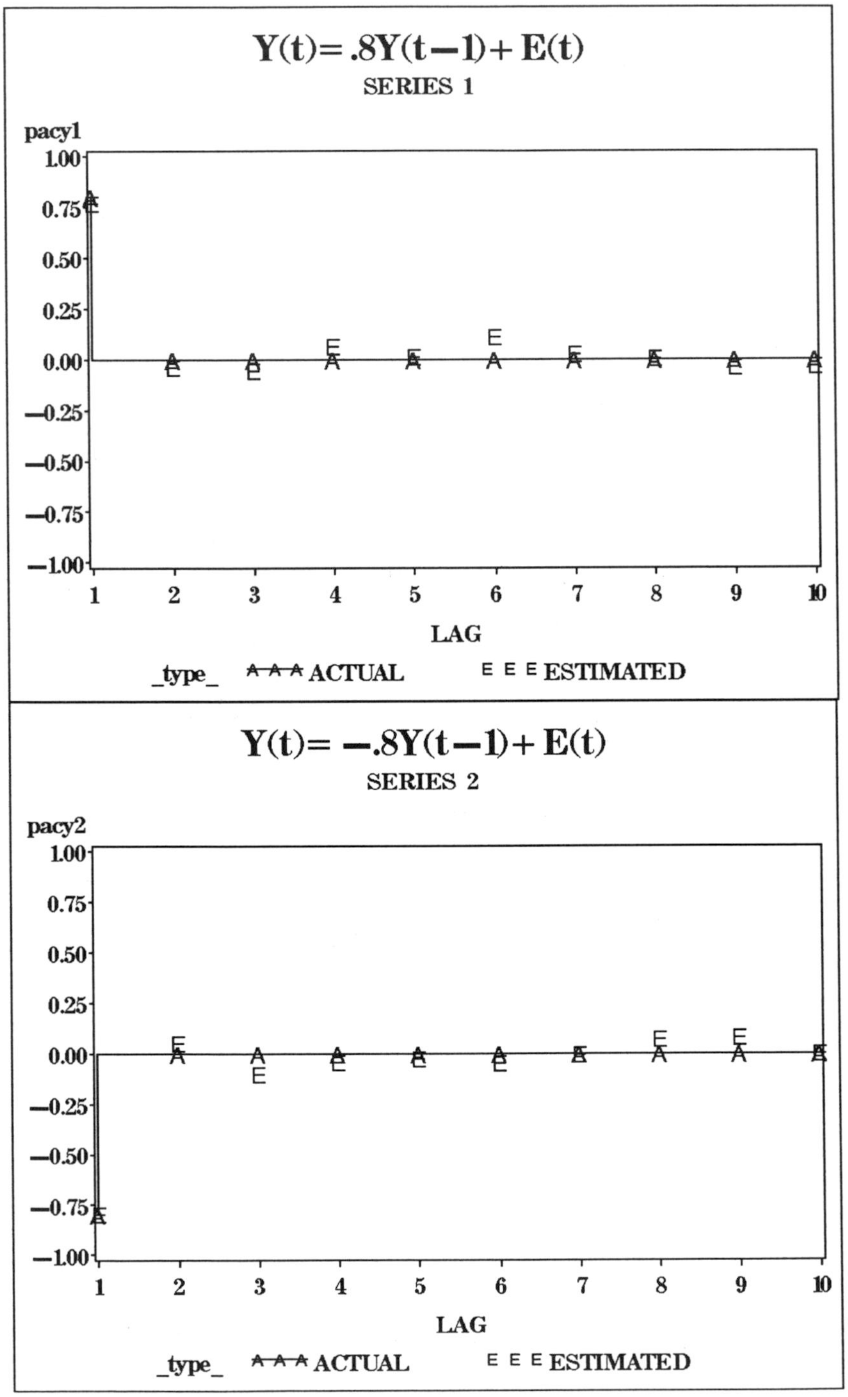

Output 3.4
Plotting Actual and Estimated Partial Autocorrelations for Series 1–8 (continued)

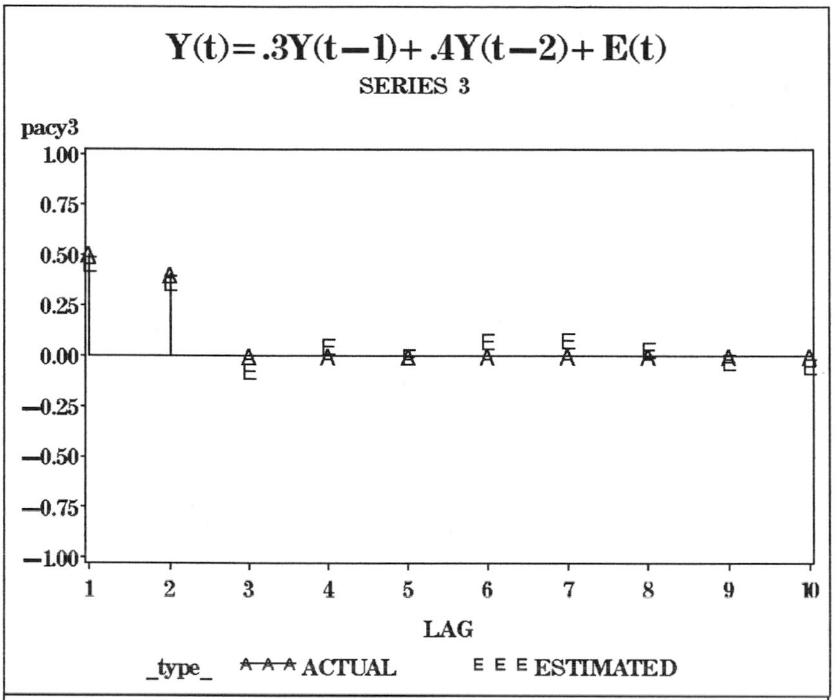

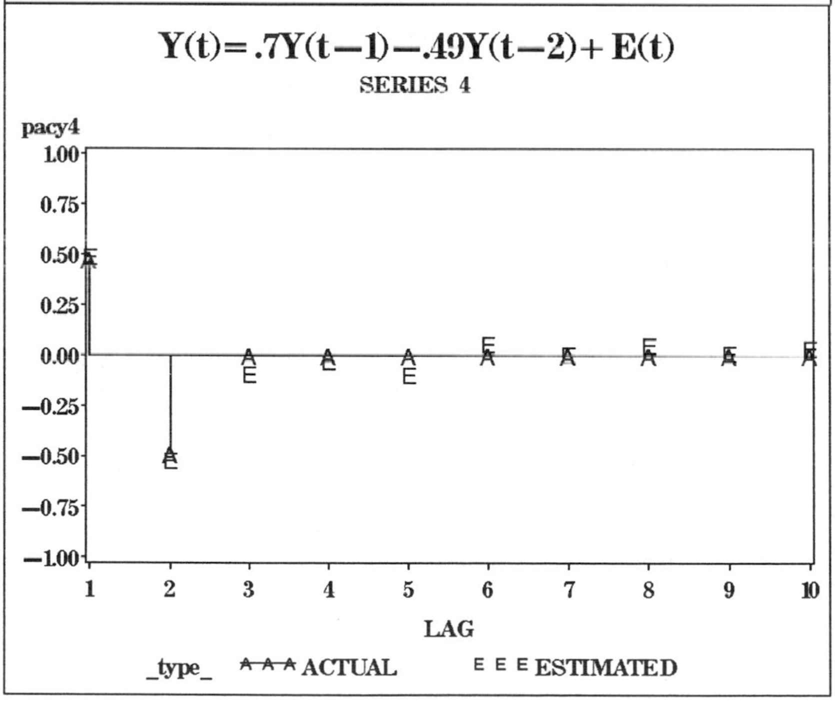

Output 3.4
Plotting Actual and Estimated Partial Autocorrelations for Series 1–8 (continued)

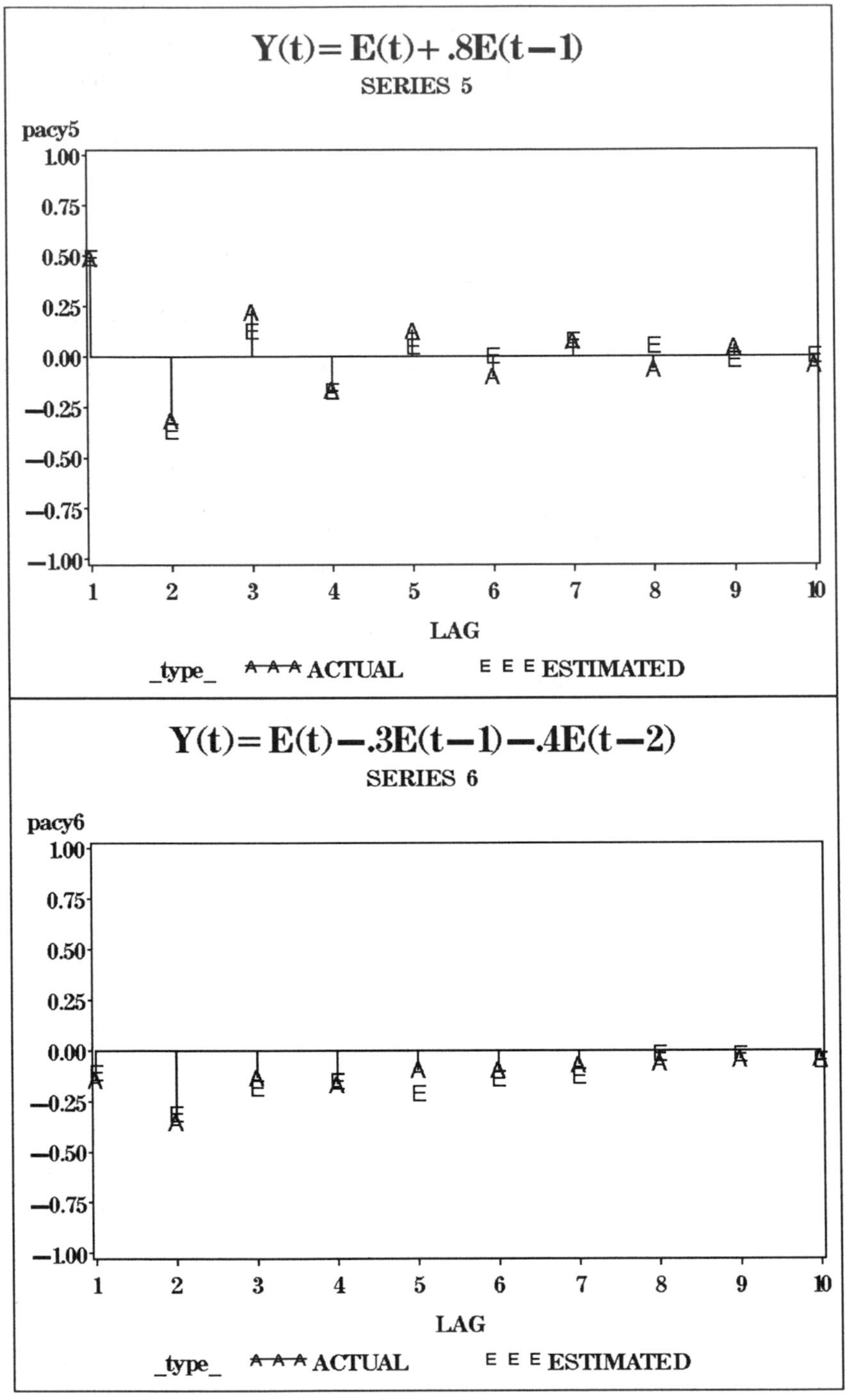

Output 3.4
Plotting Actual and Estimated Partial Autocorrelations for Series 1–8 (continued)

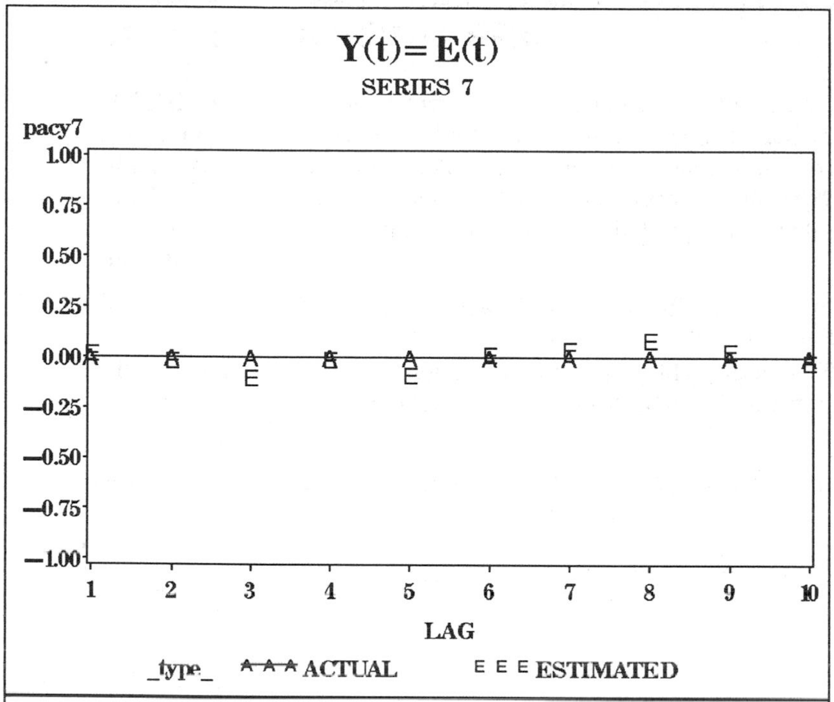

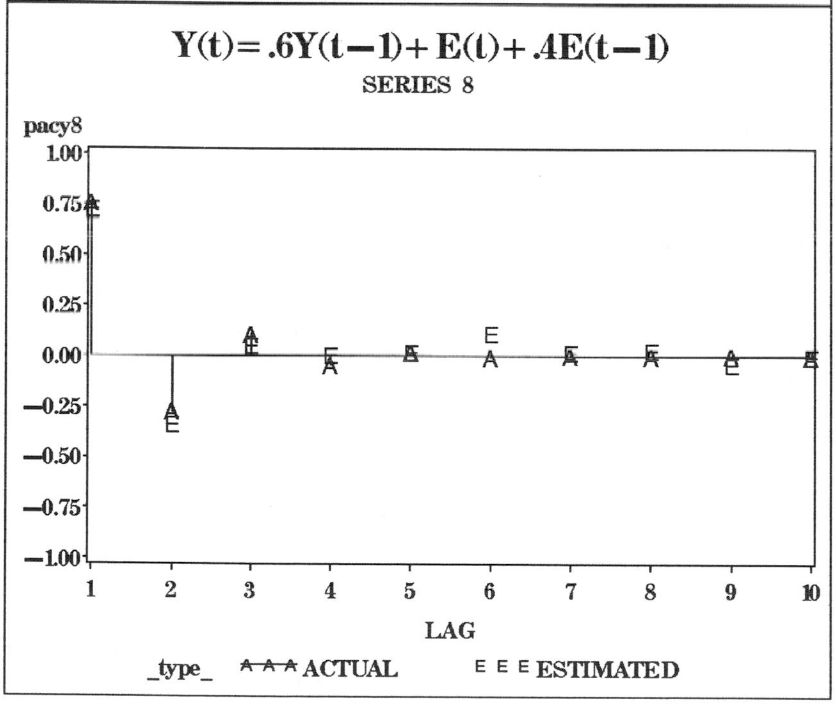

3.3.2.6 IACF

The IACF of an ARMA(*p*,*q*) model is defined as the ACF of the ARMA(*q*,*p*) model you obtain if you switch sides with the MA and AR operators. Thus, the inverse autocorrelation of

$$(1-.8B)(Y_t-\mu)=e_t$$

is defined as the ACF of

$$Y_t-\mu = e_t-.8e_{t-1}$$

In the catalog of Series 1–8, for example, the IACF of Series 3 is the same as the ACF of Series 6 and vice versa.

3.3.2.7 Estimated IACF

Suppose you know that a series comes from an AR(3) process. Fit an AR(3) model to obtain estimated coefficients—for example,

$$Y_t-\mu = .300(Y_{t-1}-\mu)+.340(Y_{t-2}-\mu)-.120(Y_{t-3}-\mu)+e_t$$

The inverse model is the moving average

$$Y_t-\mu = e_t-.300e_{t-1}-.340e_{t-2}+.120e_{t-3}$$

The inverse autocovariances are estimated by

$$(1+.300^2+.340^2+.120^2)\sigma^2$$

at lag 0,

$$(-.300+(.300)(.340)-(.340)(.120))\sigma^2$$

at lag 1,

$$(-.340-(.300)(.120))\sigma^2$$

at lag 2, and $.120\sigma^2$ at lag 3.

In general, you do not know the order p of the process, nor do you know the form (it may be MA or ARMA). Use the fact (see **Section 3.3.1**) that any invertible ARMA series can be represented as an infinite-order AR and therefore can be approximated by an AR(p) with p large.

Set p to the minimum of the NLAG value and one-half the number of observations after differencing. Then do the following:

- Fit AR(p) to the data.
- Using the estimated coefficients, compute covariances for corresponding MA series as illustrated above for p=3.
- Assign standard errors of $n^{-1/2}$ to the resulting estimates.

3.3.3 Chi-Square Check of Residuals

In the identification stage, PROC ARIMA uses the autocorrelations to form a statistic whose approximate distribution is chi-square under the null hypothesis that the series is white noise. The test is the Ljung modification of the Box-Pierce Q statistic. Both Q statistics are described in Box, Jenkins, and Riensel (1994) and the Ljung modification in Ljung and Box (1978, p. 297). The formula for this statistic is

$$n(n+2)\,\Sigma_{j=1}^{k}\, r^2(j)\,/\,(n-j)$$

where $r(j)$ is the estimated autocorrelation at lag j and k can be any positive integer. In PROC ARIMA several ks are used.

Later in the modeling stage, PROC ARIMA calculates the same statistic on the model residuals to test the hypothesis that they are white noise. The statistic is compared to critical values from a chi-square distribution. If your model is correct, the residuals should be white noise and the chi-square statistic should be small (the PROB value should be large). A significant chi-square statistic indicates that your model does not fit well.

3.3.4 Summary of Model Identification

At the identification stage, you compute the ACF, PACF, and IACF. Behavior of the estimated functions is the key to model identification. The behavior of functions for different processes is summarized in the following table:

***Table 3.1** Summary of Model Identification*

	MA(q)	AR(p)	ARMA(p, q)	White noise
ACF	D(q)	T	T	0
PACF	T	D(p)	T	0
IACF	T	D(p)	T	0

where

D(q) means the function drops off to 0 after lag q

T means the function tails off exponentially

0 means the function is 0 at all nonzero lags.

3.4 Examples and Instructions

The following pages contain results for 150 observations generated from each of the eight sample series discussed earlier. Thus, the ACFs correspond to the Es in **Output 3.3**. Even with 150 observations, considerable variation occurs.

To obtain all of the output shown for the first series Y1, use these SAS statements:

```
PROC ARIMA DATA=SERIES;
   IDENTIFY VAR=Y1 NLAG=10;
RUN;
```

The VAR= option is required. The NLAG= option gives the number of autocorrelations to be computed and defaults to 24. When you fit an ARIMA(p,d,q), NLAG+1 must be greater than p+d+q to obtain initial parameter estimates. For the ARMA(p,q) models discussed so far, d is 0.

The following options can also be used:

NOPRINT
: suppresses printout. This is useful because you must use an IDENTIFY statement prior to an ESTIMATE statement. If you have seen the output on a previous run, you may want to suppress it with this option.

CENTER
: subtracts the series mean from each observation prior to the analysis.

DATA=*SASdataset*
: specifies the SAS data set to be analyzed (the default is the most recently created SAS data set).

3.4.1 IDENTIFY Statement for Series 1–8

The following SAS statements, when used on the generated data, produce **Output 3.5**:

```
PROC ARIMA DATA=SERIES;
   IDENTIFY VAR=Y1 NLAG=10;
   IDENTIFY VAR=Y2 NLAG=10;
   more SAS statements
   IDENTIFY VAR=Y8 NLAG=10;
RUN;
```

Try to identify all eight of these series. These are presented in **Section 3.3.2.1**, so you can check your diagnosis against the actual model. For example, look at Y6. First, observe that the calculated Q statistic ❶ is 17.03, which would be compared to a chi-square distribution with six degrees of freedom. The 5% critical value is 12.59, so you have significant evidence against the null hypothesis that the considered model is adequate. Because no model is specified, this Q statistic simply tests the hypothesis that the original data are white noise. The number 0.0092 ❷ is the area under the chi-square distribution to the right of the calculated 17.03. Because 0.0092 is less than .05, without recourse to a chi-square table, you see that 17.03 is to the right of the 5% critical value. Either way, you decide that Y6 is not a white noise series. Contrast this with Y7, where the calculated statistic 2.85 ❸ has an area 0.8269 ❹ to its right; 2.85 is far to the left of the critical value and nowhere near significance. Therefore, you decide that Y7 is a white noise series.

A model is needed for Y6. The PACF and IACF are nonzero through several lags, which means that an AR diagnosis requires perhaps seven lags. A model with few parameters is preferable. The ACF is near 0 after two lags, indicating that you may choose an MA(2). Because an MA model has a persistently nonzero PACF and IACF, the MA(2) diagnosis seems appropriate. At this stage, you have identified the form of the model and can assign the remainder of the analysis to PROC ARIMA. You must identify the model because PROC ARIMA does not do it automatically.

The generated series has 150 observations; note the width of the standard error bands on the autocorrelations. Even with 150 observations, reading fine detail from the ACF is unlikely. Your goal is to use these functions to limit your search to a few plausible models rather than to pinpoint one model at the identification stage.

Output 3.5 *Using the IDENTIFY Statement for Series 1–8: PROC ARIMA*

```
                                      The ARIMA Procedure

                                    Name of Variable = Y1

                              Mean of Working Series    -0.83571
                              Standard Deviation        1.610893
                              Number of Observations         150

                                       Autocorrelations

Lag    Covariance    Correlation    -1 9 8 7 6 5 4 3 2 1 0 1 2 3 4 5 6 7 8 9 1     Std Error

  0      2.594976        1.00000    |                    |********************|             0
  1      1.993518        0.76822    |                 .  |***************     |      0.081650
  2      1.493601        0.57557    |               .    |************        |      0.120563
  3      1.063870        0.40997    |              .     |********            |      0.137669
  4      0.819993        0.31599    |              .     |******              |      0.145581
  5      0.652487        0.25144    |              .     |*****.              |      0.150084
  6      0.644574        0.24839    |              .     |*****.              |      0.152866
  7      0.637198        0.24555    |              .     |*****.              |      0.155534
  8      0.609458        0.23486    |              .     |*****.              |      0.158097
  9      0.504567        0.19444    |              .     |**** .              |      0.160406
 10      0.372414        0.14351    |              .     |***  .              |      0.161970

                                   Inverse Autocorrelations

              Lag    Correlation    -1 9 8 7 6 5 4 3 2 1 0 1 2 3 4 5 6 7 8 9 1

                1       -0.48474    |          **********|  .                 |
                2       -0.03700    |                 . *|  .                 |
                3        0.07242    |                 .  |* .                 |
                4       -0.06915    |                 . *|  .                 |
                5        0.08596    |                 .  |**.                 |
                6       -0.04379    |                 . *|  .                 |
                7        0.00593    |                 .  |  .                 |
                8       -0.02651    |                 . *|  .                 |
                9       -0.00946    |                 .  |  .                 |
               10        0.02004    |                 .  |  .                 |

                                   Partial Autocorrelations

              Lag    Correlation    -1 9 8 7 6 5 4 3 2 1 0 1 2 3 4 5 6 7 8 9 1

                1        0.76822    |                 .  |***************     |
                2       -0.03560    |                 . *|  .                 |
                3       -0.05030    |                 . *|  .                 |
                4        0.06531    |                 .  |* .                 |
                5        0.01599    |                 .  |  .                 |
                6        0.11264    |                 .  |**.                 |
                7        0.02880    |                 .  |* .                 |
                8        0.00625    |                 .  |  .                 |
                9       -0.03668    |                 . *|  .                 |
               10       -0.03308    |                 . *|  .                 |

                            Autocorrelation Check for White Noise

 To        Chi-             Pr >
Lag      Square     DF     ChiSq    -------------------Autocorrelations------------------

  6      202.72      6    <.0001     0.768     0.576     0.410     0.316     0.251     0.248
```

Output 3.5 *Using the IDENTIFY Statement for Series 1–8: PROC ARIMA (continued)*

```
                              The ARIMA Procedure
                             Name of Variable = Y2

                      Mean of Working Series    -0.07304
                      Standard Deviation        1.740946
                      Number of Observations         150

                               Autocorrelations

Lag    Covariance    Correlation    -1 9 8 7 6 5 4 3 2 1 0 1 2 3 4 5 6 7 8 9 1      Std Error

  0      3.030893        1.00000    |                    |********************|             0
  1     -2.414067        -.79649    |    ****************|   .                |      0.081650
  2      1.981819        0.65387    |               .    |*************       |      0.122985
  3     -1.735348        -.57255    |         ***********|     .              |      0.144312
  4      1.454755        0.47998    |               .    |**********          |      0.158735
  5     -1.242813        -.41005    |            ********|      .             |      0.168132
  6      1.023028        0.33753    |              .     |*******             |      0.174672
  7     -0.844730        -.27871    |              .******|      .            |      0.178968
  8      0.790137        0.26069    |              .     |*****  .            |      0.181838
  9     -0.623423        -.20569    |              .  ****|      .            |      0.184313
 10      0.494691        0.16322    |              .     |***    .            |      0.185837

                          Inverse Autocorrelations

              Lag    Correlation    -1 9 8 7 6 5 4 3 2 1 0 1 2 3 4 5 6 7 8 9 1

                1        0.50058    |                 .  |**********          |
                2        0.08442    |                 .  |**.                 |
                3        0.10032    |                 .  |**.                 |
                4        0.03955    |                 .  |* .                 |
                5        0.02433    |                 .  |  .                 |
                6       -0.01680    |                 .  |  .                 |
                7       -0.10020    |                 .**|  .                 |
                8       -0.12236    |                 .**|  .                 |
                9       -0.05245    |                 . *|  .                 |
               10       -0.00211    |                 .  |  .                 |

                          Partial Autocorrelations

              Lag    Correlation    -1 9 8 7 6 5 4 3 2 1 0 1 2 3 4 5 6 7 8 9 1

                1       -0.79649    |    ****************|  .                 |
                2        0.05329    |                 .  |* .                 |
                3       -0.10166    |                 .**|  .                 |
                4       -0.03975    |                 . *|  .                 |
                5       -0.02340    |                 .  |  .                 |
                6       -0.04177    |                 . *|  .                 |
                7       -0.00308    |                 .  |  .                 |
                8        0.07566    |                 .  |**.                 |
                9        0.08029    |                 .  |**.                 |
               10        0.00344    |                 .  |  .                 |

                    Autocorrelation Check for White Noise

 To        Chi-             Pr >
Lag      Square     DF     ChiSq    --------------------Autocorrelations--------------------

  6      294.24      6    <.0001    -0.796     0.654    -0.573     0.480    -0.410     0.338
```

Output 3.5 *Using the IDENTIFY Statement for Series 1–8: PROC ARIMA (continued)*

```
                                The ARIMA Procedure
                               Name of Variable = Y3

                         Mean of Working Series    -0.55064
                         Standard Deviation        1.237272
                         Number of Observations         150

                                  Autocorrelations

Lag   Covariance   Correlation   -1 9 8 7 6 5 4 3 2 1 0 1 2 3 4 5 6 7 8 9 1      Std Error

  0     1.530842       1.00000   |                    |********************|              0
  1     0.693513       0.45303   |                  . |*********           |       0.081650
  2     0.756838       0.49439   |                 .  |**********          |       0.096970
  3     0.395653       0.25845   |                .   |*****               |       0.112526
  4     0.417928       0.27301   |                .   |*****               |       0.116416
  5     0.243252       0.15890   |                .   |***  .              |       0.120609
  6     0.311005       0.20316   |                .   |****.               |       0.121997
  7     0.274850       0.17954   |                .   |****.               |       0.124232
  8     0.295125       0.19279   |                .   |****.               |       0.125950
  9     0.212710       0.13895   |                .   |***  .              |       0.127902
 10     0.154864       0.10116   |                .   |**   .              |       0.128904

                            Inverse Autocorrelations

               Lag   Correlation   -1 9 8 7 6 5 4 3 2 1 0 1 2 3 4 5 6 7 8 9 1

                 1      -0.17754   |                ****|  .                 |
                 2      -0.30226   |              ******|  .                 |
                 3       0.04442   |                 .  |* .                 |
                 4      -0.02891   |                 . *|  .                 |
                 5       0.07177   |                 .  |* .                 |
                 6       0.00400   |                 .  |  .                 |
                 7      -0.04363   |                 . *|  .                 |
                 8      -0.06394   |                 . *|  .                 |
                 9      -0.00347   |                 .  |  .                 |
                10       0.03944   |                 .  |* .                 |

                            Partial Autocorrelations

               Lag   Correlation   -1 9 8 7 6 5 4 3 2 1 0 1 2 3 4 5 6 7 8 9 1

                 1       0.45303   |                 .  |*********           |
                 2       0.36383   |                 .  |*******             |
                 3      -0.07085   |                 . *|  .                 |
                 4       0.04953   |                 .  |* .                 |
                 5      -0.00276   |                 .  |  .                 |
                 6       0.07517   |                 .  |**.                 |
                 7       0.07687   |                 .  |**.                 |
                 8       0.03254   |                 .  |* .                 |
                 9      -0.02622   |                 . *|  .                 |
                10      -0.04933   |                 . *|  .                 |

                      Autocorrelation Check for White Noise

 To        Chi-             Pr >
Lag      Square     DF     ChiSq    -------------------Autocorrelations------------------

  6      101.56      6    <.0001     0.453     0.494     0.258     0.273     0.159     0.203
```

Output 3.5 *Using the IDENTIFY Statement for Series 1–8: PROC ARIMA (continued)*

```
                                  The ARIMA Procedure
                                 Name of Variable = Y4

                             Mean of Working Series     -0.21583
                             Standard Deviation         1.381192
                             Number of Observations          150

                                    Autocorrelations

 Lag    Covariance    Correlation    -1 9 8 7 6 5 4 3 2 1 0 1 2 3 4 5 6 7 8 9 1      Std Error

   0      1.907692        1.00000    |                    |********************|            0
   1      0.935589        0.49043    |                .   |**********          |     0.081650
   2     -0.297975        -.15620    |              .***|   .                  |     0.099366
   3     -0.810601        -.42491    |          ********|   .                  |     0.100990
   4     -0.546360        -.28640    |            ******|   .                  |     0.112278
   5     -0.106682        -.05592    |              .   *|   .                 |     0.117047
   6      0.237817        0.12466    |              .    |**  .                |     0.117225
   7      0.324887        0.17030    |              .    |*** .                |     0.118105
   8      0.241111        0.12639    |              .    |*** .                |     0.119731
   9      0.055065        0.02886    |              .    |*   .                |     0.120617
  10     -0.073198        -.03837    |              .   *|   .                 |     0.120663
                                 Inverse Autocorrelations

                Lag    Correlation    -1 9 8 7 6 5 4 3 2 1 0 1 2 3 4 5 6 7 8 9 1

                  1       -0.57307    |        ***********|  .                 |
                  2        0.22767    |                .  |*****               |
                  3        0.10911    |                .  |**.                 |
                  4       -0.09954    |                .**|  .                 |
                  5        0.09918    |                .  |**.                 |
                  6       -0.06612    |                . *|  .                 |
                  7        0.03872    |                .  |* .                 |
                  8       -0.05099    |                . *|  .                 |
                  9        0.02259    |                .  |  .                 |
                 10       -0.01998    |                .  |  .                 |

                                 Partial Autocorrelations

                Lag    Correlation    -1 9 8 7 6 5 4 3 2 1 0 1 2 3 4 5 6 7 8 9 1

                  1        0.49043    |                .  |**********          |
                  2       -0.52236    |         **********|  .                 |
                  3       -0.09437    |                .**|  .                 |
                  4       -0.02632    |                . *|  .                 |
                  5       -0.09683    |                .**|  .                 |
                  6        0.05711    |                .  |* .                 |
                  7        0.00687    |                .  |  .                 |
                  8        0.05190    |                .  |* .                 |
                  9        0.00919    |                .  |  .                 |
                 10        0.03505    |                .  |* .                 |

                            Autocorrelation Check for White Noise

  To        Chi-               Pr >
 Lag      Square      DF      ChiSq    ------------------Autocorrelations------------------

   6       84.33       6     <.0001     0.490    -0.156    -0.425    -0.286    -0.056     0.125
```

Output 3.5 *Using the IDENTIFY Statement for Series 1–8: PROC ARIMA (continued)*

```
                                      The ARIMA Procedure
                                     Name of Variable = Y5

                             Mean of Working Series     -0.30048
                             Standard Deviation         1.316518
                             Number of Observations          150

                                        Autocorrelations

Lag    Covariance    Correlation    -1 9 8 7 6 5 4 3 2 1 0 1 2 3 4 5 6 7 8 9 1      Std Error

  0      1.733219        1.00000    |                    |********************|             0
  1      0.852275        0.49173    |                  . |**********          |      0.081650
  2     -0.055217        -.03186    |                 .  *|  .                 |      0.099452
  3     -0.200380        -.11561    |                 . **|  .                 |      0.099520
  4     -0.203287        -.11729    |                 . **|  .                 |      0.100411
  5     -0.144763        -.08352    |                 . **|  .                 |      0.101320
  6      0.011068        0.00639    |                 .   |  .                 |      0.101778
  7      0.163554        0.09436    |                 .   |**  .               |      0.101781
  8      0.234861        0.13551    |                 .   |***.                |      0.102363
  9      0.141452        0.08161    |                 .   |**  .               |      0.103551
 10      0.0013709       0.00079    |                 .   |  .                 |      0.103979
                                   Inverse Autocorrelations

               Lag    Correlation    -1 9 8 7 6 5 4 3 2 1 0 1 2 3 4 5 6 7 8 9 1

                 1       -0.69306    |      **************|  .                 |
                 2        0.44973    |                  . |*********           |
                 3       -0.24954    |               *****|  .                 |
                 4        0.13893    |                  . |***                 |
                 5       -0.04150    |                  . *|  .                |
                 6        0.00044    |                  .  |  .                |
                 7        0.01466    |                  .  |  .                |
                 8       -0.04225    |                  . *|  .                |
                 9        0.01048    |                  .  |  .                |
                10       -0.00262    |                  .  |  .                |

                                   Partial Autocorrelations

               Lag    Correlation    -1 9 8 7 6 5 4 3 2 1 0 1 2 3 4 5 6 7 8 9 1

                 1        0.49173    |                  . |**********          |
                 2       -0.36093    |             *******|  .                 |
                 3        0.12615    |                  . |***                 |
                 4       -0.17086    |                 ***|  .                 |
                 5        0.05473    |                  . |*  .                |
                 6        0.00681    |                  . |  .                 |
                 7        0.08204    |                  . |**.                 |
                 8        0.05577    |                  . |*  .                |
                 9       -0.01303    |                  . |  .                 |
                10        0.00517    |                  . |  .                 |

                            Autocorrelation Check for White Noise

  To        Chi-                 Pr >
 Lag      Square      DF        ChiSq     -------------------Autocorrelations------------------

   6       42.48       6       <.0001      0.492    -0.032    -0.116    -0.117    -0.084    0.006
```

Output 3.5 *Using the IDENTIFY Statement for Series 1–8: PROC ARIMA (continued)*

```
                                The ARIMA Procedure
                               Name of Variable = Y6

                         Mean of Working Series    -0.04253
                         Standard Deviation        1.143359
                         Number of Observations         150

                                  Autocorrelations

 Lag    Covariance    Correlation    -1 9 8 7 6 5 4 3 2 1 0 1 2 3 4 5 6 7 8 9 1      Std Error

   0      1.307271        1.00000    |                    |********************|              0
   1     -0.137276        -.10501    |                 .**|  .                 |       0.081650
   2     -0.385340        -.29477    |              ******|  .                 |       0.082545
   3     -0.118515        -.09066    |                . **|   .                |       0.089287
   4     0.0083104        0.00636    |                .   |   .                |       0.089899
   5     -0.084843        -.06490    |                .  *|   .                |       0.089902
   6      0.011812        0.00904    |                .   |   .                |       0.090214
   7      0.045677        0.03494    |                .   |*  .                |       0.090220
   8      0.119262        0.09123    |                .   |** .                |       0.090310
   9      0.018882        0.01444    |                .   |   .                |       0.090922
  10     -0.083572        -.06393    |                .  *|   .                |       0.090937

                             Inverse Autocorrelations

               Lag    Correlation    -1 9 8 7 6 5 4 3 2 1 0 1 2 3 4 5 6 7 8 9 1

                 1        0.54503    |                 .  |***********         |
                 2        0.57372    |                 .  |***********         |
                 3        0.44261    |                 .  |*********           |
                 4        0.33790    |                 .  |*******             |
                 5        0.28173    |                 .  |******              |
                 6        0.16162    |                 .  |***                 |
                 7        0.10979    |                 .  |**.                 |
                 8        0.03715    |                 .  |* .                 |
                 9        0.02286    |                 .  |  .                 |
                10        0.02606    |                 .  |* .                 |

                             Partial Autocorrelations

               Lag    Correlation    -1 9 8 7 6 5 4 3 2 1 0 1 2 3 4 5 6 7 8 9 1

                 1       -0.10501    |                 .**|  .                 |
                 2       -0.30920    |              ******|  .                 |
                 3       -0.18297    |                ****|  .                 |
                 4       -0.14923    |                 ***|  .                 |
                 5       -0.20878    |                ****|  .                 |
                 6       -0.13688    |                 ***|  .                 |
                 7       -0.12493    |                 .**|  .                 |
                 8       -0.00862    |                 .  |  .                 |
                 9       -0.01255    |                 .  |  .                 |
                10       -0.04518    |                 . *|  .                 |

                       Autocorrelation Check for White Noise
             ❶                   ❷
 To         Chi-                Pr >
 Lag       Square      DF      ChiSq    -------------------Autocorrelations-------------------

   6        17.03       6     0.0092    -0.105    -0.295    -0.091     0.006    -0.065     0.009
```

***Output 3.5** Using the IDENTIFY Statement for Series 1–8: PROC ARIMA (continued)*

```
                                 The ARIMA Procedure
                                Name of Variable = Y7

                          Mean of Working Series     -0.15762
                          Standard Deviation         1.023007
                          Number of Observations          150

                                   Autocorrelations

Lag    Covariance    Correlation   -1 9 8 7 6 5 4 3 2 1 0 1 2 3 4 5 6 7 8 9 1     Std Error

  0      1.046543        1.00000   |                    |********************|             0
  1      0.019680        0.01880   |                .   |   .                |      0.081650
  2     -0.012715        -.01215   |                .   |   .                |      0.081679
  3     -0.107313        -.10254   |                .** |   .                |      0.081691
  4     -0.012754        -.01219   |                .   |   .                |      0.082544
  5     -0.085250        -.08146   |                .** |   .                |      0.082556
  6      0.023489        0.02244   |                .   |   .                |      0.083090
  7      0.048176        0.04603   |                .   |*  .                |      0.083131
  8      0.106544        0.10181   |                .   |** .                |      0.083300
  9      0.033337        0.03185   |                .   |*  .                |      0.084126
 10     -0.026272        -.02510   |                .  *|   .                |      0.084206

                               Inverse Autocorrelations

              Lag    Correlation   -1 9 8 7 6 5 4 3 2 1 0 1 2 3 4 5 6 7 8 9 1

                1       -0.00490   |                .   |   .                |
                2        0.02012   |                .   |   .                |
                3        0.08424   |                .   |** .                |
                4       -0.00793   |                .   |   .                |
                5        0.06817   |                .   |*  .                |
                6       -0.02208   |                .   |   .                |
                7       -0.03545   |                .  *|   .                |
                8       -0.08197   |                .** |   .                |
                9       -0.03324   |                .  *|   .                |
               10        0.02062   |                .   |   .                |

                               Partial Autocorrelations

              Lag    Correlation   -1 9 8 7 6 5 4 3 2 1 0 1 2 3 4 5 6 7 8 9 1

                1        0.01880   |                .   |   .                |
                2       -0.01251   |                .   |   .                |
                3       -0.10213   |                .** |   .                |
                4       -0.00867   |                .   |   .                |
                5       -0.08435   |                .** |   .                |
                6        0.01474   |                .   |   .                |
                7        0.04159   |                .   |*  .                |
                8        0.08562   |                .   |** .                |
                9        0.03364   |                .   |*  .                |
               10       -0.02114   |                .   |   .                |

                         Autocorrelation Check for White Noise
          ❸                 ❹
 To      Chi-                Pr >
Lag     Square     DF      ChiSq     --------------------Autocorrelations--------------------

  6       2.85      6     0.8269     0.019     -0.012     -0.103     -0.012     -0.081     0.022
```

Output 3.5 *Using the IDENTIFY Statement for Series 1–8: PROC ARIMA (continued)*

```
                          The ARIMA Procedure
                          Name of Variable = Y8

                   Mean of Working Series      -0.57405
                   Standard Deviation          1.591833
                   Number of Observations           150

                             Autocorrelations
```

Lag	Covariance	Correlation	-1 9 8 7 6 5 4 3 2 1 0 1 2 3 4 5 6 7 8 9 1	Std Error
0	2.533932	1.00000	| |********************|	0
1	1.848193	0.72938	| . |*************** |	0.081650
2	0.946216	0.37342	| . |******* |	0.117303
3	0.352595	0.13915	| . |*** . |	0.124976
4	0.086093	0.03398	| . |* . |	0.126005
5	0.025473	0.01005	| . | . |	0.126066
6	0.150883	0.05955	| . |* . |	0.126071
7	0.295444	0.11659	| . |** . |	0.126259
8	0.359400	0.14183	| . |*** . |	0.126975
9	0.279258	0.11021	| . |** . |	0.128026
10	0.142827	0.05637	| . |* . |	0.128657

```
                          Inverse Autocorrelations
```

Lag	Correlation	-1 9 8 7 6 5 4 3 2 1 0 1 2 3 4 5 6 7 8 9 1
1	-0.65293	| *************| . |
2	0.20188	| . |**** |
3	0.00095	| . | . |
4	-0.06691	| . *| . |
5	0.08986	| . |** . |
6	-0.00873	| . *| . |
7	0.04555	| . |* . |
8	-0.04728	| . *| . |
9	0.01718	| . | . |
10	0.00064	| . | . |

```
                          Partial Autocorrelations
```

Lag	Correlation	-1 9 8 7 6 5 4 3 2 1 0 1 2 3 4 5 6 7 8 9 1
1	0.72938	| . |*************** |
2	-0.33883	| *******| . |
3	0.05222	| . |* . |
4	0.00753	| . | . |
5	0.01918	| . | . |
6	0.11087	| . |** . |
7	0.02107	| . | . |
8	0.02566	| . |* . |
9	-0.03977	| . *| . |
10	-0.00138	| . | . |

```
                      Autocorrelation Check for White Noise
```

To Lag	Chi-Square	DF	Pr > ChiSq	-----------------Autocorrelations-----------------					
6	106.65	6	<.0001	0.729	0.373	0.139	0.034	0.010	0.060

3.4.2 Example: Iron and Steel Export Analysis

The U.S. iron and steel export yearly series (Fairchild Publications 1981) graphed in **Output 3.6** is a good illustration of model identification.

Output 3.6
Plotting a Yearly Series

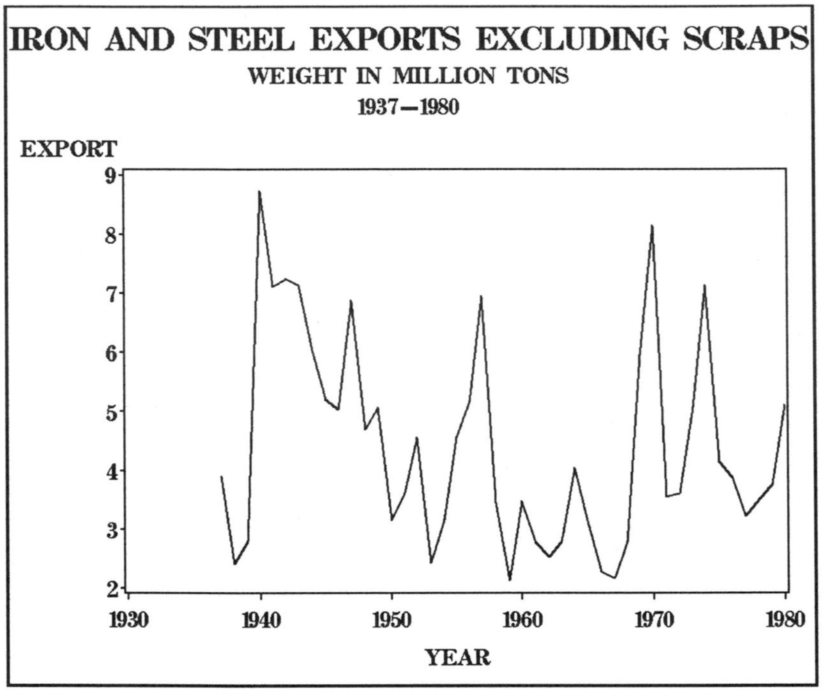

The following statements produce the results in **Output 3.7**:

```
PROC ARIMA DATA=STEEL;
   IDENTIFY VAR=EXPORT NLAG=10;
RUN;
```

Although the Q statistic ❶ fails by a slim margin to be significant, the lag 1 autocorrelation 0.47193 ❷ is beyond the two standard error bands. Thus, you want to fit a model despite the Q value. From the ACF, it appears that an MA(1) is appropriate. From the PACF and IACF, an AR(1) also appears consistent with these data. You can fit both and select the one with the smallest error mean square. To fit the MA(1) model, use the statement

```
ESTIMATE Q=1;
```

For the AR(1) model use the statement

```
ESTIMATE P=1;
```

Output 3.7 *Identifying a Model Using the IDENTIFY Statement*

```
                          IRON AND STEEL EXPORTS EXCLUDING SCRAPS
                                  WEIGHT IN MILLION TONS

                                        1937-1980

                                  The ARIMA Procedure

                               Name of Variable = EXPORT

                          Mean of Working Series     4.418182
                          Standard Deviation          1.73354
                          Number of Observations           44

                                     Autocorrelations

Lag    Covariance    Correlation    -1 9 8 7 6 5 4 3 2 1 0 1 2 3 4 5 6 7 8 9 1      Std Error

  0      3.005160        1.00000    |                    |********************|              0
  1      1.418238        0.47193    | ❷               .  |*********           |       0.150756
  2      0.313839        0.10443    |                .   |**     .            |       0.181248
  3      0.133835        0.04453    |                .   |*      .            |       0.182611
  4      0.310097        0.10319    |                .   |**     .            |       0.182858
  5      0.296534        0.09867    |                .   |**     .            |       0.184176
  6      0.024517        0.00816    |                .   |       .            |       0.185374
  7     -0.159424        -.05305    |                .  *|       .            |       0.185382
  8     -0.299770        -.09975    |                . **|       .            |       0.185727
  9     -0.247158        -.08224    |                . **|       .            |       0.186940
 10     -0.256881        -.08548    |               .  **|        .           |       0.187761

                                 Inverse Autocorrelations

              Lag    Correlation    -1 9 8 7 6 5 4 3 2 1 0 1 2 3 4 5 6 7 8 9 1

                1       -0.48107    |          **********|     .              |
                2        0.14768    |                .   |***  .              |
                3       -0.01309    |                .   |     .              |
                4       -0.03053    |                .  *|     .              |
                5       -0.05510    |                .  *|     .              |
                6        0.04941    |                .   |*    .              |
                7       -0.04857    |                .  *|     .              |
                8        0.07991    |                .   |**   .              |
                9       -0.03744    |                .  *|     .              |
               10        0.04236    |                .   |*    .              |
```

***Output** 3.7 Identifying a Model Using the IDENTIFY Statement (continued)*

```
                     IRON AND STEEL EXPORTS EXCLUDING SCRAPS
                             WEIGHT IN MILLION TONS
                                   1937-1980

                               The ARIMA Procedure

                            Partial Autocorrelations

          Lag    Correlation    -1 9 8 7 6 5 4 3 2 1 0 1 2 3 4 5 6 7 8 9 1

            1       0.47193     |                    .   |*********          |
            2      -0.15218     |                    .  ***|       .         |
            3       0.07846     |                    .     |**     .         |
            4       0.08185     |                    .     |**     .         |
            5       0.01053     |                    .     |       .         |
            6      -0.05594     |                    .    *|       .         |
            7      -0.03333     |                    .    *|       .         |
            8      -0.08310     |                    .   **|       .         |
            9      -0.01156     |                    .     |       .         |
           10      -0.05715     |                    .    *|       .         |


                       Autocorrelation Check for White Noise
             ❶
  To       Chi-              Pr >
 Lag      Square     DF     ChiSq     --------------------Autocorrelations--------------------
                                         ❷
   6       12.15      6    0.0586     0.472     0.104     0.045     0.103     0.099     0.008
```

Suppose you overfit, using an MA(2) as an initial step. Specify these statements:

```
PROC ARIMA DATA=STEEL;
   IDENTIFY VAR=EXPORT NOPRINT;
   ESTIMATE Q=2;
RUN;
```

Any ESTIMATE statement must be preceded with an IDENTIFY statement. In this example, NOPRINT suppresses the printout of ACF, IACF, and PACF. Note that the Q statistics ❶ in **Output 3.8** are quite small, indicating a good fit for the MA(2) model. However, when you examine the parameter estimates and their t statistics ❷, you see that more parameters were fit than necessary. An MA(1) model is appropriate because the t statistic for the lag 2 parameter is only –0.85. Also, it is wise to ignore the fact that the previous Q was insignificant due to the large t value, –3.60, associated with the lag 1 coefficient. In **Output 3.7** the Q was calculated from six autocorrelations ❶, and the large lag 1 autocorrelation's effect ❷ was diminished by the other five small autocorrelations.

Output 3.8 *Fitting an MA(2) Model with the ESTIMATE Statement*

```
                    IRON AND STEEL EXPORTS EXCLUDING SCRAPS
                            WEIGHT IN MILLION TONS
                                   1937-1980

                             The ARIMA Procedure

                     Conditional Least Squares Estimation

                                   Standard                 Approx
          Parameter      Estimate        Error    t Value   Pr > |t|     Lag

          MU              4.43400      0.39137      11.33     <.0001       0
          MA1,1          -0.56028      0.15542 ❷    -3.60     0.0008       1
          MA1,2          -0.13242      0.15535      -0.85     0.3990       2


                        Constant Estimate       4.433999
                        Variance Estimate       2.433068
                        Std Error Estimate      1.559829
                        AIC                     166.8821
                        SBC                     172.2347
                        Number of Residuals           44
                    * AIC and SBC do not include log determinant.


                      Correlations of Parameter Estimates

                    Parameter          MU      MA1,1      MA1,2

                    MU              1.000     -0.013     -0.011
                    MA1,1          -0.013      1.000      0.492
                    MA1,2          -0.011      0.492      1.000


                      Autocorrelation Check of Residuals
           ❶
 To       Chi-             Pr >
Lag     Square     DF     ChiSq     -------------------Autocorrelations-------------------

  6       0.58      4    0.9653    -0.002    -0.006     0.006     0.060     0.081    -0.032
 12       2.81     10    0.9855     0.005    -0.077    -0.035     0.008    -0.163     0.057
 18       6.24     16    0.9853     0.066    -0.005     0.036    -0.098     0.123    -0.125
 24      12.10     22    0.9553    -0.207    -0.086    -0.102    -0.068     0.025    -0.060


                         Model for Variable EXPORT

                         Estimated Mean     4.433999


                           Moving Average Factors

                  Factor 1:  1 + 0.56028 B**(1) + 0.13242 B**(2)
```

You now fit an MA(1) model using these statements:

```
PROC ARIMA DATA=STEEL;
   IDENTIFY VAR=EXPORT NOPRINT;
   ESTIMATE Q=1;
RUN;
```

The results are shown in **Output 3.9**. The Q statistics ❶ are still small, so you have no evidence of a lack of fit for the order 1 MA model. The estimated model is now ❷

$$Y_t = 4.421 + e_t + .4983e_{t-1}$$

Output 3.9 *Fitting an MA(1) Model with the ESTIMATE Statement*

```
                    IRON AND STEEL EXPORTS EXCLUDING SCRAPS
                            WEIGHT IN MILLION TONS
                                  1937-1980

                             The ARIMA Procedure

                      Conditional Least Squares Estimation

                                   Standard                 Approx
          Parameter      Estimate      Error    t Value    Pr > |t|     Lag

          MU              4.42102    0.34703      12.74      <.0001       0
          MA1,1          -0.49827    0.13512      -3.69      0.0006       1

                      Constant Estimate       4.421016
                      Variance Estimate       2.412583
                      Std Error Estimate      1.553249
                      AIC                     165.5704
                      SBC                     169.1388
                      Number of Residuals           44
                 * AIC and SBC do not include log determinant.

                          Correlations of Parameter
                                  Estimates

                       Parameter          MU       MA1,1

                       MU              1.000      -0.008
                       MA1,1          -0.008       1.000

                     Autocorrelation Check of Residuals
          ❶
 To      Chi-             Pr >
Lag     Square     DF     ChiSq    -------------------Autocorrelations------------------

  6       1.31      5    0.9336     0.059     0.094    -0.028     0.085     0.075    -0.020
 12       3.23     11    0.9873    -0.006    -0.079    -0.052    -0.013    -0.146     0.039
 18       6.68     17    0.9874     0.063    -0.001     0.044    -0.092     0.096    -0.149
 24      14.00     23    0.9268    -0.206    -0.135    -0.114    -0.084     0.014    -0.072

                     ❷ Model for Variable EXPORT
                       Estimated Mean     4.421016

                          Moving Average Factors
                        Factor 1:  1 + 0.49827 B**(1)
```

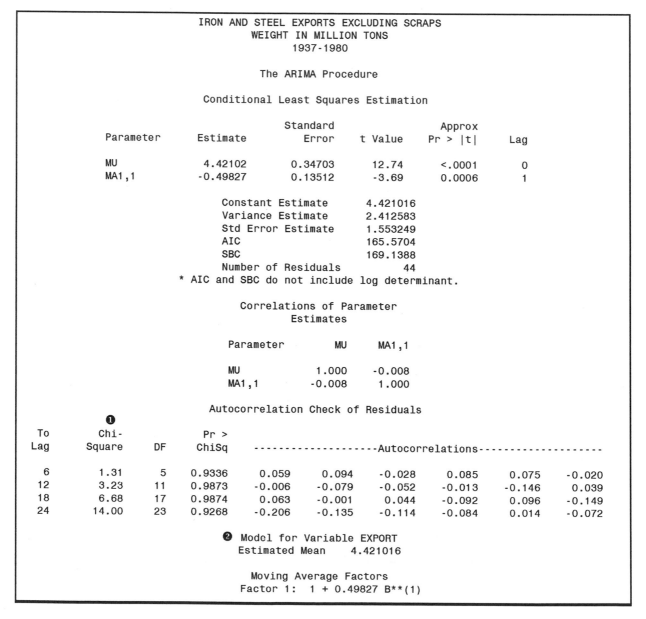

3.4.3 Estimation Methods Used in PROC ARIMA

How does PROC ARIMA estimate this MA coefficient? As in the AR case, three techniques are available:

- conditional least squares (CLS)
- unconditional least squares (ULS)
- maximum likelihood (ML).

In the CLS method you attempt to minimize

$$\Sigma_{t=p+1}^{n} \; e_t^2$$

where p is the order of the AR part of the process and e_t is a residual. In the example,

$$e_t = Y_t - \hat{\mu} + \hat{\beta}\left(e_{t-1}\right)$$

where $\hat{\mu}$ and $\hat{\beta}$ are parameter estimates. Begin by assuming e_0=0. ARIMA computations indicate that $\hat{\mu} = 4.421$ and $\hat{\beta} = -.4983$ provide the minimum for the iron export data.

To illustrate further, suppose you are given data $Y_1, Y_2, \ldots, Y_6$, where you assume

$$Y_t = e_t - \beta e_{t-1}$$

Suppose you want to estimate β from the data given below:

							Sum of squares
Y_t	−12	−3	7	9	4	−7	
$\hat{e}_t(-0.29)$	−12	0.48	6.86	7.01	1.97	−7.57	301.62
$\hat{e}_t(-0.30)$	−12	0.60	6.82	6.95	1.91	−7.57	300.13
$W_t(-0.30)$	0	12	−4	−6	−6	0	

You find that

$$\hat{\gamma}(0) = 57.89$$

and

$$\hat{\gamma}(1) = 15.26$$

Solving

$$\hat{\rho}(1) = -\beta / \left(1 + \beta^2\right) = .2636$$

yields the initial estimate $\beta = -.29$. Compute

$$\hat{e}_t = Y_t - .29\hat{e}_{t-1}$$

Starting with e_0=0, values of $\hat{e}_t$ are listed under the Y_t values. Thus,

$$\hat{e}_1 = Y_1 - .29(0) = -12$$

$$\hat{e}_2 = -3 - .29(-12) = .48$$

$$\hat{e}_3 = 7 - .29(.48) = 6.86$$

and thus

$$\hat{e}_1^2 + \hat{e}_2^2 + \ldots + \hat{e}_6^2 = 301.62$$

Perhaps you can improve upon $\hat{\beta} = -.29$. For example, using $\hat{\beta} = -.30$, you can add a second row of e_t values to the previous list and thus compute

$$\hat{e}_1^2 + \hat{e}_2^2 + \ldots + \hat{e}_6^2 = 300.13$$

The larger $\hat{\beta}$ gives a smaller sum of squares, so you would like to continue increasing $\hat{\beta}$, but by how much? Letting β_0 be the true value of the parameter, you can use Taylor's series expansion to write

$$e_t(\beta_0) = e_t(\hat{\beta}) - W_t(\hat{\beta})(\beta_0 - \hat{\beta}) + R_t \qquad (3.2)$$

where $-W_t$ is the derivative of e_t with respect to β and R_t is a remainder term. Rearranging equation 3.2 and ignoring the remainder yields

$$e_t(\hat{\beta}) = W_t(\hat{\beta})(\beta_0 - \hat{\beta}) + e_t(\beta_0)$$

Because $e_t(\beta_0)$ is white noise, this looks like a regression equation that you can use to estimate $\beta_0 - \hat{\beta}$. You need to compute the derivative W_t. Derivatives are defined as limits—for example,

$$-W_t(\hat{\beta}) = \lim_{\delta \to 0} \left(e_t(\hat{\beta} + \delta) - e_t(\hat{\beta})\right) / \delta$$

You have now computed $e_t(-.29)$ and $e_t(-.30)$, so you can approximate W_t by

$$-(e_t(-.29) - e_t(-.30)) / .01$$

as in the third row of the table above, where $\delta = .01$ and $\hat{\beta} = -.30$. In PROC ARIMA, $\delta = .001$ unless otherwise specified. Now, regressing $e_t(-.30)$ on $W_t(-.30)$ gives a coefficient

$$\begin{aligned}&((.60)(12) - (6.82)(4) - (6.95)(6) - (1.91)(6) - (7.57)(0))\\ &/\ ((12)^2 + (4)^2 + (6)^2 + (6)^2 + (0)) = -0.3157\end{aligned}$$

This is an estimate of

$$\beta_0 - \hat{\beta} = \beta_0 + .30$$

so you compute a new estimate of β by

$$-.30 - 0.3157 = -.6157$$

This estimate of β results in a lower sum of squares,

$$\Sigma \hat{e}_t^2(-.6157) = 271.87$$

Using $\hat{\beta} = -.6157$ as an initial value, you can again compute an improvement. Continue iterating the estimation improvement technique until the changes $\Delta\hat{\beta}$ become small. For this data set, $\hat{\beta} = -.6618$ appears to minimize the sum of squares at 271.153.

You can extend this method to higher-order and mixed processes. The technique used in PROC ARIMA is more sophisticated than the one given here, but it operates under the same principle. The METHOD=ULS technique more accurately computes prediction error variances and finite sample predictions than METHOD=CLS. METHOD=CLS assumes a constant variance and the same linear combination of past values as the optimum prediction. Also, when you specify METHOD=ML, the quantity to be minimized is not the sum of squares; instead, it is the negative log of the likelihood function. Although CLS, ULS, and ML should give similar results for reasonably large data sets, studies comparing the three methods indicate that ML is the most accurate. Initial values are computed from the Yule-Walker equations for the first round of the iterative procedure as in the example above. See also **Section 2.2.1**.

3.4.4 ESTIMATE Statement for Series 8

Finally, reexamine the generated series Y8,

$$Y_t - .6Y_{t-1} = e_t + .4e_{t-1}$$

The following statements produce **Output 3.10**:

```
PROC ARIMA DATA=SERIES;
   IDENTIFY VAR=Y8 NOPRINT;
   ESTIMATE P=1 Q=1 PRINTALL GRID;
   ESTIMATE P=2 Q=2;
RUN;
```

The PRINTALL option shows the iterations. Because the iterations stop when the changes in parameter estimates are small, you have no guarantee that the final parameter estimates have minimized the residual sum of squares (or maximized the likelihood). To check this, use the GRID option to evaluate the sum of squares (or likelihood) on a grid surrounding the final parameter estimates. Examine the grids ❶ ❷ ❸ in **Output 3.10** and verify that the middle sum of squares, 164.77, is the smallest of the nine tabulated values. For example, increasing the AR estimate .52459 to .52959 and decreasing the MA estimate –.32122 to –.32622 increases the sum of squares from 164.77 to 164.79. A message ❹ associated with the last command indicates that the procedure could not find estimates that minimized the error sum of squares because excess lags are specified on both sides of the ARMA model.

Output 3.10 *Using the ESTIMATE Statement for Series 8: PROC ARIMA*

The ARIMA Procedure

Preliminary Estimation

Initial Autoregressive Estimates

	Estimate
1	0.53588

Initial Moving Average Estimates

	Estimate
1	-0.27217

Constant Term Estimate	-0.17285
White Noise Variance Est	1.117013

Conditional Least Squares Estimation

Iteration	SSE	MU	MA1,1	AR1,1	Constant	Lambda	R Crit
0	165.16	-0.37242	-0.27217	0.53588	-0.17285	0.00001	1
1	164.78	-0.29450	-0.31415	0.52661	-0.13941	1E-6	0.04534
2	164.77	-0.28830	-0.31998	0.52519	-0.13689	1E-7	0.005519
3	164.77	-0.28762	-0.32102	0.52469	-0.13671	1E-8	0.000854
4	164.77	-0.28756	-0.32122	0.52459	-0.13671	1E-9	0.000151

ARIMA Estimation Optimization Summary

Estimation Method	Conditional Least Squares
Parameters Estimated	3
Termination Criteria	Maximum Relative Change in Estimates
Iteration Stopping Value	0.001
Criteria Value	0.000604
Alternate Criteria	Relative Change in Objective Function
Alternate Criteria Value	4.782E-8
Maximum Absolute Value of Gradient	0.020851
R-Square Change from Last Iteration	0.000151
Objective Function	Sum of Squared Residuals
Objective Function Value	164.7716
Marquardt's Lambda Coefficient	1E-9
Numerical Derivative Perturbation Delta	0.001
Iterations	4

***Output 3.10** Using the ESTIMATE Statement for Series 8: PROC ARIMA (continued)*

Conditional Least Squares Estimation

Parameter	Estimate	Standard Error	t Value	Approx Pr > |t|	Lag
MU	-0.28756	0.23587	-1.22	0.2247	0
MA1,1	-0.32122	0.10810	-2.97	0.0035	1
AR1,1	0.52459	0.09729	5.39	<.0001	1

Constant Estimate	-0.13671
Variance Estimate	1.120895
Std Error Estimate	1.058724
AIC	445.7703
SBC	454.8022
Number of Residuals	150

* AIC and SBC do not include log determinant.

Correlations of Parameter Estimates

Parameter	MU	MA1,1	AR1,1
MU	1.000	0.016	0.054
MA1,1	0.016	1.000	0.690
AR1,1	0.054	0.690	1.000

Autocorrelation Check of Residuals

To Lag	Chi-Square	DF	Pr > ChiSq	Autocorrelations					
6	2.08	4	0.7217	-0.006	0.003	0.039	-0.067	0.082	-0.022
12	3.22	10	0.9758	-0.059	-0.029	0.010	-0.037	0.034	-0.015
18	5.79	16	0.9902	0.004	0.106	0.051	0.038	0.002	0.004
24	11.75	22	0.9623	-0.035	0.099	-0.050	0.007	0.001	-0.140
30	13.19	28	0.9920	-0.043	-0.059	0.004	0.034	-0.011	-0.034

SSE Surface on Grid Near
Estimates: MA1,1 (y8)

MU (y8)	-0.32622	-0.32122	-0.31622
-0.29256	164.78	164.77	164.78
-0.28756	164.78	164.77	164.78
-0.28256	164.78	164.77	164.78 ❶

Output 3.10 *Using the ESTIMATE Statement for Series 8: PROC ARIMA (continued)*

```
                         SSE Surface on Grid Near
                          Estimates: AR1,1 (y8)

                MU (y8)       0.51959    0.52459    0.52959

                  -0.29256     164.78     164.77     164.78
                  -0.28756     164.78     164.77     164.78
                  -0.28256     164.78     164.77     164.78   ❷

                         SSE Surface on Grid Near
                          Estimates: AR1,1 (y8)

                MA1,1 (y8)    0.51959    0.52459    0.52959

                  -0.32622     164.77     164.78     164.79
                  -0.32122     164.78     164.77     164.78   ❸
                  -0.31622     164.79     164.78     164.77


                          Model for Variable y8

                        Estimated Mean     -0.28756


                          Autoregressive Factors

                       Factor 1:  1 - 0.52459 B**(1)


                          Moving Average Factors

                       Factor 1:  1 + 0.32122 B**(1)

WARNING: The model defined by the new estimates is unstable. The iteration process has been
         terminated.

WARNING: Estimates may not have converged.  ❹


                      ARIMA Estimation Optimization Summary

         Estimation Method                                  Conditional Least Squares
         Parameters Estimated                                                       5
         Termination Criteria                          Maximum Relative Change in Estimates
         Iteration Stopping Value                                               0.001
         Criteria Value                                                      1.627249
         Maximum Absolute Value of Gradient                                  144.3651
         R-Square Change from Last Iteration                                 0.118066
         Objective Function                                      Sum of Squared Residuals
         Objective Function Value                                            164.6437
         Marquardt's Lambda Coefficient                                       0.00001
         Numerical Derivative Perturbation Delta                                0.001
         Iterations                                                                20
         Warning Message                                 Estimates may not have converged.
```

Output 3.10 *Using the ESTIMATE Statement for Series 8: PROC ARIMA (continued)*

Conditional Least Squares Estimation

Parameter	Estimate	Standard Error	t Value	Approx Pr > |t|	Lag
MU	-0.30535	0.43056	-0.71	0.4793	0
MA1,1	0.68031	0.36968	1.84	0.0678	1
MA1,2	0.31969	0.17448	1.83	0.0690	2
AR1,1	1.52519	0.35769	4.26	<.0001	1
AR1,2	-0.52563	0.18893	-2.78	0.0061	2

Constant Estimate	-0.00013
Variance Estimate	1.135473
Std Error Estimate	1.065586
AIC	449.6538
SBC	464.707
Number of Residuals	150

* AIC and SBC do not include log determinant.

Correlations of Parameter Estimates

Parameter	MU	MA1,1	MA1,2	AR1,1	AR1,2
MU	1.000	-0.258	-0.155	-0.230	0.237
MA1,1	-0.258	1.000	0.561	0.974	-0.923
MA1,2	-0.155	0.561	1.000	0.633	-0.442
AR1,1	-0.230	0.974	0.633	1.000	-0.963
AR1,2	0.237	-0.923	-0.442	-0.963	1.000

Autocorrelation Check of Residuals

To Lag	Chi-Square	DF	Pr > ChiSq	Autocorrelations					
6	2.07	2	0.3549	-0.007	0.001	0.038	-0.068	0.082	-0.022
12	3.22	8	0.9198	-0.059	-0.029	0.010	-0.037	0.034	-0.016
18	5.78	14	0.9716	0.004	0.106	0.051	0.037	0.002	0.003
24	11.76	20	0.9242	-0.035	0.099	-0.050	0.007	0.002	-0.141
30	13.20	26	0.9821	-0.043	-0.059	0.004	0.034	-0.011	-0.035

Model for Variable y8

Estimated Mean -0.30535

Autoregressive Factors

Factor 1: 1 - 1.52519 B**(1) + 0.52563 B**(2)

Moving Average Factors

Factor 1: 1 - 0.68031B**(1) - 0.31969 B**(2)

To understand the failure to converge, note that

$$Y_t - .6Y_{t-1} = e_t + .4e_{t-1}$$

implies that

$$Y_{t-1} - .6Y_{t-2} = e_{t-1} + .4e_{t-2}$$

Now multiply this last equation on both sides by φ and add to the first equation, obtaining

$$Y_t + (\varphi - .6)Y_{t-1} - .6\varphi Y_{t-2} = e_t + (\varphi + .4)e_{t-1} + .4\varphi e_{t-2}$$

Every φ yields a different ARMA(2,2), each equivalent to the original Y8. Thus, the procedure could not find one ARMA(2,2) model that seemed best. Although you sometimes overfit and test coefficients for significance to select a model (as illustrated with the iron and steel data), the example above shows that this method fails when you overfit on both sides of the ARMA equation at once. Notice that $(1-1.525B+.525B^2)(y_t-\mu)=(1-.6803B+.3197)e_t$ is the same as $(1-B)(1-.525B)(y_t-\mu)=(1-B)(1+.3197B)e_t$ or, eliminating the common factor, $(1-.525B)(y_t-\mu)=(1+.3197B)e_t$.

3.4.5 Nonstationary Series

The theory behind PROC ARIMA requires that a series be stationary. Theoretically, the stationarity of a series

$$(1 - \alpha_1 B - \alpha_2 B^2 - \ldots - \alpha_p B^p)(Y_t - \mu)$$
$$= (1 - \beta_1 B - \beta_2 B^2 - \ldots - \beta_q B^q)e_t$$

hinges on the solutions M of the characteristic equation

$$1 - \alpha_1 M - \alpha_2 M^2 - \ldots - \alpha_p M^p = 0$$

If all Ms that satisfy this equation have $|M| > 1$, the series is stationary. For example, the series

$$(1 - 1.5B + .64B^2)(Y_t - \mu) = (1 + .8B)e_t$$

is stationary, but the following series is not:

$$(1 - 1.5B + .5B^2)(Y_t - \mu) = (1 + .8B)e_t$$

The characteristic equation for the nonstationary example above is

$$1 - 1.5M + .5M^2 = 0$$

with solutions M=1 and M=2. These solutions are called roots of the characteristic polynomial, and because one of them is 1 the series is nonstationary. This unit root nonstationarity has several implications, which are explored below. The overfit example at the end of the previous section ended when the common factor $(1-\varphi B)$ neared $(1-B)$, an "unstable" value.

First, expanding the model gives

$$Y_t - 1.5Y_{t-1} + .5Y_{t-2} + (1 - 1.5 + .5)\mu = e_t + .8_{t-1}$$

which shows that μ drops out of the equation. As a result, series forecasts do not tend to return to the historic series mean. This is in contrast to stationary series, where μ is estimated and where forecasts *always* approach this estimated mean.

In the nonstationary example, Y_t is the series level and

$$W_t = Y_t - Y_{t-1}$$

is the first difference or change in the series. By substitution,

$$W_t - .5W_{t-1} = e_t + .8e_{t-1}$$

so when the levels Y_t satisfy an equation with a single unit root nonstationarity, the first differences W_t satisfy a stationary equation, often with mean 0. Similarly, you can eliminate a double unit root as in

$$\left(1 - 2B + B^2\right)\left(Y_t - \mu\right) = e_t + .8e_{t-1}$$

by computing and then analyzing the second difference

$$W_t - W_{t-1} = Y_t - 2Y_{t-1} + Y_{t-2}$$

The first and second differences are often written ∇Y_t and $\nabla^2 Y_t$. For nonseasonal data, you rarely difference more than twice.

Because you do not know the model, how do you know when to difference? You decide by examining the ACF or performing a test as in **Section 3.4.8**. If the ACF dies off very slowly, a unit root is indicated. The slow dying off may occur after one or two substantial drops in the ACF. Note that the sequence 1, .50, .48, .49, .45, .51, .47, . . . is considered to die off slowly in this context even though the initial drop from 1 to .5 is large and the magnitude of the autocorrelation is not near 1.

Using the IDENTIFY statement, you can accomplish differencing easily. The statement

```
IDENTIFY VAR=Y(1);
```

produces the correlation function for W_t, where

$$W_t = Y_t - Y_{t-1}$$

A subsequent ESTIMATE statement operates on W_t, so the NOCONSTANT option is normally used. The statement

```
IDENTIFY VAR=Y(1,1);
```

specifies analysis of the second difference, or

$$\left(Y_t - Y_{t-1}\right) - \left(Y_{t-1} - Y_{t-2}\right)$$

The default is no differencing for the variables. Assuming a nonzero mean in the differenced data is equivalent to assuming a deterministic trend in the original data because $(\alpha + \beta t) - (\alpha + \beta(t-1)) = \beta$. You can fit this β easily by omitting the NOCONSTANT option.

3.4.6 Effect of Differencing on Forecasts

PROC ARIMA provides forecasts and 95% upper and lower confidence bounds for predictions for the general ARIMA model. If you specify differencing, modeling is done on the differenced series, but predictions are given for the original series levels. Also, when you specify a model with differencing, prediction error variances increase without bound as you predict further into the future.

In general, by using estimated parameters and by estimating σ^2 from the model residuals, you can easily derive the forecasts and their variances from the model. PROC ARIMA accomplishes this task for you automatically.

For example, in the model

$$Y_t - 1.5Y_{t-1} + .5Y_{t-2} = e_t$$

note that

$$(Y_t - Y_{t-1}) - .5(Y_{t-1} - Y_{t-2}) = e_t$$

Thus, the first differences

$$W_t = Y_t - Y_{t-1}$$

are stationary. Given data $Y_1, Y_2, \ldots, Y_n$ from this series, you predict future values by first predicting future values of W_{n+j}, using $.5^j W_n$ as the prediction. Now

$$Y_{n+j} - Y_n = W_{n+1} + W_{n+2} + \ldots + W_{n+j}$$

so the forecast of Y_{n+j} is

$$Y_n + \Sigma_{i=1}^{j} (.5)^i W_n$$

To illustrate further, the following computation of forecasts shows a few values of Y_t, W_t, and predictions $\hat{Y}_j$:

	Actual			Forecast			
t	98	99	100(*n*)	101	102	103	104
Y_t	475	518	550	566	574	578	580
W_t	28	43	32	16	8	4	2

Note that

$$\Sigma_{i=1}^{j} (.5)^i$$

approaches 1 as *j* increases, so the forecasts converge to

$$550 + (1)(32) = 582$$

Forecast errors can be computed from the forecast errors of the Ws—for example,

$$Y_{n+2} = Y_n + W_{n+1} + W_{n+2}$$

and

$$\hat{Y}_{n+2} = Y_n + .5W_n + .25W_n$$

Rewriting

$$Y_{n+2} = Y_n + (.5W_n + e_{n+1}) + (.25W_n + .5e_{n+1} + e_{n+2})$$

yields the forecast error

$$1.5e_{n+1} + e_{n+2}$$

with the variance $3.25\sigma^2$.

3.4.7 Examples: Forecasting IBM Series and Silver Series

An example that obviously needs differencing is the IBM stock price series reported by Box and Jenkins (1976). In this example, the data are analyzed with PROC ARIMA and are forecast 15 periods ahead. Box and Jenkins report values of daily closing prices of IBM stock. You read in the series and check the ACF:

```
DATA IBM;
   INPUT PRICE @@;
   T+1;
   CARDS;
data lines
;
RUN;
PROC ARIMA DATA=IBM;
   IDENTIFY VAR=PRICE CENTER NLAG=15;
   IDENTIFY VAR=PRICE(1) NLAG=15;
RUN;
```

The plot of the original data is shown in **Output 3.11**, and the IDENTIFY results in **Output 3.12**.

Output 3.11
Plotting the Original Data

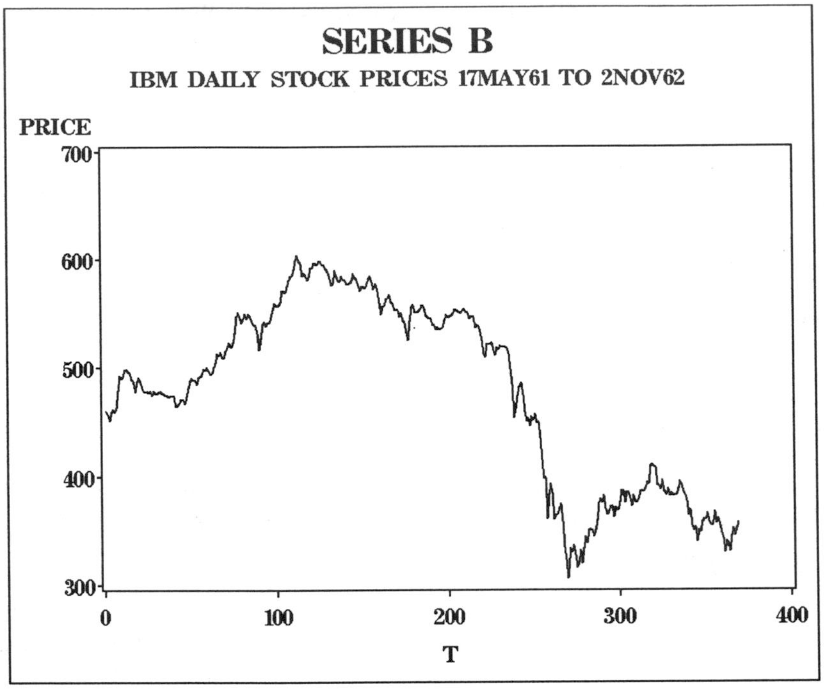

Output 3.12 *Identifying the IBM Price Series*

```
                                   SERIES B
                    IBM DAILY STOCK PRICES 17MAY61 TO 2NOV62

                              The ARIMA Procedure

                            Name of Variable = PRICE

                       Mean of Working Series            0
                       Standard Deviation         84.10504
                       Number of Observations          369


                                Autocorrelations
                                                                    ❶
Lag   Covariance   Correlation   -1 9 8 7 6 5 4 3 2 1 0 1 2 3 4 5 6 7 8 9 1      Std Error

  0     7073.658       1.00000    |                    |********************|             0
  1     7026.966       0.99340    |                  . |********************|      0.052058
  2     6973.914       0.98590    |                .   |********************|      0.089771
  3     6918.629       0.97808    |               .    |********************|      0.115443
  4     6868.433       0.97099    |               .    |******************* |      0.136059
  5     6817.810       0.96383    |              .     |******************* |      0.153695
  6     6763.587       0.95617    |             .      |******************* |      0.169285
  7     6705.771       0.94799    |             .      |******************* |      0.183337
  8     6645.401       0.93946    |            .       |******************* |      0.196172
  9     6580.448       0.93028    |            .       |******************* |      0.208008
 10     6522.985       0.92215    |           .        |******************  |      0.218993
 11     6466.010       0.91410    |           .        |******************  |      0.229274
 12     6407.497       0.90583    |          .         |******************  |      0.238947
 13     6348.092       0.89743    |          .         |******************  |      0.248078
 14     6289.664       0.88917    |          .         |******************  |      0.256726
 15     6230.941       0.88087    |         .          |******************  |      0.264940

                          "." marks two standard errors


                            Inverse Autocorrelations

              Lag   Correlation   -1 9 8 7 6 5 4 3 2 1 0 1 2 3 4 5 6 7 8 9 1

                1      -0.51704    |          **********| .                  |
                2      -0.02791    |                  .*| .                  |
                3       0.07838    |                  . |**                  |
                4      -0.01677    |                  . | .                  |
                5      -0.03290    |                  .*| .                  |
                6       0.02005    |                  . | .                  |
                7       0.01682    |                  . | .                  |
                8      -0.09039    |                  **| .                  |
                9       0.10983    |                  . |**                  |
               10      -0.02725    |                  .*| .                  |
               11      -0.02739    |                  .*| .                  |
               12       0.00859    |                  . | .                  |
               13       0.01911    |                  . | .                  |
               14      -0.01985    |                  . | .                  |
               15       0.00673    |                  . | .                  |
```

Output 3.12 *Identifying the IBM Price Series (continued)*

```
                                   SERIES B
                    IBM DAILY STOCK PRICES 17MAY61 TO 2NOV62

                              The ARIMA Procedure

                           Partial Autocorrelations
                                                                     ❷
           Lag    Correlation    -1 9 8 7 6 5 4 3 2 1 0 1 2 3 4 5 6 7 8 9 1

             1        0.99340    |                    .|********************|
             2       -0.07164    |                   .*|  .                 |
             3       -0.02325    |                    .|  .                 |
             4        0.05396    |                    .|*.                  |
             5       -0.01535    |                    .|  .                 |
             6       -0.04422    |                   .*|  .                 |
             7       -0.03394    |                   .*|  .                 |
             8       -0.02613    |                   .*|  .                 |
             9       -0.05364    |                   .*|  .                 |
            10        0.08124    |                    .|**                  |
            11       -0.00845    |                    .|  .                 |
            12       -0.02830    |                   .*|  .                 |
            13        0.00187    |                    .|  .                 |
            14        0.01204    |                    .|  .                 |
            15       -0.01469    |                    .|  .                 |


                      Autocorrelation Check for White Noise

 To       Chi-              Pr >
Lag     Square      DF     ChiSq    --------------------Autocorrelations-------------------

  6    2135.31       6    <.0001     0.993     0.986     0.978     0.971     0.964     0.956
 12    4097.40      12    <.0001     0.948     0.939     0.930     0.922     0.914     0.906


                         Name of Variable = PRICE

               Period(s) of Differencing                         1
               Mean of Working Series                     -0.27989
               Standard Deviation                         7.248345
               Number of Observations                          368
               Observation(s) eliminated by differencing         1


                              Autocorrelations
                                                                 ❸
Lag    Covariance    Correlation    -1 9 8 7 6 5 4 3 2 1 0 1 2 3 4 5 6 7 8 9 1    Std Error

  0     52.538509        1.00000    |                     |********************|           0
  1      4.496014    ❺  0.08558    |                    .|**                  |    0.052129
  2     -0.072894        -.00139    |                    .|  .                 |    0.052509
  3     -2.853759        -.05432    |                   .*|  .                 |    0.052509
  4     -1.820817        -.03466    |                   .*|  .                 |    0.052662
  5     -1.261461        -.02401    |                    .|  .                 |    0.052723
  6      6.350064        0.12086    |                    .|**                  |    0.052753
```

Output 3.12 *Identifying the IBM Price Series (continued)*

```
   7     3.585725      0.06825   |                 .  |*.                 |     0.053500
   8     1.871606      0.03562   |                 .  |*.                 |     0.053736
   9    -3.483286      -.06630   |                 .*|  .                 |     0.053801
  10     1.149218      0.02187   |                 .  |  .                |     0.054022
  11     4.043788      0.07697   |                 .  |**                 |     0.054046
  12     2.816399      0.05361   |                 .  |*.                 |     0.054343
  13    -2.508704      -.04775   |                 .*|  .                 |     0.054487
  14     3.445101      0.06557   |                 .  |*.                 |     0.054600
  15    -3.470001      -.06605   |                 .*|  .                 |     0.054814

                         "." marks two standard errors


                           Inverse Autocorrelations

          Lag    Correlation    -1 9 8 7 6 5 4 3 2 1 0 1 2 3 4 5 6 7 8 9 1

            1       -0.08768     |                 **|  .                 |
            2       -0.01236     |                 .  |  .                |
            3        0.02663     |                 .  |*.                 |
            4        0.04032     |                 .  |*.                 |
            5        0.04148     |                 .  |*.                 |
            6       -0.10091     |                 **|  .                 |
            7       -0.05960     |                 .*|  .                 |
            8       -0.02455     |                 .  |  .                |
            9        0.05083     |                 .  |*.                 |
           10       -0.02460     |                 .  |  .                |
           11       -0.07939     |                 **|  .                 |
           12       -0.03140     |                 .*|  .                 |
           13        0.07809     |                 .  |**                 |
           14       -0.07742     |                 **|  .                 |
           15        0.05362     |                 .  |*.                 |


                           Partial Autocorrelations

          Lag    Correlation    -1 9 8 7 6 5 4 3 2 1 0 1 2 3 4 5 6 7 8 9 1

            1        0.08558     |                 .  |**                 |
            2       -0.00877     |                 .  |  .                |
            3       -0.05385     |                 .*|  .                 |
            4       -0.02565     |                 .*|  .                 |
            5       -0.01940     |                 .  |  .                |
            6        0.12291     |                 .  |**                 |
            7        0.04555     |                 .  |*.                 |
            8        0.02375     |                 .  |  .                |
            9       -0.06241     |                 .*|  .                 |
           10        0.04501     |                 .  |*.                 |
           11        0.08667     |                 .  |**                 |
           12        0.02638     |                 .  |*.                 |
           13       -0.07034     |                 .*|  .                 |
           14        0.07191     |                 .  |*.                 |
           15       -0.05660     |                 .*|  .                 |
```

Output 3.12 *Identifying the IBM Price Series (continued)*

Autocorrelation Check for White Noise

To Lag	❹ Chi-Square	DF	Pr > ChiSq	--------Autocorrelations--------					❼
6	9.98	6	0.1256	0.086	-0.001	-0.054	-0.035	-0.024	0.121
12	17.42	12	0.1344	0.068	0.036	-0.066	0.022	0.077	0.054

The ACF ❶ dies off very slowly. The PACF ❷ indicates a very high coefficient, 0.99340, in the regression of Y_t on Y_{t-1}. The ACF of the differenced series ❸ looks like white noise. In fact, the *Q* statistics 9.98 and 17.42 ❹ are not significant. For example, the probability of a value larger than 9.98 in a χ^2_6 distribution is .126, so 9.98 is to the left of the critical value and, therefore, is not significant. The *Q* statistics are computed with the first six (9.98) and first twelve (17.42) autocorrelations of the differenced series. With a first difference, it is common to find an indication of a lag 1 MA term. The first autocorrelation is 0.08558 ❺ with a standard error of about $1/(368)^{1/2}$=.052.

Next, suppress the printout with the IDENTIFY statement (you have already looked at it but still want PROC ARIMA to compute initial estimates) and estimate the model:

```
PROC ARIMA DATA=IBM;
   IDENTIFY VAR=PRICE(1) NOPRINT;
   ESTIMATE Q=1 NOCONSTANT;
RUN;
```

The results are shown in **Output 3.13**.

Output 3.13 *Analyzing Daily Series with the ESTIMATE Statement: PROC ARIMA*

SERIES B
IBM DAILY STOCK PRICES 17MAY61 TO 2NOV62

The ARIMA Procedure

Conditional Least Squares Estimation

Parameter	Estimate	Standard ❻ Error	t Value	Approx Pr > |t|	Lag
MA1,1	-0.08658	0.05203	-1.66	0.0970	1

Variance Estimate	52.36132
Std Error Estimate	7.236112
AIC	2501.943
SBC	2505.851

Output 3.13 *Analyzing Daily Series with the ESTIMATE Statement: PROC ARIMA (continued)*

```
Number of Residuals        368
* AIC and SBC do not include log determinant.

Autocorrelation Check of Residuals
```

	To Lag	Chi-Square	DF	Pr > ChiSq	Autocorrelations					
	6	6.99	5	0.2217	0.001	0.005	-0.051	-0.026	-0.030	0.120
	12	13.94	11	0.2365	0.056	0.039	-0.070	0.024	0.072	0.054
❽	18	31.04	17	0.0198	-0.057	0.079	-0.081	0.118	0.113	0.040
	24	39.05	23	0.0196	0.041	0.072	-0.089	-0.027	0.066	0.025
	30	49.83	29	0.0094	0.028	-0.100	-0.055	0.051	0.028	0.099
	36	56.47	35	0.0122	0.072	-0.074	-0.063	-0.007	0.022	0.035
	42	64.42	41	0.0112	0.066	-0.085	0.059	-0.060	0.018	0.017
	48	76.33	47	0.0044	-0.116	-0.037	0.073	0.005	0.069	0.057

```
Model for Variable PRICE

Period(s) of Differencing    1

No mean term in this model.

Moving Average Factors

Factor 1:   1 + 0.08658 B**(1)
```

Although the evidence is not strong enough to indicate that the series has a nonzero first-order autocorrelation, you nevertheless fit the MA(1) model. The t statistic –1.66 ❻ is significant at the 10% level.

More attention should be paid to the lower-order and seasonal autocorrelations than to the others. In this example, you ignore an autocorrelation 0.121 ❼ at lag 6 that was even bigger than the lag 1 autocorrelation. Similarly, residuals from the final fitted model show a Q statistic 31.04 ❽ that attains significance because of autocorrelations .118 and .113 at lags 16 and 17. Ignore this significance in favor of the more parsimonious MA(1) model.

The model appears to fit; therefore, make a third run to forecast:

```
PROC ARIMA DATA=IBM;
   IDENTIFY VAR=PRICE(1) NOPRINT;
   ESTIMATE Q=1 NOCONSTANT NOPRINT;
   FORECAST LEAD=15;
RUN;
```

See the forecasts in **Output 3.14**.

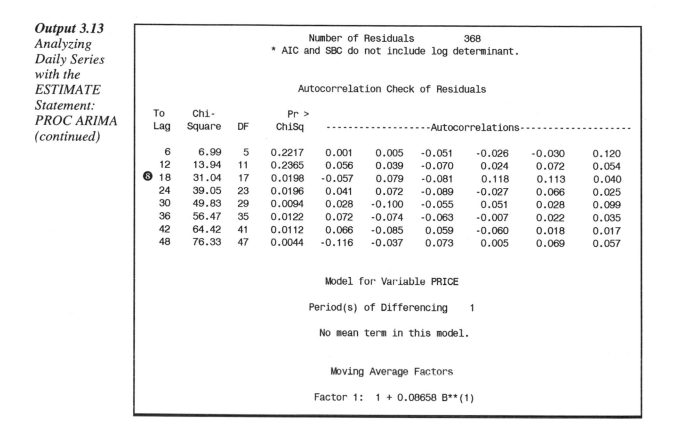

Output 3.14
Forecasting Daily Series: PROC ARIMA

```
                    SERIES B
      IBM DAILY STOCK PRICES 17MAY61 TO 2NOV62

               The ARIMA Procedure

             Model for variable PRICE

           Period(s) of Differencing    1

            No mean term in this model.


              Moving Average Factors

           Factor 1:  1 + 0.08658 B**(1)

            Forecasts for variable PRICE

Obs      Forecast    Std Error      95% Confidence Limits

370      357.3837       7.2361      343.2012      371.5662
371      357.3837      10.6856      336.4403      378.3270
372      357.3837      13.2666      331.3817      383.3857

  [more output lines]

382      357.3837      28.1816      302.1487      412.6187
383      357.3837      29.2579      300.0392      414.7282
384      357.3837      30.2960      298.0047      416.7627
```

If

$$Y_t - Y_{t-1} = e_t - \beta e_{t-1}$$

as in the IBM example, then by repeated back substitution

$$e_t = (Y_t - Y_{t-1}) + \beta(Y_{t-1} - Y_{t-2}) + \beta^2(Y_{t-2} - Y_{t-3}) + \ldots$$

or

$$Y_t = e_t + (1-\beta)\left[Y_{t-1} + \beta Y_{t-2} + \beta^2 Y_{t-3} + \ldots\right]$$

so that

$$\hat{Y}_t = (1-\beta)(Y_{t-1} + \beta Y_{t-2} + \beta^2 Y_{t-3} + \ldots)$$

Forecasting Y_t by such an exponentially weighted sum of past Ys is called single exponential smoothing. Higher degrees of differencing plus the inclusion of more MA terms is equivalent to higher-order exponential smoothing. PROC ARIMA, however, unlike PROC FORECAST with METHOD=EXPO, estimates the parameters from the data.

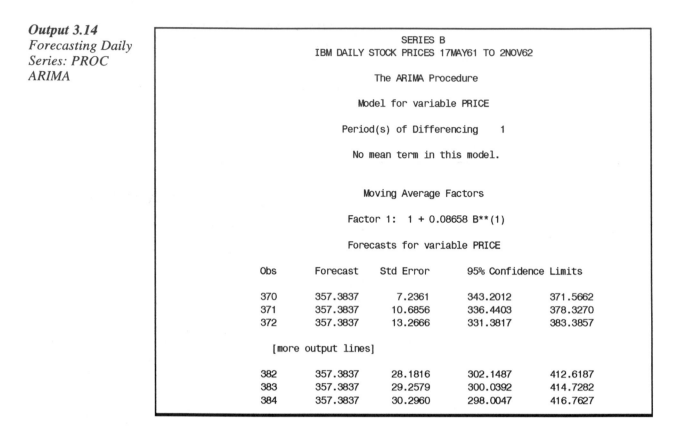

Dickey and Fuller (1979) give a formal test of the null hypothesis that an AR series has a unit root nonstationarity versus the alternative that it is stationary. Said and Dickey (1984) extend the test to ARIMA models. The test involves a regression of ∇Y_t (where $\nabla Y_t = Y_t - Y_{t-1}$) on $Y_{t-1} - \bar{Y}$, $\nabla Y_{t-1}, \ldots, \nabla Y_{t-p}$ where p is at least as large as the order of the AR process or, in the case of the mixed process, is large enough to give a good approximation to the model. The t test on $Y_{t-1} - \bar{Y}$ is called τ_μ because it does not have a Student's t distribution and must be compared to tables provided by Fuller (1996, p. 642). The silver series from Chapter 2, "Simple Models: Autoregression," is used as an illustration in the next section.

3.4.8 Models for Nonstationary Data

You can formally test for unit root nonstationarity with careful modeling and special distributions. Any autoregressive model like the AR(2) model $Y_t - \mu = \alpha_1(Y_{t-1} - \mu) + \alpha_2(Y_{t-2} - \mu) + e_t$ can be written in terms of differences and the lagged level term ($Y_{t-1} - \mu$). With a little algebra, the AR(2) becomes

$$Y_t - Y_{t-1} = -(1 - \alpha_1 - \alpha_2)(Y_{t-1} - \mu) - \alpha_2(Y_{t-1} - Y_{t-2}) + e_t$$

Stationarity depends on the roots of the characteristic equation $1 - \alpha_1 M - \alpha_2 M^2 = 0$, so if $M = 1$ is a root, then $(1 - \alpha_1 - \alpha_2) = 0$. So the $(Y_{t-1} - \mu)$ term drops out of the model and forecasts do not revert to the mean. This discussion suggests a least squares regression of

$Y_t - Y_{t-1}$ on Y_{t-1} and $(Y_{t-1} - Y_{t-2})$ with an intercept and the use of the resulting coefficient or t test on the Y_{t-1} term as a test of the null hypothesis that the series has a unit root nonstationarity. If all roots M exceed 1 in magnitude, the coefficient of $(Y_{t-1} - \mu)$ will be negative, suggesting a one-tailed test to the left if stationarity is the alternative. There is, however, one major problem with this idea: neither the estimated coefficient of $(Y_{t-1} - \mu)$ nor its t test has a standard distribution, even when the sample size becomes very large. This does not mean the test cannot be done, but it does require the tabulation of a new distribution for the test statistics.

Dickey and Fuller (1979, 1981) studied the distributions of estimators and t statistics in autoregressive models with unit roots. The leftmost column of the following tables shows the regressions they studied. Here $Y_t - Y_{t-1} = \nabla Y_t$ denotes a first difference.

Regress ∇Y_t on these:	AR(1) in deviations form
$Y_{t-1}, \nabla Y_{t-1} \cdots \nabla Y_{t-k}$	$Y_t = \rho Y_{t-1} + e_t$
$Y_{t-1}, 1, \nabla Y_{t-1} \cdots \nabla Y_{t-k}$	$Y_t - \mu = \rho(Y_{t-1} - \mu) + e_t$
$Y_{t-1}, 1, t, \nabla Y_{t-1} \cdots \nabla Y_{t-k}$	$Y_t - \alpha - \beta t = \rho(Y_{t-1} - \alpha - \beta(t-1)) + e_t$

AR(1) in regression form	$H_0: \rho = 1$
$\nabla Y_t = (\rho - 1)Y_{t-1} + e_t$	$\nabla Y_t = e_t$
$\nabla Y_t = (\rho - 1)\mu + (\rho - 1)Y_{t-1} + e_t$	$\nabla Y_t = e_t$
$\nabla Y_t = (1 - \rho)(\alpha + \beta t) + \beta + (\rho - 1)Y_{t-1} + e_t$	$\nabla Y_t = \beta + e_t$

The lagged differences are referred to as "augmenting lags" and the tests as "Augmented Dickey-Fuller" or "ADF" tests. The three regression models allow for three kinds of trends. For illustration a lag 1 autoregressive model with autoregressive parameter ρ is shown in the preceding table both in deviations form and in the algebraically equivalent regression form. The deviations form is most instructive. It shows that if $|\rho|<1$ and if we have appropriate starting values, then the expected value of Y_t is 0, μ, or $\alpha+\beta t$ depending on which model is assumed. Fit the first model only if you know the mean of your data is 0 (for example, Y_t might already be a difference of some observed variable). Use the third model if you suspect a regular trend up or down in your data. If you fit the third model when β is really 0, your tests will be valid, but not as powerful as those from the second model. The parameter β represents a trend slope when $|\rho|<1$ and is called a "drift" when $\rho=1$. Note that for known parameters and *n* data points, the forecast of Y_{n+L} would be $\alpha+\beta(n+L)+\rho^L(Y_n-\alpha-\beta n)$ for $|\rho|<1$ with forecast error variance $(1+\rho^2+\cdots+\rho^{2L-2})\sigma^2$. As L increases, the forecast error variance approaches $\sigma^2/(1-\rho^2)$, the variance of Y around the trend. However, if $\rho=1$ the L step ahead forecast is $Y_n+\beta L$ with forecast error variance $L\sigma^2$, so that the error variance increases without bound in this case. In both cases, the forecasts have a component that increases at the linear rate β.

For the regression under discussion, the distributions for the coefficients of Y_{t-1}, 1, and *t* are all nonstandard. Tables of critical values and discussion of the theory are given in Fuller (1996). One very nice feature of these regressions is that the coefficients of the lagged differences ∇Y_{t-j} have normal distributions in the limit. Thus a standard F test to see if a set of these lagged differences can be omitted is justified in large samples, as are the *t* statistics for the individual lagged difference coefficients. They converge to standard normal distributions. The coefficients of Y_{t-1} and the associated *t* tests have distributions that differ among the three regressions and are nonstandard. Fortunately, however, the *t* test statistics have the same limit distributions no matter how many augmenting lags are used.

As an example, stocks of silver on the New York Commodities Exchange were analyzed in Chapter 2 of this book. We reanalyze the data here using DEL to denote the difference, DELi for its *i*th lag, and LSILVER for the lagged level of silver. The WHERE PART=1; statement restricts analysis to the data used in the first edition.

```
PROC REG DATA=SILVER;
   MODEL DEL=LSILVER DEL1 DEL2 DEL3 DEL4 /NOPRINT;
   TEST DEL2=0, DEL3=0, DEL4=0;
   WHERE PART=1;
RUN;
PROC REG DATA=SILVER;
   MODEL DEL=LSILVER DEL1;
   WHERE PART=1;
RUN;
```

Some output follows. First you have the result of the test statement for the model with four augmenting lags in **Output 3.15**.

Output 3.15
Test of Augmenting Lags

Test 1 Results for Dependent Variable DEL

Source	DF	Mean Square	F Value	Pr > F
Numerator	3	1152.19711	1.32	0.2803
Denominator	41	871.51780		

Because this test involves only the lagged differences, the F distribution is justified in large samples. Although the sample size here is not particularly large, the p-value 0.2803 is not even close to 0.05, thus providing no evidence against leaving out all but the first augmenting lag. The second PROC REG produces **Output 3.16**.

Output 3.16
PROC REG on Silver Data

Parameter Estimates

Variable	DF	Parameter Estimate	Standard Error	t Value	Pr > |t|
Intercept	1	75.58073	27.36395	2.76	0.0082
LSILVER	1	-0.11703	0.04216	-2.78	0.0079
DEL1	1	0.67115	0.10806	6.21	<.0001

Because the printed p-value 0.0079 is less than 0.05, the uninformed user might conclude that there is strong evidence against a unit root in favor of stationarity. This is an error because all p-values from PROC REG are computed from the *t* distribution whereas, under the null hypothesis of a unit root, this statistic has the distribution tabulated by Dickey and Fuller. The appropriate 5% left tail critical value of the limit distribution is –2.86 (Fuller 1996, p. 642), so the statistic is not far enough below 0 to reject the unit root null hypothesis. Nonstationarity cannot be rejected. This test is also available in PROC ARIMA starting with Version 6 and can be obtained as follows

```
PROC ARIMA DATA=SILVER;
   I VAR = SILVER STATIONARITY=(ADF=(1)) OUTCOV=ADF;
RUN;
```

Output 3.17 contains several tests.

Output 3.17
Unit Root Tests, Silver Data

Augmented Dickey-Fuller Unit Root Tests

Type	Lags	Rho	Pr < Rho	Tau	Pr < Tau	F	Pr > F
Zero Mean	1	-0.2461	0.6232	-0.28	0.5800		
Single Mean	1	-17.7945	0.0121	-2.78	0.0689	3.86	0.1197
Trend	1	-15.1102	0.1383	-2.63	0.2697	4.29	0.3484

Every observed data point exceeds 400, so any test from a model that assumes a 0 mean can be ignored. Also, the PROC REG output strongly indicated that one lagged difference was required. Thus the tests with no lagged differences can also be ignored and are not requested here. The output shows coefficient (or "normalized bias") unit root tests that would be computed as $n(\hat{\rho}-1)$ in an AR(1) model with coefficient ρ. For the AR(2) model with roots ρ and m, the regression model form

$$Y_t - Y_{t-1} = -(1-\alpha_1-\alpha_2)(Y_{t-1}-\mu) - \alpha_2(Y_{t-1}-Y_{t-2}) + e_t$$

becomes

$$Y_t - Y_{t-1} = -(1-m)(1-\rho)(Y_{t-1}-\mu) + m\rho(Y_{t-1}-Y_{t-2}) + e_t$$

so that the coefficient of Y_{t-1} is $-(1-\rho)(1-m)$ in terms of the roots. If $\rho=1$, it is seen that the coefficient of $(Y_{t-1}-Y_{t-2})$, 0.671152 in the silver example, is an estimate of m, so it is not surprising that an adjustment using that statistic is required to get a test statistic that behaves like $n(\hat{\rho}-1)$ under $H_0:\rho=1$. Specifically you divide the lag 1 coefficient (-0.117034) by $(1-.671152)$, then multiply by n. Similar adjustments can be made in higher-order processes. For the silver data, $50(-0.117034)/(1-.671152) = -17.7945$ is shown in the printout and has a p-value (.0121) less than 0.05. However, based on simulated size and power results (Dickey 1984), the tau tests are preferable to these normalized bias tests. Furthermore, the adjustment for lagged differences is motivated by large sample theory and $n=50$ is not particularly large. The associated tau test, -2.78, has a p-value exceeding 0.05 and hence fails to provide significant evidence at the usual 0.05 level against the unit root null hypothesis. The F type statistics are discussed in Dickey and Fuller (1981). If interest lies only in inference about ρ, there is no advantage to using the F statistics, which include restrictions on the intercept and trend as a part of H_0. Simulations indicate that the polynomial deterministic trend should have as low a degree as is consistent with the data, in order to get good power. The 50 observations studied thus far do not display any noticeable trend, so the model with a constant mean seems reasonable, although tests based on the model with linear trend would be valid and would guard against any unrecognized linear trend. These tests are seen to provide even less evidence against the unit root. In summary, then, getting a test with validity and good statistical power requires appropriate decisions about the model, in terms of lags and trends. This is no surprise, as any statistical hypothesis test requires a realistic model for the data.

The data analyzed here were used in the first edition of this book. Since then, more data on this series have been collected. The full set of data make it clear that the series is not stationary, in agreement with the tau statistic. In **Output 3.18**, the original series of 50 is plotted along with forecasts and confidence bands from an AR(2) that assumes stationarity in levels (solid lines), and an AR(1) fit to the differenced data (dashed lines). The more recent data are appended to the original 50. It is seen that for a few months into the forecast the series stays within the solid line bands, and it appears that the analyst who chooses stationarity is the better forecaster. He also has much tighter forecast bands. However, a little further ahead, the observations burst through his bands, never to return. The unit root forecast, though its bands may seem unpleasantly wide, does seem to give a more realistic assessment of the uncertainty inherent in this series.

Output 3.18
Silver Series, Stationary and Nonstationary Models

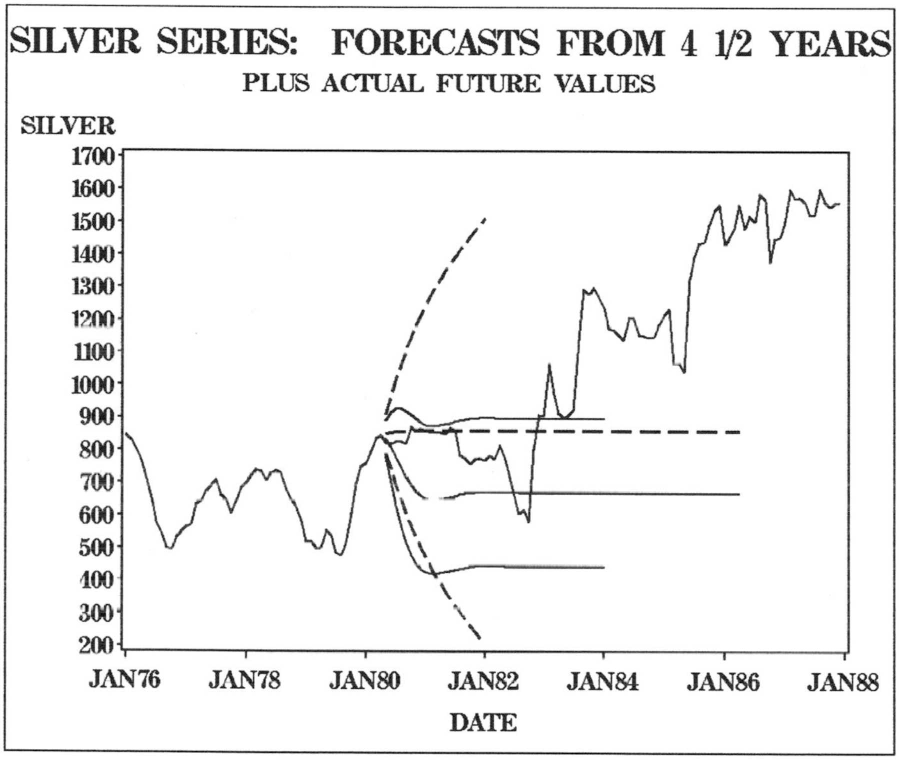

To illustrate the effects of trends, **Output 3.19** shows the logarithm of the closing price of Amazon.com stock. The data were downloaded from the stock reports available through the Web search engine Yahoo! The closing prices are fairly tightly clustered around a linear trend as displayed in the top part of the figure. The ACF, IACF, and PACF of the series are displayed just below the series plot and those of the differenced series just below that. Notice that the ACF of the original series dies off very slowly. This could be due to a deterministic trend, a unit root, or both. The three plots along the bottom seem to indicate that differencing has reduced the series to stationarity.

Output 3.19
Amazon Closing Prices

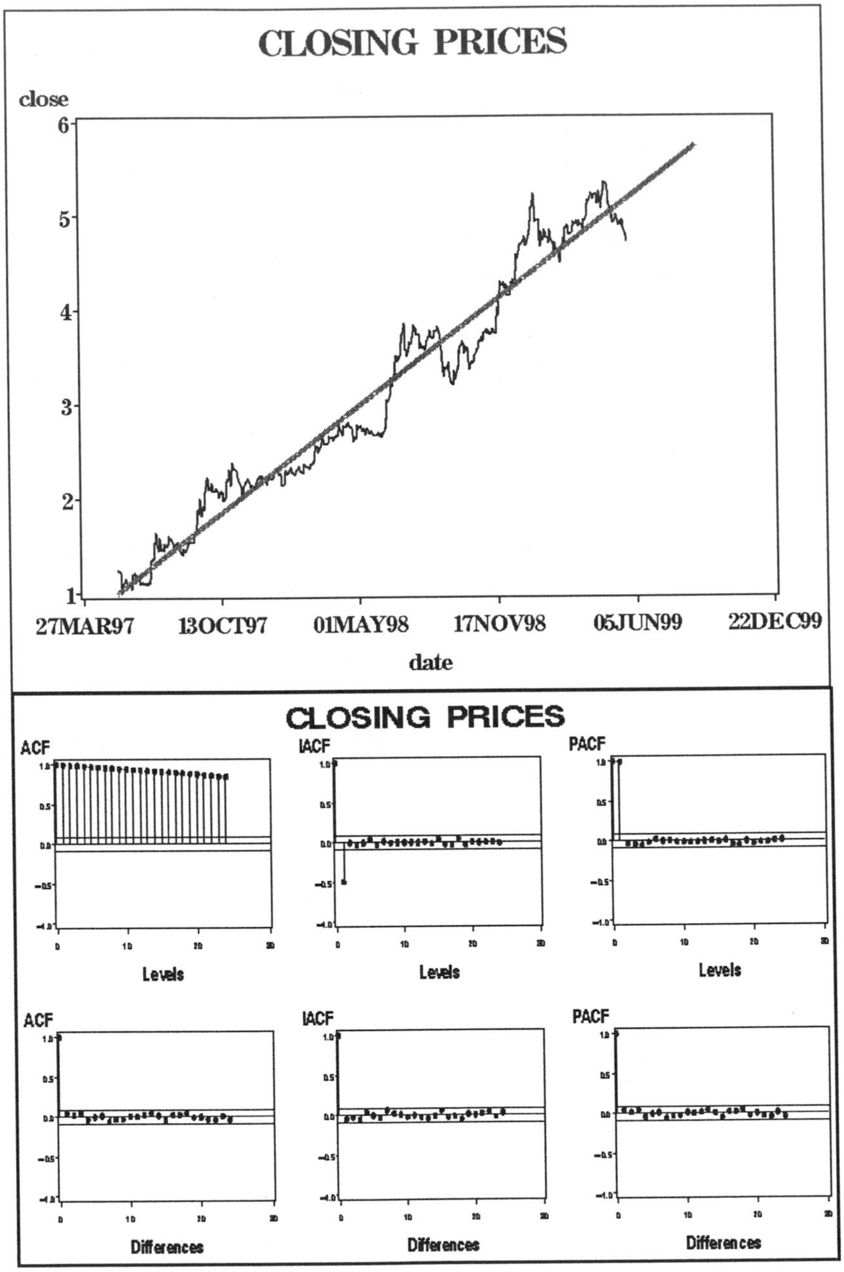

In contrast, **Output 3.20** shows the volume of the same Amazon.com stocks. These too show a trend, but notice the IACF of the differenced series. If a series has a unit root on the moving average side, the IACF will die off slowly. This is in line with what you've learned about unit roots on the autoregressive side. For the model $Y_t = e_t - \rho e_{t-1}$, the dual model obtained by switching the backshift operator to the AR side is $(1-\rho B)Y_t = e_t$, so that if ρ is (near) 1 you expect the IACF to behave like the ACF of a (near) unit root process—that is, to die off slowly.

Output 3.20
Amazon Volume

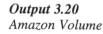

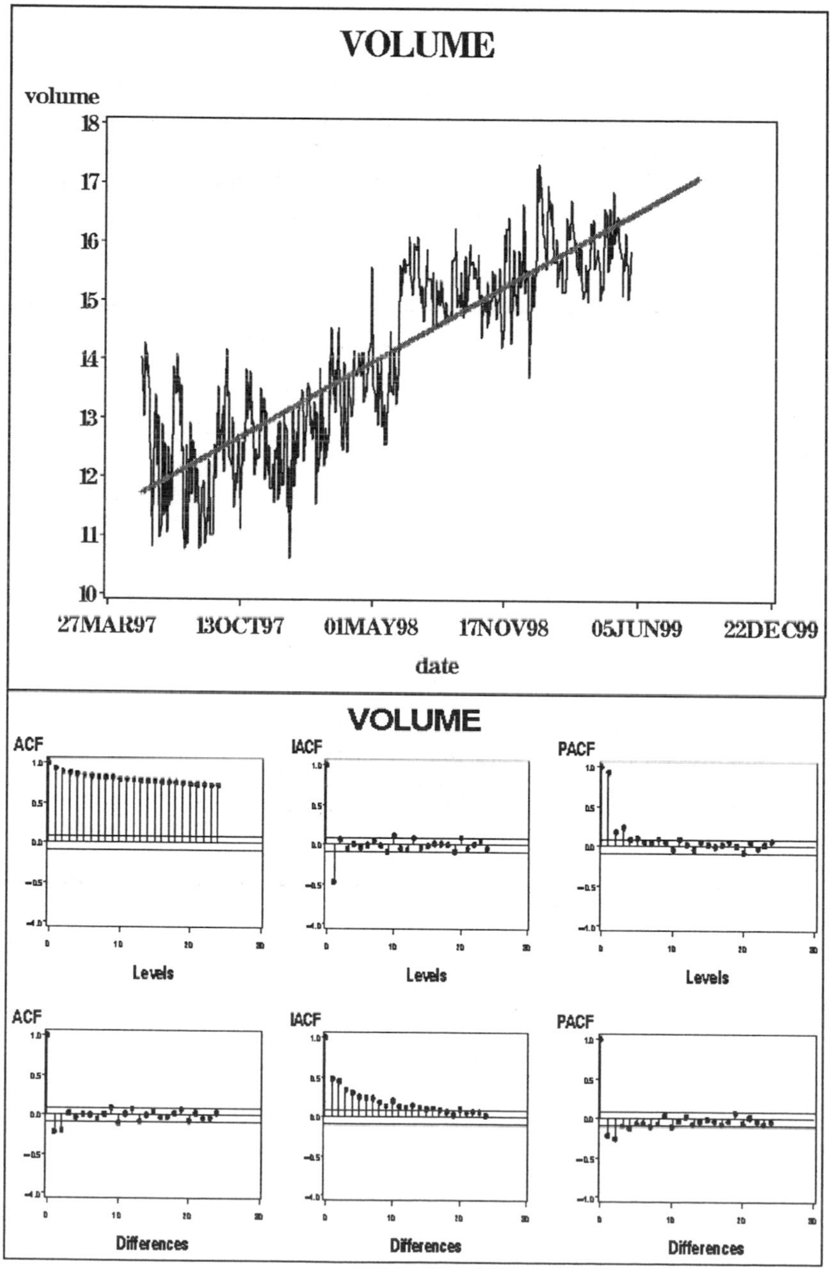

This behavior is expected anytime Y_t is the difference of an originally stationary series. Chang and Dickey (1993) give a detailed proof of what happens to the IACF when such overdifferencing occurs. They find that an essentially linear descent in the IACF is consistent with overdifferencing. This can follow an initial drop-off, as appears to happen in the volume data. Notice that a linear trend is reduced to a constant by first differencing so such a trend will not affect the behavior of the IACF of the differenced series. Of course a linear trend in the data will make the ACF of the *levels* appear to die off very slowly, as is also apparent in the volume data. The apparent mixed message-differencing indicated by the levels' ACF and too much differencing indicated by the differences' IACF is not really so inconsistent. You just need to think a little outside the class of ARIMA models to models with time trends and ARIMA errors.

Regression of differences on 1, *t*, a lagged level, and lagged differences indicated that no lagged differences were needed for the log transformed closing price series and two were needed for volume. Using the indicated models, the parameter estimates from PROC REG using the differenced series as a response, DATE as the time variable, LAGC and LAGV as the lag levels of closing price and volume, respectively, and lagged differences DV1 and DV2 for volume are shown in **Output 3.21**.

Output 3.21
Closing Price and Volume—Unit Root Test

Parameter Estimates

Variable	DF	Parameter Estimate	Standard Error	t Value	Pr > |t|	Type I SS
Intercept	1	-2.13939	0.87343	-2.45	0.0146	0.02462
date	1	0.00015950	0.00006472	2.46	0.0141	0.00052225
LAGC	1	-0.02910	0.01124	-2.59	0.0099	0.02501

Parameter Estimates

Variable	DF	Parameter Estimate	Standard Error	t Value	Pr > |t|	Type I SS
Intercept	1	-17.43463	3.11590	-5.60	<.0001	0.01588
date	1	0.00147	0.00025318	5.80	<.0001	0.00349
LAGV	1	-0.22354	0.03499	-6.39	<.0001	25.69204
DV1	1	-0.13996	0.04625	-3.03	0.0026	1.04315
DV2	1	-0.16621	0.04377	-3.80	0.0002	4.16502

As before, these tests can be automated using the IDENTIFY statement in PROC ARIMA. For these examples, clearly only the linear trend tests are to be considered. Although power is gained by using a lower-order polynomial when it is consistent with the data, the assumption that the trend is simply a constant is clearly inappropriate here.

The tau statistics (see Fuller 1996) are -2.59 (LAGC) for closing price and -6.39 (LAGV) for volume. Using the large *n* critical values -3.13 at significance level 0.10, -3.41 at 0.05, and -3.96 at 0.01, it is seen that unit roots are rejected even at the 0.01 level for volume. Thus the volume series displays stationary fluctuations around a linear trend. There is not evidence for stationarity in closing prices even at the 0.10 level, so even though the series seems to hug the linear trend line pretty closely, the deviations cannot be distinguished from a unit root process whose variance grows without bound.

An investment strategy based on an assumption of reversion of log transformed closing prices to the linear trend line does not seem to be supported here. That is not to refute the undeniable upward trend in the data—it comes out in the intercept or "drift" term (estimate 0.0068318) of the model for the differenced series. The model (computations not shown) is

$$\nabla Y_t = 0.0068318 + e_t + 0.04547 e_{t-1}$$

The differences, ∇Y_t, have this positive drift term as their average, so it implies a positive change on average with each passing unit of time. A daily increase of 0.0068318 in the logarithm implies a multiplicative $e^{0.0068318} = 1.00686$ or 0.68% daily increase, which compounds to a $e^{260(0.0068318)} = e^{1.78} = 6$-fold increase over the roughly 260 trading days in a year. This was a period of phenomenal growth for many such technology stocks, with this data going from about 3.5 to about 120 over two years' time, roughly the predicted 36-fold increase.

The top panel of **Output 3.22** shows closing price forecasts and intervals for the unit root with drift model (forecast rising almost linearly from the last observation and outermost bands) and for a model with stationary residuals from a linear trend (forecast converging to trend line and interior bands) for the log scale data. The plot below, in which each of these has been transformed back to the original scale by exponentiation, deserves some comments. First, note the strong effect of the logarithmic transformation. Any attempt to model on the original scale would have to account for the obviously unequal variation in the data and would require a somewhat complex trend function, whereas once logs are taken, a rather simple model, random walk with drift, seems to suffice. There is a fairly long string of values starting around January 1999 that are pretty far above the trend curve. Recall that this trend curve is simply an exponentiation of the linear trend on the log scale and hence approximates a median, not a mean. This 50% probability number, the median, may be a more easily understood number for an investment strategist than the mean in a highly skewed distribution such as this. Also note that the chosen model, random walk with drift, does not even use this curve, so a forecast beginning on February 1, 1999, for example, would emanate from the February 1, 1999, data point and follow a path approximately parallel to this trend line. The residuals from this trend line would not represent forecasting errors from either model. Even for the model that assumes stationary but strongly correlated errors, the forecast consists of the trend plus an adjustment based on the error correlation structure.

Output 3.22 *Amazon Closing Price (two Models, two Scales)*

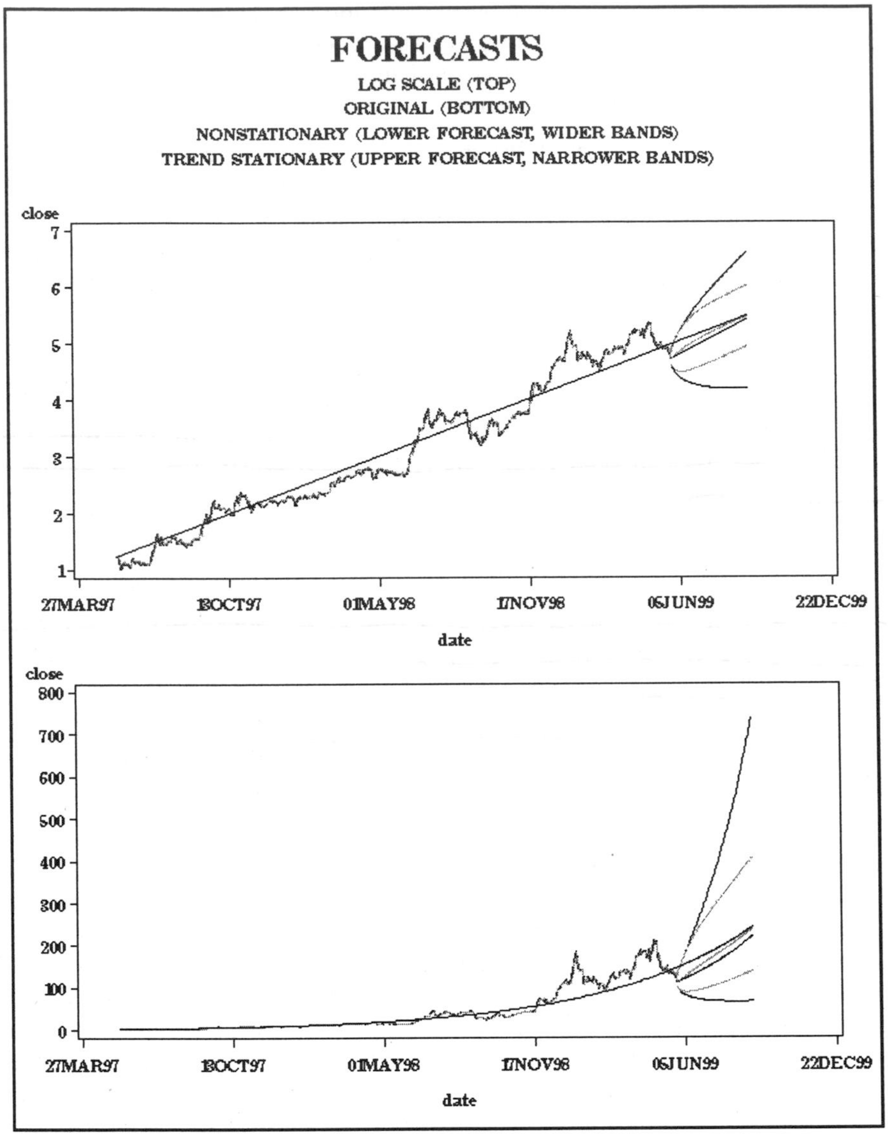

In fact the plot actually contains forecasts throughout the historic series from both models but they overlay the data so closely as to be hardly distinguishable from it. Note also that the combination of logs and differencing, while it makes the transformed series behave nicely statistically, produces very wide forecast intervals on the original scale. While this may disappoint the analyst, it might nevertheless be a reasonable assessment of uncertainty, given that 95% confidence is required and that this is a volatile series.

In summary, ignorance of unit roots and deterministic trends in time series can lead to clearly inappropriate mean reverting forecasts, while careful modeling of unit roots and deterministic trends can lead to quite reasonable and informative forecasts. Note that p-values produced under the assumption of stationarity can be quite misleading when unit roots are in fact present as shown in the silver and stock closing price examples. Both of these show inappropriately small p-values when the p-values are computed from the t rather than from the Dickey-Fuller distributions. In the regression of differences on trend terms, lagged level, and lagged differences, the usual (t and F) distributions are appropriate in large samples for inference on the lagged differences. To get tests with the proper behavior, carefully deciding on the number of lagged differences is important. Hall (1992) studies several methods and finds that overfitting lagged differences then testing to leave some out is a good method. This was illustrated in the silver example and was done for all examples here. Dickey, Bell, and Miller (1986) in their appendix show that the addition of seasonal dummy variables to a model does not change the large sample (limit) behavior of the unit root tests discussed here.

Some practitioners are under the false impression that differencing is justified anytime data appear to have a trend. In fact, such differencing may or may not be appropriate. This is discussed next.

3.4.9 Differencing to Remove a Linear Trend

Occasionally, practitioners difference data to remove a linear trend. Note that if Y_t has a linear trend

$$\alpha + \beta t$$

then the differenced series

$$W_t = Y_t - Y_{t-1}$$

involves only the constant. For example, suppose

$$Y_t = \alpha + \beta t + e_t$$

where e_t is white noise. Then

$$W_t = \beta + e_t - e_{t-1}$$

which does not have a trend but, unfortunately, is a noninvertible moving average. Thus, the data have been overdifferenced. Now the IACF of W looks like the ACF of a time series with a unit root nonstationarity; that is, the IACF of W dies off very slowly. You can detect overdifferencing this way.

The linear trend plus white noise model presented above is interesting. The ACF of the original data dies off slowly because of the trend. You respond by differencing, and then the IACF of the differenced series indicates that you have overdifferenced. This mixed signaling by the diagnostic functions simply tells you that the data do not fit an ARMA model on the original levels scale or on the differences scale. You can obtain the correct analysis in this particular case by regressing Y on t using PROC REG or PROC GLM. The situation is different if the error series e_t is not white noise but is instead a nonstationary time series whose difference

$$e_t - e_{t-1}$$

is stationary. In that case, a model in the differences is appropriate and has an intercept estimating β. This scenario seems to hold in the publishing and printing data that produce the plot (U.S. Bureau of Labor 1977) shown in **Output 3.23**. The data are the percentages of nonproduction workers in the industry over several years.

Output 3.23
Plotting the Original Series

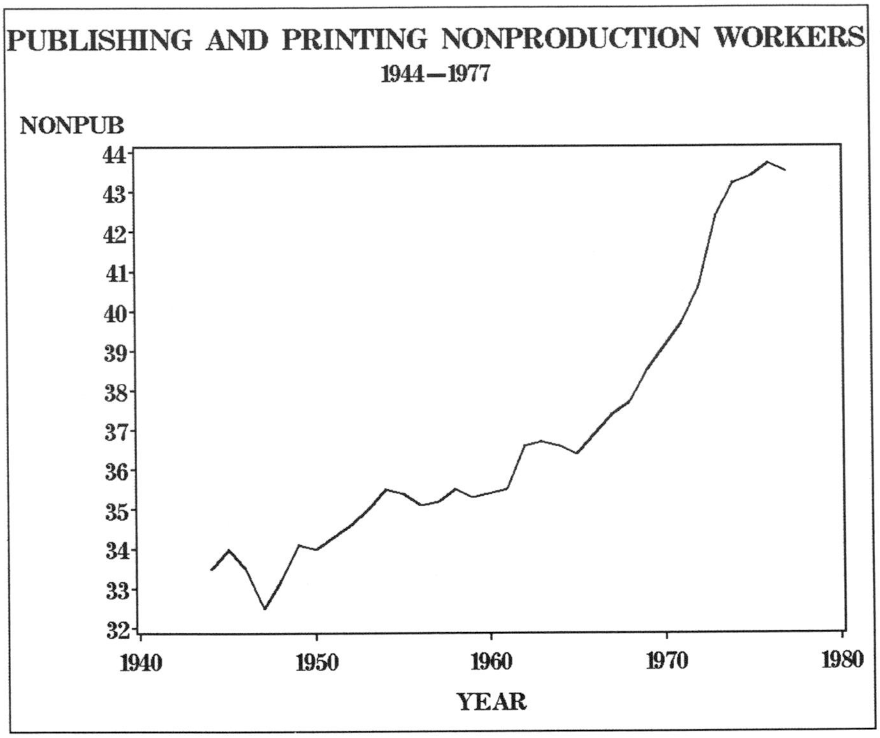

The ACF shown in **Output 3.24** is obtained by specifying the following statements:

```
PROC ARIMA DATA=WORKERS;
   IDENTIFY VAR=NONPUB(1) NLAG=10;
   TITLE 'PUBLISHING AND PRINTING NONPRODUCTION WORKERS';
   TITLE2 '1944-1977';
RUN;
```

Because the ACF ❶ looks like that of an MA(1) and because it is very common to fit an MA(1) term when a first difference is taken, you do that fitting by specifying these statements:

```
PROC ARIMA DATA=WORKERS;
   IDENTIFY VAR=NONPUB(1) NOPRINT;
   ESTIMATE Q=1;
   FORECAST LEAD=10;
RUN;
```

The output shows a good fit based on the *Q* statistics ❷ the parameter estimates, and their *t* statistics ❸. Note that the MU (0.3033) ❹ estimate is statistically significant and is roughly the slope in the plot of the data. Also, the MA coefficient is not near 1; in fact, it is a negative number. Thus, you have little evidence of overdifferencing. With only 33 observations, you have a lot of sampling variability (for example, look at the two standard error marks on the ACF). The number 0.3033 is sometimes called drift.

Output 3.24 *Modeling and Forecasting with the IDENTIFY, ESTIMATE, and FORECAST Statements: PROC ARIMA*

```
                    PUBLISHING AND PRINTING NONPRODUCTION WORKERS
                                      1944-1977

                                 The ARIMA Procedure

                              Name of Variable = NONPUB

                  Period(s) of Differencing                         1
                  Mean of Working Series                      0.30303
                  Standard Deviation                         0.513741
                  Number of Observations                           33
                  Observation(s) eliminated by differencing         1

                                  Autocorrelations
                                                             ❶
Lag   Covariance   Correlation   -1 9 8 7 6 5 4 3 2 1 0 1 2 3 4 5 6 7 8 9 1     Std Error

  0     0.263930       1.00000   |                    |********************|            0
  1     0.082387       0.31216   |             .      |******.             |     0.174078
  2    -0.025565       -.09686   |            .     **|       .            |     0.190285
  3    0.0079422       0.03009   |            .       |*      .            |     0.191774
  4     0.034691       0.13144   |            .       |***    .            |     0.191917
  5     0.010641       0.04032   |            .       |*      .            |     0.194626
  6    -0.014492       -.05491   |            .      *|       .            |     0.194879
  7    -0.028101       -.10647   |            .     **|       .            |     0.195347
  8    -0.018074       -.06848   |            .      *|       .            |     0.197098
  9     0.024708       0.09362   |            .       |**     .            |     0.197817
 10   -0.0016373       -.00620   |            .       |       .            |     0.199155
```

Output 3.24 *Modeling and Forecasting with the IDENTIFY, ESTIMATE, and FORECAST Statements: PROC ARIMA (continued)*

```
                          Inverse Autocorrelations

 Lag    Correlation    -1 9 8 7 6 5 4 3 2 1 0 1 2 3 4 5 6 7 8 9 1

   1       -0.43944     |           *********|      .             |
   2        0.22666     |             .      |***** .             |
   3       -0.08116     |             .    **|      .             |
   4       -0.06102     |             .     *|      .             |
   5        0.01986     |             .      |      .             |
   6        0.03297     |             .      |*     .             |
   7       -0.04359     |             .     *|      .             |
   8        0.14827     |             .      |***   .             |
   9       -0.15649     |             .   ***|      .             |
  10        0.07695     |             .      |**    .             |


                          Partial Autocorrelations

 Lag    Correlation    -1 9 8 7 6 5 4 3 2 1 0 1 2 3 4 5 6 7 8 9 1

   1        0.31216     |             .      |******.             |
   2       -0.21528     |             .  ****|      .             |
   3        0.15573     |             .      |***   .             |
   4        0.05219     |             .      |*     .             |
   5       -0.00997     |             .      |      .             |
   6       -0.03898     |             .     *|      .             |
   7       -0.09630     |             .    **|      .             |
   8       -0.02738     |             .     *|      .             |
   9        0.12247     |             .      |**    .             |
  10       -0.10058     |             .    **|      .             |


                   Autocorrelation Check for White Noise

  To       Chi-              Pr >
 Lag     Square     DF     ChiSq    --------------------Autocorrelations-------------------

   6       4.79      6    0.5716     0.312    -0.097     0.030     0.131     0.040    -0.055
```

Output 3.24 *Modeling and Forecasting with the IDENTIFY, ESTIMATE, and FORECAST Statements: PROC ARIMA (continued)*

PUBLISHING AND PRINTING NONPRODUCTION WORKERS
1944-1977

The ARIMA Procedure

Conditional Least Squares Estimation

Parameter	Estimate	Standard Error ❸	t Value	Approx Pr > |t|	Lag
MU	❹ 0.30330	0.12294	2.47	0.0193	0
MA1,1	-0.46626	0.16148	-2.89	0.0070	1

Constant Estimate	0.3033
Variance Estimate	0.238422
Std Error Estimate	0.488284
AIC	48.27419
SBC	51.2672
Number of Residuals	33

* AIC and SBC do not include log determinant.

Correlations of Parameter Estimates

Parameter	MU	MA1,1
MU	1.000	0.006
MA1,1	0.006	1.000

Autocorrelation Check of Residuals

To Lag	❷ Chi-Square	DF	Pr > ChiSq	Autocorrelations					
6	1.01	5	0.9619	-0.033	-0.089	0.020	0.119	-0.032	-0.036
12	3.80	11	0.9754	-0.054	-0.114	0.157	-0.093	0.072	-0.057
18	7.41	17	0.9776	0.064	0.001	-0.175	-0.108	-0.027	0.085
24	10.07	23	0.9909	0.007	-0.023	0.067	-0.003	-0.123	0.057

Model for Variable NONPUB

Estimated Mean	0.3033
Period(s) of Differencing	1

Output 3.24 *Modeling and Forecasting with the IDENTIFY, ESTIMATE, and FORECAST Statements: PROC ARIMA (continued)*

```
                         Moving Average Factors

                       Factor 1:   1 + 0.46626 B**(1)

                       Forecasts for variable NONPUB

      Obs        Forecast    Std Error          95% Confidence Limits

       35        43.5635       0.4883        42.6065         44.5206
       36        43.8668       0.8666        42.1683         45.5654
       37        44.1701       1.1241        41.9669         46.3733
       38        44.4734       1.3327        41.8613         47.0855
       39        44.7767       1.5129        41.8116         47.7419
       40        45.0800       1.6737        41.7996         48.3605
       41        45.3833       1.8204        41.8154         48.9513
       42        45.6866       1.9562        41.8526         49.5206
       43        45.9899       2.0831        41.9072         50.0727
       44        46.2932       2.2027        41.9761         50.6104
```

3.4.10 Other Identification Techniques

In addition to the ACF, IACF, and PACF, three methods called ESACF, SCAN, and MINIC are available for simultaneously identifying both the autoregressive and moving average orders. These consist of tables with rows labeled AR 0, AR 1, etc. and columns MA 0, MA 1, etc. You look at the table entries to find the row and column whose labels give the correct *p* and *q*. Tsay and Tiao (1984, 1985) develop the ESACF and SCAN methods and show they even work when the autoregressive operator has roots on the unit circle, in which case $p+d$ rather than *p* is found. For $(\mathrm{Y}_t-\mathrm{Y}_{t-2})-0.7(\mathrm{Y}_{t-1}-\mathrm{Y}_{t-3})=e_t$ ESACF and SCAN should give 3 as the autoregressive order. The key to showing their results is that standard estimation techniques give consistent estimators of the autoregressive operator coefficients even in the presence of unit roots.

These methods can be understood through an ARMA(1,1) example. Suppose you have the ARMA(1,1) process $\mathrm{Z}_t-\alpha\mathrm{Z}_{t-1}=e_t-\beta e_{t-1}$, where Z_t is the deviation from the mean at time *t*. The autocorrelations $\rho(j)$ are $\rho(0)=1$, $\rho(1)=[(\alpha-\beta)(1-\alpha\beta)]/[1-\alpha^2+(\beta-\alpha)^2]$, and $\rho(j)=\alpha\rho(j-1)$ for $j>1$.

The partial autocorrelations are motivated by the problem of finding the best linear predictor of Z_t based on $\mathrm{Z}_{t-1},\ldots,\mathrm{Z}_{t-k}$. That is, you want to find coefficients ϕ_{kj} for which $\mathrm{E}\{(\mathrm{Z}_t-\phi_{k1}\mathrm{Z}_{t-1}-\phi_{k2}\mathrm{Z}_{t-2}-\cdots-\phi_{kk}\mathrm{Z}_{t-k})^2\}$ is minimized. This is sometimes referred to as "performing a theoretical regression" of Z_t on Z_{t-1}, $\mathrm{Z}_{t-2},\ldots,\mathrm{Z}_{t-k}$ or "projecting" Z_t onto the space spanned by Z_{t-1}, $\mathrm{Z}_{t-2},\ldots,\mathrm{Z}_{t-k}$. It is accomplished by solving the matrix system of equations

$$\begin{pmatrix} 1 & \rho(1) & \cdots & \rho(k-1) \\ \rho(1) & 1 & \cdots & \rho(k-2) \\ \vdots & \vdots & \ddots & \vdots \\ \rho(k-1) & \rho(k-2) & \cdots & 1 \end{pmatrix} \begin{pmatrix} \phi_{k1} \\ \phi_{k2} \\ \vdots \\ \phi_{kk} \end{pmatrix} = \begin{pmatrix} \rho(1) \\ \rho(2) \\ \vdots \\ \rho(k) \end{pmatrix}$$

Letting $\pi_k = \phi_{kk}$ for $k = 1, 2, \ldots$ produces the sequence π_k of partial autocorrelations. (See **Section 3.3.2.3.**)

At $k = 1$ in the ARMA(1,1) example, you note that
$\phi_{11} = \pi_1 = \rho(1) = [(\alpha - \beta)(1 - \alpha\beta)]/[1 - \alpha^2 + (\beta - \alpha)^2]$, which does not in general equal α. Therefore $Z_t - \phi_{11}Z_{t-1}$ is not $Z_t - \alpha Z_{t-1}$ and thus does not equal $e_t - \beta e_{t-1}$. The autocorrelations of $Z_t - \phi_{11}Z_{t-1}$ would not drop to 0 beyond the moving average order. Increasing *k* beyond 1 will not solve the problem. Still, it is clear that there is some linear combination of Z_t and Z_{t-1}, namely $Z_t - \alpha Z_{t-1}$, whose autocorrelations theoretically identify the order of the moving average part of your model. In general neither the π_k sequence nor any ϕ_{kj} sequence contains the autoregressive coefficients unless the process is a pure autoregression. You are looking for a linear combination
$Z_t - C_1Z_{t-1} - C_2Z_{t-2} - \cdots - C_pZ_{t-p}$ whose autocorrelation is 0 for *j* exceeding the moving average order *q* (1 in our example). The trick is to discover *p* and the C_j s from the data.

The lagged residual from the theoretical regression of Z_t on Z_{t-1} is $R_{1,t-1} = Z_{t-1} - \phi_{11}Z_{t-2}$, which is a linear combination of Z_{t-1} and Z_{t-2}, so regressing Z_t on Z_{t-1} and $R_{1,t-1}$ produces regression coefficients, say C_{21} and C_{22}, which give the same fit, or projection, as regressing Z_t on Z_{t-1} and Z_{t-2}. That is, $C_{21}Z_{t-1} + C_{22}R_{1,t-1} = C_{21}Z_{t-1} + C_{22}(Z_{t-1} - \phi_{11}Z_{t-2}) = \phi_{21}Z_{t-1} + \phi_{22}Z_{t-2}$. Thus it must be that $\phi_{21} = C_{21} + C_{22}$ and $\phi_{22} = -\phi_{11}C_{22}$. In matrix form

$$\begin{pmatrix}\phi_{21}\\ \phi_{22}\end{pmatrix} = \begin{pmatrix}1 & 1\\ 0 & -\phi_{11}\end{pmatrix}\begin{pmatrix}C_{21}\\ C_{22}\end{pmatrix}$$

Noting that $\rho(2) = \alpha\rho(1)$, the ϕ_{2j} coefficients satisfy

$$\begin{pmatrix}1 & \rho(1)\\ \rho(1) & 1\end{pmatrix}\begin{pmatrix}\phi_{21}\\ \phi_{22}\end{pmatrix} = \rho(1)\begin{pmatrix}1\\ \alpha\end{pmatrix}$$

Relating this to the Cs and noting that $\rho(1) = \phi_{11}$, you have

$$\begin{pmatrix}1 & \phi_{11}\\ \phi_{11} & 1\end{pmatrix}\begin{pmatrix}1 & 1\\ 0 & -\phi_{11}\end{pmatrix}\begin{pmatrix}C_{21}\\ C_{22}\end{pmatrix} = \phi_{11}\begin{pmatrix}1\\ \alpha\end{pmatrix}$$

or

$$\begin{pmatrix}C_{21}\\ C_{22}\end{pmatrix} = \frac{\phi_{11}}{\phi_{11}(\phi_{11}^2 - 1)}\begin{pmatrix}0 & \phi_{11}^2 - 1\\ -\phi_{11} & 1\end{pmatrix}\begin{pmatrix}1\\ \alpha\end{pmatrix} = \begin{pmatrix}\alpha\\ \dfrac{\alpha - \phi_{11}}{\phi_{11}^2 - 1}\end{pmatrix}$$

You now "filter" Z using only $C_{21} = \alpha$; that is, you compute $Z_t - C_{21}Z_{t-1}$, which is just $Z_t - \alpha Z_{t-1}$, and this in turn is a moving average of order 1. Its lag 1 autocorrelation (it is nonzero) will appear in the AR 1 row and MA 0 column of the ESACF table. Let the residual from this regression be denoted $R_{2,t}$. The next step is to regress Z_t on Z_{t-1}, $R_{1,t-2}$, and $R_{2,t-1}$. In this regression, the theoretical coefficient of Z_{t-1} will again be α, but its estimate may differ somewhat from the one obtained previously. Notice the use of the lagged value of $R_{2,t}$ and the second lag of the first round residual $R_{1,t-2} = Z_{t-2} - \phi_1 Z_{t-3}$. The lag 2 autocorrelation of $Z_t - \alpha Z_{t-1}$, which is 0, will be written in the MA 1 column of the AR 1 row. For the ESACF of a general ARMA(p,q) in the AR p row, once your regression has at least q lagged residuals, the first p theoretical C_{kj} will be the p autoregressive coefficients and the filtered series will be a MA(q), so its autocorrelations will be 0 beyond lag q.

The entries in the AR k row of the ESACF table are computed as follows:

(1) Regress Z_t on $Z_{t-1}, Z_{t-2}, \ldots, Z_{t-k}$ with residual $R_{1,t}$
Coefficients: $C_{11}, C_{12}, \ldots, C_{1k}$

(2) Regress Z_t on $Z_{t-1}, Z_{t-2}, \ldots, Z_{t-k}, R_{1,t-1}$ with residual $R_{2,t}$
Second-round coefficients: $C_{21}, \ldots, C_{2k}$, (and $C_{2,k+1}$)
Record in MA 0 column, the lag 1 autocorrelation of
$Z_t - C_{21}Z_{t-1} - C_{22}Z_{t-2} - \cdots - C_{2k}Z_{t-k}$

(3) Regress Z_t on $Z_{t-1}, Z_{t-2}, \ldots, Z_{t-k}, R_{1,t-2}, R_{2,t-1}$ with residual $R_{3,t}$
Third-round coefficients: $C_{31}, \ldots, C_{3k}$, (and $C_{3,k+1}, C_{3,k+2}$)
Record in MA 1 column the lag 2 autocorrelation of
$Z_t - C_{31}Z_{t-1} - C_{32}Z_{t-2} - \cdots - C_{3k}Z_{t-k}$
etc.

Notice that at each step, you lag all residuals that were previously included as regressors and add the lag of the most recent residual to your regression. The estimated C coefficients and resulting filtered series differ at each step. Looking down the ESACF table of an AR(p,q), theoretically row p should be the first row in which a string of 0s appears and it should start at the MA q column. Finding that row and the first 0 entry in it puts you in row p column q of the ESACF. The model is now identified.

Here is a theoretical ESACF table for an ARMA(1,1) with "X" for nonzero numbers:

	MA 0	MA 1	MA 2	MA 3	MA 4	MA 5
AR 0	X	X	X	X	X	X
AR 1	X	0*	0	0	0	0
AR 2	X	X	0	0	0	0
AR 3	X	X	X	0	0	0
AR 4	X	X	X	X	0	0

The string of 0s slides to the right as the AR row number moves beyond p, so there appears a triangular array of 0s whose "point" 0* is at the correct (p,q) combination.

In practice, the theoretical regressions are replaced by least squares regressions, so the ESACF table will only have numbers near 0 where the theoretical ESACF table has 0s. A recursive algorithm is used to quickly compute the needed coefficients without having to compute so many actual regressions. PROC ARIMA will also use asymptotically valid standard errors based on Bartlett's formula to deliver a table of approximate p-values for the ESACF entries and will suggest values of p and q as a tentative identification. See Tsay and Tiao (1984) for further details.

Tsay and Tiao (1985) suggest a second table called SCAN. It is computed using canonical correlations. For the ARMA(1,1) model, recall that the autocovariances are $\gamma(0)$, $\gamma(1)$, $\gamma(2)=\alpha\gamma(1)$, $\gamma(3)=\alpha^2\gamma(1)$, $\gamma(4)=\alpha^3\gamma(1)$, etc., so the covariance matrix of $Y_t, Y_{t-1}, \ldots, Y_{t-5}$ is

$$\Gamma = \begin{pmatrix} \gamma(0) & \gamma(1) & \alpha\gamma(1) & \alpha^2\gamma(1) & \alpha^3\gamma(1) & \alpha^4\gamma(1) \\ \gamma(1) & \gamma(0) & \gamma(1) & \alpha\gamma(1) & \alpha^2\gamma(1) & \alpha^3\gamma(1) \\ [\alpha\gamma(1)] & [\gamma(1)] & \gamma(0) & \gamma(1) & \alpha\gamma(1) & \alpha^2\gamma(1) \\ [\alpha^2\gamma(1)] & [\alpha\gamma(1)] & \gamma(1) & \gamma(0) & \gamma(1) & \alpha\gamma(1) \\ \alpha^3\gamma(1) & \alpha^2\gamma(1) & \alpha\gamma(1) & \gamma(1) & \gamma(0) & \gamma(1) \\ \alpha^4\gamma(1) & \alpha^3\gamma(1) & \alpha^2\gamma(1) & \alpha\gamma(1) & \gamma(1) & \gamma(0) \end{pmatrix}$$

The entries in square brackets form the 2×2 submatrix of covariances between the vectors (Y_t, Y_{t-1}) and (Y_{t-2}, Y_{t-3}). That submatrix $\mathbf{A}$, the variance matrix $\mathbf{C}_{11}$ of (Y_t, Y_{t-1}), and the variance matrix $\mathbf{C}_{22}$ of (Y_{t-2}, Y_{t-3}) are

$$\mathbf{A} = \gamma(1)\begin{pmatrix} \alpha & 1 \\ \alpha^2 & \alpha \end{pmatrix} \quad \mathbf{C}_{11} = \mathbf{C}_{22} = \begin{pmatrix} \gamma(0) & \gamma(1) \\ \gamma(1) & \gamma(0) \end{pmatrix}$$

The best linear predictor of $(Y_t, Y_{t-1})'$ based on $(Y_{t-2}, Y_{t-3})'$ is $\mathbf{A}'\mathbf{C}_{22}^{-1}(Y_{t-2}, Y_{t-3})'$ with prediction error variance matrix $\mathbf{C}_{11} - \mathbf{A}'\mathbf{C}_{22}^{-1}\mathbf{A}$. Because matrix $\mathbf{C}_{11}$ represents the variance of (Y_t, Y_{t-1}), the matrix $\mathbf{C}_{11}^{-1}\mathbf{A}'\mathbf{C}_{22}^{-1}\mathbf{A}$ is analogous to a regression R^2 statistic. Its eigenvalues are called squared canonical correlations between $(Y_t, Y_{t-1})'$ and $(Y_{t-2}, Y_{t-3})'$.

Recall that, for a square matrix $\mathbf{M}$, if a column vector $\mathbf{H}$ exists such that $\mathbf{MH} = b\mathbf{H}$, then $\mathbf{H}$ is called an *eigenvector* and the scalar b is the corresponding *eigenvalue* of matrix $\mathbf{M}$. Using $\mathbf{H} = (1, -\alpha)'$, you see that $\mathbf{AH} = (0,0)'$, so $\mathbf{C}_{11}^{-1}\mathbf{A}'\mathbf{C}_{22}^{-1}\mathbf{AH} = 0\mathbf{H}$; that is, $\mathbf{C}_{11}^{-1}\mathbf{A}'\mathbf{C}_{22}^{-1}\mathbf{A}$ has an eigenvalue 0. The number of 0 eigenvalues of $\mathbf{A}$ is the same as the number of 0 eigenvalues of $\mathbf{C}_{11}^{-1}\mathbf{A}'\mathbf{C}_{22}^{-1}\mathbf{A}$. This is true for general time series covariance matrices.

The matrix **A** has first column that is α times the second, which implies these equivalent statements:

(1) The 2×2 matrix **A** is not of full rank (its rank is 1).

(2) The 2×2 matrix **A** has at least one eigenvalue 0.

(3) The 2×2 matrix $\mathbf{C}_{11}^{-1}\mathbf{A}'\mathbf{C}_{22}^{-1}\mathbf{A}$ has at least one eigenvalue 0.

(4) The vectors (Y_t, Y_{t-1}) and (Y_{t-2}, Y_{t-3}) have at least one squared canonical correlation that is 0.

The fourth of these statements is easily seen. The linear combinations $Y_t - \alpha Y_{t-1}$ and its second lag $Y_{t-2} - \alpha Y_{t-3}$ have correlation 0 because each is an MA(1). The smallest canonical correlation is obtained by taking linear combinations of (Y_t, Y_{t-1}) and (Y_{t-2}, Y_{t-3}) and finding the pair with correlation closest to 0. Since there exist linear combinations in the two sets that are *uncorrelated*, the smallest canonical correlation must be 0. Again you have a method of finding a linear combination whose autocorrelation sequence is 0 beyond the moving average lag q.

In general, construct an arbitrarily large covariance matrix of $Y_t, Y_{t-1}, Y_{t-2}, \ldots,$ and let $\mathbf{A}_{j,m}$ be the $m \times m$ matrix whose upper-left element is in row $j+1$, column 1 of the original matrix. In this notation, the **A** with square bracketed elements is denoted $\mathbf{A}_{2,2}$ and the bottom left 3×3 matrix of Γ is $\mathbf{A}_{3,3}$. Again there is a full-rank 3×2 matrix **H** for which $\mathbf{A}_{3,3}\mathbf{H}$ has all 0 elements, namely

$$\mathbf{A}_{3,3}\mathbf{H} = \begin{pmatrix} \alpha^2\gamma(1) & \alpha\gamma(1) & \gamma(1) \\ \alpha^3\gamma(1) & \alpha^2\gamma(1) & \alpha\gamma(1) \\ \alpha^4\gamma(1) & \alpha^3\gamma(1) & \alpha^2\gamma(1) \end{pmatrix} \begin{pmatrix} 1 & 0 \\ -\alpha & 1 \\ 0 & -\alpha \end{pmatrix} = \begin{pmatrix} 0 & 0 \\ 0 & 0 \\ 0 & 0 \end{pmatrix}$$

showing that matrix $\mathbf{A}_{3,3}$ has (at least) 2 eigenvalues that are 0 with the columns of **H** being the corresponding eigenvectors. Similarly, using $\mathbf{A}_{3,2}$ and $\mathbf{H} = (1, -\alpha)$

$$\mathbf{A}_{3,2}\mathbf{H} = \begin{pmatrix} \alpha^2\gamma(1) & \alpha\gamma(1) \\ \alpha^3\gamma(1) & \alpha^2\gamma(1) \end{pmatrix} \begin{pmatrix} 1 \\ -\alpha \end{pmatrix} = \begin{pmatrix} 0 \\ 0 \end{pmatrix}$$

so $\mathbf{A}_{3,2}$ has (at least) one 0 eigenvalue, as does $\mathbf{A}_{j,2}$ for all $j > 1$. In fact all $\mathbf{A}_{j,m}$ with $j > 1$ and $m > 1$ have at least one 0 eigenvalue for this example. For general ARIMA (p,q) models, all $\mathbf{A}_{j,m}$ with $j > q$ and $m > p$ have at least one 0 eigenvalue. This provides the key to the SCAN table. If you make a table whose *m*th row, *j*th column entry is the smallest canonical correlation derived from $\mathbf{A}_{j,m}$, you have this table for the current example:

	$m=1$	$m=2$	$m=3$	$m=4$
$j=1$	X	X	X	X
$j=2$	X	0	0	0
$j=3$	X	0	0	0
$j=4$	X	0	0	0

	$p=0$	$p=1$	$p=2$	$p=3$
$q=0$	X	X	X	X
$q=1$	X	0	0	0
$q=2$	X	0	0	0
$q=3$	X	0	0	0

where the Xs represent nonzero numbers. Relabeling the rows and columns with $q = j - 1$ and $p = m - 1$ gives the SCAN (smallest canonical correlation) table. It has a rectangular array of 0s whose upper left corner is at the p and q corresponding to the correct model, ARMA(1,1) for the current example. The first column of the SCAN table consists of the autocorrelations and the first row consists of the partial autocorrelations.

In PROC ARIMA, entries of the 6×6 variance-covariance matrix $\mathbf{\Gamma}$ above would be replaced by estimated autocovariances. To see why the 0s appear for an ARMA (p,q) whose autoregressive coefficients are α_i, you notice from the Yule-Walker equations that $\gamma(j) - \alpha_1\gamma(j-1) - \alpha_2\gamma(j-2) - \cdots - \alpha_p\gamma(j-p)$ is zero for $j > q$. Therefore, in the variance covariance matrix for such a process, any $m \times m$ submatrix with $m > p$ whose upper-left element is at row j, column 1 of the original matrix will have at least one 0 eigenvalue with eigenvector $(1, -\alpha_1, -\alpha_2, \ldots, -\alpha_p, 0, 0, \ldots, 0)'$ if $j > q$. Hence 0 will appear in the theoretical table whenever $m > p$ and $j > q$. Approximate standard errors are obtained by applying Bartlett's formula to the series filtered by the autoregressive coefficients, which in turn can be extracted from the **H** matrix (eigenvectors). An asymptotically valid test, again making use of Bartlett's formula, is available and PROC ARIMA displays a table of the resulting p-values.

The MINIC method simply attempts to fit models over a grid of p and q choices, and records the SBC information criterion for each fit in a table. The Schwartz Bayesian Information Criterion is $\text{SBC} = n\ln(s^2) + (p+q)\ln(n)$, where p and q are the autoregressive and moving average orders of the candidate model and s^2 is an estimate of the innovations variance. Some sources refer to Schwartz's criterion, perhaps normalized by n, as BIC. Here, the symbol SBC is used so that Schwartz's criterion will not be confused with the BIC criterion of Sawa (1978). Sawa's BIC, used as a model selection tool in PROC REG, is $n\ln(s^2) + 2[(k+2)\frac{n}{n-k} - (\frac{n}{n-k})^2]$ for a full regression model with n observations and k parameters. The MINIC technique chooses p and q giving the smallest SBC. It is possible, of course, that the fitting will fail due to singularities in which case the SBC is set to missing.

The fitting of models in computing MINIC follows a clever algorithm suggested by Hannan and Rissanen (1982) using ideas dating back to Durbin (1960). First, using the Yule-Walker equations, a long autoregressive model is fit to the data. For the ARMA(1,1) example of this section it is seen that

$$Y_t = (\alpha - \beta)[Y_{t-1} + \beta Y_{t-2} + \beta^2 Y_{t-3} + \cdots] + e_t$$

and as long as $|\beta| < 1$, the coefficients on lagged Y will die off quite quickly, indicating that a truncated version of this infinite autoregression will approximate the e_t process well. To the extent that this is true, the Yule-Walker equations for a length k (k large) autoregression can be solved to give estimates, say $\hat{b}_j$, of the coefficients of the Y_{t-j} terms and a residual series $\hat{e}_t = Y_t - \hat{b}_1 Y_{t-1} - \hat{b}_2 Y_{t-2} - \cdots - \hat{b}_k Y_{t-k}$ that is close to the actual e_t series. Next, for a candidate model of order p,q, regress Y_t on $Y_{t-1},\ldots,Y_{t-p},\hat{e}_{t-1},\hat{e}_{t-2},\ldots,\hat{e}_{t-q}$. Letting $\hat{\sigma}^2_{pq}$ be $1/n$ times the error sum of squares for this regression, pick p and q to minimize the SBC criterion $\text{SBC} = n\ln(\hat{\sigma}^2_{pq}) + (p+q)\ln(n)$. The length of the autoregressive model for the $\hat{e}_t$ series can be selected by minimizing the AIC criterion.

To illustrate, 1000 observations on an ARMA(1,1) with $\alpha = .8$ and $\beta = .4$ are generated and analyzed. The following code generates **Output 3.25**:

```
PROC ARIMA DATA=A;
   I VAR=Y NLAG=1 MINIC P=(0:5) Q=(0:5);
   I VAR=Y NLAG=1 ESACF P=(0:5) Q=(0:5);
   I VAR=Y NLAG=1 SCAN  P=(0:5) Q=(0:5);
RUN;
```

Output 3.25 *ESACF, SCAN and MINIC Displays*

Minimum Information Criterion

Lags	MA 0	MA 1	MA 2	MA 3	MA 4	MA 5
AR 0	0.28456	0.177502	0.117561	0.059353	0.028157	0.003877
AR 1	-0.0088	-0.04753	-0.04502	-0.0403	-0.03565	-0.03028
AR 2	-0.03958	-0.04404	-0.04121	-0.0352	-0.03027	-0.02428
AR 3	-0.04837	-0.04168	-0.03537	-0.02854	-0.02366	-0.01792
AR 4	-0.04386	-0.03696	-0.03047	-0.02372	-0.01711	-0.01153
AR 5	-0.03833	-0.03145	-0.02461	-0.0177	-0.01176	-0.00497

Error series model: AR(9)
Minimum Table Value: BIC(3,0) = -0.04837

Extended Sample Autocorrelation Function

Lags	MA 0	MA 1	MA 2	MA 3	MA 4	MA 5
AR 0	0.5055	0.3944	0.3407	0.2575	0.2184	0.1567
AR 1	-0.3326	-0.0514	0.0564	-0.0360	0.0417	-0.0242
AR 2	-0.4574	-0.2993	0.0197	0.0184	0.0186	-0.0217
AR 3	-0.1207	-0.2357	0.1902	0.0020	0.0116	0.0006
AR 4	-0.4074	-0.1753	0.1042	-0.0132	0.0119	0.0015
AR 5	0.4836	0.1777	-0.0733	0.0336	0.0388	-0.0051

ESACF Probability Values

Lags	MA 0	MA 1	MA 2	MA 3	MA 4	MA 5
AR 0	0.0001	0.0001	0.0001	0.0001	0.0001	0.0010
AR 1	0.0001	0.1489	0.1045	0.3129	0.2263	0.4951
AR 2	0.0001	0.0001	0.5640	0.6013	0.5793	0.6003
AR 3	0.0001	0.0001	0.0001	0.9598	0.7634	0.9874
AR 4	0.0001	0.0001	0.0001	0.7445	0.7580	0.9692
AR 5	0.0001	0.0001	0.0831	0.3789	0.2880	0.8851

ARMA(p+d,q) Tentative Order Selection Tests

(5% Significance Level)

ESACF	p+d	q
	1	1
	4	3
	5	2

***Output 3.25** ESACF, SCAN and MINIC Display (continued)*

Squared Canonical Correlation Estimates

Lags	MA 0	MA 1	MA 2	MA 3	MA 4	MA 5
AR 0	0.2567	0.1563	0.1170	0.0670	0.0483	0.0249
AR 1	0.0347	0.0018	0.0021	0.0008	0.0011	0.0003
AR 2	0.0140	0.0023	0.0002	0.0002	0.0002	0.0010
AR 3	0.0002	0.0007	0.0002	0.0001	0.0002	0.0001
AR 4	0.0008	0.0010	0.0002	0.0002	0.0002	0.0002
AR 5	0.0005	0.0001	0.0002	0.0001	0.0002	0.0004

SCAN Chi-Square[1] Probability Values

Lags	MA 0	MA 1	MA 2	MA 3	MA 4	MA 5
AR 0	0.0001	0.0001	0.0001	0.0001	0.0001	0.0010
AR 1	0.0001	0.2263	0.1945	0.4097	0.3513	0.5935
AR 2	0.0002	0.1849	0.7141	0.6767	0.7220	0.3455
AR 3	0.6467	0.4280	0.6670	0.9731	0.6766	0.9877
AR 4	0.3741	0.3922	0.6795	0.6631	0.7331	0.7080
AR 5	0.4933	0.8558	0.7413	0.9111	0.6878	0.6004

ARMA(p+d,q) Tentative Order Selection Tests
(5% Significance Level)

SCAN	p+d	q
	1	1
	3	0

The tentative order selections in ESACF and SCAN simply look at all triangles (rectangles) for which every element is insignificant at the specified level (0.05 by default). These are listed in descending order of size (below the tables), size being the number of elements in the triangle or rectangle. In our example ESACF (previous page) and SCAN (above) list the correct (1,1) order at the top of the list. The MINIC criterion uses $k = 9$, a preliminary AR(9) model, to create the estimated white noise series, then selects $(p,q) = (3,0)$ as the order, this also being one choice given by the SCAN option. The second smallest SBC, $-.04753$, occurs at the correct $(p,q) = (1,1)$.

As a check on the relative merits of these methods, 50 ARMA(1,1) series each of length 500 are generated for each of the 12 (α,β) pairs obtained by choosing α and β from $\{-.9,-.3,.3,.9\}$ such that $\alpha \neq \beta$. This gives 600 series. For each, the ESACF, SCAN, and MINIC methods are used, the results are saved, and the estimated p and q are extracted for each method. The whole experiment is repeated with series of length 50. A final set of 600 runs for $Y_t = .5Y_{t-4} + e_t + .3e_{t-1}$ using $n = 50$ gives the last three columns. Asterisks indicate the correct model.

pq	|<- ARMA(1,1) n=500 ->|			|<- ARMA(1,1) n=50 ->|			|<- ARMA(4,1) n=50 ->|	
pq	BIC	ESACF	SCAN	BIC	ESACF	SCAN	BIC	ESACF	SCAN
00	2	1	1	25	40	25	69	64	35
01	0	0	0	48	146	126	28	46	33
02	0	0	0	17	21	8	5	9	11
03	0	0	0	7	4	16	4	6	2
04	0	0	0	7	3	2	41	20	35
05	0	0	0	6	1	0	14	0	2
10	1	0	0	112	101	145	28	15	38
11	* 252 ***	441 ***	461 ***	53 **	165 **	203 *	5	47	78
12	13	23	8	16	7	9	1	10	30
13	13	18	8	12	0	2	0	1	3
14	17	5	3	2	0	1	3	0	0
15	53	6	0	5	0	0	4	0	0
20	95	6	12	91	41	18	26	16	19
21	9	6	25	9	22	14	2	42	25
22	24	46	32	4	8	7	3	62	121
23	1	0	1	1	2	1	1	2	8
24	4	0	2	3	0	1	2	1	0
25	10	0	1	6	0	0	0	0	0
30	35	2	9	50	6	8	30	9	21
31	5	3	11	1	10	3	3	23	27
32	3	6	1	3	4	0	3	21	7
33	3	15	13	0	2	0	1	16	2
34	5	2	0	1	0	0	0	0	0
35	4	0	0	2	0	0	0	0	0
40	5	0	0	61	6	6	170	66	98
41	3	0	5	3	4	0	* 10 ***	52 ***	0 *
42	2	4	2	1	2	0	4	24	0
43	5	3	0	0	0	0	0	22	0
44	1	4	1	1	0	0	0	0	0
45	6	0	0	5	0	0	1	0	0
50	5	0	0	32	3	2	116	6	5
51	0	1	2	10	1	0	18	13	0
52	3	0	1	2	0	0	2	6	0
53	9	2	0	2	0	0	5	0	0
54	6	1	0	2	1	0	1	0	0
55	6	0	0	0	0	0	0	0	0
totals	600	595	599	600	600	597	600	599	600

It is reassuring that the methods almost never underestimate p or q when n is 500. For the ARMA(1,1) with parameters in this range, it appears that SCAN does slightly better than ESACF, with both being superior to MINIC. The SCAN and ESACF columns do not always add to 600 because, for some cases, no rectangle or triangle can be found with all elements insignificant. Because SCAN compares the *smallest* normalized squared canonical correlation to a distribution (χ_1^2) that is appropriate for a randomly selected one, it is also very conservative. By analogy, even if 5% of men exceed 6 feet in height, finding a random sample of 10 men whose shortest member exceeds 6 feet in height would be extremely rare. Thus the appearance of a significant bottom-right-corner element in the SCAN table, which would imply no rectangle of insignificant values, happens rarely—not the 30 times you would expect from $600(.05) = 30$.

The conservatism of the test also implies that for moderately large p and q there is a fairly good chance that a rectangle (triangle) of "insignificant" terms will appear by chance having p or q too small. Indeed for 600 replicates of the model $Y_t = .5Y_{t-4} + e_t + .3e_{t-1}$ using $n = 50$, we see that $(p,q) = (4,1)$ is rarely chosen by any technique with SCAN giving no correct choices. There does not seem to be a universally preferable choice among the three.

As a real data example, **Output 3.26** shows monthly interbank loans in billions of dollars. The data were downloaded from the Federal Reserve Web site. Also shown are the differences (upper-right corner) and the corresponding log scale graphs. The data require differencing and the right-side graphs seem to indicate the need for logarithms to stabilize the variance.

Output 3.26
Loans

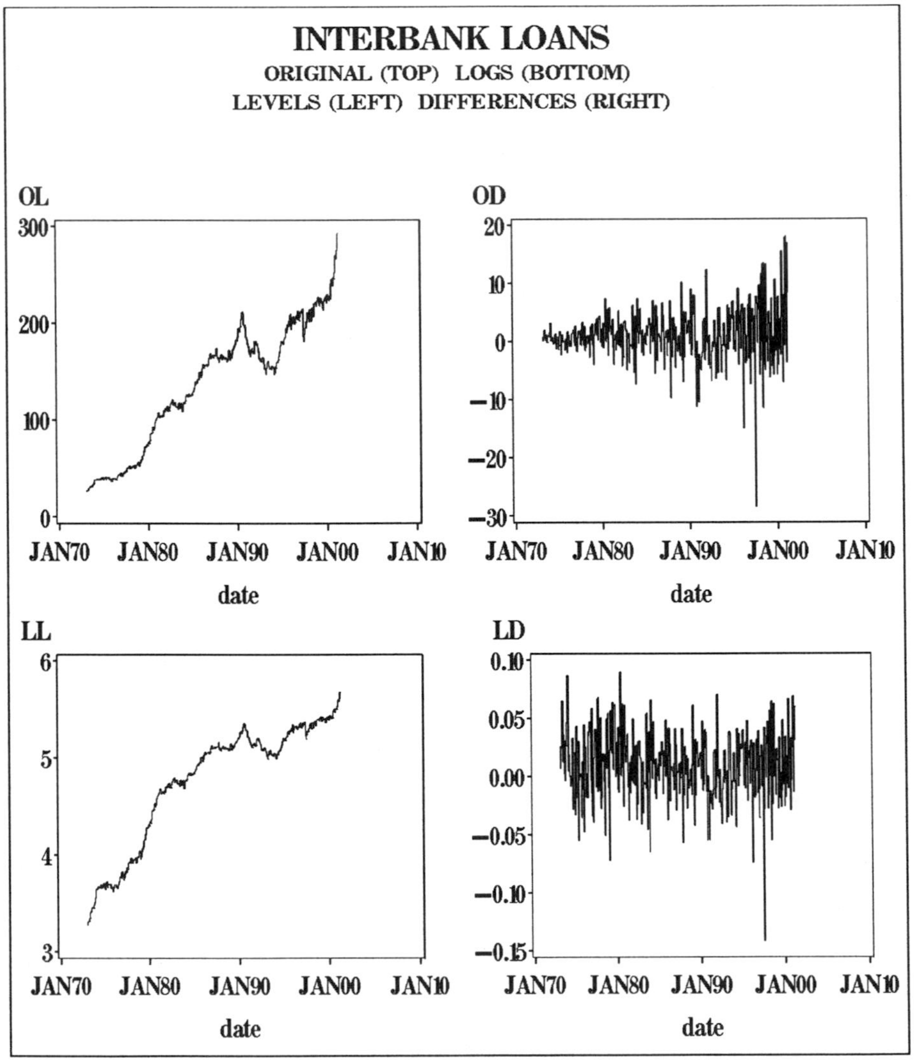

To identify the log transformed variable, called LOANS in the data set, use this code to get the SCAN table.

```
PROC ARIMA DATA=IBL;
   IDENTIFY VAR=LOANS SCAN P=(0:5) Q=(0:5);
RUN;
```

Output 3.27 shows the SCAN results. They indicate several possible models.

Output 3.27
SCAN Table for Interbank Loans

Squared Canonical Correlation Estimates

Lags	MA 0	MA 1	MA 2	MA 3	MA 4	MA 5
AR 0	0.9976	0.9952	0.9931	0.9899	0.9868	0.9835
AR 1	<.0001	0.0037	0.0397	0.0007	0.0024	0.0308
AR 2	0.0037	0.0003	0.0317	0.0020	<.0001	0.0133
AR 3	0.0407	0.0309	0.0274	0.0126	0.0134	0.0125
AR 4	0.0004	0.0053	0.0076	0.0004	0.0022	0.0084
AR 5	0.0058	0.0003	0.0067	0.0022	<.0001	0.0078

SCAN Chi-Square[1] Probability Values

Lags	MA 0	MA 1	MA 2	MA 3	MA 4	MA 5
AR 0	<.0001	<.0001	<.0001	<.0001	<.0001	<.0001
AR 1	0.9125	0.2653	0.0003	0.6474	0.3936	0.0019
AR 2	0.2618	0.7467	0.0033	0.4419	0.9227	0.0940
AR 3	0.0002	0.0043	0.0136	0.0856	0.0881	0.1302
AR 4	0.7231	0.1942	0.1562	0.7588	0.4753	0.1589
AR 5	0.1613	0.7678	0.1901	0.4709	0.9708	0.1836

The ARIMA Procedure

ARMA(p+d,q)
Tentative Order Selection Tests

----SCAN---	
p+d	q
4	0
2	3

(5% Significance Level)

The SCAN table was computed on log transformed, undifferenced data. Therefore, the listed number $p+d$ represents $p+1$ and SCAN suggests ARIMA(3,1,0) or ARIMA(1,1,3).

```
IDENTIFY VAR=LOANS(1) NOPRINT;
   ESTIMATE P=3 ML;
   ESTIMATE P=1 Q=3 ML;
RUN;
```

The chi-square checks for both of these models are insignificant at all lags, indicating both models fit well. Both models have some insignificant parameters and could be refined by omitting some lags if desired (output not shown).

3.5 Summary

The steps for analyzing nonseasonal univariate series are outlined below.

1. Check for nonstationarity using

   - data plot to monitor slow level shifts in the data (as in IBM example)
   - ACF to monitor very slow decay (IBM or publishing and printing example)
   - Dickey and Fuller test for stationarity (silver example).

   If any of these tests indicate nonstationarity, difference the series using VAR=Y(1) in the IDENTIFY statement and repeat step 1. If necessary, difference again by specifying VAR=Y(1,1).

2. Check the Q statistic (CHI SQUARE) at the bottom of the printout. If Q is small (in other words, PROB is fairly large) and if the first few autocorrelations are small, you may want to assume that your (possibly differenced) series is just white noise.

3. Check the ACF, IACF, and PACF to identify a model. If the ACF drops to 0 after q lags, this indicates an MA(q) model. If the IACF or PACF drops to 0 after p lags, this indicates an AR(p) model. If you have differenced the series once or twice, one or two MA lags are likely to be indicated.

4. You can use the SCAN, ESACF, and MINIC tables to determine initial starting models to try in an ESTIMATE statement.

5. Using the ESTIMATE statement, specify the model you picked (or several candidate models). For example, you fit the model

   $$\left(Y_t - Y_{t-1}\right) = \left(1 - \theta_1 B\right)e_t$$

   by specifying these statements:

```
PROC ARIMA DATA=SASDS;
   IDENTIFY VAR=Y(1);
   ESTIMATE Q=1 NOCONSTANT;
RUN;
```

6. Check the Q statistic (CHI SQUARE) at the bottom of the ESTIMATE printout. If it is insignificant, your model fits reasonably well according to this criterion. Otherwise, return to the original ACF, IACF, and PACF of your (possibly differenced) data to determine if you have missed something. This is generally more advisable than plotting the ACF of the residuals from this misspecified model.

   If you have differenced, the mean is often (IBM data), but not always (publishing and printing data), 0. Use the NOCONSTANT option to suppress the fitting of a constant.

   Fitting extra lags and excluding insignificant lags in an attempt to bypass identification causes unstable parameter estimates and possible convergence problems if you overfit on both sides (AR and MA) at once. Correlations of parameter estimates are extremely high in this case (if, in fact, the estimation algorithm converges). Overfitting on one side at a time to check the model is no problem.

7. Use the FORECAST statement with LEAD=k to produce forecasts from the fitted model. It is a good idea to specify BACK=b to start the forecast b steps before the end of the series. You can then compare the last b forecasts to data values at the end of the series. If you note a large discrepancy, you may want to adjust your forecasts. You omit the BACK= option on your final forecast. It is used only as a diagnostic tool.

8. Examine plots of residuals and possibly use PROC UNIVARIATE to examine the distribution and PROC SPECTRA to test the white noise assumption further. (See Chapter 7, "Spectral Analysis," for more information.)

Chapter 4 The ARIMA Model: Introductory Applications

4.1 Seasonal Time Series

4.1.1 Introduction to Seasonal Modeling

The first priority in seasonal modeling is to specify correct differencing and appropriate transformations. This topic is discussed first, followed by model identification. The potential behavior of autocorrelation functions (ACFs) for seasonal models is not easy to characterize, but ACFs are given for a few seasonal models. You should find a pattern that matches your data among these diagnostic plots.

Consider the model

$$Y_t - \mu = \alpha\left(Y_{t-12} - \mu\right) + e_t$$

where e_t is white noise. This model is applied to monthly data and expresses this December's Y, for example, as μ plus a proportion of last December's deviation from μ. If $\mu = 100$, $\alpha = .8$, and last December's Y=120, the model forecasts this December's Y as 100+.8(20)=116. The forecast for the next December's Y is 100+.64(20), and the forecast for *j* Decembers ahead is $100+.8^j\,(20)$.

The model responds to change in the series because it uses only the most recent December to forecast the future. This approach contrasts with the indicator variables in the regression approach discussed in Chapter 1, "Overview of Time Series," where the average of all December values goes into the forecast for this December. For the autoregressive (AR) seasonal model above, the further into the future you forecast, the closer your forecast is to the mean μ. Suppose you allow α to be 1 in the AR seasonal model. Your model is nonstationary and reduces to

$$Y_t = Y_{t-12} + e_t$$

This model uses last December's Y as the forecast for next December (and for any other future December). The difference

$$Y_t - Y_{t-12}$$

is stationary (white noise, in this case) and is specified using the PROC ARIMA statement

```
IDENTIFY VAR=Y(12);
```

This is called a span 12 difference. The forecast does not tend to return to the historical series mean, as evidenced by the lack of a μ term in the model.

When you encounter a span 12 difference, often the differenced series is not white noise but is instead a moving average of the form

$$e_t - \beta e_{t-12}$$

For example, if

$$Y_t - Y_{t-12} = e_t - .5e_{t-12}$$

you see that

$$\begin{aligned} e_{t-12} &= Y_{t-12} - Y_{t-24} + .5e_{t-24} \\ &= Y_{t-12} - Y_{t-24} + .5\left(Y_{t-24} - Y_{t-36} + .5e_{t-36}\right) \end{aligned}$$

If you continue in this fashion, you can express Y_t as e_t plus an infinite weighted sum of past Y values, namely

$$Y_t = e_t + .5(Y_{t-12} + .5Y_{t-24} + \ldots)$$

Thus, the forecast for any future December is a weighted sum of past December values, with weights decreasing exponentially as you move further into the past. Although the forecast involves many past Decembers, the decreasing weights make it respond to recent changes.

Differencing over seasonal spans is indicated when the ACF at the seasonal lags dies off very slowly. Often this behavior is masked in the original ACF, which dies off slowly at all lags. In that case you should difference, as the ACF seems to indicate, by specifying the PROC ARIMA statement

```
IDENTIFY VAR=Y(1);
```

Now look at the ACF of the differenced series, considering only seasonal lags (12, 24, 36, and so on). If these ACF values die off very slowly, you want to take a span 12 difference in addition to the first difference. You accomplish this by specifying

```
IDENTIFY VAR=Y(1,12);
```

Note how the differencing specification works. For example,

```
IDENTIFY VAR=Y(1,1);
```

specifies a second difference

$$(Y_t - Y_{t-1}) - (Y_{t-1} - Y_{t-2}) = Y_t - 2Y_{t-1} + Y_{t-2}$$

whereas the specification

```
IDENTIFY VAR=Y(2);
```

creates the span 2 difference

$$Y_t - Y_{t-2}$$

Calling the span 1 and span 12 differenced series V_t, you create

$$V_t = (Y_t - Y_{t-1}) - (Y_{t-12} - Y_{t-13})$$

and consider models for V_t.

4.1.2 Model Identification

If V_t appears to be white noise, the model becomes

$$Y_t = Y_{t-1} + (Y_{t-12} - Y_{t-13}) + e_t$$

Thus, with data through this November, you forecast this December's Y_t as the November value (Y_{t-1}) plus last year's November-to-December change ($Y_{t-12} - Y_{t-13}$).

More commonly, you find that the differenced series V_t satisfies

$$V_t = (1 - \theta_1 B)(1 - \theta_2 B^{12}) e_t$$

This is called a seasonal multiplicative moving average. The meaning of a product of backshift factors like this is simply

$$V_t = e_t - \theta_1 e_{t-1} - \theta_2 e_{t-12} - \delta e_{t-13}$$

where $\delta = -\theta_1\theta_2$. If you are not sure about the multiplicative structure, you can specify

```
ESTIMATE Q=(1,12,13);
```

and check to see if the third estimated moving average (MA) coefficient δ is approximately the negative of the product of the other two ($\theta_1\theta_2$). To specify the multiplicative structure, issue the PROC ARIMA statement

```
ESTIMATE Q=(1)(12);
```

After differencing, the intercept is probably 0, so you can use the NOCONSTANT option. You can fit seasonal multiplicative factors on the AR side also. For example, specifying

```
ESTIMATE P=(1,2)(12) NOCONSTANT;
```

causes the model

$$\left(1-\alpha_1 B-\alpha_2 B^2\right)\left(1-\alpha_3 B^{12}\right)V_t = e_t$$

to be fit to the data.

Consider the monthly number of U.S. masonry and electrical construction workers in thousands (U.S. Bureau of Census 1982). You issue the following SAS statements to plot the data and compute the ACF for the original series, first differenced series, and first and seasonally differenced series:

```
PROC GPLOT DATA=CONST;
   PLOT CONSTRCT*DATE/HMINOR=0 VMINOR=0;
   TITLE 'CONSTRUCTION REVIEW';
   TITLE2 'CONSTRUCTION WORKERS IN THOUSANDS';
   SYMBOL1 L=1 I=JOIN C=BLACK V=NONE;
RUN;
PROC ARIMA DATA=CONST;
   IDENTIFY VAR=CONSTRCT NLAG=36;
   IDENTIFY VAR=CONSTRCT(1) NLAG=36;
   IDENTIFY VAR=CONSTRCT(1,12) NLAG=36;
RUN;
```

The plot is shown in **Output 4.1**. The ACFs are shown in **Output 4.2**. The plot of the data displays nonstationary behavior (nonconstant mean). The original ACF ❶ shows slow decay, indicating a first differencing. The ACF of the first differenced series ❷ shows slow decay at the seasonal lags, indicating a span 12 difference. The Q statistics ❸ on the CONSTRCT(1,12) differenced variable indicate that no AR or MA terms are needed.

Output 4.1
Plotting the Original Data

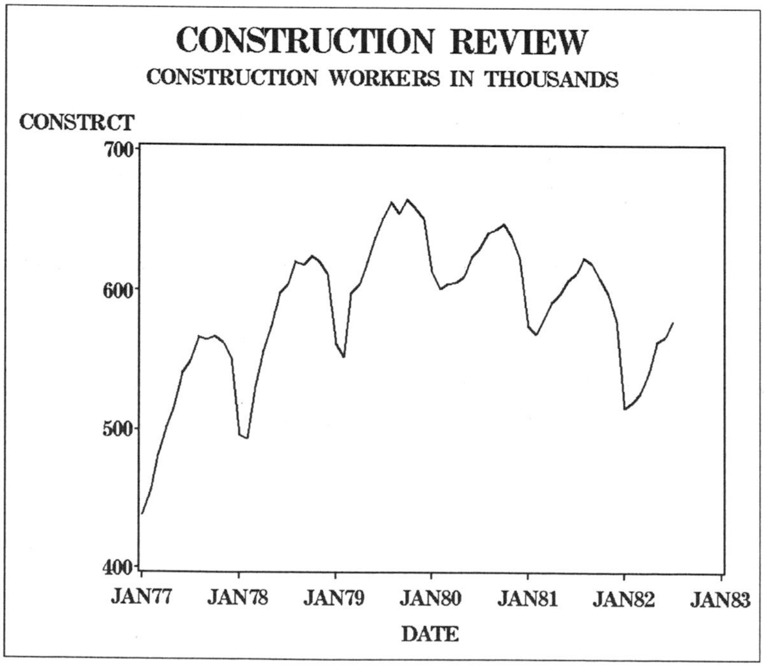

Output 4.2 *Computing the ACF with the IDENTIFY Statement: PROC ARIMA*

```
                              CONSTRUCTION REVIEW
                        CONSTRUCTION WORKERS IN THOUSANDS

                              The ARIMA Procedure


                         Name of Variable = CONSTRCT

                    Mean of Working Series     585.4149
                    Standard Deviation         50.65318
                    Number of Observations           67


                                Autocorrelations
                                                                      ❶
Lag    Covariance    Correlation    -1 9 8 7 6 5 4 3 2 1 0 1 2 3 4 5 6 7 8 9 1      Std Error

  0      2565.745        1.00000    |                    |********************|              0
  1      2213.592        0.86275    |               .    |*****************   |       0.122169
  2      1746.293        0.68062    |            .       |**************      |       0.192729
  3      1296.936        0.50548    |           .        |**********          |       0.225771
  4       907.000        0.35350    |          .         |*******  .          |       0.242074
  5       598.110        0.23311    |          .         |*****    .          |       0.249660
  6       443.389        0.17281    |          .         |***      .          |       0.252887
  7       418.306        0.16304    |          .         |***      .          |       0.254644
  8       541.150        0.21091    |          .         |****     .          |       0.256197
  9       712.443        0.27767    |          .         |******   .          |       0.258776
 10       922.830        0.35967    |         .          |*******   .         |       0.263185
 11      1126.964        0.43923    |         .          |********* .         |       0.270422
 12      1201.001        0.46809    |         .          |********* .         |       0.280868
 13       894.577        0.34866    |        .           |*******    .        |       0.292280
 14       461.050        0.18087    |        .           |****       .        |       0.298423
 15     78.980204        0.03078    |        .           |*          .        |       0.300055


                           Inverse Autocorrelations

                 Lag    Correlation    -1 9 8 7 6 5 4 3 2 1 0 1 2 3 4 5 6 7 8 9 1

                   1       -0.48370    |          **********|    .               |
                   2       -0.02286    |               .    |    .               |
                   3       -0.00329    |               .    |    .               |
                   4        0.01486    |               .    |    .               |
                   5        0.01432    |               .    |    .               |
                   6       -0.02317    |               .    |    .               |
                   7        0.04451    |               .    |*   .               |
                   8       -0.04284    |               .   *|    .               |
                   9        0.01813    |               .    |    .               |
                  10        0.00084    |               .    |    .               |
                  11        0.10220    |               .    |**  .               |
                  12       -0.23143    |               *****|    .               |
                  13        0.07845    |               .    |**  .               |
                  14        0.02590    |               .    |*   .               |
                  15        0.03720    |               .    |*   .               |
```

Output 4.2 *Computing the ACF with the IDENTIFY Statement: PROC ARIMA (continued)*

```
                                  Partial Autocorrelations

     Lag     Correlation    -1 9 8 7 6 5 4 3 2 1 0 1 2 3 4 5 6 7 8 9 1

       1         0.86275    |               .    |*****************   |
       2        -0.24922    |               *****|    .               |
       3        -0.05442    |               .   *|    .               |
       4        -0.03252    |               .   *|    .               |
       5        -0.00610    |               .    |    .               |
       6         0.11891    |               .    |**  .               |
       7         0.08462    |               .    |**  .               |
       8         0.17698    |               .    |****.               |
       9         0.06457    |               .    |*   .               |
      10         0.13938    |               .    |*** .               |
      11         0.11326    |               .    |**  .               |
      12        -0.05743    |               .   *|    .               |
      13        -0.46306    |           *********|    .               |
      14        -0.11123    |               .  **|    .               |
      15         0.03114    |               .    |*   .               |
```

Autocorrelation Check for White Noise

To Lag	Chi-Square	DF	Pr > ChiSq	--------Autocorrelations--------					
6	119.03	6	<.0001	0.863	0.681	0.505	0.354	0.233	0.173
12	175.54	12	<.0001	0.163	0.211	0.278	0.360	0.439	0.468
18	199.05	18	<.0001	0.349	0.181	0.031	-0.095	-0.196	-0.248
24	215.31	24	<.0001	-0.261	-0.227	-0.177	-0.111	-0.040	-0.004
30	296.18	30	<.0001	-0.078	-0.193	-0.284	-0.368	-0.436	-0.468
36	376.88	36	<.0001	-0.463	-0.414	-0.349	-0.259	-0.157	-0.082

Name of Variable = CONSTRCT

Period(s) of Differencing	1
Mean of Working Series	2.113636
Standard Deviation	19.56132
Number of Observations	66
Observation(s) eliminated by differencing	1

Output 4.2 *Computing the ACF with the IDENTIFY Statement: PROC ARIMA (continued)*

```
                                        Autocorrelations
                                                                            ❷
 Lag    Covariance    Correlation    -1 9 8 7 6 5 4 3 2 1 0 1 2 3 4 5 6 7 8 9 1      Std Error

   0      382.645        1.00000    |                    |********************|             0
   1      150.101        0.39227    |               .    |********            |      0.123091
   2    34.261654        0.08954    |               .    |**  .               |      0.140764
   3   -17.049788        -.04456    |               .   *|    .               |      0.141624
   4   -50.307780        -.13147    |              .  ***|     .              |      0.141836
   5     -104.359        -.27273    |              .*****|     .              |      0.143671
   6     -113.459        -.29651    |              ******|     .              |      0.151312
   7     -104.599        -.27336    |              .*****|     .              |      0.159874
   8   -53.315553        -.13933    |             .   ***|      .             |      0.166805
   9   -34.163118        -.08928    |             .    **|      .             |      0.168559
  10    -5.892301        -.01540    |             .      |      .             |      0.169274
  11      104.746        0.27374    |             .      |***** .             |      0.169296
  12      258.268        0.67495    |             .      |*************       |      0.175874
  13      114.671        0.29968    |            .       |****** .            |      0.211510
  14    31.767495        0.08302    |           .        |**      .           |      0.217849
  15    -9.281516        -.02426    |           .        |        .           |      0.218328
```

```
                                  Inverse Autocorrelations

                 Lag    Correlation    -1 9 8 7 6 5 4 3 2 1 0 1 2 3 4 5 6 7 8 9 1

                   1       -0.22458    |               .****|    .               |
                   2       -0.08489    |               .  **|    .               |
                   3       -0.02789    |               .   *|    .               |
                   4        0.08315    |               .    |**  .               |
                   5        0.02735    |               .    |*   .               |
                   6       -0.02083    |               .    |    .               |
                   7        0.07075    |               .    |*   .               |
                   8       -0.07942    |               .  **|    .               |
                   9        0.06801    |               .    |*   .               |
                  10        0.00071    |               .    |    .               |
                  11        0.07826    |               .    |**  .               |
                  12       -0.44727    |           *********|    .               |
                  13        0.06835    |               .    |*   .               |
                  14        0.08653    |               .    |**  .               |
                  15       -0.04643    |               .   *|    .               |
```

```
                                  Partial Autocorrelations

                 Lag    Correlation    -1 9 8 7 6 5 4 3 2 1 0 1 2 3 4 5 6 7 8 9 1

                   1        0.39227    |               .    |********            |
                   2       -0.07604    |               .  **|    .               |
                   3       -0.06244    |               .   *|    .               |
                   4       -0.10017    |               .  **|    .               |
                   5       -0.21696    |               .****|    .               |
                   6       -0.14352    |               . ***|    .               |
                   7       -0.15335    |               . ***|    .               |
                   8       -0.03507    |               .   *|    .               |
                   9       -0.11707    |               .  **|    .               |
                  10       -0.06943    |               .   *|    .               |
                  11        0.23978    |               .    |*****               |
                  12        0.56027    |               .    |***********         |
                  13       -0.20675    |               .****|    .               |
                  14       -0.05794    |               .   *|    .               |
                  15       -0.00776    |               .    |    .               |
```

Output 4.2 *Computing the ACF with the IDENTIFY Statement: PROC ARIMA (continued)*

Autocorrelation Check for White Noise

To Lag	Chi-Square	DF	Pr > ChiSq	---------------	--------------	Autocorrelations	--------------	---------------	---------------
6	24.63	6	0.0004	0.392	0.090	-0.045	-0.131	-0.273	-0.297
12	76.44	12	<.0001	-0.273	-0.139	-0.089	-0.015	0.274	0.675
18	93.66	18	<.0001	0.300	0.083	-0.024	-0.071	-0.202	-0.226
24	124.70	24	<.0001	-0.211	-0.124	-0.081	-0.058	0.181	0.442
30	137.06	30	<.0001	0.214	0.091	-0.022	-0.016	-0.148	-0.172
36	171.23	36	<.0001	-0.224	-0.161	-0.117	-0.125	0.140	0.341

Name of Variable = CONSTRCT

Period(s) of Differencing	1,12
Mean of Working Series	-1.70926
Standard Deviation	9.624434
Number of Observations	54
Observation(s) eliminated by differencing	13

Autocorrelations

```
Lag   Covariance   Correlation   -1 9 8 7 6 5 4 3 2 1 0 1 2 3 4 5 6 7 8 9 1   Std Error

  0    92.629729     1.00000     |                    |********************|          0
  1    -4.306435     -.04649     |               .   *|    .               |   0.136083
  2     1.797140     0.01940     |               .    |    .               |   0.136377
  3    -0.505176     -.00545     |               .    |    .               |   0.136428
  4   -12.210895     -.13182     |               . ***|    .               |   0.136432
  5    -4.617964     -.04985     |              .    *|     .              |   0.138770
  6    -9.764872     -.10542     |              .   **|     .              |   0.139102
  7     1.989406     0.02148     |              .     |     .              |   0.140573
  8    -6.583051     -.07107     |              .    *|     .              |   0.140634
  9     1.441982     0.01557     |              .     |     .              |   0.141298
 10     5.785042     0.06245     |              .     |*    .              |   0.141329
 11     1.738408     0.01877     |              .     |     .              |   0.141840
 12   -10.793891     -.11653     |              .   **|     .              |   0.141886
 13    -7.469517     -.08064     |              .   **|     .              |   0.143647
 14     7.038939     0.07599     |              .     |**   .              |   0.144483
 15     7.022723     0.07582     |              .     |**   .              |   0.145221
```

Output 4.2 *Computing the ACF with the IDENTIFY Statement: PROC ARIMA (continued)*

```
                            Inverse Autocorrelations

Lag    Correlation    -1 9 8 7 6 5 4 3 2 1 0 1 2 3 4 5 6 7 8 9 1

  1        0.13117    |              .    |*** .              |
  2       -0.09234    |              .  **|    .              |
  3       -0.03146    |              .   *|    .              |
  4        0.11189    |              .    |**  .              |
  5        0.11239    |              .    |**  .              |
  6        0.12587    |              .    |*** .              |
  7        0.05514    |              .    |*   .              |
  8        0.07675    |              .    |**  .              |
  9       -0.03391    |              .   *|    .              |
 10       -0.04485    |              .   *|    .              |
 11        0.09691    |              .    |**  .              |
 12        0.22039    |              .    |****.              |
 13        0.08367    |              .    |**  .              |
 14       -0.09861    |              .  **|    .              |
 15       -0.07616    |              .  **|    .              |

                            Partial Autocorrelations

Lag    Correlation    -1 9 8 7 6 5 4 3 2 1 0 1 2 3 4 5 6 7 8 9 1

  1       -0.04649    |              .   *|    .              |
  2        0.01728    |              .    |    .              |
  3       -0.00377    |              .    |    .              |
  4       -0.13291    |              . ***|    .              |
  5       -0.06308    |              .   *|    .              |
  6       -0.10861    |              .  **|    .              |
  7        0.00986    |              .    |    .              |
  8       -0.08828    |              .  **|    .              |
  9       -0.01180    |              .    |    .              |
 10        0.03282    |              .    |*   .              |
 11        0.01461    |              .    |    .              |
 12       -0.15265    |              . ***|    .              |
 13       -0.10793    |              .  **|    .              |
 14        0.06736    |              .    |*   .              |
 15        0.10476    |              .    |**  .              |

                     Autocorrelation Check for White Noise

            ❸
 To        Chi-             Pr >
Lag      Square     DF     ChiSq    --------------------Autocorrelations--------------------

  6        2.05      6    0.9149    -0.046     0.019    -0.005    -0.132    -0.050    -0.105
 12        3.70     12    0.9883     0.021    -0.071     0.016     0.062     0.019    -0.117
 18        5.97     18    0.9963    -0.081     0.076     0.076    -0.037    -0.090     0.040
 24       19.59     24    0.7196     0.104     0.084    -0.011     0.045    -0.249    -0.240
 30       26.63     30    0.6424     0.135    -0.182     0.051     0.053     0.055     0.066
 36       28.38     36    0.8134    -0.017    -0.027     0.021    -0.038     0.084    -0.035
```

To forecast the seasonal data, use the following statements:

```
PROC ARIMA DATA=CONST;
   IDENTIFY VAR=CONSTRCT(1,12) NOPRINT;
   ESTIMATE NOCONSTANT METHOD=ML;
   FORECAST LEAD=12 INTERVAL=MONTH ID=DATE OUT=OUTF;
RUN;
```

The results are shown in **Output 4.3**.

The model

$$(1-B)(1-B^{12})Y_t = (1-\theta_1 B)(1-\theta_1 B^{12})e_t$$

is known as the airline model. Its popularity started when Box and Jenkins (1976) used it to model sales of international airline tickets on a logarithmic scale. **Output 4.4** shows plots of the original and log scale data from Box and Jenkins's text.

Output 4.3 Forecasting Seasonal Data with the IDENTIFY, ESTIMATE, and FORECAST Statements: PROC ARIMA

```
                              CONSTRUCTION REVIEW
                        CONSTRUCTION WORKERS IN THOUSANDS

                              The ARIMA Procedure

                         Variance Estimate        95.5513
                         Std Error Estimate      9.775034
                         AIC                     399.4672
                         SBC                     399.4672
                         Number of Residuals           54

                       Autocorrelation Check of Residuals

 To       Chi-             Pr >
 Lag     Square    DF     ChiSq     -------------------Autocorrelations------------------

   6       0.93     6    0.9880    -0.019     0.048     0.031    -0.089    -0.011    -0.061
  12       2.62    12    0.9977     0.057    -0.033     0.052     0.093     0.050    -0.079
  18       5.22    18    0.9985    -0.046     0.108     0.102    -0.003    -0.054     0.078
  24      16.49    24    0.8695     0.135     0.112     0.020     0.072    -0.209    -0.195

                           Model for Variable CONSTRCT

                        Period(s) of Differencing    1,12

                           No mean term in this model.

                         Forecasts for Variable CONSTRCT

              Obs       Forecast     Std Error        95% Confidence Limits

               68       588.9000        9.7750        569.7413       608.0587
               69       585.0000       13.8240        557.9055       612.0945
               70       574.6000       16.9309        541.4161       607.7839

             [more output lines]

               77       529.5000       30.9114        468.9148       590.0852
               78       532.8000       32.4201        469.2577       596.3423
               79       543.6000       33.8617        477.2323       609.9677
```

Output 4.4
Plotting the Original and Log Transformed Box and Jenkins Airline Data

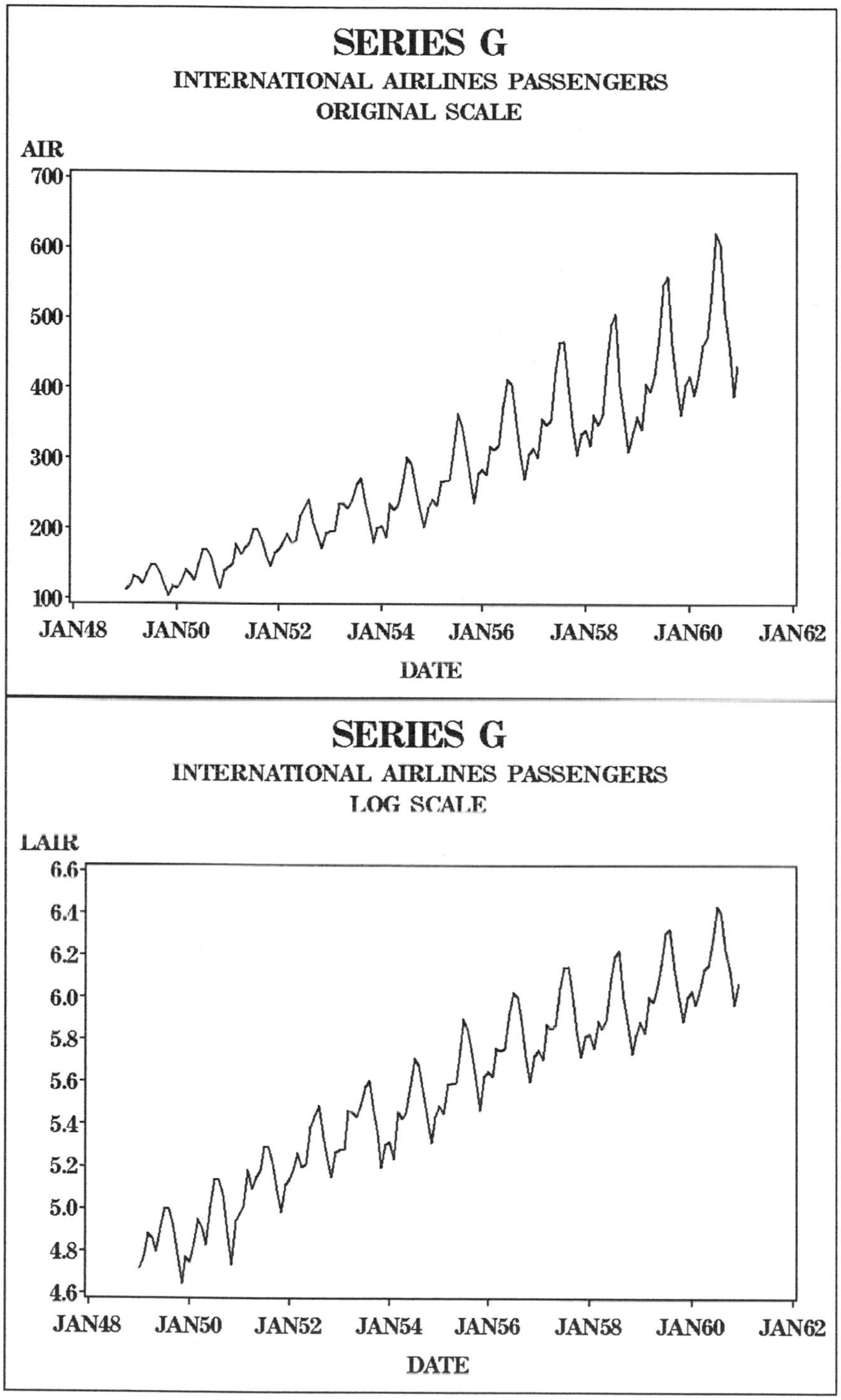

Now analyze the logarithms, which have the more stable seasonal pattern, using these SAS statements:

```
PROC ARIMA DATA=AIRLINE;
   IDENTIFY VAR=LAIR;
   IDENTIFY VAR=LAIR(1);
   TITLE 'SERIES G';
   TITLE2 'INTERNATIONAL AIRLINES PASSENGERS';
RUN;
```

The results are shown in **Output 4.5**. It is hard to detect seasonality in the ACF of the original series ❶ because all the values are so near 1. The slow decay is much more evident here than in the construction example. Once you take the first difference, you obtain the ACF ❷. Looking at the seasonal lags (12,24), you see little decay, indicating you should consider a span 12 difference. To create the variable

$$V_t = (Y_t - Y_{t-1}) - (Y_{t-12} - Y_{t-13})$$

and its ACF, inverse autocorrelation function (IACF), and partial autocorrelation function (PACF), issue the following SAS statements:

```
PROC ARIMA DATA=AIRLINE;
   IDENTIFY VAR=LAIR(1,12);
```

The model is identified from the autocorrelations. Identification depends on pattern recognition in the plot of the ACF values against the lags. The nonzero ACF values are called spikes to draw to mind the plots PROC ARIMA produces in the IDENTIFY stage. For the airline model, if $\theta_1 > 0$ and $\theta_2 > 0$, the theoretical autocorrelations of the series

$$V_t = (1 - \theta_1 B)(1 - \theta_2 B^{12}) e_t$$

should have

- a (negative) spike at lag 1
- a (negative) spike at lag 12
- equal (and positive) spikes at lags 11 and 13 called side lobes of the lag 12 spike
- all other lag correlations 0.

Output 4.5 *Identifying the Logarithms with the IDENTIFY Statement: PROC ARIMA*

```
                                   SERIES G
                        INTERNATIONAL AIRLINES PASSENGERS

                             The ARIMA Procedure

                          Name of Variable = LAIR

                     Mean of Working Series     5.542176
                     Standard Deviation         0.439921
                     Number of Observations          144

                               Autocorrelations
                                                                 ❶
Lag   Covariance   Correlation   -1 9 8 7 6 5 4 3 2 1 0 1 2 3 4 5 6 7 8 9 1    Std Error

  0     0.193530      1.00000    |                    |********************|          0
  1     0.184571      0.95370    |                  . |*******************   |   0.083333
  2     0.173968      0.89892    |                .   |******************    |   0.139918
  3     0.164656      0.85080    |               .    |*****************     |   0.175499
  4     0.156455      0.80843    |              .     |****************      |   0.202123
  5     0.150741      0.77890    |             .      |****************      |   0.223452
  6     0.146395      0.75644    |            .       |***************       |   0.241572
  7     0.142748      0.73760    |            .       |***************       |   0.257496
  8     0.140722      0.72713    |           .        |***************       |   0.271773
  9     0.141983      0.73365    |           .        |***************       |   0.284963
 10     0.144036      0.74426    |          .         |***************       |   0.297791
 11     0.146701      0.75803    |          .         |***************       |   0.310440
 12     0.147459      0.76194    |         .          |***************       |   0.323038
 13     0.138665      0.71650    |         .          |**************        |   0.335286
 14     0.128319      0.66304    |        .           |*************.        |   0.345756
 15     0.119672      0.61836    |        .           |************ .        |   0.354475


                          Inverse Autocorrelations

             Lag    Correlation    -1 9 8 7 6 5 4 3 2 1 0 1 2 3 4 5 6 7 8 9 1

               1       -0.50387    |          **********|  .                 |
               2       -0.00329    |                  . |  .                 |
               3        0.02976    |                  . |* .                 |
               4       -0.01239    |                  . |  .                 |
               5       -0.02237    |                  . |  .                 |
               6        0.01070    |                  . |  .                 |
               7       -0.00970    |                  . |  .                 |
               8        0.01922    |                  . |  .                 |
               9        0.01018    |                  . |  .                 |
              10       -0.00820    |                  . |  .                 |
              11        0.15344    |                  . |***                 |
              12       -0.34183    |             *******|  .                 |
              13        0.14016    |                  . |***                 |
              14        0.07699    |                  . |**.                 |
              15       -0.06626    |                  .*|  .                 |
```

Output 4.5 *Identifying the Logarithms with the IDENTIFY Statement: PROC ARIMA (continued)*

```
                            Partial Autocorrelations
          Lag     Correlation     -1 9 8 7 6 5 4 3 2 1 0 1 2 3 4 5 6 7 8 9 1

            1        0.95370      |                    .  |*******************  |
            2       -0.11757      |                   .**|  .                   |
            3        0.05423      |                    .  |* .                  |
            4        0.02376      |                    .  |  .                  |
            5        0.11582      |                    .  |**.                  |
            6        0.04437      |                    .  |* .                  |
            7        0.03803      |                    .  |* .                  |
            8        0.09962      |                    .  |**.                  |
            9        0.20410      |                    .  |****                 |
           10        0.06391      |                    .  |* .                  |
           11        0.10604      |                    .  |**.                  |
           12       -0.04247      |                    . *|  .                  |
           13       -0.48543      |          **********|  .                    |
           14       -0.03435      |                    . *|  .                  |
           15        0.04222      |                    .  |* .                  |

                        Autocorrelation Check for White Noise

 To        Chi-             Pr >
Lag      Square     DF     ChiSq     --------------------Autocorrelations-------------------

  6      638.37      6    <.0001      0.954      0.899      0.851      0.808      0.779      0.756
 12     1157.62     12    <.0001      0.738      0.727      0.734      0.744      0.758      0.762
 18     1521.94     18    <.0001      0.717      0.663      0.618      0.576      0.544      0.519
 24     1785.32     24    <.0001      0.501      0.490      0.498      0.506      0.517      0.520


                              Name of Variable = LAIR

                  Period(s) of Differencing                           1
                  Mean of Working Series                        0.00944
                  Standard Deviation                           0.106183
                  Number of Observations                            143
                  Observation(s) eliminated by differencing           1

                                  Autocorrelations
                                                                          ❷
Lag    Covariance     Correlation    -1 9 8 7 6 5 4 3 2 1 0 1 2 3 4 5 6 7 8 9 1        Std Error

  0      0.011275        1.00000     |                    |********************|               0
  1     0.0022522        0.19975     |                  . |****                |        0.083624
  2    -0.0013542        -.12010     |                 .**|  .                 |        0.086897
  3    -0.0016999        -.15077     |                .***|  .                 |        0.088050
  4    -0.0036313        -.32207     |              ******|  .                 |        0.089837
  5    -0.0009468        -.08397     |                . **|  .                 |        0.097578
  6    0.00029065        0.02578     |                .   |* .                 |        0.098082
  7    -0.0012511        -.11096     |                . **|  .                 |        0.098130
  8    -0.0037965        -.33672     |             *******|  .                 |        0.099003
  9    -0.0013032        -.11559     |                . **|  .                 |        0.106712
 10    -0.0012320        -.10927     |                . **|  .                 |        0.107584
 11     0.0023209        0.20585     |                .   |****               |        0.108357
 12     0.0094870        0.84143     |                .   |*****************   |        0.111058
 13     0.0024251        0.21509     |              .     |****  .             |        0.149118
 14    -0.0015734        -.13955     |              .   ***|     .            |        0.151272
 15    -0.0013078        -.11600     |              .    **|     .            |        0.152169
```

Output 4.5 *Identifying the Logarithms with the IDENTIFY Statement: PROC ARIMA (continued)*

```
                          Inverse Autocorrelations

 Lag    Correlation    -1 9 8 7 6 5 4 3 2 1 0 1 2 3 4 5 6 7 8 9 1

   1        0.17650    |                    .  |****                 |
   2        0.09526    |                    .  |**.                  |
   3        0.34062    |                    .  |*******              |
   4        0.28364    |                    .  |******               |
   5        0.05975    |                    .  |* .                  |
   6        0.13322    |                    .  |***                  |
   7        0.24104    |                    .  |*****                |
   8        0.11930    |                    .  |**.                  |
   9       -0.04769    |                    . *|  .                  |
  10        0.19042    |                    .  |****                 |
  11        0.10362    |                    .  |**.                  |
  12       -0.27362    |                 *****|  .                  |
  13        0.02062    |                    .  |  .                  |
  14        0.15054    |                    .  |***                  |
  15       -0.03827    |                    . *|  .                  |


                          Partial Autocorrelations

 Lag    Correlation    -1 9 8 7 6 5 4 3 2 1 0 1 2 3 4 5 6 7 8 9 1

   1        0.19975    |                    .  |****                 |
   2       -0.16665    |                    ***|  .                  |
   3       -0.09588    |                    .**|  .                  |
   4       -0.31089    |                 ******|  .                  |
   5        0.00778    |                    .  |  .                  |
   6       -0.07455    |                    . *|  .                  |
   7       -0.21028    |                   ****|  .                  |
   8       -0.49476    |             **********|  .                  |
   9       -0.19229    |                   ****|  .                  |
  10       -0.53188    |            ***********|  .                  |
  11       -0.30229    |                 ******|  .                  |
  12        0.58604    |                    .  |************         |
  13        0.02598    |                    .  |* .                  |
  14       -0.18119    |                   ****|  .                  |
  15        0.12004    |                    .  |**.                  |
```

Output 4.5 *Identifying the Logarithms with the IDENTIFY Statement: PROC ARIMA (continued)*

```
                          Autocorrelation Check for White Noise

 To        Chi-            Pr >
Lag       Square    DF     ChiSq     --------------------Autocorrelations--------------------

  6        27.95     6    <.0001      0.200    -0.120    -0.151    -0.322    -0.084     0.026
 12       169.89    12    <.0001     -0.111    -0.337    -0.116    -0.109     0.206     0.841
 18       195.75    18    <.0001      0.215    -0.140    -0.116    -0.279    -0.052     0.012
 24       321.53    24    <.0001     -0.114    -0.337    -0.107    -0.075     0.199     0.737

                                        SERIES G
                               INTERNATIONAL AIRLINES PASSENGERS

                                   The ARIMA Procedure

                                 Name of Variable = LAIR

                      Period(s) of Differencing                     1,12
                      Mean of Working Series                    0.000291
                      Standard Deviation                        0.045673
                      Number of Observations                         131
                      Observation(s) eliminated by differencing       13


                                     Autocorrelations

Lag    Covariance    Correlation    -1 9 8 7 6 5 4 3 2 1 0 1 2 3 4 5 6 7 8 9 1      Std Error

  0     0.0020860       1.00000    |                    |********************|              0
  1    -0.0007116       -.34112    |             *******|  .                 |       0.087370
  2    0.00021913       0.10505    |                .   |** .                |       0.097006
  3    -0.0004217       -.20214    |                ****|   .                |       0.097870
  4    0.00004456       0.02136    |                .   |   .                |       0.101007
  5    0.00011610       0.05565    |                .   |*  .                |       0.101042
  6    0.00006426       0.03080    |                .   |*  .                |       0.101275
  7    -0.0001159       -.05558    |                .  *|   .                |       0.101347
  8    -1.5867E-6       -.00076    |                .   |   .                |       0.101579
  9    0.00036791       0.17637    |                .   |****                |       0.101579
 10    -0.0001593       -.07636    |                . **|   .                |       0.103891
 11    0.00013431       0.06438    |                .   |*  .                |       0.104318
 12    -0.0008065       -.38661    |            ********|   .                |       0.104621
 13    0.00031624       0.15160    |               .    |*** .               |       0.115011
 14    -0.0001202       -.05761    |               .   *|    .               |       0.116526
 15    0.00031200       0.14957    |               .    |*** .               |       0.116744
```

Output 4.5 *Identifying the Logarithms with the IDENTIFY Statement: PROC ARIMA (continued)*

```
                             Inverse Autocorrelations

        Lag    Correlation    -1 9 8 7 6 5 4 3 2 1 0 1 2 3 4 5 6 7 8 9 1

          1        0.32632    |                    .  |*******           |
          2        0.09594    |                    .  |**.               |
          3        0.09992    |                    .  |**.               |
          4        0.10889    |                    .  |**.               |
          5       -0.12127    |                    .**|  .               |
          6       -0.16601    |                    ***|  .               |
          7       -0.05979    |                    . *|  .               |
          8        0.02949    |                    .  |* .               |
          9       -0.08480    |                    .**|  .               |
         10        0.01413    |                    .  |  .               |
         11        0.10508    |                    .  |**.               |
         12        0.37985    |                    .  |********          |
         13        0.12446    |                    .  |**.               |
         14        0.05655    |                    .  |* .               |
         15        0.05144    |                    .  |* .               |

                             Partial Autocorrelations

        Lag    Correlation    -1 9 8 7 6 5 4 3 2 1 0 1 2 3 4 5 6 7 8 9 1

          1       -0.34112    |               *******|  .               |
          2       -0.01281    |                    .  |  .               |
          3       -0.19266    |                  ****|  .               |
          4       -0.12503    |                   ***|  .               |
          5        0.03309    |                    .  |* .               |
          6        0.03468    |                    .  |* .               |
          7       -0.06019    |                    . *|  .               |
          8       -0.02022    |                    .  |  .               |
          9        0.22550    |                    .  |*****             |
         10        0.04307    |                    .  |* .               |
         11        0.04659    |                    .  |* .               |
         12       -0.33869    |               *******|  .               |
         13       -0.10918    |                    .**|  .               |
         14       -0.07684    |                    .**|  .               |
         15       -0.02175    |                    .  |  .               |

                        Autocorrelation Check for White Noise

 To       Chi-             Pr >
Lag     Square     DF     ChiSq    --------------------Autocorrelations--------------------

  6      23.27      6    0.0007    -0.341     0.105    -0.202     0.021     0.056     0.031
 12      51.47     12    <.0001    -0.056    -0.001     0.176    -0.076     0.064    -0.387
 18      62.44     18    <.0001     0.152    -0.058     0.150    -0.139     0.070     0.016
 24      74.27     24    <.0001    -0.011    -0.117     0.039    -0.091     0.223    -0.018
```

The pattern that follows represents the ACF of

$$V_t = (1-\theta_1 B)(1-\theta_2 B^{12})e_t$$

or the IACF of

$$(1-\theta_1 B)(1-\theta_2 B^{12})V_t = e_t$$

```
    |
1   *
    *
    *
    *                     *   *
    *                     *   *
    +-1-2-3-4-5-6-7-8-9-0-1-2-3-4-5-6-7-8-9-0-1-2-3-4-5-6---> Lag
      *                     *
      *                     *
      *                     *
      *
```

When you compare this pattern to the ACF of the LAIR(1,12) variable, you find reasonable agreement. If the signs of the parameters are changed, the spikes and side lobes have different signs but remain at the same lags. The spike and side lobes at the seasonal lag are characteristic of seasonal multiplicative models. Note that if the multiplicative factor is on the AR side, this pattern appears in the IACF instead of in the ACF. In that case, the IACF and PACF behave differently and the IACF is easier to interpret.

If the model is changed to

$$V_t - \alpha V_{t-12} = (1 - \theta_1 B)(1 - \theta_2 B^{12}) e_t$$

the spike and side lobes are visible at the seasonal lag (for example, 12) and its multiples (24, 36, and so on), but the magnitudes of the spikes at the multiples decrease exponentially at rate α. If the decay is extremely slow, an additional seasonal difference is needed $(\alpha = 1)$. If the pattern appears in the IACF, the following model is indicated:

$$(1 - \theta_1 B)(1 - \theta_2 B^{12}) V_t = (1 - \alpha B^{12}) e_t$$

The SAS code for the airline data is

```
PROC ARIMA DATA=AIRLINE;
   IDENTIFY VAR=LAIR(1,12) NOPRINT;
   ESTIMATE Q=(1)(12) NOCONSTANT;
   FORECAST LEAD=12 OUT=FORE ID=DATE INTERVAL=MONTH;
RUN;
PROC GPLOT DATA=FORE(FIRSTOBS=120);
   PLOT (LAIR FORECAST L95 U95)*DATE / OVERLAY HMINOR=0;
   SYMBOL1 V=A L=1 I=JOIN C=BLACK;
   SYMBOL2 V=F L=2 I=JOIN C=BLACK;
   SYMBOL3 V=L C=BLACK I=NONE;
   SYMBOL4 V=U C=BLACK I=NONE;
RUN;
DATA FORE;
   SET FORE;
   IF RESIDUAL NE .;
RUN;
PROC SPECTRA P WHITETEST DATA=FORE OUT=RESID;
VAR RESIDUAL;
RUN;
PROC GPLOT DATA=RESID;
PLOT P_01*FREQ/HMINOR=0;
SYMBOL1 F=TRIPLEX V=* I=JOIN C=BLACK;
RUN;
```

The results are shown in **Output 4.6** and **Output 4.7**.

Output 4.6 *Fitting the Airline Model: PROC ARIMA*

SERIES G
INTERNATIONAL AIRLINES PASSENGERS

The ARIMA Procedure

Conditional Least Squares Estimation

Parameter	Estimate	Standard Error	t Value	Approx Pr > |t|	Lag
MA1,1	0.37727	0.08196	4.60	<.0001	1
MA2,1	0.57236	0.07802	7.34	<.0001	12

Variance Estimate	0.00141
Std Error Estimate	0.037554
AIC	-486.133
SBC	-480.383
Number of Residuals	131

* AIC and SBC do not include log determinant.

Correlations of Parameter Estimates

Parameter	MA1,1	MA2,1
MA1,1	1.000	-0.091
MA2,1	-0.091	1.000

Autocorrelation Check of Residuals

To Lag	Chi-Square	DF	Pr > ChiSq	Autocorrelations					
6	5.15	4	0.2723	0.010	0.028	-0.119	-0.100	0.081	0.077
12	7.89	10	0.6400	-0.049	-0.023	0.114	-0.045	0.025	-0.023
18	11.98	16	0.7452	0.012	0.036	0.064	-0.136	0.055	0.011
24	22.56	22	0.4272	-0.098	-0.096	-0.031	-0.021	0.214	0.013

Model for Variable LAIR

Period(s) of Differencing 1,12

No mean term in this model.

Moving Average Factors

Factor 1: 1 - 0.37727 B**(1)
Factor 2: 1 - 0.57236 B**(12)

Output 4.6 *Fitting the Airline Model: PROC ARIMA (continued)*

```
                    SERIES G
        INTERNATIONAL AIRLINES PASSENGERS

               The ARIMA Procedure

          Forecasts for Variable LAIR

Obs      Forecast   Std Error       95% Confidence Limits

145        6.1095      0.0376       6.0359         6.1831
146        6.0536      0.0442       5.9669         6.1404
147        6.1728      0.0500       6.0747         6.2709

[more output lines]

154        6.2081      0.0796       6.0521         6.3641
155        6.0631      0.0829       5.9005         6.2256
156        6.1678      0.0862       5.9989         6.3367

                    SERIES G
        INTERNATIONAL AIRLINES PASSENGERS

              The SPECTRA Procedure

     Test for White Noise for Variable RESIDUAL

               M          =          65
               Max(P(*))         0.0102
               Sum(P(*))       0.181402


      Fisher's Kappa: M*MAX(P(*))/SUM(P(*))

                Kappa     3.655039


     Bartlett's Kolmogorov-Smirnov Statistic:
   Maximum absolute difference of the standardized
   partial sums of the periodogram and the CDF of a
           uniform(0,1) random variable.

      Test Statistic                          0.089019
      Approximate P-Value                       0.6816
```

Output 4.7
Plotting the Forecasts and the Periodogram: PROC ARIMA and PROC SPECTRA

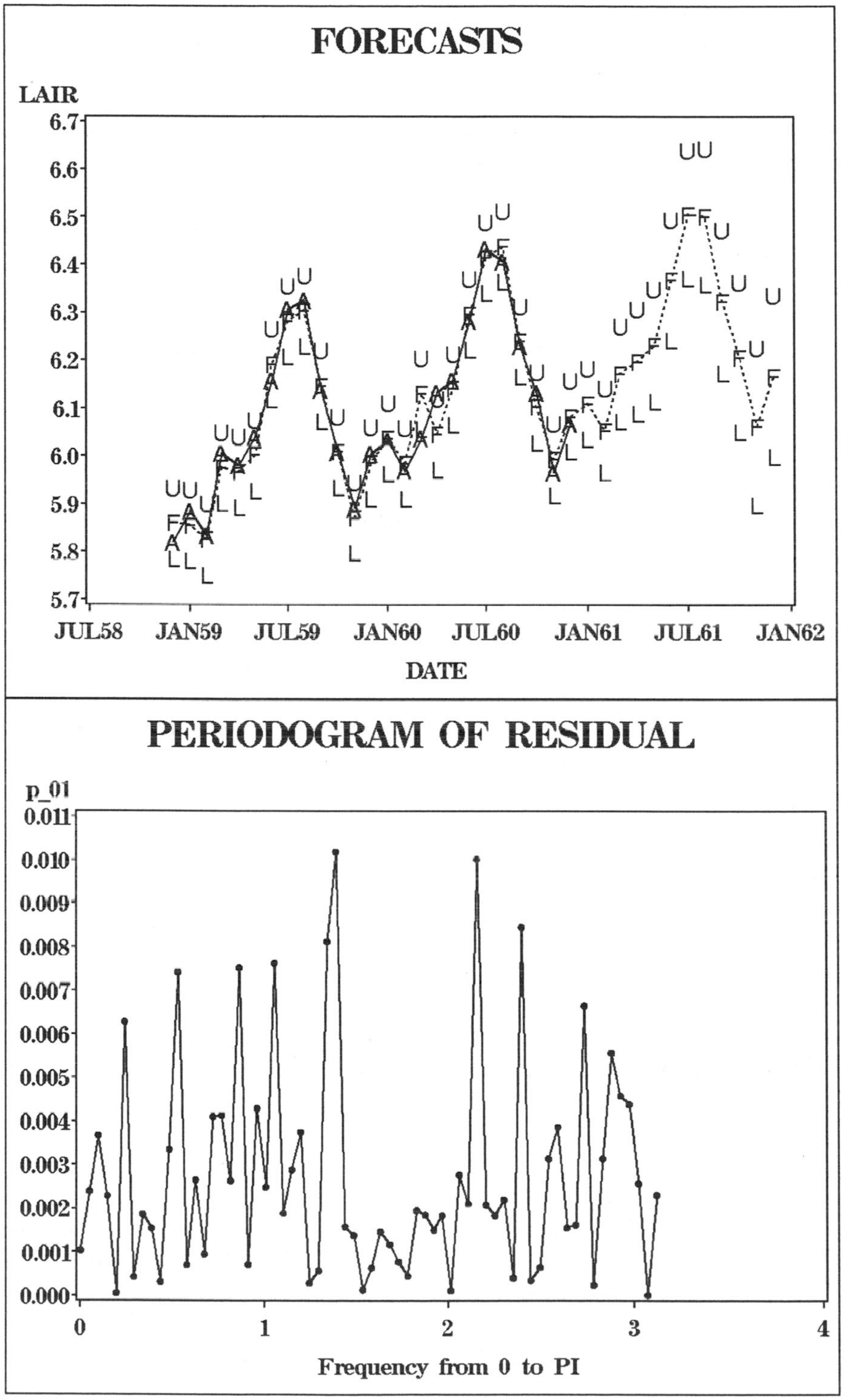

PROC SPECTRA is also used to search for hidden periodicities in the airline residuals. No periodicities are indicated in the periodogram plot or in the white noise tests produced by PROC SPECTRA. Refer to Chapter 7, "Spectral Analysis," for more information on PROC SPECTRA.

4.2 Models with Explanatory Variables

Sometimes you can improve forecasts by relating the series of interest to other explanatory variables. Obviously, forecasting in such situations requires knowledge (or at least forecasts) of future values of such variables. The nature of the explanatory variables and of the model relating them to the target series determines the optimal forecasting method. Explanatory variables are addressed in Chapter 2, "Simple Models: Autoregression." There, they are deterministic, meaning that their future values are determined without error. Seasonal indicator variables and time t are deterministic. Explanatory variables like interest rates and unemployment are not deterministic because their future values are unknown.

Chapter 2 assumes that the relationship between the target series Y_t and the explanatory series X_{1t}, X_{2t}, . . . , X_{kt} satisfies the usual regression model assumptions

$$Y_t = \beta_0 + \beta_1 X_{1t} + . . . + \beta_k X_{kt} + e_t$$

where e_t is white noise. The Durbin-Watson statistic is used in Chapter 2 to detect departures from the assumptions on e_t. The following methods are appropriate when the Durbin-Watson statistic from PROC REG or PROC GLM shows significant autocorrelation. Recall that if the regression analysis from PROC REG or PROC GLM shows no autocorrelation and if known future values (as opposed to forecasts) are available for all Xs, you can forecast with appropriate prediction intervals by

- supplying future Xs and missing values (.) for future Ys
- regressing Y on the Xs with the CLI option in the MODEL statement or the keywords U95=, L95= in the OUTPUT statement.

This chapter combines regression with time series errors to provide a richer class of forecasting models. Three cases are delineated below, presented in order of increasing complexity. Examples are included, and special cases are highlighted.

4.2.1 Case 1: Regression with Time Series Errors

The model is

$$Y_t = \beta_0 + \beta_1 X_{1t} + \beta_2 X_{2t} + ... + \beta_k X_{kt} + Z_t$$

where Z_t is an ARIMA time series. This is a typical regression except that you allow for autocorrelation in the error term Z. The Y series does not depend on lagged values of the Xs. If the error series is purely autoregressive of order p, the SAS code

```
PROC AUTOREG DATA=EXAMP;
   MODEL Y=X1 X2 X3 / NLAG=P;
RUN;
```

properly fits a model to k=3 explanatory variables. Because PROC ARIMA can do this and can also accommodate mixed models and differencing, it is used instead of PROC AUTOREG in the analyses below.

In case 1, forecasts of Y and forecast intervals are produced whenever future values of the Xs are supplied. If these future Xs are user-supplied forecasts, the procedure cannot incorporate the uncertainty of these future Xs into the intervals around the forecasts of Y. Thus, the Y forecast intervals are too narrow. Valid intervals are produced when you supply future values of deterministic Xs or when PROC ARIMA forecasts the Xs in a transfer function setting as in cases 2 and 3.

4.2.2 Case 1A: Intervention

If one of the X variables is an indicator variable (each value 1 or 0), the modeling above is called intervention analysis. The reason for this term is that X usually changes from 0 to 1 during periods of expected change in the level of Y, such as strikes, power outages, and war. For example, suppose Y is the daily death rate from automobile accidents in the United States. Suppose that on day 50 the speed limit is reduced from 65 mph to 55 mph. Suppose you have another 100 days of data after this intervention. In that case, designate X_t as 0 before day 50 and as 1 on and following day 50. The model

$$Y_t = \beta_0 + \beta_1 X_t + Z_t$$

explains Y in terms of two means (plus the error term). Before day 50 the mean is

$$\beta_0 + (\beta_1)(0) = \beta_0$$

and on and following day 50 the mean is $\beta_0 + \beta_1$. Thus, β_1 is the effect of a lower speed limit, and its statistical significance can be judged based on the t test for $H_0 : \beta_1 = 0$.

If the model is fit by ordinary regression but the Zs are autocorrelated, this t test is not valid. Using PROC ARIMA to fit the model allows a valid test; supplying future values of the deterministic X produces forecasts with valid forecast intervals.

The 1s and 0s can occur in any meaningful place in X. For example, if the speed limit reverts to 65 mph on day 70, you set X back to 0 starting on day 70.

If a data point is considered an outlier, you can use an indicator variable that is 1 only for that data point in order to eliminate its influence on the ARMA parameter estimates. Deleting the point results in a missing value (.) in the series; closing the gap with a DELETE statement makes the lags across the gap incorrect. You can avoid these problems with the indicator variable approach. PROC ARIMA also provides an outlier detection routine.

4.2.3 Case 2: Simple Transfer Function

In this case, the model is

$$Y_t = \beta_0 + \beta_1 X_t + Z_t$$

where X_t and Z_t are independent ARIMA processes. Because X is an ARIMA process, you can estimate a model for X in PROC ARIMA and use it to forecast future Xs. The algorithm allows you to compute forecast error variances for these future Xs, which are automatically incorporated later into the Y forecast intervals. First, however, you must identify a model and fit it to the Z series. You accomplish this by studying the ACF, IACF, and PACF of residuals from a regression of Y on X. In fact, you can accomplish this entire procedure within PROC ARIMA. Once you have identified and fit models for X and Z, you can produce forecasts and associated intervals easily.

You can use several explanatory variables, but for proper forecasting they should be independent of one another. If the explanatory variables contain arbitrary correlations, use the STATESPACE procedure, which takes advantage of these correlations to produce forecast intervals.

4.2.4 Case 3: General Transfer Function

In case 3, you allow the target series Y_t to depend on current and past values of the explanatory variable X. The model is

$$Y_t = \alpha + \Sigma_{j=0}^{\infty} \beta_j X_{t-j} + Z_t$$

where X and Z are independent ARIMA time series. Because it is impossible to fit an infinite number of unrestricted βs to a finite data set, you restrict the βs to have certain functional forms depending on only a few parameters. The appropriate form for a given data set is determined by an identification process for the βs that is very similar to the usual identification process with the ACFs. Instead of inspecting autocorrelations, you inspect cross-correlations; but you are looking for the same patterns as in univariate ARIMA modeling. The βs are called transfer function weights or impulse-response weights.

You can use several explanatory Xs, but they should be independent of one another for proper forecasting and identification of the βs. Even if you can identify the model properly, correlation among explanatory variables causes incorrect forecast intervals because the procedure assumes independence when it computes forecast error variances.

Because you need forecasts of explanatory variables to forecast the target series, it is crucial that X does not depend on past values of Y. Such a dependency is called feedback. Feedback puts you in a circular situation where you need forecasts of X to forecast Y and forecasts of Y to forecast X. You can use PROC STATESPACE to model a series with arbitrary forms of feedback and cross-correlated inputs. Strictly AR models, including feedback, can be fit by multiple regression as proved by Fuller (1996). A general approach to AR modeling by nonlinear regression is also given by Fuller (1986).

4.2.5 Case 3A: Leading Indicators

Suppose in the model above you find that

$$\beta_0 = \beta_1 = 0, \ \beta_2 \neq 0$$

Then Y responds two periods later to movements in X. X is called a leading indicator for Y because its movements allow you to predict movements in Y two periods ahead. The lead of two periods is also called a shift or a pure delay in the response of Y to X. Such models are highly desirable for forecasting.

4.2.6 Case 3B: Intervention

You can use an indicator variable as input in case 3B, as was suggested in case 1A. However, you identify the pattern of the βs differently than in case 3. In case 3 cross-correlations are the key to identifying the β pattern, but in case 3B cross-correlations are virtually useless.

4.3 Methodology and Example

4.3.1 Case 1: Regression with Time Series Errors

In this example, a manufacturer of building supplies monitors sales (S) for one of his product lines in terms of disposable income (D), U.S. housing starts (H), and mortgage rates (M). The data are obtained quarterly. Plots of the four series are given in **Output 4.8.**

The first task is to determine the differencing desired. Each series has a fairly slowly decaying ACF, and you decide to use a differenced series. Each first differenced series has an ACF consistent with the assumption of stationarity. The D series has differences that display a slight, upward trend. This trend is not of concern unless you plan to model D. Currently, you are using it just as an explanatory variable. The fact that you differenced all the series (including sales) implies an assumption about the error term. Your model in the original levels of the variable is

$$S_t = \beta_0 + \beta_1 D_t + \beta_2 H_t + \beta_3 M_t + \eta_t$$

When you lag by 1, you get

$$S_{t-1} = \beta_0 + \beta_1 D_{t-1} + \beta_2 H_{t-1} + \beta_3 M_{t-1} + \eta_{t-1}$$

When you subtract, you get

$$\nabla S_t = 0 + \beta_1 \nabla D_t + \beta_2 \nabla H_t + \beta_3 \nabla M_t + \nabla \eta_t$$

Output 4.8
Plotting Building- and Manufacturing-Related Quarterly Data

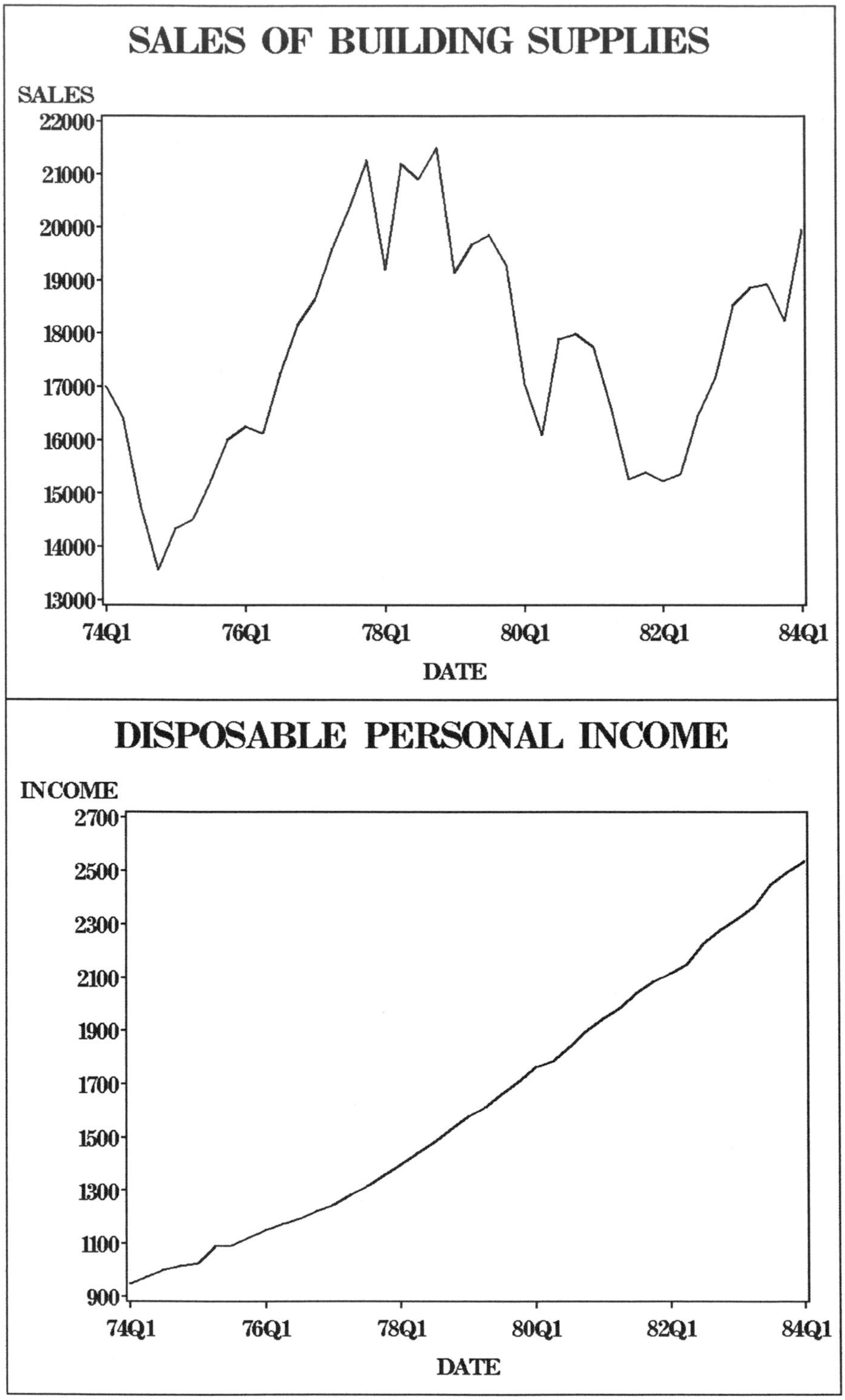

Output 4.8
Plotting Building- and Manufacturing-Related Quarterly Data (continued)

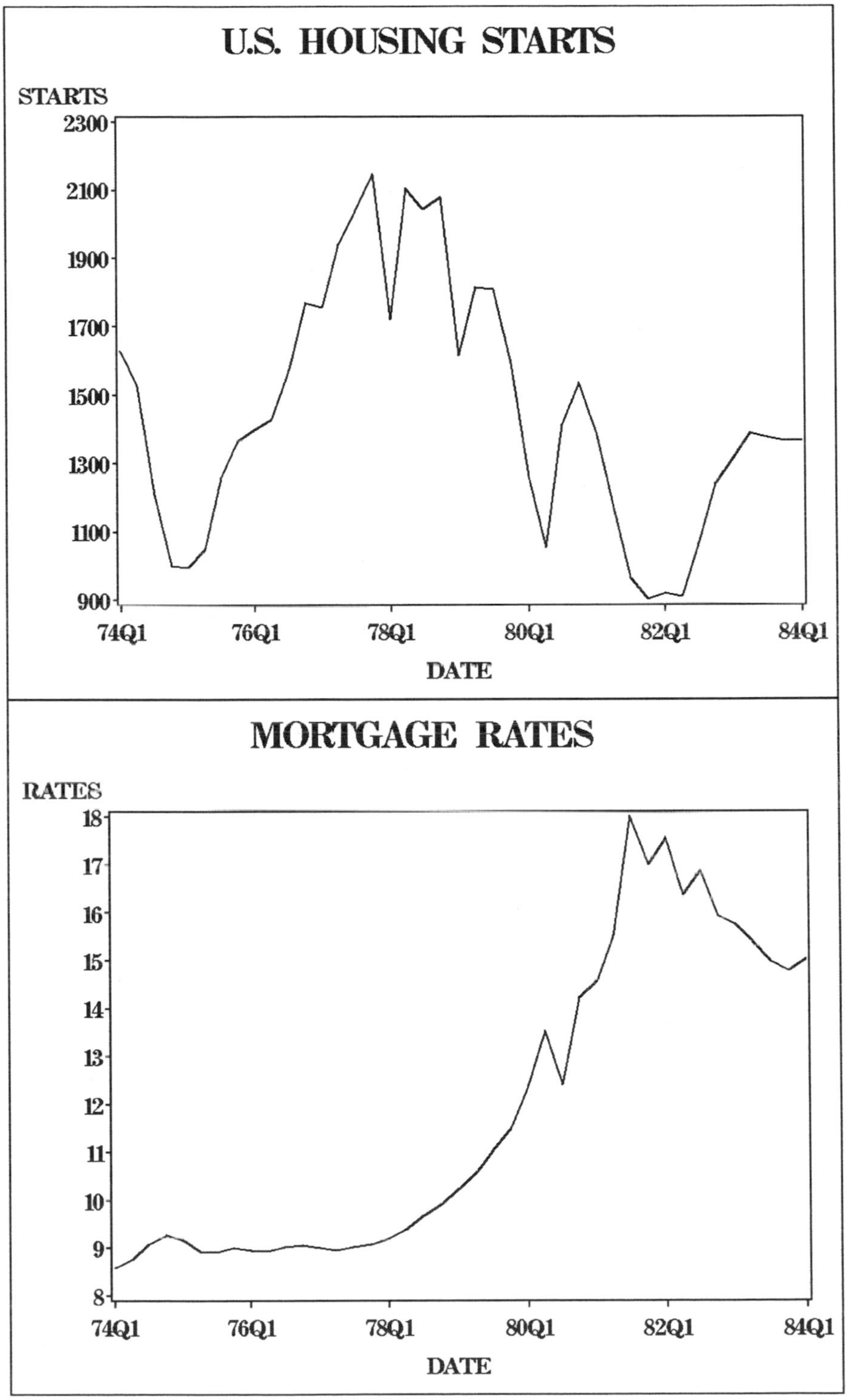

Thus, differencing implies that η_t had a unit root nonstationarity, so the differenced error series is stationary. This assumption, unlike assumptions about the explanatory series, is crucial. If you do not want to make this assumption, you can model the series in the original levels. Also, in the development above, you assume a simple intercept β_0 that canceled out of the differenced model. If, in fact, a trend $\beta_0 + \psi t$ is present, the differenced series has intercept ψ. If you had decided to fit the model in the original levels and to allow only AR error structures, PROC AUTOREG or Fuller's PROC NLIN method (1986) would have been an appropriate tool for the fitting.

Assuming differencing is appropriate, your next task is to output the residuals from regression and to choose a time series model for the error structure $\nabla\eta_t$. To accomplish this in PROC ARIMA, you must modify your IDENTIFY and ESTIMATE statements. The IDENTIFY statement is used to call in all explanatory variables of interest and to declare the degree of differencing for each. The CROSSCOR= option accomplishes this goal. You specify the following SAS statements:

```
PROC ARIMA DATA=HOUSING;
   TITLE 'MODEL IN FIRST DIFFERENCES';
   IDENTIFY VAR=SALES(1) CROSSCOR=(MORT(1) DPIC(1) STARTS(1)) NOPRINT;
RUN;
```

The NOPRINT option eliminates the printing of the cross-correlation function. Because you assume a contemporaneous relationship between sales and the explanatory variables, you do not check the cross-correlation function for dependence of sales on lagged values of the explanatory variables. If you want to check for lagged dependencies, you need to model the explanatory series to perform prewhitening. This is the only way you can get clear information from the cross-correlations.

To run a regression of SALES(1) on MORT(1), DPIC(1), and STARTS(1), add the following statement to your PROC ARIMA code:

```
ESTIMATE INPUT=(MORT DPIC STARTS) PLOT METHOD=ML;
RUN;
```

The INPUT= option denotes which variables in the CROSSCOR= list are to be used in the regression. Specifying differencing in the INPUT= option is not allowed. The order of differencing in the CROSSCOR= list is the order used. The PLOT option creates and plots the ACF, IACF, and PACF of the residuals. The results are shown in **Output 4.9**.

Output 4.9 *Using the INPUT= Option of the ESTIMATE Statement to Run a Regression: PROC ARIMA*

```
                              The ARIMA Procedure

                         Maximum Likelihood Estimation

                           Standard       ❺          Approx
Parameter      Estimate       Error    t Value    Pr > |t|     Lag    Variable    Shift

MU            170.03857   181.63744       0.94      0.3492       0    SALES           0
NUM1         -151.07400   112.12506      -1.35      0.1779       0    MORT            0
NUM2           -1.00212     4.22489      -0.24      0.8125       0    DPIC            0
NUM3            4.93009     0.39852      12.37      <.0001       0    STARTS          0


                           Constant Estimate        170.0386
                           Variance Estimate        200686.5
                           Std Error Estimate       447.9805
                           AIC                      605.6806
                           SBC                      612.4362
                           Number of Residuals            40


                        Correlations of Parameter Estimates

                Variable              SALES      MORT       DPIC     STARTS
                Parameter                MU      NUM1       NUM2       NUM3

                SALES          MU     1.000    -0.028     -0.916      0.016
                MORT         NUM1    -0.028     1.000     -0.071      0.349
                DPIC         NUM2    -0.916    -0.071      1.000     -0.039
                STARTS       NUM3     0.016     0.349     -0.039      1.000


                        Autocorrelation Check of Residuals
             ❻
 To         Chi-              Pr >
Lag       Square      DF     ChiSq    -------------------Autocorrelations-------------------

  6         8.90       6    0.1793    -0.397    -0.028    -0.012     0.128    -0.088     0.138
 12        10.27      12    0.5923    -0.099     0.035    -0.033    -0.099     0.053     0.026
 18        16.98      18    0.5245    -0.011    -0.122     0.143    -0.189     0.014     0.156
 24        23.37      24    0.4981    -0.033    -0.169     0.174    -0.043    -0.039     0.093

                              The ARIMA Procedure

                         Autocorrelation Plot of Residuals

Lag    Covariance    Correlation    -1 9 8 7 6 5 4 3 2 1 0 1 2 3 4 5 6 7 8 9 1      Std Error

  0        200687        1.00000    |                    |********************|            0
  1    -79669.125        -.39698    | ❼          ********|      .             |     0.158114
  2     -5682.473        -.02832    |           .       *|      .             |     0.181328
  3     -2373.543        -.01183    |           .        |      .             |     0.181438
  4     25596.562        0.12754    |           .        |***   .             |     0.181458
  5    -17570.433        -.08755    |           .      **|      .             |     0.183685
  6     27703.666        0.13804    |           .        |***   .             |     0.184725
  7    -19780.522        -.09856    |           .      **|      .             |     0.187287
  8      7007.339        0.03492    |          .         |*      .            |     0.188579
  9     -6640.664        -.03309    |          .        *|       .            |     0.188741
 10    -19897.518        -.09915    |          .       **|       .            |     0.188886
```

Output 4.9 *Using the INPUT= Option of the ESTIMATE Statement to Run a Regression: PROC ARIMA (continued)*

```
                         Inverse Autocorrelations   ❽

 Lag    Correlation    -1 9 8 7 6 5 4 3 2 1 0 1 2 3 4 5 6 7 8 9 1

   1        0.50240     |               .    |**********           |
   2        0.20274     |               .    |****  .              |
   3       -0.05699     |               .   *|      .              |
   4       -0.23103     |               .*****|     .              |
   5       -0.18394     |               . ****|     .              |
   6       -0.15639     |               .  ***|     .              |
   7        0.03430     |               .    |*     .              |
   8        0.11237     |               .    |**    .              |
   9        0.15648     |               .    |***   .              |
  10        0.13664     |               .    |***   .              |


                         Partial Autocorrelations

 Lag    Correlation    -1 9 8 7 6 5 4 3 2 1 0 1 2 3 4 5 6 7 8 9 1

   1       -0.39698     |           ********|       .              |
   2       -0.22069     |               . ****|     .              |
   3       -0.14121     |               .  ***|     .              |
   4        0.07283     |               .    |*     .              |
   5       -0.00035     |               .    |      .              |
   6        0.16471     |               .    |***   .              |
   7        0.03533     |               .    |*     .              |
   8        0.02720     |               .    |*     .              |
   9       -0.02400     |               .    |      .              |
  10       -0.20475     |               . ****|     .              |


                          The ARIMA Procedure
           ❶           Model for Variable SALES

                 Estimated Intercept             170.0386
                 Period(s) of Differencing              1

                             Input Number 1

           ❸     Input Variable                      MORT
                 Period(s) of Differencing              1
                 Overall Regression Factor       -151.074

                             Input Number 2

           ❹     Input Variable                      DPIC
                 Period(s) of Differencing              1
                 Overall Regression Factor       -1.00212

                             Input Number 3

           ❷     Input Variable                    STARTS
                 Period(s) of Differencing              1
                 Overall Regression Factor       4.930094
```

Output from the ESTIMATE statement for the sales data indicates that sales ❶ are positively related to housing starts ❷ but negatively related to mortgage rates ❸ and disposable personal income ❹. In terms of significance, only the *t* statistic ❺ for housing starts exceeds 2.

However, unless you fit the correct model, the *t* statistics are meaningless. The correct model includes specifying the error structure, which you have not yet done. For the moment, ignore these *t* statistics. You may argue based on the chi-square checks ❻ that the residuals are not autocorrelated. However, because the first chi-square statistic uses six correlations, the influence of a reasonably large correlation at lag 1 may be lessened to such an extent by the other five small correlations that significance is lost. Look separately at the first few autocorrelations, and remember that differencing is often accompanied by an MA term. Thus, you fit a model to the error series and wait to judge the significance of your *t* statistics until all important variables (including lagged error values) have been incorporated into the model. You use the same procedure here as in regression settings, where you do not use the *t* statistic for a variable in a model with an important explanatory variable omitted.

Based on the ACF of the differenced series, you fit an MA(1) model to the errors. You interpret the ACF of the differenced series as having a nonzero value (–0.39698) ❼ at lag 1 and a near-zero value at the other lags. Also, check the IACF ❽ to see if you have overdifferenced the series. If you have, the IACF dies off very slowly. Suppose you decide the IACF dies off rapidly enough and that you were correct to difference. Note that if

$$Y = \alpha + \beta X + \eta$$

where X and η are unit root processes, regression of Y on X produces an inconsistent estimate of β. This makes it impossible for you to use the PLOT option in a model in the original levels of the series to determine if you should difference. Residuals from the model may not resemble the true errors in the series because the estimate of β is inconsistent. Because the explanatory series seems to require differencing, you decide to model the SALES series in differences also and then to check for overdifferencing with the PLOT option. Overdifferencing also results in an MA coefficient that is an estimate of 1.

The next step, then, is to fit the regression model with an MA error term. You can accomplish this in PROC ARIMA by replacing the ESTIMATE statement above with

```
ESTIMATE INPUT=(MORT DPIC STARTS) Q=1 METHOD=ML;
RUN;
```

The results are shown in **Output 4.10**.

Output 4.10 *Fitting the Regression Model with an MA Error Term: PROC ARIMA*

```
                         The ARIMA Procedure
                     MODEL IN FIRST DIFFERENCES

     Estimation Method                                  Maximum Likelihood
     Parameters Estimated                                                5
     Termination Criteria                  Maximum Relative Change in Estimates
     Iteration Stopping Value                                        0.001
     Criteria Value                                               27.51793
     Maximum Absolute Value of Gradient                           234254.1
     R-Square Change from Last Iteration                          0.153278
     Objective Function                              Log Gaussian Likelihood
     Objective Function Value                                     -289.068
     Marquardt's Lambda Coefficient                                   1E-8
     Numerical Derivative Perturbation Delta                         0.001
     Iterations                                                         13
     Warning Message                       Estimates may not have converged.
```

Maximum Likelihood Estimation

Parameter	Estimate	Standard Error	❷ t Value	Approx Pr > |t|	Lag	Variable	Shift
MU	91.38149	47.30628	1.93	0.0534	0	SALES	0
MA1,1 ❶	0.99973	30.48149	0.03	0.9738	1	SALES	0
NUM1	-202.26240	60.63966	-3.34	0.0009	0	MORT	0
NUM2	0.89566	1.14910	0.78	0.4357	0	DPIC	0
NUM3	5.13054	0.28083	18.27	<.0001	0	STARTS	0

Constant Estimate	91.38149
Variance Estimate	115435.7
Std Error Estimate	339.7582
AIC	588.1353
SBC	596.5797
Number of Residuals	40

Correlations of Parameter Estimates

Variable	Parameter	SALES MU	SALES MA1,1	MORT NUM1	DPIC NUM2	STARTS NUM3
SALES	MU	1.000	0.360	-0.219	-0.958	-0.612
SALES	MA1,1	0.360	1.000	0.159	-0.402	-0.314
MORT	NUM1	-0.219	0.159	1.000	-0.052	0.651
DPIC	NUM2	-0.958	-0.402	-0.052	1.000	0.457
STARTS	NUM3	-0.612	-0.314	0.651	0.457	1.000

Output 4.10 *Fitting the Regression Model with an MA Error Term: PROC ARIMA (continued)*

```
                          Autocorrelation Check of Residuals

 To        Chi-             Pr >
Lag      Square     DF     ChiSq     --------------------Autocorrelations--------------------

  6        2.08      5    0.8382     0.017     0.079     0.076     0.147    -0.066     0.080
 12        6.50     11    0.8383    -0.128    -0.079    -0.144    -0.186    -0.037    -0.044
 18       11.43     17    0.8336    -0.120    -0.183     0.024    -0.137     0.040     0.074
 24       14.65     23    0.9067    -0.073    -0.045     0.108    -0.053    -0.073     0.088


                             Model for Variable SALES

                        Estimated Intercept          91.38149
                        Period(s) of Differencing           1


                             Moving Average Factors

                          Factor 1:  1 - 0.99973 B**(1)


                                 Input Number 1

                        Input Variable                   MORT
                        Period(s) of Differencing           1
                        Overall Regression Factor    -202.262


                                 Input Number 2

                        Input Variable                   DPIC
                        Period(s) of Differencing           1
                        Overall Regression Factor    0.895659


                                 Input Number 3

                        Input Variable                 STARTS
                        Period(s) of Differencing           1
                        Overall Regression Factor    5.130543
```

You have used the generally more accurate maximum-likelihood (ML) method of estimation on the differenced series. Remember that the IDENTIFY statement determines the degree of differencing used. You should note that the MA parameter 0.99973 ❶ is not significant (p-value >.05). The calculated *t* statistics ❷ on the explanatory variables have changed from the values they had in the regression with no model for the error series. Also note that PROC AUTOREG, another SAS procedure for regression with time series errors, cannot be used here because it does not allow for differencing (a problem that can be alleviated in the DATA step but could be very cumbersome for handling the forecasts and standard errors) and because it works only with AR error terms.

Something has happened here that can happen in practice and is worth noting. The moving average parameter estimate is almost 1. A moving average parameter of 1 is exactly what would be expected if the original regression model in series levels had a white noise error term. This in turn indicates that just an ordinary regression would suffice to fit the model without any differencing being required. Further inspection of the printout, however, reveals that this number may in fact not be a good estimate of the true moving average parameter, this coming from the message about estimates not converging. Decisions made on the basis of this number can thus not be supported.

It is worth noting that since the first edition of this book, in which the example first appeared, some relevant developments have taken place. If a regression model with stationary errors is appropriate for data in which the variables themselves appear to be nonstationary, then these errors are a stationary linear combination of nonstationary variables. The variables, independent and dependent, are then said to be *cointegrated*. Tests for cointegration are available in PROC VARMAX, discussed in Chapter 5, Section 5.2. It will be seen that elimination of some seemingly unimportant input variables in the example results in a model that does not show this problem, and this is the route that will be taken here. However, a test for cointegration could also be used to make a more informed decision as to whether the differencing was appropriate.

Any model that can be fit in PROC AUTOREG can also be fit in PROC ARIMA, which makes PROC ARIMA more generally applicable than PROC AUTOREG. The only advantage of PROC AUTOREG in this setting is its automatic selection of an AR model and, starting with Version 8 of SAS, its ability to handle strings of missing data.

A final modeling step is to delete insignificant explanatory variables. Do not calculate SALES forecasts based on forecasts of unrelated series. If you do, the forecast error variance is unnecessarily large because the forecast then responds to fluctuations in irrelevant variables. Is it acceptable to eliminate simultaneously all variables with insignificant t statistics? No, it is not acceptable. Eliminating a single insignificant regressor, like DPIC, can change the t statistics on all remaining parameters.

In the example above DPIC drifts upward, along with SALES. A nonzero MU in the differenced model also corresponds to drift in the original levels. The t statistic on MU is currently insignificant because DPIC takes over as the explainer of drift if MU is removed. Similarly, MU takes over if DPIC is removed. However, if you remove both terms from the model, the fit deteriorates significantly. DPIC and MU have the lowest t statistics. Remove DPIC and leave MU in the model because it is much easier to forecast than DPIC.

When DPIC is removed from the INPUT= list in your ESTIMATE statement, what happens then to the t test for MU? Omitting the insignificant DPIC results in a t statistic 3.86 (not shown) on MU. Also note that the other t statistics change but that the mortgage rates are still not statistically significant. Removing the mortgage rates from the INPUT= list results in a fairly simple model. Review the progression of your modeling thus far:

- You noticed that the inputs and the dependent variable SALES were nonstationary.
- You checked the residuals from a regression of differenced SALES on differenced DPIC, STARTS, and MORT. The residuals seemed stationary and reasonably invertible (in other words, the IACF died down reasonably fast).
- You used the PLOT option to identify an error term model that was MA(1). This term was problematic in that its estimate was near 1, it had a huge standard error, and the estimation procedure may not have converged.
- You used t statistics to sequentially remove insignificant terms and obtain

$$\nabla S_t = \nabla H_t + \psi + e_t - \beta e_{t-1}$$

where

∇ indicates a first difference

S_t is sales at time t

H_t is U.S. housing starts at time *t*

ψ is a constant (drift) that corresponds to the slope in a plot of the undifferenced series against time.

The final MA estimate 0.60397 ❸ is not particularly close to 1, giving you some confidence that you have not overdifferenced. No convergence problems remain at this point. (See **Output 4.11**).

Consider two scenarios for forecasting this series. First, suppose you are supplied with future values of housing starts H_{t+1} from some source. You incorporate these into your data set along with missing values for the unknown future values of SALES, and you call for a forecast. You do not supply information about the forecast accuracy of future housing start values, nor can the procedure use such information. It simply treats these futures as known values. In the second scenario, you model housing starts and then forecast them from within PROC ARIMA. This, then, provides an example of a case 2 problem.

For the first scenario, imagine you have been given future values of U.S. housing starts (the values are actually those that would be forecast from PROC ARIMA, giving you an opportunity to see the effect of treating forecasts as perfectly known values). The first step is to create a data set with future values for DATE and STARTS and missing values for SALES. This data set is then concatenated to the original data set. The combined data set COMB has eight values of future STARTS. Use the following SAS statements:

```
PROC ARIMA DATA=COMB;
   IDENTIFY VAR=SALES(1) CROSSCOR=(STARTS(1)) NOPRINT;
   ESTIMATE Q=1 INPUT=(STARTS) METHOD=ML;
   FORECAST LEAD=8 ID=DATE INTERVAL=QTR OUT=FOR1;
   TITLE 'DATA WITH FORECASTS OF STARTS APPENDED AND SALES=.';
RUN;
```

The results are shown in **Output 4.11**.

***Output 4.11** Forecasting with Future Input Values and Missing Future Sales Values*

DATA WITH FORECASTS OF STARTS APPENDED AND SALES=.

The ARIMA Procedure

Maximum Likelihood Estimation

Parameter	Estimate	Standard Error	t Value	Approx Pr > |t|	Lag	Variable	Shift
MU	91.99669	25.60324	3.59	0.0003	0	SALES	0
MA1,1	0.60397 ❸	0.15203	3.97	<.0001	1	SALES	0
NUM1	5.45100	0.26085	20.90	<.0001	0	STARTS	0

Constant Estimate	91.99669
Variance Estimate	152042.2
Std Error Estimate	389.9259
AIC	594.1269
SBC	599.1936
Number of Residuals	40

Output 4.11 *Forecasting with Future Input Values and Missing Future Sales Values (continued)*

Correlations of Parameter Estimates

Variable		SALES	SALES	STARTS
Parameter		MU	MA1,1	NUM1
SALES	MU	1.000	-0.170	0.010
SALES	MA1,1	-0.170	1.000	0.068
STARTS	NUM1	0.010	0.068	1.000

Autocorrelation Check of Residuals

To Lag	Chi-Square	DF	Pr > ChiSq	--------Autocorrelations--------					
6	2.07	5	0.8388	-0.100	0.030	0.142	0.112	-0.044	0.011
12	6.81	11	0.8140	-0.134	-0.057	-0.092	-0.229	-0.061	-0.029
18	11.77	17	0.8140	-0.187	-0.130	0.007	-0.078	0.059	0.115
24	19.29	23	0.6844	-0.025	0.051	0.218	-0.010	-0.072	0.155

Model for Variable SALES

Estimated Intercept	91.99669
Period(s) of Differencing	1

Moving Average Factors

Factor 1: 1 - 0.60397 B**(1)

Forecasts for Variable SALES

Obs	Forecast	Std Error	95% Confidence Limits	
42	19322.8041	389.9259	18558.5634	20087.0447
43	19493.4413	419.3912	18671.4496	20315.4329
44	19484.7198	446.9181	18608.7764	20360.6631
45	19576.7165	472.8452	18649.9570	20503.4759
46	19679.9629	497.4227	18705.0324	20654.8935
47	19794.3720	520.8417	18773.5409	20815.2030
48	19857.6642	543.2521	18792.9096	20922.4189
49	19949.6609	564.7740	18842.7242	21056.5976

The estimation is exactly the same as in the original data set because SALES has missing values for all future quarters, and thus these points cannot be used in the estimation. Because future values are available for all inputs, forecasts are generated. A request of LEAD=10 also gives only eight forecasts because only eight future STARTS are supplied. Note that future values were supplied to and not generated by the procedure. Forecast intervals are valid if you can guarantee the future values supplied for housing starts. Otherwise, they are too small. **Section 4.3.2** displays a plot of the forecasts from this procedure and also displays a similar plot in which PROC ARIMA is used to forecast the input variable. See **Output 4.13**.

Predicted SALES are the same (recall that future values of STARTS in this example are the same as those produced in PROC ARIMA), but forecast intervals differ considerably. Note that the general increase in predicted SALES is caused by including the drift term ψ.

4.3.2 Case 2: Simple Transfer Functions

In case 2, housing starts H_t are used as an explanatory variable for a company's sales. Using fitting and diagnostic checking, you obtain the model

$$\nabla S_t = \psi + \beta \nabla H_t + \eta_t$$

where η is the moving average

$$\eta_t = e_t - \theta e_{t-1}$$

In case 1, you supplied future values of H_t to PROC ARIMA and obtained forecasts and forecast intervals. The forecasts were valid, but the intervals were not large enough because future values of housing starts were forecasts. In addition, you have the problem of obtaining these future values for housing starts. PROC ARIMA correctly incorporates the uncertainty of future housing start values into the sales forecast.

Step 1 in this methodology identifies and estimates a model for the explanatory variable H_t, U.S. housing starts. The data are quarterly and, based on the usual criteria, the series should be differenced. The differenced series ∇H_t shows some correlation at lag 4 but not enough to warrant a span 4 difference. Use an AR factor to handle the seasonality of this series.

Diagnostic checking was done on the STARTS series H_t. The model

$$\left(1 - \alpha B^4\right) \nabla H_t = \left(1 - \theta B^3\right) e_t$$

fits well. In **Section 4.3.1**, the series was forecast eight periods ahead to obtain future values. You do not need to request forecasts of your inputs (explanatory series) if your goal is only to forecast target series (SALES, in this case). The procedure automatically generates forecasts of inputs that it needs, but you do not see them unless you request them.

In step 2, an input series is used in an input option to identify and estimate a model for the target series S_t. This part of the SAS code is the same as that in the previous example. The two steps must be together in a single PROC ARIMA segment. The entire set of code is shown below and some of the output is shown in **Output 4.12**. Some of the output has been suppressed (it was displayed earlier). Also, forecast intervals are wider than in case 1, where forecasts of H_t were taken from this run and concatenated to the end of the data set instead of being forecast by the procedure. This made it impossible to incorporate forecast errors for H_t into the forecast of S_t. The SAS code follows:

```
PROC ARIMA DATA=HOUSING;
   TITLE 'FORECASTING STARTS AND SALES';
   IDENTIFY VAR=STARTS(1) NOPRINT;
   ESTIMATE P=(4) Q=(3) METHOD=ML NOCONSTANT;
   FORECAST LEAD=8;
   IDENTIFY VAR=SALES(1) CROSSCOR=(STARTS(1)) NOPRINT;
   ESTIMATE Q=1 INPUT=(STARTS) METHOD=ML NOPRINT;
   FORECAST LEAD=8 ID=DATE INTERVAL=QTR OUT=FOR2 NOPRINT;
RUN;
```

Output 4.12 *Estimating Using Maximum Likelihood: PROC ARIMA*

FORECASTING STARTS AND SALES

The ARIMA Procedure

Maximum Likelihood Estimation

Parameter	Estimate	Standard Error	t Value	Approx Pr > |t|	Lag
MA1,1	0.42332	0.15283	2.77	0.0056	3
AR1,1	0.28500	0.15582	1.83	0.0674	4

Variance Estimate	30360.5
Std Error Estimate	174.2426
AIC	529.229
SBC	532.6068
Number of Residuals	40

Correlations of Parameter Estimates

Parameter	MA1,1	AR1,1
MA1,1	1.000	0.193
AR1,1	0.193	1.000

Autocorrelation Check of Residuals

To Lag	Chi-Square	DF	Pr > ChiSq	--------------------Autocorrelations-------------------					
6	2.55	4	0.6351	0.197	0.106	0.034	0.053	0.004	-0.063
12	9.90	10	0.4496	-0.021	-0.019	-0.033	-0.232	-0.167	-0.208
18	14.58	16	0.5558	-0.153	-0.117	-0.114	-0.116	-0.058	-0.060
24	18.62	22	0.6686	-0.096	0.118	0.109	0.068	0.066	-0.047

***Output 4.12** Estimating Using Maximum Likelihood: PROC ARIMA (continued)*

```
                    Model for Variable STARTS

                  Period(s) of Differencing    1

                    No mean term in this model.


                      Autoregressive Factors

                    Factor 1:   1 - 0.285 B**(4)

                      Moving Average Factors

                  Factor 1:   1 - 0.42332 B**(3)


                  Forecasts for variable STARTS

Obs      Forecast    Std Error       95% Confidence Limits

 42     1373.2426     174.2426     1031.7333      1714.7519
 43     1387.6693     246.4163      904.7022      1870.6364
 44     1369.1927     301.7971      777.6812      1960.7041
 45     1369.1927     318.0853      745.7569      1992.6284
 46     1371.2568     351.7394      681.8602      2060.6534
 47     1375.3683     382.4434      625.7930      2124.9437
 48     1370.1026     410.8593      564.8332      2175.3719
 49     1370.1026     430.6707      526.0035      2214.2016
```

You can now merge the data sets FOR1 and FOR2 from the previous two examples and plot the forecasts and intervals on the same graph. This is illustrated in **Output 4.13** to indicate the difference in interval widths for these data.

The first graph gives forecast intervals that arose from using PROC ARIMA to forecast housing starts. The second plot gives these forecast intervals as a solid line along with intervals from the previous analysis (broken line), where the same future values for housing starts are read into the data set rather than being forecast by PROC ARIMA. Note how the broken line drastically underestimates the uncertainty in the forecasts. The narrower interval is questionable in light of the downturn in SALES at the end of the series.

Output 4.13 *Plotting Forecast Intervals*

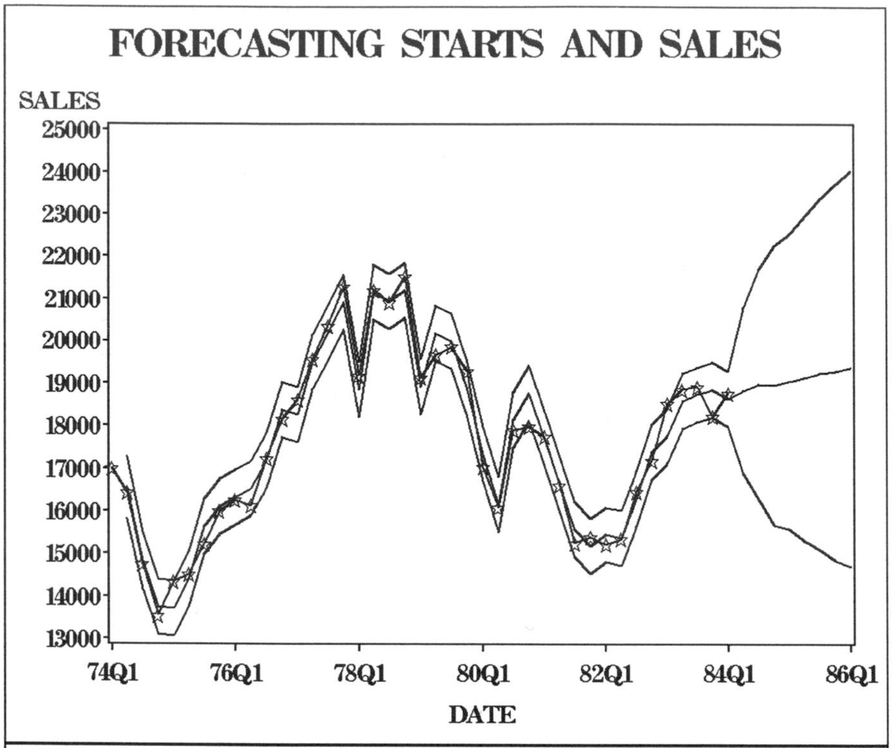

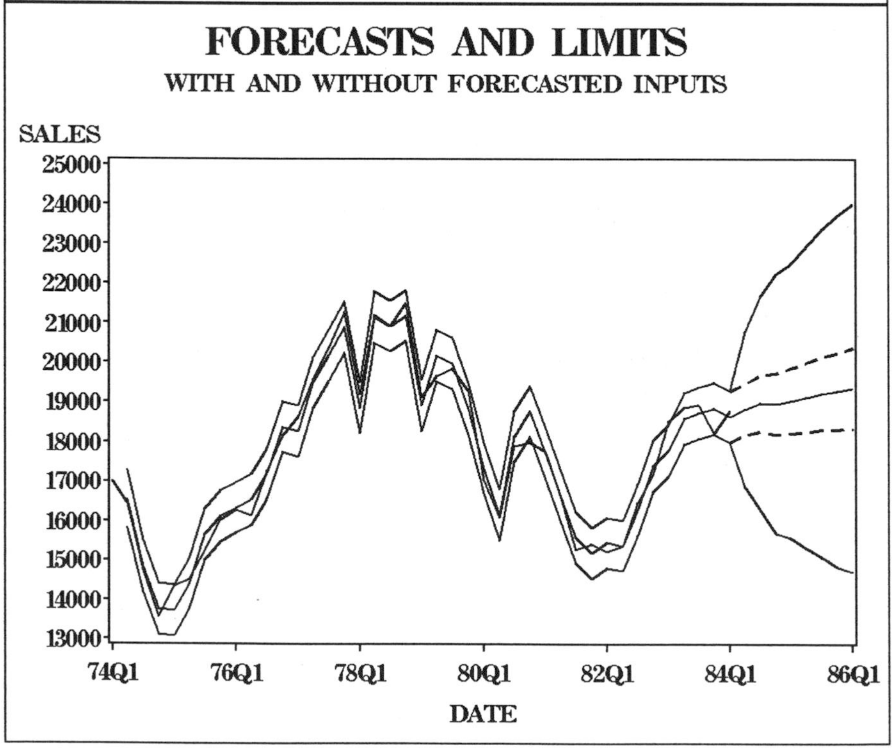

4.3.3 Case 3: General Transfer Functions

4.3.3.1 Model Identification

You have specified the ARMA model with backshift operators. For example, you can write the ARMA(1,1) model

$$Y_t - \alpha Y_{t-1} = e_t - \theta e_{t-1}$$

as

$$(1-\alpha B)Y_t = (1-\theta B)e_t$$

or as

$$Y_t = (1-\theta B)/(1-\alpha B)e_t$$

or, finally, as

$$Y_t = e_t + (\alpha - \theta)e_{t-1} + \alpha(\alpha - \theta)e_{t-2} + \alpha^2(\alpha - \theta)e_{t-3} + \ldots$$

The pattern of the weights (coefficients on the e_ts) determines that the process has one AR and one MA parameter in the same way the ACF does. For example, if

$$Y_t = e_t + 1.2e_{t-1} + .6e_{t-2} + .3e_{t-3} + .15e_{t-4} + .075e_{t-5} + \ldots$$

the weights are 1, 1.2, .6, .3, .15, .075, ... The pattern is characterized by one arbitrary change (from 1 to 1.2) followed by exponential decay at the rate .5 (.6= (.5)(1.2), .3=(.5)(.6), . . .). The exponential decay tells you to put a factor (1−.5B) in the denominator of the expression multiplying e_t (in other words, $\alpha = .5$).

Because $1.2 = \alpha - \theta$, you see that $\theta = -.7$. The model, then, is

$$Y_t = (1 + .7B)/(1 - .5B)e_t$$

What have you learned from this exercise? First, you see that you can write any ARMA model by setting Y_t equal to a ratio of polynomial factors in the backshift operator B operating on e_t. Next, you see that if you can estimate the sequence of weights on the e_ts, you can determine how many AR and MA lags you need. Finally, in this representation, you see that the numerator polynomial corresponds to MA factors and the denominator corresponds to AR factors.

If you can apply a ratio of backshift polynomials to an unobserved error series e_t, why not apply one to an observable input? This is exactly what you do in case 3. For example, suppose you write

$$Y_t - .8Y_{t-1} = 3(X_{t-1} - .4X_{t-2}) + \eta_t$$

where η_t is the moving average

$$\eta_t = e_t + .6e_{t-1}$$

You then obtain

$$(1 - .8B)Y_t = 3(1 - .4B)X_{t-1} + (1 + .6B)e_t$$

or

$$Y_t = 0 + 3(1 - .4B) / (1 - .8B)X_{t-1} + (1 + .6B) / (1 - .8B)e_t$$

This is called a transfer function. Y_t is modeled as a function of lagged values of the input series X_t and current and lagged values of the shocks e_t. Usually, the intercept is not 0, although for simplicity 0 is used in the preceding example.

You now have a potentially useful model, but how is it used? With real data, how will you know the form of the backshift expression that multiplies X_{t-1}? The answer is in the cross-correlations. Define the cross-covariance as

$$\gamma_{YX}(j) = \text{cov}(Y_t, X_{t+j})$$

and

$$\gamma_{XY}(j) = \text{cov}(X_t, Y_{t+j}) = \text{cov}(X_{t-j}, Y_t) = \gamma_{YX}(-j)$$

Estimate $\gamma_{XY}(j)$ by

$$C_{XY}(j) = \Sigma\ (X_t - \bar{X})(Y_{t+j} - \bar{Y})\ /\ n$$

Define the cross-correlation as

$$\rho_{XY}(j) = \gamma_{XY}(j) / (\gamma_{XX}(0)\gamma_{YY}(0))^{.5}$$

Estimate this by

$$r_{XY}(j) = C_{XY}(j) / (C_{XX}(0)C_{YY}(0))^{.5}$$

To illustrate the theoretical cross-covariances for a transfer function, assume that X_t is a white noise process independent of the error series η_t. The cross-covariances are computed below and are direct multiples of $\gamma_{XX}(0)$, the variance of X (this holds only when X is white noise):

$$Y_t - .8Y_{t-1} = 3X_{t-1} - 1.2X_{t-2} + \eta_t$$

and

$$Y_t = 3X_{t-1} + 1.2X_{t-2} + .96X_{t-3} + .768X_{t-4} + \ .\ .\ . + \text{noise}$$

Multiplying both sides by X_{t-j}, j=0, 1, 2, 3, and computing expected values gives

$$\gamma_{XY}(0) = E(X_t Y_t) = 0$$

$$\gamma_{XY}(1) = E(X_t Y_{t+1}) = E(X_{t-1} Y_t) = 3\gamma_{XX}(0)$$

$$\gamma_{XY}(2) = E(X_t Y_{t+2}) = E(X_{t-2} Y_t) = 1.2\gamma_{XX}(0)$$

and

$$\gamma_{XY}(3) = E(X_t Y_{t+3}) = .96\gamma_{XX}(0)$$

When you divide each term in the cross-covariance sequence by $\gamma_{XX}(0)$, you obtain the weights β_j:

LAG J	−1	0	1	2	3	4	5
WEIGHT β_j	0	0	3	1.2	.96	$(1.2)(.8)^2$	$(1.2)(.8)^3$

Note that the model involves

$$3(1-.4B)/(1-.8B)X_{t-1} = 0X_t + 3X_{t-1} + 1.2X_{t-2} + .96X_{t-3} + \ldots$$

so if X is white noise, the cross-covariances are proportional to the transfer function weights β_j.

These weights β_j are also known as the impulse-response function. The reason for this name is clear if you ignore the error term in the model and let X_t be a pulse; that is, X_t=0 except at t=10 where X_{10}=1. Ignoring the white noise term, you have

$$Y_t = 3X_{t-1} + 1.2X_{t-2} + .96X_{t-3} + \ldots$$

so

$$Y_0 = Y_1 = Y_2 = \ldots = Y_{10} = 0$$

$$Y_{11} = 3X_{10} + 1.2X_9 + .96X_8 + \ldots = 3$$

$$Y_{12} = 3(0) + 1.2(1) + .96(0) + \ldots = 1.2$$

and

$$Y_{13} = .96$$

The weights are the expected responses to a pulse input. The pulse is delayed by one period. Its effect continues to be felt starting with t=11 but diminishes quickly because of the stationary denominator (in other words, AR-type operator $(1-.8B)^{-1}$ on X_{t-1}).

The crucial point is that if you can obtain the cross-correlations, you have the impulse-response weight pattern, which you can then analyze by the same rules used for the ACFs. In the example above, the 0 on X_t indicates a pure delay. The arbitrary jump from 3 to 1.2, followed by exponential decay at rate .8, indicates that the multiplier on X_{t-1} has one numerator (MA) lag and one denominator (AR) lag. The only problem is the requirement that X_t be white noise, which is addressed below.

Suppose you have the same transfer function, but X_t is AR(1) with parameter α. You have

$$Y_t = 0 + 3(1-.4B)/(1-.8B)X_{t-1} + (1+.6B)/(1-.8B)e_t$$

and

$$X_t = \alpha X_{t-1} + \varepsilon_t$$

where the X_ts are independent of the e_ts and where e_t and ε_t are two (independent) white noise sequences.

Note that

$$Y_t = 3X_{t-1} + 1.2X_{t-2} + .96X_{t-3} + \ldots + \text{noise}_t$$

so

$$\alpha Y_{t-1} = 3\alpha X_{t-2} + 1.2\alpha X_{t-3} + .96\alpha X_{t-4} + \ldots + \alpha(\text{noise}_{t-1})$$

and

$$Y_t - \alpha Y_{t-1} = 3(X_{t-1} - \alpha X_{t-2}) + 1.2(X_{t-2} - \alpha X_{t-3}) + .96(X_{t-3} - \alpha X_{t-4}) + \ldots + N'_t$$

where N'_t is a noise term. Set

$$Y_t - \alpha Y_{t-1} = Y'_t$$

and note that

$$X_t - \alpha X_{t-1} = \varepsilon_t$$

is a white noise sequence, so the expression above becomes

$$Y'_t = 3\varepsilon_{t-1} + 1.2\varepsilon_{t-2} + .96\varepsilon_{t-3} + \ldots + N'_t$$

The impulse-response function is exactly what you want, and

$$X_t - \alpha X_{t-1} = \varepsilon_t$$

is a white noise sequence.

You want to model X and use that model to estimate Y'_t and ε_t. This process is known as prewhitening, although it really only whitens X. Next, compute the cross-correlations of the prewhitened X and Y (in other words, the estimated Y'_t and ε_t). Note that the prewhitened variables are used only to compute the cross-correlations. The parameter estimation in PROC ARIMA is always performed on the original variables.

4.3.3.2 Statements for Transfer Function Modeling in the IDENTIFY Stage

Use the IDENTIFY and ESTIMATE statements in PROC ARIMA to model X. A subsequent IDENTIFY statement for Y with the CROSSCOR=(X) option automatically prewhitens X and Y, using the previously estimated model for X. For this example, you specify the following SAS statements:

```
PROC ARIMA DATA=TRANSFER;
   TITLE 'FITTING A TRANSFER FUNCTION';
   IDENTIFY VAR=X;
   ESTIMATE P=1;
   IDENTIFY VAR=Y CROSSCOR=(X) NLAG=10;
RUN;
```

The results are shown in **Output 4.14**.

Output 4.14 *Fitting a Transfer Function with the IDENTIFY and ESTIMATE Statements: PROC ARIMA*

```
                              FITTING A TRANSFER FUNCTION

                                 The ARIMA Procedure

                                Name of Variable = X

                         Mean of Working Series     5.000159
                         Standard Deviation         1.168982
                         Number of Observations          500

                                  Autocorrelations

 Lag    Covariance    Correlation    -1 9 8 7 6 5 4 3 2 1 0 1 2 3 4 5 6 7 8 9 1      Std Error

   0      1.366519        1.00000    |                    |********************|            0
   1      0.694896        0.50851    |                  . |**********          |     0.044721
   2      0.362484        0.26526    |                  . |*****               |     0.055085
   3      0.225201        0.16480    |                  . |***                 |     0.057583
   4      0.187569        0.13726    |                  . |***                 |     0.058519
   5      0.115383        0.08444    |                  . |**                  |     0.059159
   6      0.117494        0.08598    |                  . |**                  |     0.059400
   7      0.094599        0.06923    |                  . |*.                  |     0.059648
   8     -0.010062        -.00736    |                  . | .                  |     0.059808
   9     -0.038517        -.02819    |                  .*| .                  |     0.059810
  10     -0.029129        -.02132    |                  . | .                  |     0.059837
  11     -0.078230        -.05725    |                  .*| .                  |     0.059852
  12     -0.153233        -.11213    |                  **| .                  |     0.059961
  13     -0.107038        -.07833    |                  **| .                  |     0.060379
  14     -0.101788        -.07449    |                  .*| .                  |     0.060582
  15     -0.090632        -.06632    |                  .*| .                  |     0.060765


                             Inverse Autocorrelations

              Lag    Correlation    -1 9 8 7 6 5 4 3 2 1 0 1 2 3 4 5 6 7 8 9 1

                1       -0.39174    |            ********| .                  |
                2       -0.00306    |                  . | .                  |
                3        0.03538    |                  . |*.                  |
                4       -0.05678    |                  .*| .                  |
                5        0.02589    |                  . |*.                  |
                6       -0.01120    |                  . | .                  |
                7       -0.06020    |                  .*| .                  |
                8        0.03468    |                  . |*.                  |
                9        0.01185    |                  . | .                  |
               10       -0.02814    |                  .*| .                  |
               11       -0.02175    |                  . | .                  |
               12        0.08543    |                  . |**                  |
               13       -0.03405    |                  .*| .                  |
               14        0.00512    |                  . | .                  |
               15        0.03973    |                  . |*.                  |
```

Output 4.14 *Fitting a Transfer Function with the IDENTIFY and ESTIMATE Statements: PROC ARIMA (continued)*

```
                              Partial Autocorrelations

          Lag    Correlation    -1 9 8 7 6 5 4 3 2 1 0 1 2 3 4 5 6 7 8 9 1

            1        0.50851     |                    . |**********          |
            2        0.00900     |                    . | .                  |
            3        0.03581     |                    . |*.                  |
            4        0.05194     |                    . |*.                  |
            5       -0.01624     |                    . | .                  |
            6        0.04802     |                    . |*.                  |
            7        0.00269     |                    . | .                  |
            8       -0.07792     |                    **| .                  |
            9       -0.00485     |                    . | .                  |
           10       -0.00078     |                    . | .                  |
           11       -0.05930     |                    .*| .                  |
           12       -0.07466     |                    .*| .                  |
           13        0.02408     |                    . | .                  |
           14       -0.02906     |                    .*| .                  |
           15       -0.00077     |                    . | .                  |

                       Autocorrelation Check for White Noise

 To        Chi-             Pr >
Lag      Square     DF     ChiSq     -------------------Autocorrelations------------------

  6      196.16      6    <.0001      0.509     0.265     0.165     0.137     0.084     0.086
 12      207.41     12    <.0001      0.069    -0.007    -0.028    -0.021    -0.057    -0.112
 18      217.70     18    <.0001     -0.078    -0.074    -0.066    -0.011     0.051     0.033
 24      241.42     24    <.0001      0.041     0.071     0.104     0.115     0.100     0.066


                       Conditional Least Squares Estimation

                                    Standard                 Approx
         Parameter      Estimate       Error    t Value    Pr > |t|     Lag

         MU              5.00569     0.09149      54.71      <.0001       0
         AR1,1           0.50854     0.03858      13.18      <.0001       1


                         Constant Estimate        2.460094
                         Variance Estimate        1.017213
                         Std Error Estimate        1.00857
                         AIC                      1429.468
                         SBC                      1437.897
                         Number of Residuals           500
                       * AIC and SBC do not include log determinant.
```

Output 4.14 *Fitting a Transfer Function with the IDENTIFY and ESTIMATE Statements: PROC ARIMA (continued)*

Correlations of Parameter Estimates

Parameter	MU	AR1,1
MU	1.000	0.005
AR1,1	0.005	1.000

Autocorrelation Check of Residuals

To Lag	Chi-Square	DF	Pr > ChiSq	Autocorrelations					
6	2.92	5	0.7120	-0.005	-0.012	0.004	0.062	-0.010	0.041
12	11.37	11	0.4127	0.063	-0.041	-0.028	0.022	-0.006	-0.097
18	15.99	17	0.5246	-0.005	-0.027	-0.054	-0.008	0.072	-0.007
24	20.23	23	0.6278	-0.001	0.020	0.050	0.055	0.046	0.008
30	24.29	29	0.7144	-0.013	0.050	0.042	-0.011	-0.005	0.055
36	26.44	35	0.8506	0.016	-0.030	-0.014	-0.023	0.045	0.011
42	29.67	41	0.9055	-0.033	0.043	0.012	0.017	-0.009	0.049
48	34.37	47	0.9148	0.025	-0.021	0.018	0.024	0.038	0.071

Model for Variable X

Estimated Mean 5.005601

Autoregressive Factors

Factor 1: 1 - 0.50854 B**(1)

Name of Variable = Y

Mean of Working Series	10.05915
Standard Deviation	6.141561
Number of Observations	500

Output 4.14 *Fitting a Transfer Function with the IDENTIFY and ESTIMATE Statements: PROC ARIMA (continued)*

```
                                    Autocorrelations

 Lag    Covariance    Correlation    -1 9 8 7 6 5 4 3 2 1 0 1 2 3 4 5 6 7 8 9 1      Std Error

   0     37.718773        1.00000    |                    |********************|             0
   1     32.298318        0.85629    |                  . |*****************   |      0.044721
   2     27.167391        0.72026    |                 .  |**************      |      0.070235
   3     22.911916        0.60744    |                 .  |************        |      0.083714
   4     19.393508        0.51416    |                .   |**********          |      0.092109
   5     15.904433        0.42166    |                .   |********            |      0.097680
   6     13.223877        0.35059    |                .   |*******             |      0.101255
   7     10.558154        0.27992    |                .   |******              |      0.103655
   8      7.414747        0.19658    |                .   |****                |      0.105156
   9      5.180506        0.13735    |                .   |***.                |      0.105888
  10      3.731949        0.09894    |                .   |** .                |      0.106244


                              Inverse Autocorrelations

               Lag    Correlation    -1 9 8 7 6 5 4 3 2 1 0 1 2 3 4 5 6 7 8 9 1

                 1       -0.52149    |          **********| .                  |
                 2        0.01944    |                  . | .                  |
                 3        0.04247    |                  . |*.                  |
                 4       -0.07597    |                  **| .                  |
                 5        0.06999    |                  . |*.                  |
                 6       -0.00905    |                  . | .                  |
                 7       -0.08143    |                  **| .                  |
                 8        0.07222    |                  . |*.                  |
                 9        0.00167    |                  . | .                  |
                10       -0.01040    |                  . | .                  |


                              Partial Autocorrelations

               Lag    Correlation    -1 9 8 7 6 5 4 3 2 1 0 1 2 3 4 5 6 7 8 9 1

                 1        0.85629    |                  . |*****************   |
                 2       -0.04864    |                  .*| .                  |
                 3        0.00878    |                  . | .                  |
                 4        0.00619    |                  . | .                  |
                 5       -0.05155    |                  .*| .                  |
                 6        0.02489    |                  . | .                  |
                 7       -0.04671    |                  .*| .                  |
                 8       -0.09421    |                  **| .                  |
                 9        0.03137    |                  . |*.                  |
                10        0.01915    |                  . | .                  |
```

Output 4.14 *Fitting a Transfer Function with the IDENTIFY and ESTIMATE Statements: PROC ARIMA (continued)*

```
                        Autocorrelation Check for White Noise

 To        Chi-             Pr >
 Lag      Square     DF     ChiSq    --------------------Autocorrelations-------------------

  6      1103.03      6    <.0001      0.856     0.720     0.607     0.514     0.422     0.351


                               Correlation of Y and X

                    Number of Observations                    500
                    Variance of transformed series Y     14.62306
                    Variance of transformed series X     1.013152

                          Both series have been prewhitened.

                                  Crosscorrelations

       Lag    Covariance    Correlation    -1 9 8 7 6 5 4 3 2 1 0 1 2 3 4 5 6 7 8 9 1

       -10     -0.297935       -.07740     |                   **| .                   |
        -9     -0.147123       -.03822     |                   .*| .                   |
        -8     -0.066199       -.01720     |                    . | .                   |
        -7     -0.145114       -.03770     |                   .*| .                   |
        -6     -0.232308       -.06035     |                   .*| .                   |
        -5      0.135384       0.03517     |                    . |*.                   |
        -4      0.147029       0.03820     |                    . |*.                   |
        -3      0.023585       0.00613     |                    . | .                   |
        -2      0.249856       0.06491     |                    . |*.                   |
        -1      0.090743       0.02358     |                    . | .                   |
         0      0.089734       0.02331     |                    . | .                   |
         1      0.033887       0.00880     |                    . | .                   |
         2      3.053341       0.79327     | ❶                  . |****************    |
         3      1.275799       0.33146     | ❷                  . |*******             |
         4      0.920176       0.23906     |                    . |*****               |
         5      0.777420       0.20198     |                    . |****                |
         6      0.806808       0.20961     |                    . |****                |
         7      0.505834       0.13142     |                    . |***                 |
         8      0.577540       0.15005     |                    . |***                 |
         9      0.562041       0.14602     |                    . |***                 |
        10      0.254582       0.06614     |                    . |*.                  |
```

Output 4.14 *Fitting a Transfer Function with the IDENTIFY and ESTIMATE Statements: PROC ARIMA (continued)*

```
                         Crosscorrelation Check Between Series

 To       Chi-                Pr >
 Lag     Square     DF       ChiSq    -------------------Crosscorrelations-----------------

  5      418.85      6      <.0001     0.023     0.009     0.793     0.331     0.239     0.202

                Both variables have been prewhitened by the following filter:

                                   Prewhitening Filter


                                  Autoregressive Factors

                               Factor 1:  1 - 0.50854 B**(1)
```

Data for this example are generated from the model

$$(Y_t - 10) - .8(Y_{t-1} - 10) = 3((X_{t-2} - 5) - .4(X_{t-3} - 5)) + N_t$$

where

$$(X_t - 5) = .5(X_{t-1} - 5) + e_t$$

The cross-correlations are near 0 until you reach lag 2. You now see a spike (0.79327) ❶ followed by an arbitrary drop to 0.33146 ❷ followed by a roughly exponential decay. The one arbitrary drop corresponds to one numerator (MA) lag $(1 - \theta B)$ and the exponential decay to one denominator (AR) lag $(1 - \alpha B)$. The form of the transfer function is then

$$C(1 - \theta B) / (1 - \alpha B)X_{t-2} = (C - (C\theta)B) / (1 - \alpha B)X_{t-2}$$

Note the pure delay of two periods. The default in PROC ARIMA is to estimate the model with the C multiplied through the numerator as shown on the right. The ALTPARM option gives the factored C form as on the left.

Now review the PROC ARIMA instructions needed to run this example. In INPUT=(*form1 variable1 form2 variable2. . .*), the specification for the transfer function form is

$$S\$(L_{1,1}, L_{1,2}, \ldots)\ldots(L_{k,1} \ldots) / (L_{k+1,1} \ldots)\ldots(\ldots)$$

where

S	is the shift or pure delay (2 in the example)
lag polynomials	are written in multiplicative form
variable j	is not followed by differencing numbers (this is done in CROSSCOR).

For example,

```
INPUT=(2$(1,3)(1)/(1)X) ALTPARM;
```

indicates

$$Y_t = \theta_0 + \left(C\left(1 - \theta_1 B - \theta_2 B^3\right)\left(1 - \alpha B\right) / \left(1 - \delta B\right)\right) X_{t-2} + \text{noise}$$

Several numerator and denominator factors can be multiplied together. Note the absence of a transfer function form in the sales and housing starts example, which assumes that only contemporaneous relationships exist among sales, S_t, and the input variables.

For the current (generated data) example, the transfer function form should indicate a pure delay of two (2$), one numerator (MA) lag (2$(1)), and one denominator lag (2$(1)/(1)). Use the PLOT option to analyze the residuals and then estimate the transfer function with the noise model.

To continue with the generated data, add these SAS statements to those used earlier to identify and estimate the X model and to identify the Y model:

```
ESTIMATE INPUT=(2$(1)/(1)X) MAXIT=30
   ALTPARM PLOT METHOD=ML;
RUN;
```

The code above produces **Output 4.15**. Note the AR(1) nature of the autocorrelation plot of residuals. Continue with the following code to produce **Output 4.16**:

```
ESTIMATE P=1 INPUT=(2$(1)/(1)X)
   PRINTALL ALTPARM METHOD=ML;
FORECAST LEAD=10 OUT=OUTDATA ID=T;
RUN;
DATA NEXT;
   SET OUTDATA;
   IF T>480;
RUN;
PROC PRINT DATA=NEXT;
   TITLE 'FORECAST OUTPUT DATA SET';
RUN;
PROC GPLOT DATA=NEXT;
   PLOT L95*T U95*T FORECAST*T Y*T / OVERLAY HMINOR=0;
   SYMBOL1 V=L I=NONE C=BLACK;
   SYMBOL2 V=U I=NONE C=BLACK;
   SYMBOL3 V=F L=2 I=JOIN C=BLACK;
   SYMBOL4 V=A L=1 I=JOIN C=BLACK;
   TITLE 'FORECASTS FOR GENERATED DATA';
RUN;
```

4.3.3.3 Model Evaluation

The estimated model is

$$Y_t = -32.46 + 2.99\left(1 - .78B\right)^{-1}\left(1 - .37B\right)X_{t-2} + \left(1 - .79B\right)^{-1}\eta_t$$

as shown in **Output 4.16** ❶ Standard errors are (1.73), (.05), (.01), (.02), and (.03).

In the autocorrelation and cross-correlation checks of residuals and input, note the following facts:

- Chi-square statistics automatically printed by PROC ARIMA are like the *Q* statistics discussed earlier for standard PROC ARIMA models.
- Cross-correlation of residuals with input implies improper identification of the transfer function model. This is often accompanied by autocorrelation in residuals.

- Autocorrelation of residuals not accompanied by cross-correlation of residuals with X indicates that the transfer function is right but that the noise model is not properly identified.

See Output 4.15 ❷, from

```
ESTIMATE INPUT=(2$(1)/(1)X) . . . ;
```

versus **Output 4.16** ❸, from

```
ESTIMATE  P=1 . . . ;
```

Neither cross-correlation check ❹ ❺ indicates any problem with the transfer specification. First, the inputs are forecast and then used to forecast Y. In an example without prewhitening, future values of X must be in the original data set.

Output 4.15 *Fitting a Transfer Function: PROC ARIMA*

```
                              FITTING A TRANSFER FUNCTION

                                 The ARIMA Procedure


                              Maximum Likelihood Estimation

                               Standard                  Approx
Parameter      Estimate           Error    t Value     Pr > |t|     Lag    Variable    Shift

MU            -33.99761         0.77553     -43.84       <.0001       0    Y               0
SCALE1          3.06271         0.07035      43.53       <.0001       0    X               2
NUM1,1          0.39865         0.02465      16.17       <.0001       1    X               2
DEN1,1          0.79069       0.0079330      99.67       <.0001       1    X               2


                              Constant Estimate        -33.9976
                              Variance Estimate        2.702153
                              Std Error Estimate       1.643823
                              AIC                      1908.451
                              SBC                      1925.285
                              Number of Residuals           497
```

Output 4.15 *Fitting a Transfer Function: PROC ARIMA (continued)*

Correlations of Parameter Estimates

Variable		Y	X	X	X
Parameter		MU	SCALE1	NUM1,1	DEN1,1
Y	MU	1.000	-0.328	-0.347	-0.634
X	SCALE1	-0.328	1.000	0.689	0.291
X	NUM1,1	-0.347	0.689	1.000	0.821
X	DEN1,1	-0.634	0.291	0.821	1.000

Autocorrelation Check of Residuals

To Lag	Chi-Square	DF	Pr > ChiSq	Autocorrelations					
6	731.66	6	<.0001	0.780	0.607	0.489	0.365	0.265	0.201
12	751.66	12	<.0001	0.165	0.084	0.043	0.048	0.031	0.002
❷ 18	771.14	18	<.0001	-0.027	-0.047	-0.096	-0.096	-0.095	-0.086
24	784.80	24	<.0001	-0.087	-0.071	-0.062	-0.075	-0.059	-0.025
30	802.67	30	<.0001	-0.021	-0.019	-0.048	-0.083	-0.111	-0.107
36	862.27	36	<.0001	-0.116	-0.141	-0.147	-0.143	-0.142	-0.125
42	892.19	42	<.0001	-0.108	-0.093	-0.106	-0.088	-0.090	-0.088
48	913.92	48	<.0001	-0.041	-0.011	0.018	0.053	0.116	0.145

Autocorrelation Plot of Residuals

```
Lag   Covariance   Correlation   -1 9 8 7 6 5 4 3 2 1 0 1 2 3 4 5 6 7 8 9 1      Std Error

  0     2.702153       1.00000    |                    |********************|           0
  1     2.107649       0.77999    |                  . |****************    |    0.044856
  2     1.640982       0.60729    |                 .  |************        |    0.066785
  3     1.321663       0.48911    |                 .  |**********          |    0.077100
  4     0.985109       0.36456    |                 .  |*******             |    0.083109
  5     0.715251       0.26470    |                 .  |*****               |    0.086267
  6     0.543254       0.20104    |                .   |****                |    0.087886
  7     0.447149       0.16548    |                .   |***.               |    0.088806
  8     0.226080       0.08367    |                .   |** .               |    0.089424
  9     0.115412       0.04271    |                .   |*  .               |    0.089582
 10     0.129203       0.04781    |                .   |*  .               |    0.089623
```

Output 4.15 *Fitting a Transfer Function: PROC ARIMA (continued)*

```
                    Inverse Autocorrelations

 Lag    Correlation    -1 9 8 7 6 5 4 3 2 1 0 1 2 3 4 5 6 7 8 9 1

   1       -0.49941     |         **********| .                 |
   2        0.05096     |                 . |*.                 |
   3       -0.08471     |                 **| .                 |
   4        0.05085     |                 . |*.                 |
   5       -0.00835     |                 . | .                 |
   6        0.06962     |                 . |*.                 |
   7       -0.14938     |                ***| .                 |
   8        0.08073     |                 . |**                 |
   9        0.04528     |                 . |*.                 |
  10       -0.04108     |                 .*| .                 |

                    Partial Autocorrelations

 Lag    Correlation    -1 9 8 7 6 5 4 3 2 1 0 1 2 3 4 5 6 7 8 9 1

   1        0.77999     |                 . |****************   |
   2       -0.00280     |                 . | .                 |
   3        0.04161     |                 . |*.                 |
   4       -0.07432     |                 .*| .                 |
   5       -0.01456     |                 . | .                 |
   6        0.02003     |                 . | .                 |
   7        0.03808     |                 . |*.                 |
   8       -0.13172     |                ***| .                 |
   9        0.03341     |                 . |*.                 |
  10        0.06932     |                 . |*.                 |
```

Crosscorrelation Check of Residuals with Input X

To Lag	Chi-Square	DF	Pr > ChiSq	--------------------Crosscorrelations-------------------					
5	0.57	4	0.9668	-0.031	0.008	-0.004	-0.007	-0.006	0.001
11	1.93	10	0.9969	0.007	-0.009	0.004	-0.002	-0.018	0.048
❹ 17	7.25	16	0.9682	0.056	0.042	0.037	0.043	0.044	0.027
23	15.71	22	0.8300	0.032	0.072	0.055	0.058	0.057	0.035
29	16.35	28	0.9603	0.019	-0.003	0.016	0.012	-0.019	0.013
35	19.87	34	0.9743	-0.075	-0.022	-0.002	0.029	-0.013	0.003
41	23.89	40	0.9796	0.002	0.002	-0.019	0.026	0.070	0.047
47	26.13	46	0.9919	0.021	0.023	-0.001	-0.019	0.032	0.047

Output 4.15 *Fitting a Transfer Function: PROC ARIMA (continued)*

```
                Model for Variable Y

           Estimated Intercept    -33.9976


                  Input Number 1

        Input Variable                      X
        Shift                               2
        Overall Regression Factor    3.062708


                 Numerator Factors

            Factor 1:  1 - 0.39865 B**(1)


                Denominator Factors

            Factor 1:  1 - 0.79069 B**(1)
```

Output 4.16 *Modeling and Plotting Forecasts for Generated Data*

```
             FITTING A TRANSFER FUNCTION

                The ARIMA Procedure


              Preliminary Estimation

              Initial Autoregressive
                     Estimates

                            Estimate

                1            0.85629


        Constant Term Estimate      1.445571
        White Noise Variance Est    10.06195
```

Output 4.16 *Modeling and Plotting Forecasts for Generated Data (continued)*

Conditional Least Squares Estimation

Iteration	SSE	MU	AR1,1	SCALE1	NUM1,1	DEN1,1	Constant	Lambda	R Crit
0	3908.38	10.05915	0.85629	3.01371	0.10000	0.10000	1.445571	0.00001	1
1	2311.96	-4.07135	0.90707	2.78324	0.45056	0.75974	-0.37835	0.00001	0.909428
2	552.12	-39.3685	0.87876	2.98650	0.39526	0.81121	-4.7732	1E-6	0.881845
3	532.23	-32.7390	0.82664	3.00105	0.39186	0.78952	-5.67548	0.001	0.257986
4	516.31	-31.1812	0.77427	2.99250	0.36716	0.76939	-7.03863	0.0001	0.186947

Maximum Likelihood Estimation

Iter	Loglike	MU	AR1,1	SCALE1	NUM1,1	DEN1,1	Constant	Lambda	R Crit
0	-711.36100	-31.1812	0.77427	2.99250	0.36716	0.76939	-7.03863	0.00001	1
1	-710.55902	-32.4260	0.79415	2.99417	0.37233	0.77860	-6.67482	1E-6	0.058893
2	-710.53822	-32.4551	0.79253	2.99332	0.37251	0.77865	-6.73363	1E-7	0.009205
3	-710.53820	-32.4632	0.79263	2.99340	0.37258	0.77872	-6.73204	1E-8	0.000325

ARIMA Estimation Optimization Summary

Estimation Method	Maximum Likelihood
Parameters Estimated	5
Termination Criteria	Maximum Relative Change in Estimates
Iteration Stopping Value	0.001
Criteria Value	0.000248
Alternate Criteria	Relative Change in Objective Function
Alternate Criteria Value	3.152E-6
Maximum Absolute Value of Gradient	3.435957
R-Square Change from Last Iteration	0.000325
Objective Function	Log Gaussian Likelihood
Objective Function Value	-710.538
Marquardt's Lambda Coefficient	1E-8

ARIMA Estimation Optimization Summary

Numerical Derivative Perturbation Delta	0.001
Iterations	3

Output 4.16 *Modeling and Plotting Forecasts for Generated Data (continued)*

Maximum Likelihood Estimation

	Parameter	Estimate	Standard Error	t Value	Approx Pr > |t|	Lag	Variable	Shift
❶	MU	-32.46316	1.72566	-18.81	<.0001	0	Y	0
	AR1,1	0.79263	0.02739	28.94	<.0001	1	Y	0
	SCALE1	2.99340	0.04554	65.74	<.0001	0	X	2
	NUM1,1	0.37258	0.02286	16.30	<.0001	1	X	2
	DEN1,1	0.77872	0.01369	56.86	<.0001	1	X	2

Constant Estimate	-6.73204
Variance Estimate	1.029993
Std Error Estimate	1.014886
AIC	1431.076
SBC	1452.119
Number of Residuals	497

Correlations of Parameter Estimates

Variable		Y	Y	X	X	X
Parameter		MU	AR1,1	SCALE1	NUM1,1	DEN1,1
Y	MU	1.000	0.028	-0.288	-0.384	-0.795
Y	AR1,1	0.028	1.000	-0.021	0.002	-0.012
X	SCALE1	-0.288	-0.021	1.000	0.070	-0.015
X	NUM1,1	-0.384	0.002	0.070	1.000	0.818
X	DEN1,1	-0.795	-0.012	-0.015	0.818	1.000

Autocorrelation Check of Residuals

	To Lag	Chi-Square	DF	Pr > ChiSq	--------Autocorrelations--------					
	6	4.22	5	0.5184	-0.011	-0.040	0.063	-0.012	-0.037	-0.034
❸	12	19.06	11	0.0599	0.118	-0.071	-0.084	0.044	0.037	-0.003
	18	26.15	17	0.0718	-0.017	0.052	-0.098	-0.005	-0.029	0.017
	24	34.15	23	0.0630	-0.043	0.006	0.046	-0.073	-0.040	0.067
	30	39.57	29	0.0911	-0.006	0.072	0.010	-0.025	-0.066	0.002
	36	42.81	35	0.1710	0.019	-0.040	-0.038	-0.003	-0.049	-0.018
	42	52.32	41	0.1107	-0.007	0.036	-0.067	0.030	-0.010	-0.104
	48	58.30	47	0.1248	0.028	-0.003	-0.012	-0.045	0.085	0.026

***Output 4.16** Modeling and Plotting Forecasts for Generated Data (continued)*

Crosscorrelation Check of Residuals with Input X

To Lag	Chi-Square	DF	Pr > ChiSq	Crosscorrelations					
❺ 5	0.75	4	0.9447	-0.002	0.019	-0.020	0.005	0.018	0.020
11	6.96	10	0.7294	0.026	-0.010	0.023	-0.002	-0.015	0.105
17	7.70	16	0.9573	0.026	0.003	0.007	0.020	0.018	-0.008
23	11.08	22	0.9736	0.013	0.074	-0.003	0.024	0.019	-0.015
29	14.34	28	0.9846	-0.015	-0.029	0.035	0.001	-0.046	0.047
35	28.14	34	0.7498	-0.131	0.060	0.024	0.051	-0.056	0.027
41	33.95	40	0.7384	-0.001	-0.002	-0.028	0.067	0.079	-0.015
47	38.77	46	0.7664	-0.021	0.011	-0.033	-0.026	0.077	0.038

```
                    Model for Variable Y

              Estimated Intercept    -32.4632


                Autoregressive Factors

              Factor 1:  1 - 0.79263 B**(1)


                    Input Number 1

          Input Variable                   X
          Shift                            2
          Overall Regression Factor  2.993395


                  Numerator Factors

              Factor 1:  1 - 0.37258 B**(1)


                 Denominator Factors

              Factor 1:  1 - 0.77872 B**(1)
```

Output 4.16 *Modeling and Plotting Forecasts for Generated Data (continued)*

Forecasts for Variable Y

Obs	Forecast	Std Error	95% Confidence Limits	
501	12.7292	1.0149	10.7400	14.7183
502	10.9660	1.2950	8.4278	13.5042
503	10.7301	3.3464	4.1713	17.2889
504	10.5608	4.3680	1.9997	19.1219
505	10.4360	4.9805	0.6744	20.1976
506	10.3423	5.3556	-0.1545	20.8390
507	10.2709	5.5859	-0.6772	21.2190
508	10.2160	5.7270	-1.0088	21.4407
509	10.1735	5.8134	-1.2206	21.5675
510	10.1404	5.8661	-1.3569	21.6378

FORECAST OUTPUT DATA SET

Obs	T	Y	FORECAST	STD	L95	U95	RESIDUAL
1	481	0.7780	0.7496	1.01489	-1.2396	2.7387	0.02846
2	482	1.9415	3.3176	1.01489	1.3285	5.3068	-1.37609
3	483	5.6783	4.3649	1.01489	2.3758	6.3541	1.31336
4	484	8.1388	8.0927	1.01489	6.1036	10.0818	0.04612
5	485	9.7920	9.6122	1.01489	7.6230	11.6013	0.17978

(more output lines)

Obs	T	Y	FORECAST	STD	L95	U95	RESIDUAL
26	506	.	10.3423	5.35559	-0.1545	20.8390	.
27	507	.	10.2709	5.58587	-0.6772	21.2190	.
28	508	.	10.2160	5.72703	-1.0088	21.4407	.
29	509	.	10.1735	5.81339	-1.2206	21.5675	.
30	510	.	10.1404	5.86612	-1.3569	21.6378	.

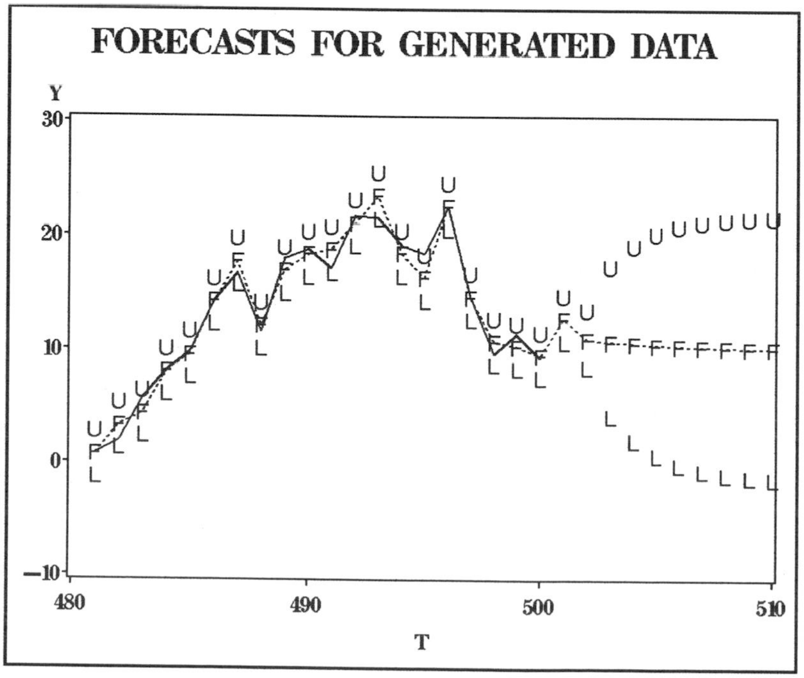

In addition to generated data, logarithms of flow rates for the Neuse River at Goldsboro, North Carolina, and 30 miles downstream at Kinston, North Carolina, are analyzed. These data include 400 daily observations. Obviously, the flow rates develop a seasonal pattern over the 365 days in a year, causing the ACF to die off slowly. Taking differences of the logarithmic observations produces ACFs that seem well behaved. The goal is to relate flow rates at Kinston to those at Goldsboro. The differenced data should suffice here even though nonstationarity is probably caused by the 365-day seasonal periodicity in flows.

You can obtain a model for the logarithms of the Goldsboro flow rates by using the following SAS statements:

```
PROC ARIMA DATA=RIVER;
   IDENTIFY VAR=LGOLD(1) NOPRINT;
   ESTIMATE Q=1 P=3 METHOD=ML NOCONSTANT MAXIT=100;
   IDENTIFY VAR=LKINS(1) CROSSCOR=(LGOLD(1));
   TITLE 'FLOW RATES OF NEUSE RIVER AT GOLDSBORO AND KINSTON';
RUN;
```

The results are shown in **Output 4.17**.

Output 4.17 *Analyzing Logarithms of Flow Data with the IDENTIFY and ESTIMATE Statements: PROC ARIMA*

FLOW RATES OF NEUSE RIVER AT GOLDSBORO AND KINSTON

The ARIMA Procedure

Maximum Likelihood Estimation

Parameter	Estimate	Standard Error	t Value	Approx Pr > |t|	Lag
MA1,1	0.87394	0.05878	14.87	<.0001	1
AR1,1	1.24083	0.07467	16.62	<.0001	1
AR1,2	-0.29074	0.08442	-3.44	0.0006	2
AR1,3	-0.11724	0.05394	-2.17	0.0297	3

Variance Estimate	0.039916
Std Error Estimate	0.199791
AIC	-148.394
SBC	-132.438
Number of Residuals	399

Correlations of Parameter Estimates

Parameter	MA1,1	AR1,1	AR1,2	AR1,3
MA1,1	1.000	0.745	-0.367	0.374
AR1,1	0.745	1.000	-0.783	0.554
AR1,2	-0.367	-0.783	1.000	-0.847
AR1,3	0.374	0.554	-0.847	1.000

Output 4.17 *Analyzing Logarithms of Flow Data with the IDENTIFY and ESTIMATE Statements: PROC ARIMA (continued)*

```
                          Autocorrelation Check of Residuals

 To        Chi-             Pr >
 Lag      Square     DF     ChiSq    --------------------Autocorrelations-------------------

   6        0.77      2    0.6819    -0.001     0.003     0.012    -0.032     0.020    -0.019
  12        8.67      8    0.3711     0.023    -0.032     0.074     0.041    -0.095    -0.040
❶ 18       14.28     14    0.4287    -0.040     0.006     0.071    -0.066     0.045     0.020
  24       19.19     20    0.5097     0.024     0.004     0.072    -0.014     0.071    -0.024
  30       22.15     26    0.6807    -0.018     0.058    -0.052    -0.021     0.007    -0.001
  36       26.18     32    0.7555     0.005     0.025     0.022    -0.089    -0.008    -0.005
  42       32.18     38    0.7347    -0.065     0.047     0.080    -0.016     0.006    -0.023
  48       34.21     44    0.8555    -0.029     0.005    -0.007    -0.010     0.042     0.041

                              Model for Variable LGOLD

                            Period(s) of Differencing     1

                              No mean term in this model.


                               Autoregressive Factors

          Factor 1:  1 - 1.24083 B**(1) + 0.29074 B**(2) + 0.11724 B**(3)

                               Moving Average Factors

                            Factor 1:  1 - 0.87394 B**(1)

                              Name of Variable = LKINS

                    Period(s) of Differencing                        1
                    Mean of Working Series                    0.006805
                    Standard Deviation                        0.152423
                    Number of Observations                         399
                    Observation(s) eliminated by differencing        1

                                   Autocorrelations

 Lag    Covariance    Correlation    -1 9 8 7 6 5 4 3 2 1 0 1 2 3 4 5 6 7 8 9 1      Std Error

   0      0.023233        1.00000    |                    |********************|              0
   1      0.012735        0.54814    |                  . |***********         |       0.050063
   2     0.0043502        0.18724    |                  . |****                |       0.063343
   3     0.0022257        0.09580    |                  . |**.                 |       0.064715
   4    -0.0002524        -.01086    |                  . |  .                 |       0.065070
   5    -0.0025054        -.10784    |                 .**|  .                 |       0.065074
   6    -0.0037296        -.16053    |                 ***|  .                 |       0.065521
   7    -0.0047520        -.20454    |                ****|  .                 |       0.066499
   8    -0.0040610        -.17480    |                 ***|  .                 |       0.068057
   9    -0.0028046        -.12072    |                 .**|  .                 |       0.069173
  10    -0.0033163        -.14274    |                 ***|  .                 |       0.069699
  11    -0.0034509        -.14853    |                 ***|  .                 |       0.070428
  12    -0.0026756        -.11517    |                 .**|  .                 |       0.071209
  13    -0.0012425        -.05348    |                 . *|  .                 |       0.071674
  14    0.00073313        0.03156    |                  . |* .                 |       0.071774
  15     0.0014717        0.06334    |                  . |* .                 |       0.071809
```

Output 4.17 *Analyzing Logarithms of Flow Data with the IDENTIFY and ESTIMATE Statements: PROC ARIMA (continued)*

```
                         Inverse Autocorrelations

 Lag    Correlation    -1 9 8 7 6 5 4 3 2 1 0 1 2 3 4 5 6 7 8 9 1

   1       -0.57039    |         ***********| .                  |
   2        0.26078    |                  . |*****               |
   3       -0.13604    |                 ***| .                  |
   4        0.05091    |                  . |*.                  |
   5        0.01526    |                  . | .                  |
   6        0.02426    |                  . | .                  |
   7       -0.00102    |                  . | .                  |
   8        0.06337    |                  . |*.                  |
   9       -0.05979    |                  .*| .                  |
  10        0.03898    |                  . |*.                  |
  11       -0.00474    |                  . | .                  |
  12        0.05098    |                  . |*.                  |
  13       -0.04073    |                  .*| .                  |
  14        0.08960    |                  . |**                  |
  15       -0.17275    |                 ***| .                  |


                         Partial Autocorrelations

 Lag    Correlation    -1 9 8 7 6 5 4 3 2 1 0 1 2 3 4 5 6 7 8 9 1

   1        0.54814    |                  . |***********         |
   2       -0.16184    |                 ***| .                  |
   3        0.09580    |                  . |**                  |
   4       -0.12466    |                  **| .                  |
   5       -0.06238    |                  .*| .                  |
   6       -0.08767    |                  **| .                  |
   7       -0.10017    |                  **| .                  |
   8       -0.00664    |                  . | .                  |
   9       -0.02847    |                  .*| .                  |
  10       -0.10749    |                  **| .                  |
  11       -0.05583    |                  .*| .                  |
  12       -0.05502    |                  .*| .                  |
  13        0.00784    |                  . | .                  |
  14        0.04108    |                  . |*.                  |
  15       -0.01643    |                  . | .                  |


                   Autocorrelation Check for White Noise

 To       Chi-             Pr >
Lag     Square     DF     ChiSq    --------------------Autocorrelations--------------------

  6     153.89      6    <.0001     0.548    0.187    0.096   -0.011   -0.108   -0.161
 12     212.40     12    <.0001    -0.205   -0.175   -0.121   -0.143   -0.149   -0.115
 18     222.41     18    <.0001    -0.053    0.032    0.063    0.002    0.076    0.101
 24     231.81     24    <.0001     0.074    0.074    0.069    0.057    0.030   -0.050
```

Output 4.17 *Analyzing Logarithms of Flow Data with the IDENTIFY and ESTIMATE Statements: PROC ARIMA (continued)*

```
                     Variable LGOLD has been differenced.


                       Correlation of LKINS and LGOLD

              Period(s) of Differencing                            1
              Number of Observations                             399
              Observation(s) eliminated by differencing            1
              Variance of transformed series LKINS          0.016725
              Variance of transformed series LGOLD          0.039507

                     Both series have been prewhitened.


                              Crosscorrelations

  Lag    Covariance    Correlation   -1 9 8 7 6 5 4 3 2 1 0 1 2 3 4 5 6 7 8 9 1

  -15    -0.0005089       -.01980    |                  . | .                  |
  -14    0.00075859       0.02951    |                  . |*.                  |
  -13    0.00003543       0.00138    |                  . | .                  |
  -12    -0.0005770       -.02245    |                  . | .                  |
  -11    -0.0004178       -.01625    |                  . | .                  |
  -10    -0.0021530       -.08376    |                  **| .                  |
   -9    -0.0000599       -.00233    |                  . | .                  |
   -8     0.0010101       0.03929    |                  . |*.                  |
   -7    0.00044281       0.01723    |                  . | .                  |
   -6    -0.0010635       -.04137    |                  .*| .                  |
   -5    -0.0002580       -.01004    |                  . | .                  |
   -4    0.00001085       0.00042    |                  . | .                  |
   -3    0.00092003       0.03579    |                  . |*.                  |
   -2     0.0042649       0.16592    |                  . |***                 |
   -1    -0.0000873       -.00339    |                  . | .                  |
    0    -0.0012901       -.05019    |                  .*| .                  |
    1      0.017885       0.69577    |                  . |**************  ❷   |
    2     0.0092795       0.36100    |                  . |*******             |
    3    0.00031429       0.01223    |                  . | .                  |
    4     0.0017646       0.06865    |                  . |*.                  |
    5    0.00069119       0.02689    |                  . |*.                  |
    6    0.00027701       0.01078    |                  . | .                  |
    7    0.00061810       0.02405    |                  . | .                  |
    8    0.00012023       0.00468    |                  . | .                  |
    9    0.00043703       0.01700    |                  . | .                  |
   10    0.00098972       0.03850    |                  . |*.                  |
   11    0.00044395       0.01727    |                  . | .                  |
   12    -0.0022054       -.08580    |                  **| .                  |
   13    -0.0013050       -.05077    |                  .*| .                  |
   14    -0.0012982       -.05050    |                  .*| .                  |
   15    0.00034243       0.01332    |                  . | .                  |
```

Output 4.17 *Analyzing Logarithms of Flow Data with the IDENTIFY and ESTIMATE Statements: PROC ARIMA (continued)*

Crosscorrelation Check Between Series

To Lag	Chi-Square	DF	Pr > ChiSq	--------------------Crosscorrelations------------------					
5	248.39	6	<.0001	-0.050	0.696	0.361	0.012	0.069	0.027
11	249.50	12	<.0001	0.011	0.024	0.005	0.017	0.039	0.017
17	263.88	18	<.0001	-0.086	-0.051	-0.051	0.013	0.136	-0.070
23	266.85	24	<.0001	0.031	0.043	-0.019	0.001	0.064	0.013

Both variables have been prewhitened by the following filter:

Prewhitening Filter

Autoregressive Factors

Factor 1: 1 - 1.24083 B**(1) + 0.29074 B**(2) + 0.11724 B**(3)

Moving Average Factors

Factor 1: 1 - 0.87394 B**(1)

The output from the ESTIMATE statement ❶ shows a reasonable fit. Cross-correlations from the second IDENTIFY statement ❷ show that a change in flow rates at Goldsboro affects the flow at Kinston one and two days later, with little other effect. This suggests $C(1-\theta B)X_{t-1}$ as a transfer function. Add the following SAS statements to the code above:

```
ESTIMATE INPUT=(1$(1)LGOLD) PLOT METHOD=ML;
RUN;
```

Results are shown in **Output 4.18**. Diagnostics from the PLOT option ❶ are used to identify an error model. The cross-correlation check ❷ looks reasonable, but the autocorrelation check ❸ indicates that the error is not white noise. This also implies that *t* statistics for the model parameters are computed from improper standard errors.

Output 4.18 *Modeling Flow Rates: Identifying an Error Model through the PLOT Option*

FLOW RATES OF NEUSE RIVER AT GOLDSBORO AND KINSTON

The ARIMA Procedure

Maximum Likelihood Estimation

Parameter	Estimate	Standard Error	t Value	Approx Pr > |t|	Lag	Variable	Shift
MU	0.0018976	0.0043054	0.44	0.6594	0	LKINS	0
NUM1	0.43109	0.02107	20.46	<.0001	0	LGOLD	1
NUM1,1	-0.22837	0.02106	-10.84	<.0001	1	LGOLD	1

Constant Estimate	0.001898
Variance Estimate	0.00735
Std Error Estimate	0.085733
AIC	-820.849
SBC	-808.897
Number of Residuals	397

Correlations of Parameter Estimates

Variable		LKINS	LGOLD	LGOLD
Parameter		MU	NUM1	NUM1,1
LKINS	MU	1.000	-0.018	0.020
LGOLD	NUM1	-0.018	1.000	0.410
LGOLD	NUM1,1	0.020	0.410	1.000

Autocorrelation Check of Residuals ❸

To Lag	Chi-Square	DF	Pr > ChiSq	Autocorrelations					
6	76.40	6	<.0001	0.315	-0.088	-0.195	-0.169	-0.106	-0.073
12	87.15	12	<.0001	-0.105	-0.081	0.062	0.042	0.054	-0.013
18	97.51	18	<.0001	-0.023	-0.005	-0.033	-0.140	-0.010	0.059
24	114.97	24	<.0001	0.092	0.150	0.059	0.036	-0.033	-0.069
30	120.35	30	<.0001	-0.044	-0.053	-0.056	-0.062	-0.023	-0.019
36	131.49	36	<.0001	0.031	0.042	0.087	0.099	0.063	-0.039
42	137.18	42	<.0001	-0.049	-0.042	-0.041	-0.004	-0.048	-0.068
48	152.24	48	<.0001	0.018	0.128	0.103	0.056	0.013	-0.054

Output 4.18 *Modeling Flow Rates: Identifying an Error Model through the PLOT Option (continued)*

```
                              Autocorrelation Plot of Residuals
                                                                       ❶
Lag    Covariance    Correlation    -1 9 8 7 6 5 4 3 2 1 0 1 2 3 4 5 6 7 8 9 1      Std Error

  0     0.0073501       1.00000     |                    |********************|             0
  1     0.0023175       0.31531     |                  . |******              |      0.050189
  2    -0.0006451       -.08776     |                  **| .                  |      0.054952
  3    -0.0014358       -.19534     |                ****| .                  |      0.055304
  4    -0.0012412       -.16886     |                 ***| .                  |      0.057016
  5    -0.0007808       -.10623     |                  **| .                  |      0.058262
  6    -0.0005350       -.07279     |                  .*| .                  |      0.058747
  7    -0.0007739       -.10530     |                  **| .                  |      0.058974
  8    -0.0005976       -.08130     |                  **| .                  |      0.059446
  9    0.00045435       0.06182     |                  . |*.                  |      0.059725
 10    0.00031116       0.04233     |                  . |*.                  |      0.059886
 11    0.00039753       0.05408     |                  . |*.                  |      0.059962
 12    -0.0000927       -.01261     |                  . | .                  |      0.060084
 13    -0.0001700       -.02312     |                  . | .                  |      0.060091
 14    -0.0000400       -.00545     |                  . | .                  |      0.060113
 15    -0.0002424       -.03298     |                  .*| .                  |      0.060115
 16    -0.0010321       -.14042     |                 ***| .                  |      0.060160
 17    -0.0000751       -.01021     |                  . | .                  |      0.060980
 18    0.00043128       0.05868     |                  . |*.                  |      0.060984
 19    0.00067524       0.09187     |                  . |**                  |      0.061126
 20     0.0011004       0.14972     |                  . |***                 |      0.061473
 21    0.00043052       0.05857     |                  . |*.                  |      0.062385
 22    0.00026751       0.03640     |                 .  |* .                 |      0.062523
 23    -0.0002427       -.03301     |                 . *|  .                 |      0.062577
 24    -0.0005082       -.06914     |                 . *|  .                 |      0.062621

                                 "." marks two standard errors

                                   Inverse Autocorrelations

                   Lag    Correlation    -1 9 8 7 6 5 4 3 2 1 0 1 2 3 4 5 6 7 8 9 1

                     1       -0.28165    |              ******| .                  |
                     2        0.20889    |                  . |****                |
                     3        0.12155    |                  . |**                  |
                     4        0.07801    |                  . |**                  |
                     5        0.10685    |                  . |**                  |
                     6        0.05649    |                  . |*.                  |
                     7        0.03509    |                  . |*.                  |
                     8        0.17435    |                  . |***                 |
                     9       -0.06121    |                  .*| .                  |
                    10        0.09295    |                  . |**                  |
                    11       -0.00055    |                  . | .                  |
                    12        0.03401    |                  . |*.                  |
                    13        0.03761    |                  . |*.                  |
                    14        0.01513    |                  . | .                  |
                    15       -0.07029    |                  .*| .                  |
                    16        0.15677    |                  . |***                 |
                    17       -0.09352    |                  **| .                  |
```

Output 4.18 *Modeling Flow Rates: Identifying an Error Model through the PLOT Option (continued)*

```
    18       0.00402    |                 .  |  .                |
    19       0.02305    |                 .  |  .                |
    20      -0.09855    |                **  |  .                |
    21       0.06347    |                 .  |* .                |
    22      -0.07294    |                 .* |  .                |
    23       0.02384    |                 .  |  .                |
    24       0.00998    |                 .  |  .                |


                        Partial Autocorrelations

   Lag   Correlation    -1 9 8 7 6 5 4 3 2 1 0 1 2 3 4 5 6 7 8 9 1

     1       0.31531    |                 .  |******             |
     2      -0.20784    |              ****  |  .                |
     3      -0.11186    |                **  |  .                |
     4      -0.09195    |                **  |  .                |
     5      -0.07202    |                 .* |  .                |
     6      -0.08240    |                **  |  .                |
     7      -0.13441    |               ***  |  .                |
     8      -0.07821    |                **  |  .                |
     9       0.04706    |                 .  |* .                |
    10      -0.08698    |                **  |  .                |
    11       0.01795    |                 .  |  .                |
    12      -0.07777    |                **  |  .                |
    13      -0.01386    |                 .  |  .                |
    14      -0.01931    |                 .  |  .                |
    15      -0.07046    |                 .* |  .                |
    16      -0.15748    |               ***  |  .                |
    17       0.07068    |                 .  |* .                |
    18      -0.03508    |                 .* |  .                |
    19       0.03301    |                 .  |* .                |
    20       0.08679    |                 .  |**                 |
    21      -0.00999    |                 .  |  .                |
    22       0.07693    |                 .  |**                 |
    23      -0.03975    |                 .* |  .                |
    24      -0.01307    |                 .  |  .                |


              Crosscorrelation Check of Residuals with Input LGOLD
```

To Lag	Chi-Square	DF	Pr > ChiSq	❷ ---------------	---------------	---------Crosscorrelations	---------------	---------------	---------------
5	4.08	5	0.5373	0.011	-0.012	-0.010	0.071	0.065	0.027
11	11.11	11	0.4345	0.036	-0.007	0.019	-0.008	-0.076	-0.101
17	21.14	17	0.2202	-0.033	-0.025	0.027	0.140	-0.007	0.056
23	24.04	23	0.4014	0.036	-0.048	-0.041	-0.015	-0.002	0.042
29	27.77	29	0.5303	-0.024	-0.065	-0.063	-0.014	-0.006	0.020
35	33.12	35	0.5593	0.023	0.007	0.011	-0.063	0.016	0.093
41	44.70	41	0.3193	0.126	0.045	-0.038	-0.067	-0.047	0.055
47	50.83	47	0.3252	-0.044	-0.057	0.062	0.072	-0.005	-0.033

Output 4.18 *Modeling Flow Rates: Identifying an Error Model through the PLOT Option (continued)*

```
                    Model for Variable LKINS

               Estimated Intercept          0.001898
               Period(s) of Differencing           1


                         Input Number 1

                Input Variable                  LGOLD
                Shift                               1
                Period(s) of Differencing           1


                        Numerator Factors

                Factor 1:  0.43109 + 0.22837 B**(1)
```

An ARMA(2,1) model fits the error term. Make the final estimation of the transfer function with noise by replacing the ESTIMATE statement (the one with the PLOT option) with

```
ESTIMATE P=2 Q=1 INPUT=(1$(1)LGOLD) METHOD=ML NOCONSTANT
   ALTPARM;
RUN;
```

Output 4.19 shows the results, and the model becomes ❶

$$\nabla \text{LKINS}_t = .49539(1+.55\text{B})\,\nabla \text{LGOLD}_{t-1} + (1-.8877\text{B})$$
$$/\left(1-1.16325\text{B}+.47963\text{B}^2\right)e_t$$

Because you encountered a pure delay, this is an example of a leading indicator, although this term is generally reserved for economic data. More insight into the effect of this pure delay is obtained through the cross-spectral analysis in Chapter 7.

Output 4.19 *Estimating the Final Transfer Function: PROC ARIMA*

FLOW RATES OF NEUSE RIVER AT GOLDSBORO AND KINSTON

The ARIMA Procedure

Maximum Likelihood Estimation

Parameter	Estimate	Standard Error	t Value	Approx Pr > |t|	Lag	Variable	Shift
MA1,1	0.88776	0.03506	25.32	<.0001	1	LKINS	0
AR1,1	1.16325	0.05046	23.05	<.0001	1	LKINS	0
AR1,2	-0.47963	0.04564	-10.51	<.0001	2	LKINS	0
SCALE1	0.49539	0.01847	26.83	<.0001	0	LGOLD	1
NUM1,1	-0.55026	0.04540	-12.12	<.0001	1	LGOLD	1

Variance Estimate	0.005838
Std Error Estimate	0.076407
AIC	-909.399
SBC	-889.48
Number of Residuals	397

Correlations of Parameter Estimates

Variable	Parameter	LKINS MA1,1	LKINS AR1,1	LKINS AR1,2	LGOLD SCALE1	LGOLD NUM1,1
LKINS	MA1,1	1.000	0.478	0.126	0.108	-0.091
LKINS	AR1,1	0.478	1.000	-0.618	0.056	-0.097
LKINS	AR1,2	0.126	-0.618	1.000	-0.121	0.042
LGOLD	SCALE1	0.108	0.056	-0.121	1.000	0.590
LGOLD	NUM1,1	-0.091	-0.097	0.042	0.590	1.000

Autocorrelation Check of Residuals

To Lag	Chi-Square	DF	Pr > ChiSq	--------------------Autocorrelations--------------------					
6	0.37	3	0.9468	0.005	-0.008	-0.010	-0.001	0.018	0.020
12	10.07	9	0.3446	-0.043	-0.094	0.092	-0.020	0.044	-0.048
18	20.04	15	0.1703	-0.030	0.004	0.036	-0.132	0.048	0.045
24	31.10	21	0.0720	0.023	0.145	0.001	0.064	-0.005	-0.025
30	33.30	27	0.1874	0.019	-0.010	-0.018	-0.056	0.002	-0.035
36	37.66	33	0.2645	0.024	-0.007	0.045	0.044	0.052	-0.052
42	39.40	39	0.4522	0.003	0.018	0.005	0.058	-0.005	-0.012
48	47.78	45	0.3605	0.028	0.111	0.054	0.035	0.035	-0.010

***Output 4.19** Estimating the Final Transfer Function: PROC ARIMA (continued)*

Crosscorrelation Check of Residuals with Input LGOLD

To Lag	Chi-Square	DF	Pr > ChiSq	--------------------Crosscorrelations					------------------
5	6.41	4	0.1705	-0.044	-0.062	0.010	0.079	0.054	0.034
11	14.95	10	0.1340	0.083	0.038	0.100	0.042	-0.031	-0.024
17	28.12	16	0.0306	0.041	-0.002	0.040	0.130	-0.051	0.102
23	29.97	22	0.1192	0.044	-0.036	0.006	0.003	0.003	0.038
29	33.48	28	0.2185	-0.047	-0.053	-0.056	-0.006	-0.026	-0.002
35	40.58	34	0.2028	-0.008	-0.018	-0.003	-0.087	0.045	0.089
41	48.60	40	0.1650	0.104	0.040	-0.008	-0.025	-0.001	0.085
47	55.14	46	0.1674	-0.069	-0.032	0.088	0.048	-0.019	-0.017

Model for Variable LKINS ❶

Period(s) of Differencing 1

No mean term in this model.

Autoregressive Factors

Factor 1: 1 - 1.16325 B**(1) + 0.47963 B**(2)

Moving Average Factors

Factor 1: 1 - 0.88776 B**(1)

Input Number 1

Input Variable	LGOLD
Shift	1
Period(s) of Differencing	1
Overall Regression Factor	0.495394

Numerator Factors

Factor 1: 1 + 0.55026 B**(1)

4.3.3.4 Summary of Modeling Strategy

Follow these steps in case 3 to complete your modeling:

1. Identify and estimate model for input X (IDENTIFY, ESTIMATE).
2. Prewhiten Y and X using model from item 1 (IDENTIFY).
3. Compute cross-correlations, $\gamma_{XY}(j)$, to identify transfer function form (IDENTIFY).
4. Fit transfer function and compute and analyze residuals (ESTIMATE, PLOT).
5. Fit transfer function with noise model (ESTIMATE).
6. Forecast X and Y (FORECAST).

4.3.4 Case 3B: Intervention

Suppose you use as an input X_t a sequence that is 0 through time 20 and 1 from time 21 onward. If the model is

$$Y_t = \alpha + \beta X_t + \text{noise}$$

you have

$$Y_t = \alpha + \text{noise}$$

through time 20 and

$$Y_t = \alpha + \beta + \text{noise}$$

after time 20. Thus, Y experiences an immediate level shift $(\text{from } \alpha \text{ to } \alpha + \beta)$ at time 21. Now change the model to

$$Y_t - \rho Y_{t-1} = \alpha + \beta X_t + \text{noise}$$

or

$$Y_t = \alpha' + \beta / (1 - \rho B) X_t + \text{noise}$$

where $\alpha' = \alpha/(1-\rho)$ (the expected value of Y when X is 0). You can also write

$$Y_t = \alpha' + \beta\left(X_t + \rho X_{t-1} + \rho^2 X_{t-2} + \ldots\right) + \text{noise}$$

At time 21, X_{21}=1 and the previous Xs are 0, so

$$Y_{21} = \alpha' + \beta + \text{noise}$$

At time 22 you get

$$Y_{22} = \alpha' + \beta(1+\rho) + \text{noise}$$

Y_t eventually approaches

$$\alpha' + \beta\left(1 + \rho + \rho^2 + \ldots\right) = \alpha' + \beta/(1-\rho)$$

if you ignore the noise term. Thus, you see that ratios of polynomials in the backshift operator B can provide interesting approaches to new levels.

When you use an indicator input, you cannot prewhiten. Therefore, impulse-response weights are not proportional to cross-covariances. You make the identification by comparing the behavior of Y_t near the intervention point with a catalog of typical behaviors for various transfer function forms. Several such response functions for X_t=1 when t>20 and 0 otherwise are shown in **Output 4.20**.

Output 4.20 *Plotting Intervention Models*

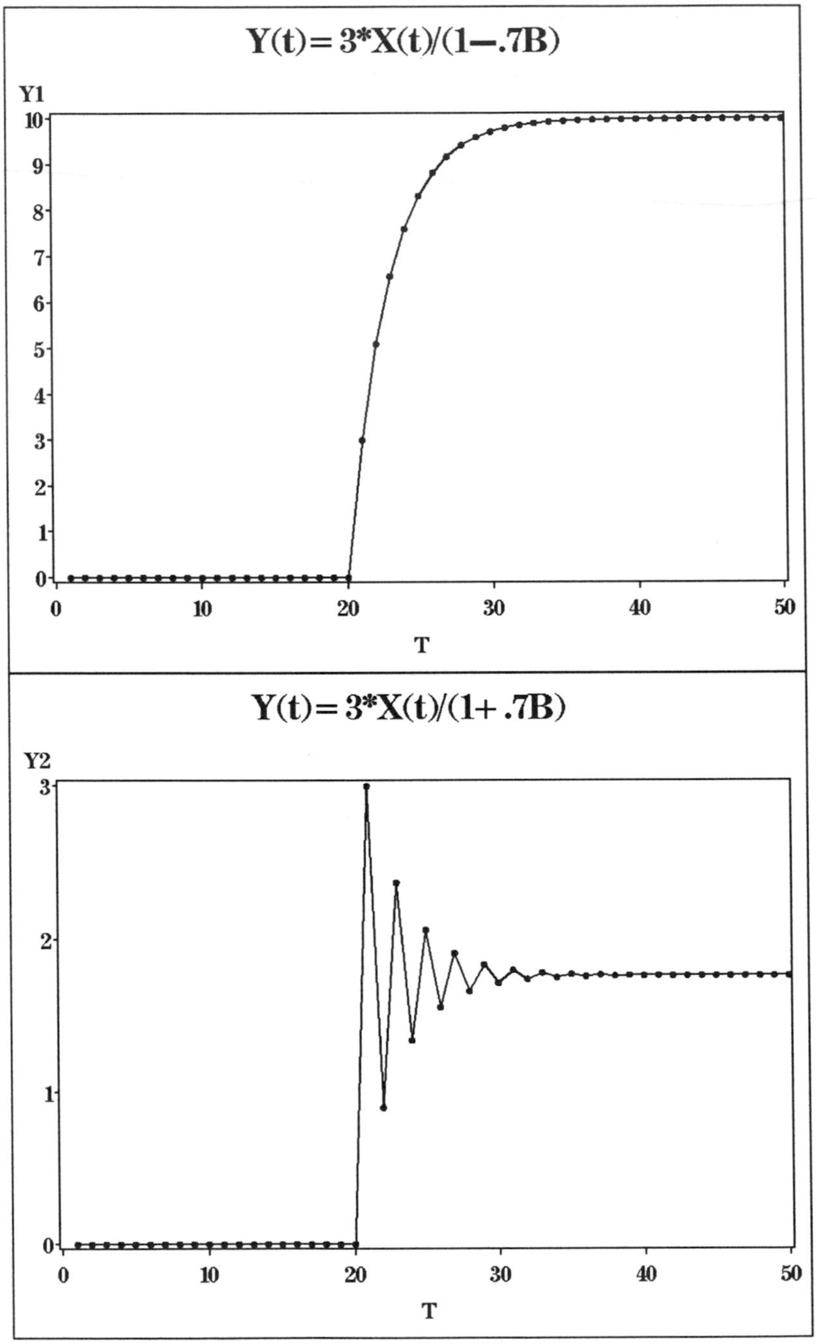

Output 4.20
Plotting Intervention Models (continued)

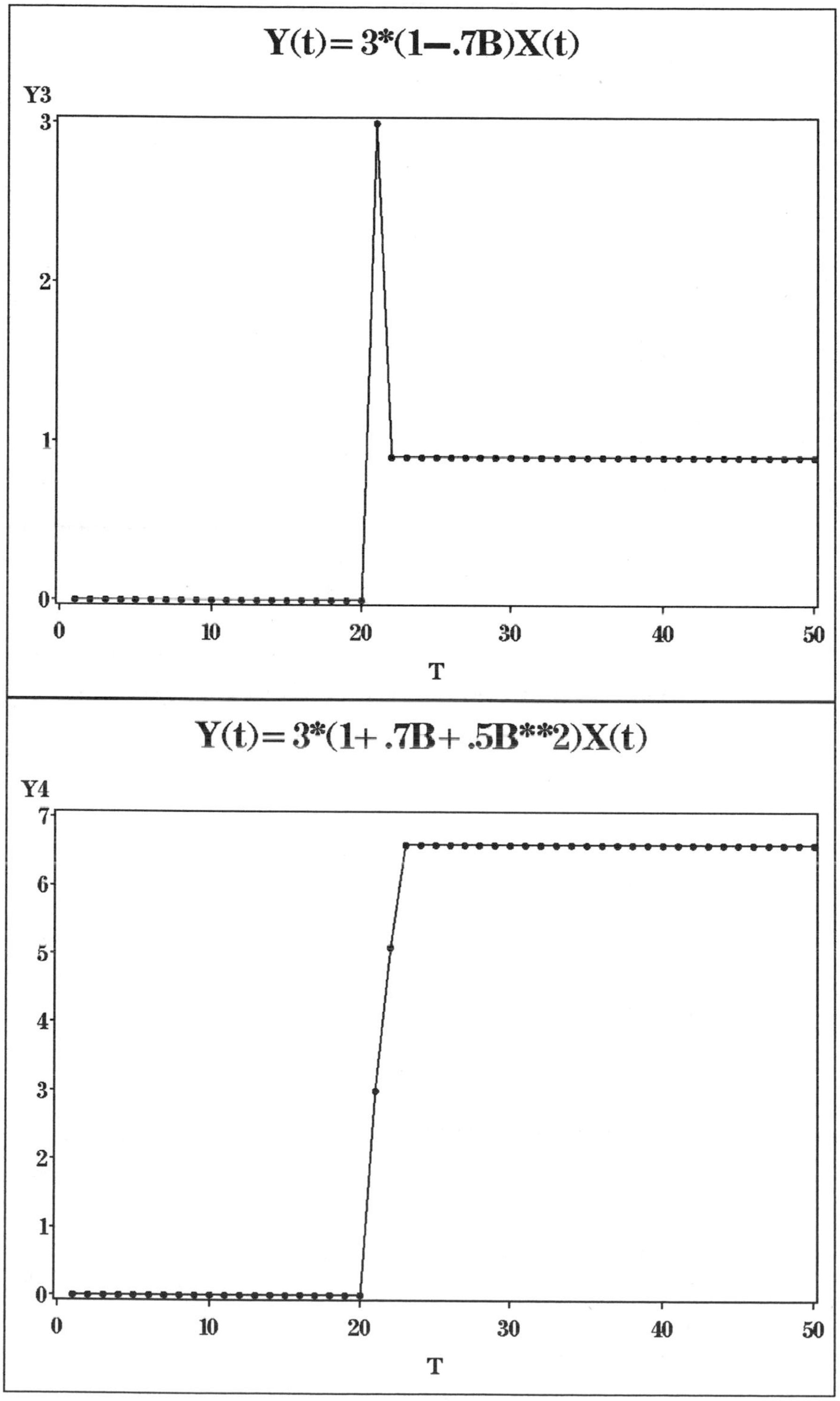

Output 4.20 *Plotting Intervention Models (continued)*

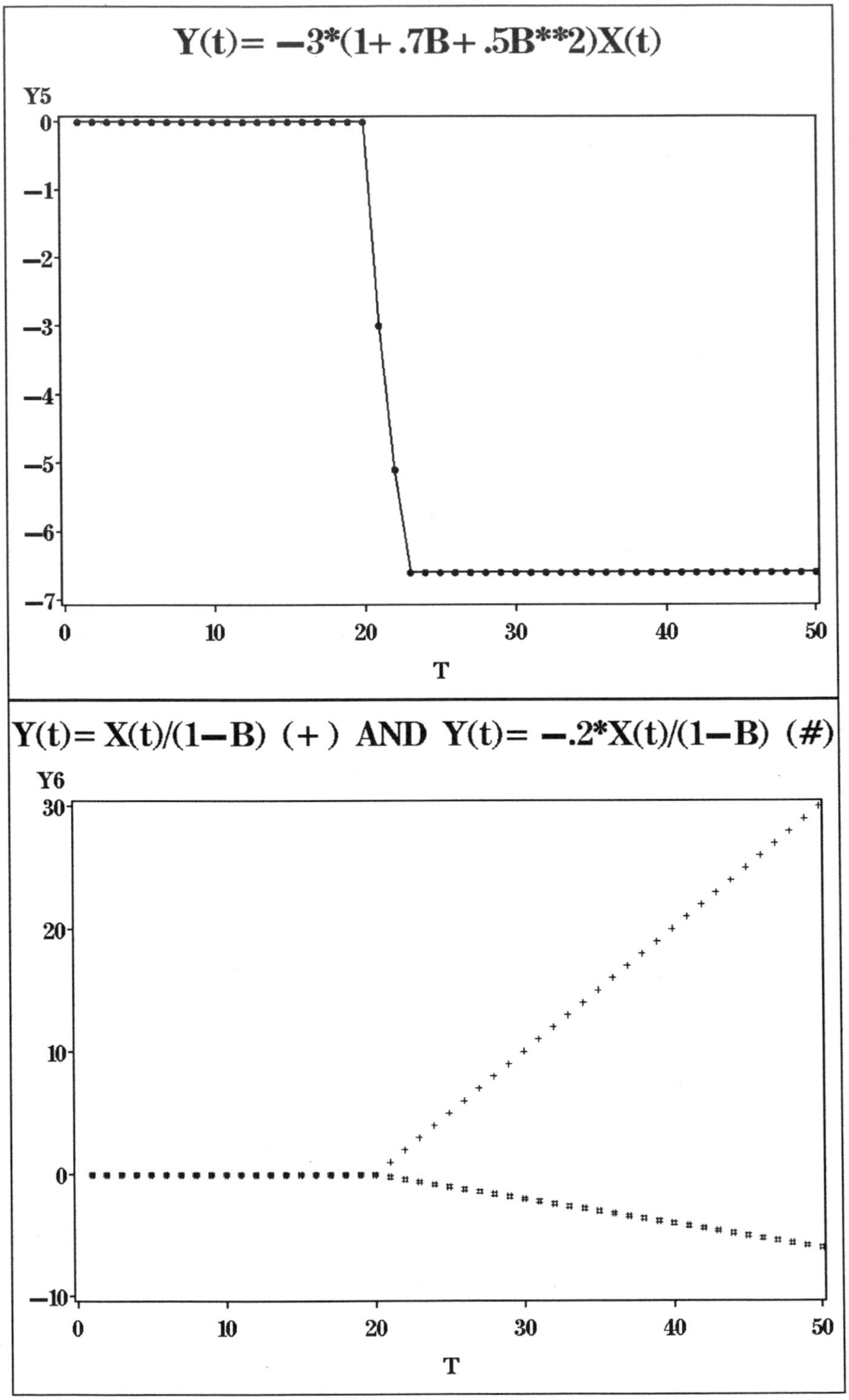

Output 4.20
Plotting Intervention Models (continued)

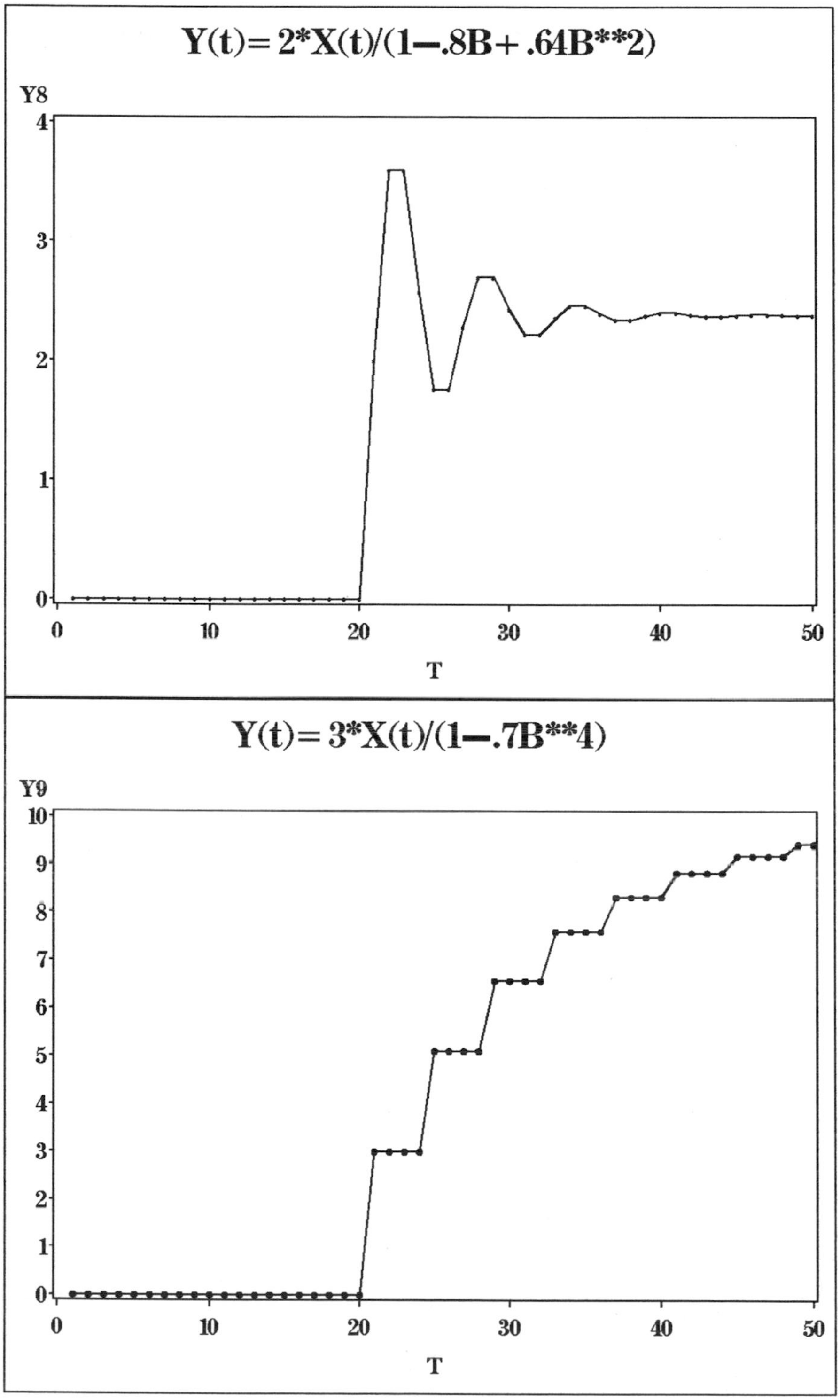

Output 4.20
Plotting Intervention Models (continued)

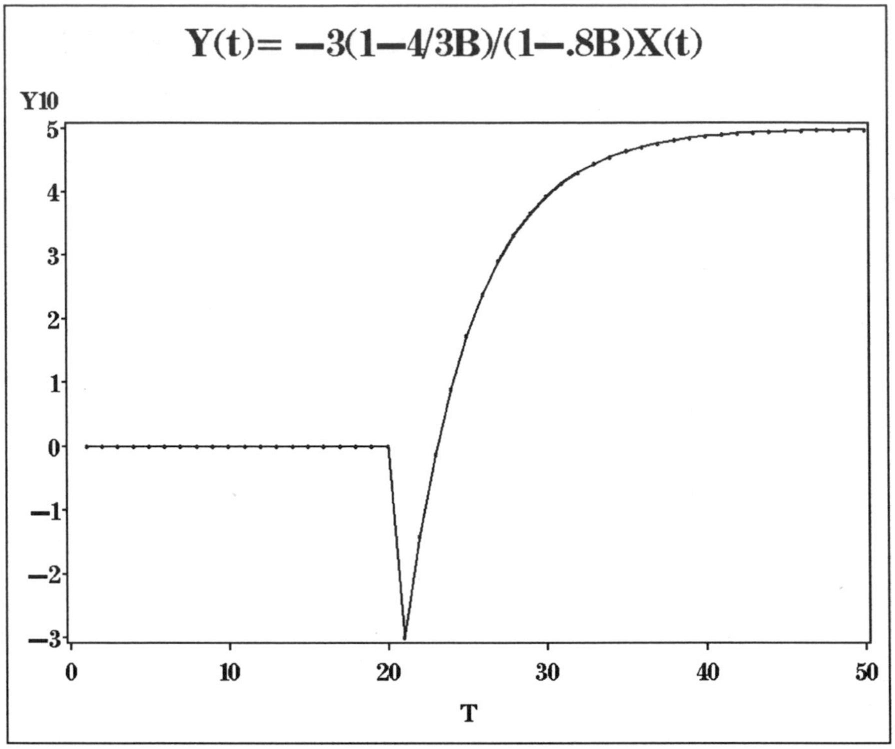

Output 4.21 shows calls for directory assistance in Cincinnati, Ohio (McSweeny, 1978).

Output 4.21
Plotting the Original Data

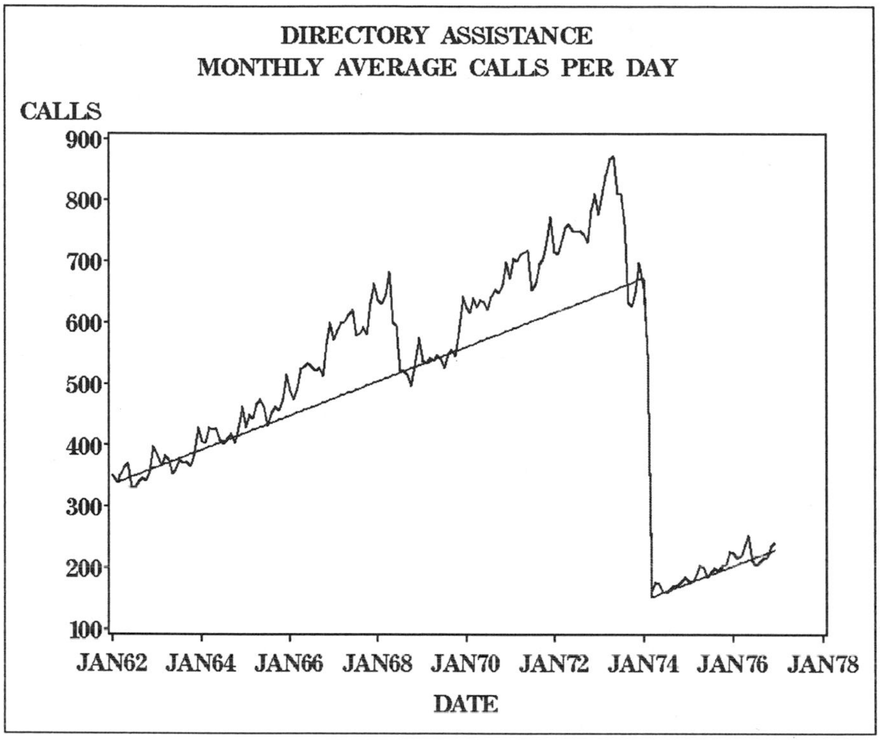

Prior to March 1974 directory assistance was free, but from that day on a charge was imposed. The data seem to show an initial falling off of demand starting in February, which may be an anticipation effect. The data clearly show an upward trend. You check the pre-intervention data for stationarity with the code

```
PROC ARIMA DATA=CALLS;
   IDENTIFY VAR=CALLS STATIONARITY = (ADF=(2,3,12,13) );
   IDENTIFY VAR=CALLS(1);
   ESTIMATE P=(12) METHOD=ML;
   WHERE DATE < '01FEB74'D;
RUN;
```

Some of the results are shown in **Output 4.22**. Only the trend tests are of interest since there is clearly a trend; however, none of the other tests could reject a unit root either. Tests with 12 and 13 lagged differences are requested in anticipation of seasonality. Below this are the chi-square checks for a seasonal AR model for the first differences. The fit is excellent and the seasonal AR parameter 0.5693 is not too close to 1. With this information you see that only the unit root tests with 12 or more lags are valid.

Output 4.22 *Unit Root Tests, Pre-intervention Calls Data*

Augmented Dickey-Fuller Unit Root Tests

Type	Lags	Rho	Pr < Rho	Tau	Pr < Tau	F	Pr > F
Trend	2	-18.1708	0.0888	-2.48	0.3378	3.37	0.5037
	3	-31.1136	0.0045	-3.04	0.1252	4.84	0.2091
	12	124.1892	0.9999	-2.96	0.1489	4.70	0.2387
	13	52.3936	0.9999	-3.18	0.0924	5.46	0.0964

The ARIMA Procedure

Autocorrelation Check of Residuals

To Lag	Chi-Square	DF	Pr > ChiSq	--------------Autocorrelations--------------					
6	2.71	5	0.7451	-0.015	-0.001	0.065	0.019	-0.111	-0.030
12	6.07	11	0.8683	-0.004	0.060	-0.039	-0.085	0.067	-0.067
18	11.10	17	0.8511	0.018	-0.067	0.029	-0.095	0.035	-0.120
24	16.64	23	0.8263	-0.002	-0.028	0.082	-0.149	-0.006	0.049

Model for Variable CALLS

```
Estimated Mean                 1.077355
Period(s) of Differencing             1

          Autoregressive Factors

     Factor 1:  1 - 0.56934 B**(12)
```

A first difference will reduce a linear trend to a constant, so calls tend to increase by 1.077 per month. The intervention variable IMPACT is created, having value 1 from February 1974 onward. Since the majority of the drop is seen in March, you fit an intervention model of the form $(\beta_0 - \beta_1 B)X_t$, where X_t is the IMPACT variable at time *t*. The first time X_t is 1, the effect is β_0, and after that both X_t and X_{t-1} will be 1 so that the effect is $\beta_0 - \beta_1$. You anticipate a negative β_0 and a larger-in-magnitude and positive β_1. A test that $\beta_0 = 0$ is a test for an anticipation effect. Motivated by the pre-intervention analysis, you try the same seasonal AR(1) error structure and check the diagnostics to see if it suffices. The code is as follows:

```
PROC ARIMA;
   IDENTIFY VAR=CALLS(1) CROSSCOR= (IMPACT(1)) NOPRINT;
   ESTIMATE INPUT = ((1)IMPACT) P=(12) METHOD=ML;
RUN;
```

It is seen in **Output 4.23** that all terms except mu are significant. The trend part of the fitted model is overlaid on the data in **Output 4.24**. Because the model has a unit root, the data can wander fairly far from this trend, and this indeed happens. It also explains why the standard error for mu is so large; that is, with random walk errors it is difficult to accurately estimate the drift term. Despite this, the model seems to capture the intervention well and seems poised to offer an accurate forecast of the next few values. The drop of –123 in calls the month prior to the charge is significant, so there was an anticipation effect. An additional drop of 400 leaves the calls at 523 below the previous levels.

Output 4.23
PROC ARIMA for Calls Data

DIRECTORY ASSISTANCE, MONTHLY AVERAGE CALLS PER DAY

The ARIMA Procedure

Maximum Likelihood Estimation

Parameter	Estimate	Standard Error	t Value	Approx Pr > |t|	Lag	Variable	Shift
MU	2.32863	2.88679	0.81	0.4199	0	CALLS	0
AR1,1	0.45045	0.06740	6.68	<.0001	12	CALLS	0
NUM1	-123.18861	20.39502	-6.04	<.0001	0	IMPACT	0
NUM1,1	400.69122	20.34270	19.70	<.0001	1	IMPACT	0

Constant Estimate	1.279694
Variance Estimate	503.7929
Std Error Estimate	22.44533
AIC	1619.363
SBC	1632.09
Number of Residuals	178

Autocorrelation Check of Residuals

To Lag	Chi-Square	DF	Pr > ChiSq	Autocorrelations					
6	3.43	5	0.6342	0.009	-0.046	0.058	0.016	-0.109	-0.030
12	7.91	11	0.7209	-0.026	0.048	-0.022	-0.110	0.015	-0.088
18	11.47	17	0.8312	-0.024	-0.093	0.021	-0.055	0.025	-0.068
24	19.83	23	0.6518	0.006	0.001	0.062	-0.162	-0.029	0.098
30	22.73	29	0.7886	-0.021	0.019	0.016	-0.025	-0.026	0.105

Output 4.24
Effect of Charge for Directory Assistance

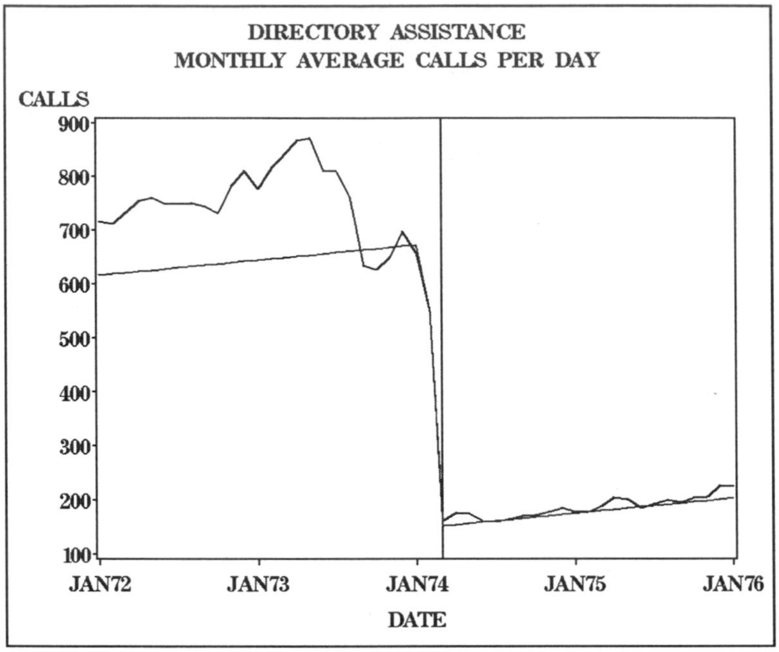

To forecast the next few months, you extend the data set with missing values for calls and set the intervention variable to 1, assuming the charge will remain in effect. The code below produces the plot in **Output 4.25**. Note how the forecasts and intervals for the historical data have been deleted from the plot. The intervals are quite wide due to the unit root structure of the errors. Recall that even the historical data have produced some notable departures from trend. Adding other predictor variables, like population or new phone installations, might help reduce the size of these intervals, but the predictors would need to be extrapolated into the future.

```
DATA EXTRA;
   DO T=1 TO 24;
      DATE = INTNX('MONTH','01DEC76'D,T);
      IMPACT=1;
   OUTPUT;
   END;
RUN;
DATA ALL;
   SET CALLS EXTRA;
RUN;
PROC ARIMA;
   IDENTIFY VAR=CALLS(1) CROSSCOR=( IMPACT(1) )    NOPRINT;
   ESTIMATE INPUT = ( (1) IMPACT ) P=(12)  METHOD=ML NOPRINT;
   FORECAST LEAD=24 OUT=GRAPH ID=DATE INTERVAL=MONTH;
RUN;
DATA GRAPH;
   SET GRAPH;
   IF CALLS NE . THEN DO;
      FORECAST=.; U95=.; L95=.;
   END;
RUN;
PROC GPLOT DATA=GRAPH;
   PLOT (CALLS FORECAST U95 L95)*DATE/OVERLAY;
   SYMBOL1 V=NONE I=JOIN R=4;
   TITLE ``FORECASTED CALLS'';
RUN;
```

Output 4.25
Forecasts from Intervention Model

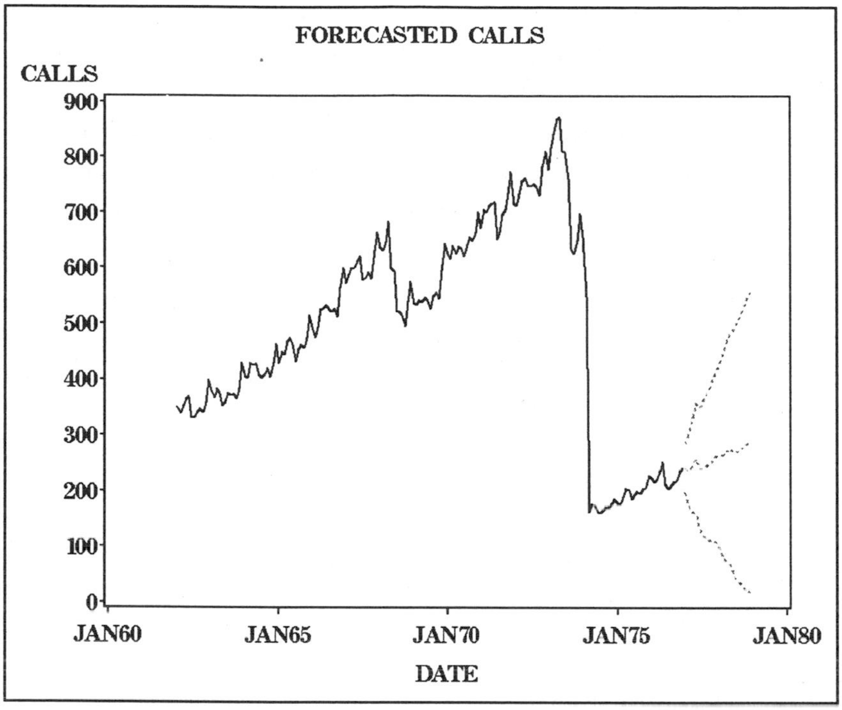

4.4 Further Examples

4.4.1 North Carolina Retail Sales

Consider again the North Carolina retail sales data investigated in Chapter 1. Recall that there the quarterly sales increases were modeled using seasonal dummy variables; that is, seasonal dummy variables were fit to the first differences of quarterly sales. The models discussed in this section potentially provide an alternative approach. Here the full monthly data (from which the quarterly numbers were computed as averages) will be used. This is an example in which the airline model seems a good choice at first, but later runs into some problems. Recall that when a first difference is found, often a moving average at lag 1 is appropriate. Likewise, a multiplicative moving average structure, specified by ESTIMATE Q = (1)(12), often works well when the first and span 12 difference, $(Y_t - Y_{t-1}) - (Y_{t-12} - Y_{t-13})$, has been taken. You can think of these moving average terms as somewhat mitigating the impact of the rather heavy-handed differencing operator. As in the IBM example in Section 3.4.7, the fitting of these moving average terms causes forecasts to be weighted averages of seasonal patterns over all past years where the weights decrease exponentially as you move further into the past. Thus the forecast is influenced somewhat by all past patterns but most substantially by those of the most recent years.

The airline model just discussed will be written here as $(1-B)(1-B^{12})Y_t = (1-\theta_{1,1}B)(1-\theta_{2,1}B^{12})e_t$, introducing double subscripts to indicate which factor and which lag within that factor is being modeled. This double-subscript notation corresponds to PROC ARIMA output. The airline model is often a good first try when seasonal data are encountered. Now if, for example, $\theta_{2,1} = 1$, then there is cancellation on both sides of the model and it reduces to $(1-B)Y_t = (1-\theta_{1,1}B)e_t$. Surprisingly, this can happen even with strongly seasonal data. If it does, as it will for the retail sales, it suggests considering a model outside the ARIMA class. Consider a model $Y_t = \mu + S_t + Z_t$ where $S_t = S_{t-12}$, and Z_t has some ARIMA structure, perhaps even having unit roots. Note that S_t forms an exactly repeating seasonal pattern, as would be modeled using dummy variables. Because of S_t the autocorrelation function will have spikes at lag 12, as will that of the ordinary first differences since $S_t - S_{t-1}$ is also periodic. However, the span 12 difference $Y_t - Y_{t-12}$ will involve $(1-B^{12})Z_t$, and unless Z_t has a unit root at lag 12, estimates of the coefficient of Z_{t-12} will be forced toward the moving average boundary. This overdifferencing often results in failure to converge.

You issue the following SAS statements to plot the data and compute the ACF of the original series, first differenced series, and first and seasonally differenced series.

```
PROC GPLOT DATA=NCRETAIL;
   PLOT SALES*DATE/HMINOR=0
      HREF='01DEC83'D '01DEC84'D '01DEC85'D '01DEC86'D
           '01DEC87'D '01DEC88'D '01DEC89'D '01DEC90'D
           '01DEC91'D '01DEC92'D '01DEC93'D '01DEC94'D;
   TITLE "NORTH CAROLINA RETAIL SALES";
   TITLE2 "IN MILLIONS";
RUN;
PROC ARIMA DATA=NCRETAIL ;
   IDENTIFY VAR=SALES OUTCOV=LEVELS NLAG=36;
   IDENTIFY VAR=SALES(1) OUTCOV=DIFF NLAG=36;
   IDENTIFY VAR=SALES(1,12) OUTCOV=SEAS NLAG=36;
RUN;
```

The data plot is shown in **Output 4.26**. The ACF, IACF, and PACF have been saved with the OUTCOV=option. **Output 4.27** uses this with SAS/GRAPH and a template to produce a matrix of plots with rows representing original, (1), and (1,12) differenced data and columns representing, from left to right, the ACF, IACF, and PACF.

Output 4.26
Plotting the Original Data

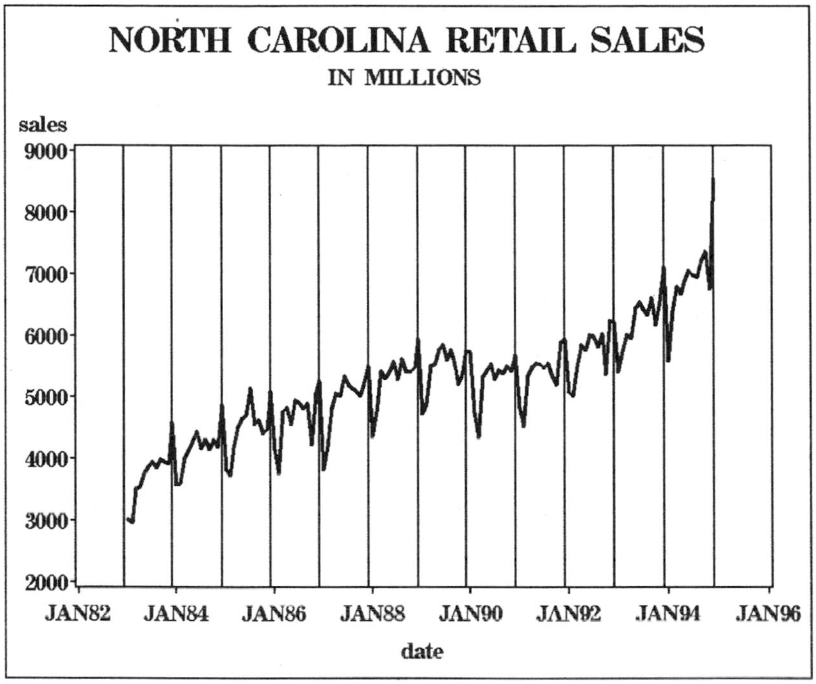

Output 4.27
Computing the ACF with the IDENTIFY Statement: PROC ARIMA

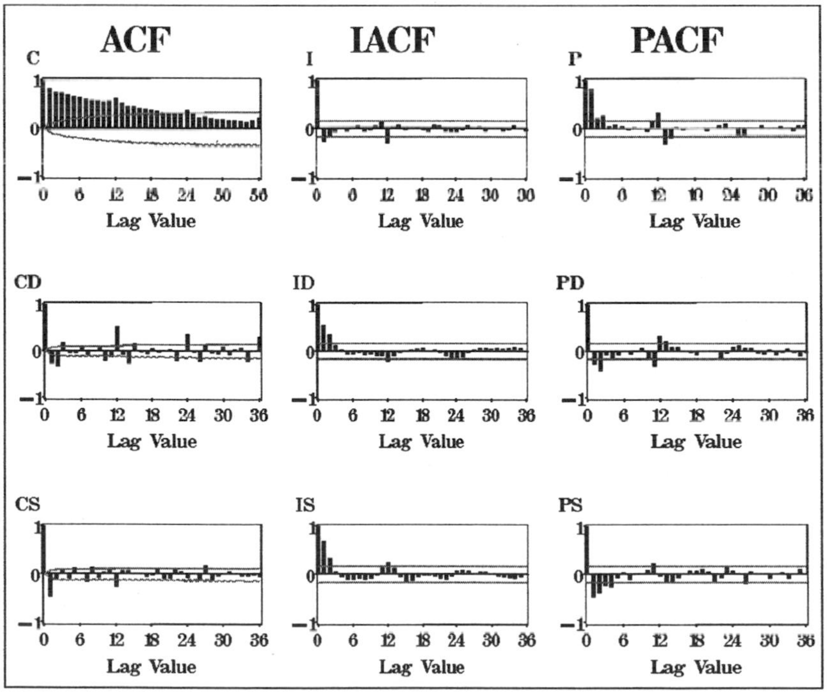

The plot of the data displays nonstationary behavior (nonconstant mean). The original ACF shows slow decay, indicating a first differencing. The ACF of the differenced series shows somewhat slow decay at the seasonal lags, indicating a possible span 12 difference. The *Q* statistics and ACF on the SALES(1,12) differenced variable indicate that some MA terms are needed, with the ACF spikes at 1 and 12 indicating MA terms at lags 1 and 12. Heeding the remarks at the beginning of this section, you try a multiplicative structure even though the expected side lobes at 11 and 13 (that such a structure implies) are not evident in the ACF. Such a structure also serves as a check on the differencing, as you will see. Adding

```
ESTIMATE Q=(1)(12) ML;
```

to the above code requests that maximum likelihood estimates of the multiplicative MA be fitted to the first and span 12 differenced data. The results are in **Output 4.28.**

Output 4.28 *Fitting the Multiplicative MA Structure*

```
WARNING: The model defined by the new estimates is unstable. The iteration
         process has been terminated.

WARNING: Estimates may not have converged.

                  ARIMA Estimation Optimization Summary

Estimation Method                                         Maximum Likelihood
Parameters Estimated                                                       3
Termination Criteria                                   Maximum Relative Change
                                                                in Estimates
Iteration Stopping Value                                               0.001
Criteria Value                                                      69.27824
Maximum Absolute Value of Gradient                                  365546.8
R-Square Change from Last Iteration                                  0.13941
Objective Function                                     Log Gaussian Likelihood
Objective Function Value                                             -923.09
Marquardt's Lambda Coefficient                                       0.00001
Numerical Derivative Perturbation Delta                                0.001
Iterations                                                                 8
Warning Message                                  Estimates may not have converged.

                         The ARIMA Procedure

                    Maximum Likelihood Estimation

                               Standard                 Approx
    Parameter      Estimate       Error    t Value    Pr > |t|     Lag

    MU             -0.65905     1.52987      -0.43      0.6666       0
    MA1,1           0.74136     0.05997      12.36      <.0001       1
    MA2,1           0.99979    83.86699       0.01      0.9905      12

                   Constant Estimate      -0.65905
                   Variance Estimate      62632.24
                   Std Error Estimate     250.2643
                   AIC                     1852.18
                   SBC                    1860.805
                   Number of Residuals         131
```

Output 4.28 *Fitting the Multiplicative MA Structure (continued)*

Correlations of Parameter Estimates

Parameter	MU	MA1,1	MA2,1
MU	1.000	0.205	-0.117
MA1,1	0.205	1.00	-0.089
MA2,1	-0.117	-0.089	1.000

Autocorrelation Check of Residuals

To Lag	Chi-Square	DF	Pr > ChiSq	--------------Autocorrelations--------------					
6	9.12	4	0.0582	-0.131	-0.056	0.175	0.046	0.115	0.032
12	16.16	10	0.0950	0.047	0.115	0.062	0.120	-0.047	0.114
18	22.96	16	0.1149	0.009	0.003	0.211	0.014	-0.001	-0.014
24	33.26	22	0.0583	0.075	-0.026	-0.074	0.163	0.087	-0.137

Model for variable sales

Estimated Mean	-0.65905
Period(s) of Differencing	1,12

Moving Average Factors

```
Factor 1:  1 - 0.74136 B**(1)
Factor 2:  1 - 0.99979 B**(12)
```

In **Output 4.28**, you see that there seems to be a problem. The procedure had trouble converging, the standard error on the lag 12 coefficient is extremely large, and the estimate itself is almost 1, indicating a possibly noninvertible model. You can think of a near 1.00 moving average coefficient at lag 12 as trying to undo the span 12 differencing. Of course, trying to make inferences when convergence has not been verified is, at best, questionable. Returning to the discussion at the opening of this section, a possible explanation is that the seasonality S_t is regular enough to be accounted for by seasonal dummy variables. That scenario is consistent with all that has been observed about these data. The first difference plus dummy variable model of section 1 did seem to fit the data pretty well.

The dummy variables can be incorporated in PROC ARIMA using techniques in **Section 4.2.** Letting $S_{1,t}$ through $S_{12,t}$ denote monthly indicator variables (dummy variables), your model is

$$Y_t = \alpha + \beta t + \delta_1 S_{1,t} + \delta_2 S_{2,t} + \cdots + \delta_{11} S_{11,t} + Z_t$$

where, from your previous modeling, Z_t seems to have a (nonseasonal) unit root. You interpret $\alpha + \beta t$ as a "December line" in that, for December, each S_{jt} is 0. For January, the expected value of Y is $(\alpha + \delta_1) + \beta t$; that is, δ_1 is a shift in the trend line that is included for all January data and similar δ_j values allow shifts for the other 10 months up through November. Because Christmas sales are always relatively high, you anticipate that all these δs, and especially δ_1, will be negative. Using ∇ to denote a first difference, write the model at time t and at time $t-1$, then subtract to get

$$\nabla Y_t = \nabla\alpha + \beta(\nabla t) + \delta_1 \nabla S_{1,t} + \delta_2 \nabla S_{2,t} + \cdots + \delta_{11} \nabla S_{11,t} + \nabla Z_t$$

Now ∇Z_t is stationary if Z_t has a unit root, $\nabla\alpha = \alpha - \alpha = 0$, and $\nabla t = t - (t-1) = 1$. Since errors should be stationary for proper modeling in PROC ARIMA, the model will be specified in first differences as

$$\nabla Y_t = \beta + \delta_1 \nabla S_{1,t} + \delta_2 \nabla S_{2,t} + \cdots + \delta_{11} \nabla S_{11,t} + \nabla Z_t$$

The parameters have the same interpretations as before. This code fits the model with ∇Z_t specified as ARMA(2,1) and plots forecasts. The data set had 24 missing values for sales at the end with seasonal indicator variables S_{jt} nonmissing. Note that the seasonal indicator variables can be generated without error and so are valid deterministic inputs.

The following code produces **Output 4.29** and **Output 4.30**.

```
PROC ARIMA DATA=NCRETAIL;
   IDENTIFY VAR=SALES(1)    CROSSCOR =
     (S1(1) S2(1) S3(1) S4(1) S5(1) S6(1) S7(1) S8(1) S9(1)
      S10(1) S11(1) )
   NOPRINT;
   ESTIMATE INPUT=(S1 S2 S3 S4 S5 S6 S7 S8 S9 S10 S11) p=2 q=1 ml;
   FORECAST LEAD=24  OUT=OUT1 ID=DATE INTERVAL= MONTH;
RUN;
PROC GPLOT DATA=OUT1;
   PLOT (SALES L95 U95 FORECAST)*DATE/
      OVERLAY HREF ='01DEC94'D; WHERE DATE> '01JAN90'D;
      SYMBOL1 V=NONE I=JOIN C=BLACK L=1 R=1 W=1;
      SYMBOL2 V=NONE I=JOIN C=BLACK L=2 R=2 W=1;
      SYMBOL3 V=NONE I=JOIN C=BLACK L=1 R=1 W=2;
RUN;
```

Output 4.29 *Seasonal Model for North Carolina Retail Sales*

NC RETAIL SALES
2 YEARS OF FORECASTS

The ARIMA Procedure

Maximum Likelihood Estimation

Parameter	Estimate	Standard Error	t Value	Approx Pr > |t|	Lag	Variable	Shift
MU	26.82314	5.89090	4.55	<.0001	0	sales	0
MA1,1	0.50693	0.14084	3.60	0.0003	1	sales	0
AR1,1	-0.39666	0.14155	-2.80	0.0051	1	sales	0
AR1,2	-0.29811	0.12524	-2.38	0.0173	2	sales	0
NUM1	-1068.4	95.36731	-11.20	<.0001	0	S1	0
NUM2	-1092.1	93.36419	-11.70	<.0001	0	S2	0
NUM3	-611.48245	82.36315	-7.42	<.0001	0	S3	0
NUM4	-476.60662	89.45164	-5.33	<.0001	0	S4	0
NUM5	-396.94536	90.59402	-4.38	<.0001	0	S5	0
NUM6	-264.63164	87.94063	-3.01	0.0026	0	S6	0
NUM7	-371.30277	90.53160	-4.10	<.0001	0	S7	0
NUM8	-424.65711	88.92377	-4.78	<.0001	0	S8	0
NUM9	-440.79196	81.05429	-5.44	<.0001	0	S9	0
NUM10	-642.50812	92.86014	-6.92	<.0001	0	S10	0
NUM11	-467.54818	94.61205	-4.94	<.0001	0	S11	0

Output 4.29 *Seasonal Model for North Carolina Retail Sales (continued)*

```
                    Constant Estimate       45.45892
                    Variance Estimate       57397.62
                    Std Error Estimate       239.578
                    AIC                     1988.001
                    SBC                     2032.443
                    Number of Residuals          143

                   Autocorrelation Check of Residuals

 To       Chi-          Pr >
Lag     Square    DF    ChiSq   --------------Autocorrelations--------------

  6       1.88     3   0.5973   -0.009   -0.021    0.008    0.001    0.108    0.016
 12       7.55     9   0.5801    0.034    0.119    0.057    0.108   -0.076    0.024
 18      16.08    15   0.3770    0.013    0.030    0.222    0.030    0.014   -0.037
 24      27.84    21   0.1446    0.013   -0.040   -0.020    0.168    0.096   -0.169

                        Model for Variable SALES

                  Estimated Intercept          26.82314
                  Period(s) of Differencing           1

                         Autoregressive Factors

              Factor 1:  1 + 0.39666 B**(1) + 0.29811 B**(2)


                         Moving Average Factors

                     Factor 1:  1 - 0.50693 B**(1)

                           Input Number 1
                  Input Variable                     S1
                  Period(s) of Differencing           1
                  Overall Regression Factor    -1068.41

                           Input Number 2
                  Input Variable                     S2
                  Period(s) of Differencing           1
                  Overall Regression Factor    -1092.13

                           Input Number 3
                  Input Variable                     S3
                  Period(s) of Differencing           1
                  Overall Regression Factor    -611.482

                           Input Number 4
                  Input Variable                     S4
                  Period(s) of Differencing           1
                  Overall Regression Factor    -476.607

                           Input Number 5
                  Input Variable                     S5
                  Period(s) of Differencing           1
                  Overall Regression Factor    -396.945
```

Output 4.29 *Seasonal Model for North Carolina Retail Sales (continued)*

```
                  Input Number 6
          Input Variable                      S6
          Period(s) of Differencing            1
          Overall Regression Factor     -264.632

                  Input Number 7
          Input Variable                      S7
          Period(s) of Differencing            1
          Overall Regression Factor     -371.303

                  Input Number 8
          Input Variable                      S8
          Period(s) of Differencing            1
          Overall Regression Factor     -424.657

                  Input Number 9
          Input Variable                      S9
          Period(s) of Differencing            1
          Overall Regression Factor     -440.792

                  Input Number 10
          Input Variable                     S10
          Period(s) of Differencing            1
          Overall Regression Factor     -642.508

                  Input Number 11
          Input Variable                     S11
          Period(s) of Differencing            1
          Overall Regression Factor     -467.548
```

Forecasts for Variable SALES

Obs	Forecast	Std Error	95% Confidence Limits	
145	6769.7677	239.5780	6300.2034	7239.3320
146	6677.4766	240.6889	6205.7350	7149.2182
147	7438.1332	243.5997	6960.6865	7915.5799
148	7527.8413	261.9620	7014.4053	8041.2774
149	7587.4027	270.8251	7056.5952	8118.2102
150	7786.6130	277.8644	7242.0088	8331.2172
151	7704.8579	287.2914	7141.7771	8267.9387
152	7667.1369	295.8507	7087.2802	8246.9935
153	7682.8322	303.6424	7087.7040	8277.9604
154	7509.2889	311.5971	6898.5697	8120.0081
155	7709.0439	319.3641	7083.1018	8334.9861
156	8203.8173	326.8401	7563.2225	8844.4121
157	7162.6783	334.1872	6507.6835	7817.6731
158	7165.4868	341.3916	6496.3715	7834.6021
159	7672.9379	348.4302	6990.0272	8355.8485
160	7834.7312	355.3315	7138.2942	8531.1682
161	7941.1827	362.1054	7231.4692	8650.8962
162	8100.3044	368.7526	7377.5626	8823.0463
163	8020.4722	375.2818	7284.9333	8756.0111
164	7993.9393	381.7001	7245.8207	8742.0578
165	8004.6236	388.0121	7244.1338	8765.1133
166	7829.7326	394.2228	7057.0701	8602.3952
167	8031.5161	400.3374	7246.8692	8816.1629
168	8525.8866	406.3599	7729.4359	9322.3374

Output 4.30 shows the resulting graph.

Output 4.30
Forecasts from Seasonal Model

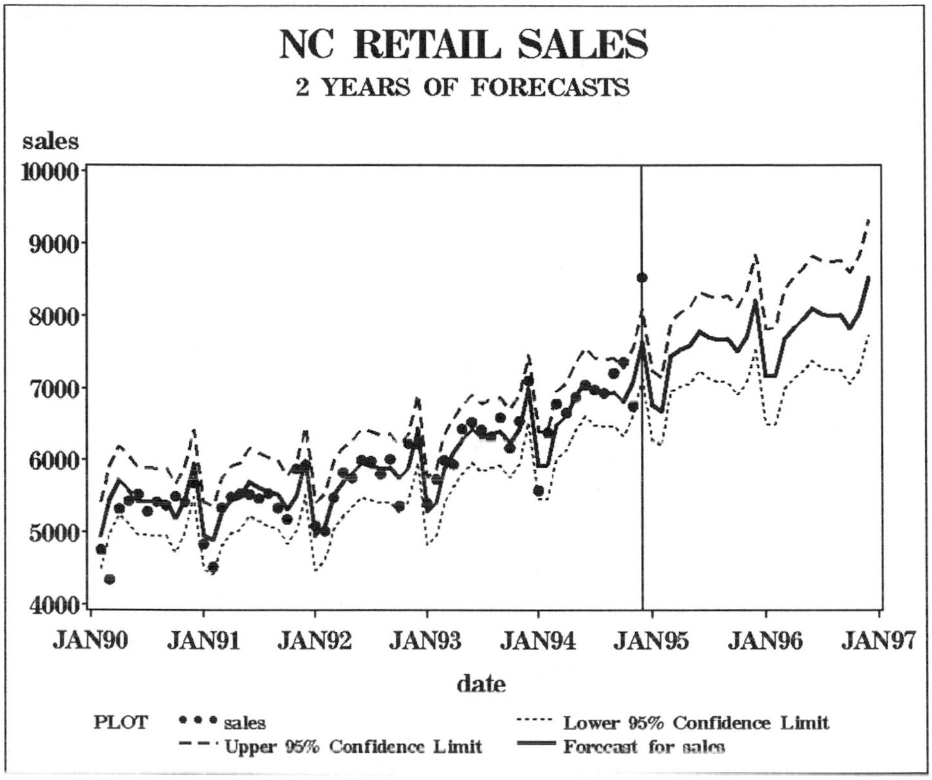

4.4.2 Construction Series Revisited

Returning to the construction worker series at the beginning of **Section 4.1.2**, you can fit two models both having a first difference. Let one incorporate a seasonal difference and the other incorporate seasonal dummy variables S1 through S12 to model the seasonal pattern. This code produces two forecast data sets, OUTDUM and OUTDIF, that have 24 forecasts from the two models. The data set ALL has the original construction data along with seasonal dummy variables S1 through S12 that extend 24 periods into the future. In **Section 4.4.1** the December indicator S12 was dropped to avoid a collinearity problem involving the intercept. An equally valid approach is to drop the intercept (NOCONSTANT) and retain all 12 seasonal indicators. That approach is used here.

```
PROC ARIMA DATA=ALL;
   IDENTIFY VAR=CONSTRCT NLAG=36 NOPRINT;
   IDENTIFY VAR=CONSTRCT(1) STATIONARITY=(ADF=(1,2,3) DLAG=12);
   IDENTIFY VAR=CONSTRCT(1) NOPRINT
      CROSSCOR = (S1(1) S2(1) S3(1) S4(1) S5(1) S6(1)
                  S7(1) S8(1) S9(1) S10(1) S11(1) S12(1) );
   ESTIMATE INPUT = (S1 S2 S3 S4 S5 S6 S7 S8 S9 S10 S11 S12 )
      NOCONSTANT METHOD=ML NOPRINT;
   FORECAST LEAD=24 ID=DATE INTERVAL=MONTH OUT=OUTDUM;
   IDENTIFY VAR=CONSTRCT(1,12) NOPRINT;
   ESTIMATE NOCONSTANT METHOD=ML NOPRINT;
   FORECAST LEAD=24 INTERVAL=MONTH ID=DATE OUT=OUTDIF NOPRINT;
RUN;
```

Output 4.31
Seasonal Dummy and Seasonal Difference Forecasts

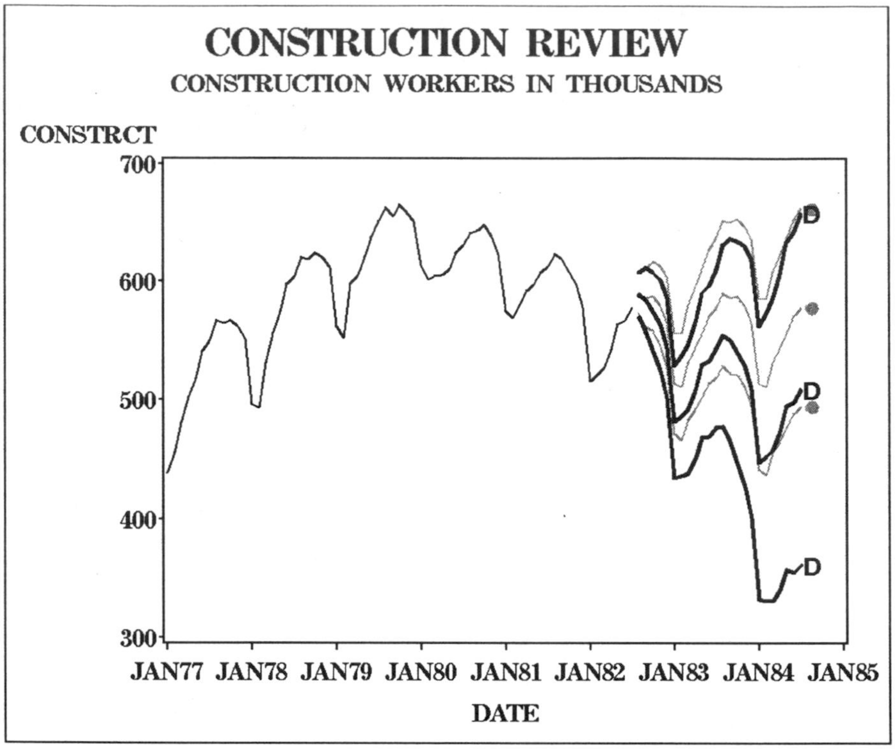

In **Output 4.31** the forecast data sets have been merged and forecasts 24 periods ahead have been plotted. The forecasts and intervals for the span 12 differenced series are shown as darker lines labeled "D," and those for the dummy variable model are shown as lighter lines with a dot label on the far right. The forecasts are quite different. The seasonally differenced series gives much wider intervals and a general pattern of decline. The seasonal dummy variables produce forecast intervals that are less pessimistic and, 24 periods into the future, are about half the width of the others. Of course, wide intervals are expected with differencing. Is there a way to see which model is more appropriate? The chi-square statistics for both models show no problems with the models. Note the code STATIONARITY=(ADF=(1,2,3) DLAG=12) for the first differenced series. This DLAG=12 option requests a *seasonal* unit root test. Dickey, Hasza, and Fuller (1984) develop this and other seasonal unit root tests. **Output 4.32** shows the results, and the tau statistics give some evidence against the null hypothesis of a seasonal unit root.

Output 4.32
Seasonal Unit Root Tests for Construction Data

Seasonal Augmented Dickey-Fuller Unit Root Tests

Type	Lags	Rho	Pr < Rho	Tau	Pr < Tau
Zero Mean	1	-6.2101	0.1810	-1.77	0.0499
	2	-7.4267	0.1389	-2.09	0.0251
	3	-7.7560	0.1291	-2.20	0.0198
Single Mean	1	-5.7554	0.2606	-1.61	0.0991
	2	-6.9994	0.2068	-1.95	0.0541
	3	-7.3604	0.1930	-2.07	0.0428

The seasonal dummy variable model does not lose as much data to differencing, is a little easier to understand, has narrower intervals, and does more averaging of past seasonal behavior. In fact, the first and span 12 difference model has forecast $\hat{Y}_t = Y_{t-1} + (Y_{t-12} - Y_{t-13})$, so the forecast for this August is just this July's value with last year's July-to-August change added in. The forecast effectively makes a copy of last year's seasonal pattern and attaches it to the end of the series as a forecast. Without moving average terms, last year's pattern alone gives the forecast. For these data, these comments along with the fact that the data themselves reject the seasonal difference suggest the use of the dummy variable model.

4.4.3 Milk Scare (Intervention)

Liu et al. (1998) discuss milk sales in Oahu, Hawaii, during a time period in which the discovery of high pesticide levels in milk was publicized. Liu (personal communication) provided the data here. The data indicate April 1982 as the month of first impact, although some tainted milk was found in March. **Output 4.33** shows a graph with March, April, and May 1982 indicated by dots. Ultimately eight recalls were issued and publicized, with over 36 million pounds of contaminated milk found. It might be reasonable to expect a resulting drop in milk sales that may or may not have a long-term effect. It appears that, with the multiple recalls and escalating publicity, the full impact was not realized until May 1982, after which recovery began.

Initially a model was fit to the data before the intervention. A seasonal pattern was detected, but no ordinary or seasonal differencing seemed necessary. A P=(1)(12) specification left a somewhat large correlation at lag 2, so Q=(2) was added and the resulting model fit the pre-intervention data nicely. The intervention response seemed to show an arbitrary value after the first drop, in fact another drop, followed by exponential increase upward. The second drop suggests a numerator lag and the exponential increase suggests a denominator lag in the transfer function operator. X is a variable that is 1 for April 1982 and 0 otherwise. The following code produces an intervention model with this pattern.

```
PROC ARIMA DATA=LIU;
   IDENTIFY VAR=SALES NOPRINT CROSSCOR=(X);
   ESTIMATE INPUT=( (1) /(1) X ) P=(1)(12) Q=(2) METHOD=ML;
RUN;
```

Output 4.33 and **Output 4.34** show the results.

Output 4.33
Effect of Tainted Milk

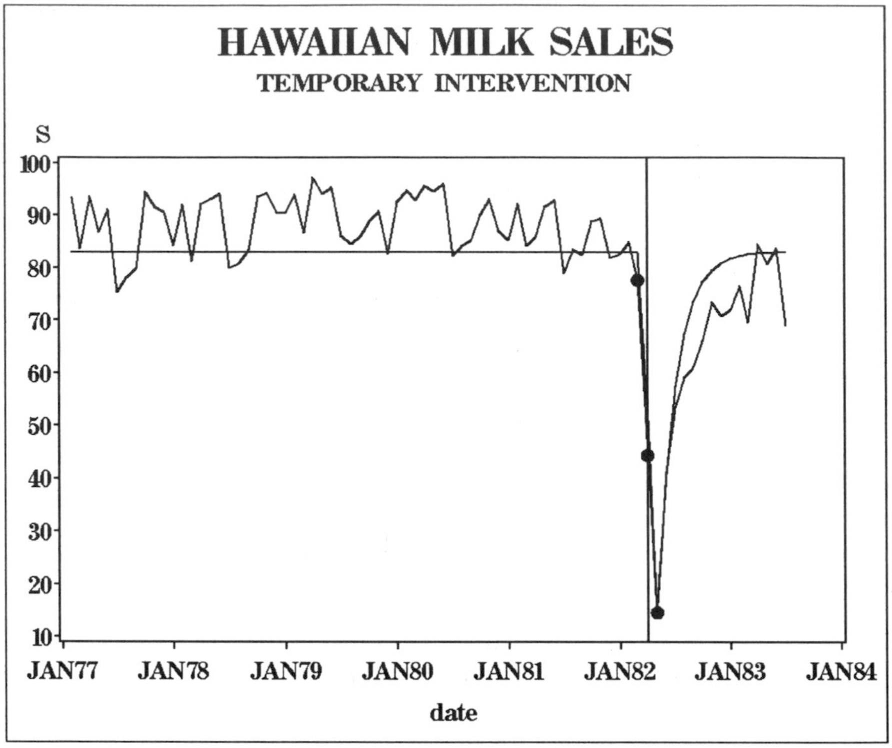

Output 4.34
Model for Milk Sales Intervention

Effect of Negative Publicity (Contamination)
on Milk Sales

The ARIMA Procedure

Maximum Likelihood Estimation

Parameter	Estimate	Standard Error	t Value	Approx Pr > |t|	Lag	Variable	Shift
MU	83.01550	3.98010	20.86	<.0001	0	sales	0
MA1,1	-0.34929	0.11980	-2.92	0.0035	2	sales	0
AR1,1	0.53417	0.10566	5.06	<.0001	1	sales	0
AR2,1	0.78929	0.06964	11.33	<.0001	12	sales	0
NUM1	-39.89641	2.83209	-14.09	<.0001	0	x	0
NUM1,1	49.55934	2.96258	16.73	<.0001	1	x	0
DEN1,1	0.61051	0.03289	18.56	<.0001	1	x	0

Output 4.34
Model for Milk Sales Intervention (continued)

```
                        Constant Estimate         8.148426
                        Variance Estimate         15.88348
                        Std Error Estimate        3.985408
                        AIC                       450.5909
                        SBC                       466.9975
                        Number of Residuals             77


                        Autocorrelation Check of Residuals

 To        Chi-             Pr >
Lag      Square     DF     ChiSq     --------------Autocorrelations--------------

  6        3.98      3    0.2636    -0.027    0.030   -0.015    0.110    0.019    0.180
 12       14.35      9    0.1104     0.116    0.141    0.149    0.000    0.235   -0.063
 18       18.96     15    0.2157     0.052    0.133   -0.006    0.005    0.154   -0.046
 24       23.01     21    0.3434     0.118    0.064    0.035    0.017    0.010   -0.131
```

By specifying INPUT=((1)/(1) X), where X is 1 for April 1982 and 0 otherwise, you are fitting an intervention model whose form is

$$(\beta_0 - \beta_1 B)/(1 - \alpha_1 B)X_t$$

Filling in the estimates, you have

$$(-40 - 50B)/(1 - 0.61B)X_t = (-40 - 50B)(1 + 0.61B + 0.61^2B^2 + \cdots)X_t$$
$$= -40X_t - 74.4X_{t-1} + 0.61(-74.4)X_{t-2} + 0.61^2(-74.4)X_{t-3} \;|$$

so when X_t is 1, the estimated effect is −40. The next month, X_{t-1} is 1 and the effect is −74.4. Two months after the intervention the estimated effect is 0.61(−74.4) as recovery begins. This model forces a return to the original level. In **Output 4.33** a horizontal line at the intercept 83 has been drawn and the intervention effects −40, −74.4, and so on, have been added in. Notice how the intercept line underestimates the pre-intervention level, and how the estimated recovery seems faster than the data suggest. Had you plotted the forecasts, including the autoregressive components, this failure of the mean structure in the model might not have been noticed. The importance of plotting cannot be overemphasized. It is a critical component of data analysis. Note also that the statistics in **Output 4.34** give no warning signs of any problems. Again one might think of the autoregressive structure as compensating for some lack of fit.

Might there be some permanent effect of this incident? The model now under consideration does not allow it. To investigate this, you add a level shift variable.

Define the variable LEVEL1 to be 1 prior to April 1982 and 0 otherwise. This will add a constant, the coefficient of the column, for the pre-intervention period. It represents the difference between the pre-intervention mean and the level to which the post-intervention trend is moving—that is, the level attained long after the intervention. If this shift is not significantly different from 0, then the model shows no permanent effect. If the shift (coefficient) is significantly larger than 0, then a permanent decrease in sales is suggested by the model. If the coefficient happens to be negative, then the pre-intervention level is less than the level toward which the data are now moving. You issue the following code to fit a model with both temporary effects (X) and a permanent level shift (LEVEL1):

```
PROC ARIMA;
   IDENTIFY VAR=SALES NOPRINT CROSSCOR=(X LEVEL1);
   ESTIMATE INPUT=((1)/(1)X LEVEL1) P=(1)(12) Q=(2) METHOD=ML;
RUN;
```

Output 4.35 and **Output 4.36** show the results.

Output 4.35
Model Allowing Permanent Effect

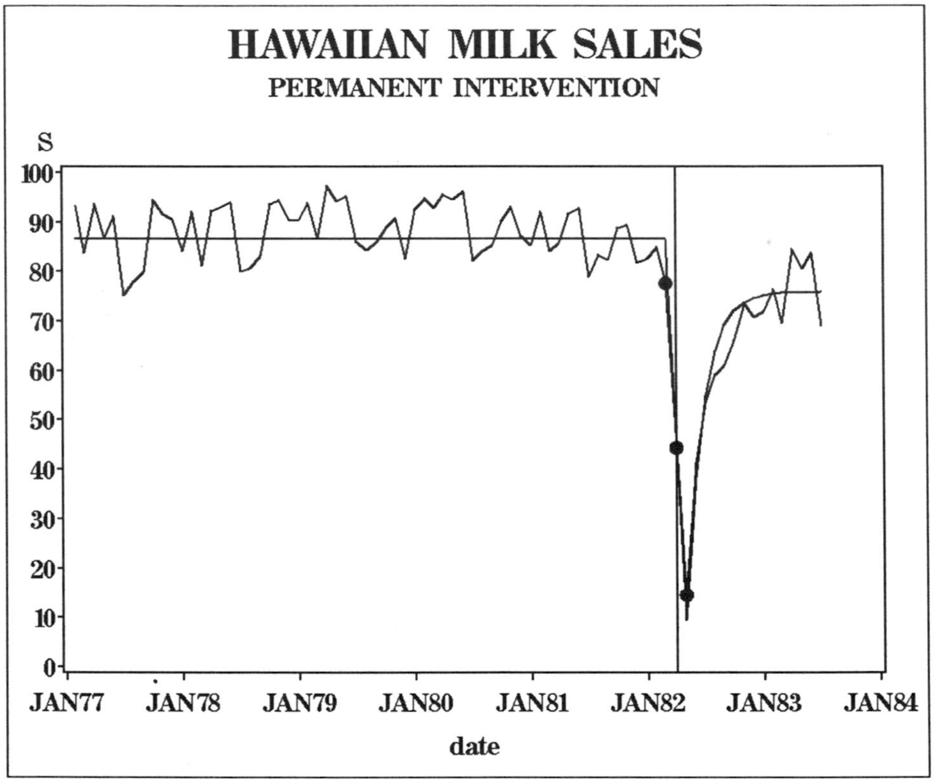

Output 4.36
Intervention Model with Permanent Shift

Maximum Likelihood Estimation

Parameter	Estimate	Standard Error	t Value	Approx Pr > |t|	Lag	Variable	Shift
MU	75.95497	2.71751	27.95	<.0001	0	sales	0
MA1,1	-0.31261	0.12036	-2.60	0.0094	2	sales	0
AR1,1	0.29442	0.11634	2.53	0.0114	1	sales	0
AR2,1	0.77634	0.07042	11.02	<.0001	12	sales	0
NUM1	-31.67677	3.24333	-9.77	<.0001	0	x	0
NUM1,1	48.55454	2.99270	16.22	<.0001	1	x	0
DEN1,1	0.56565	0.03353	16.87	<.0001	1	x	0
NUM2	10.79096	2.22677	4.85	<.0001	0	level1	0

Constant Estimate	11.98663
Variance Estimate	13.73465
Std Error Estimate	3.706029
AIC	439.2273
SBC	457.9778
Number of Residuals	77

Output 4.36
Intervention Model with Permanent Shift (continued)

Autocorrelation Check of Residuals

To Lag	Chi-Square	DF	Pr > ChiSq	Autocorrelations					
6	2.46	3	0.4831	0.020	-0.010	-0.091	-0.011	-0.090	0.110
12	9.47	9	0.3949	0.094	0.125	0.099	-0.033	0.188	-0.083
18	12.27	15	0.6582	-0.010	0.067	-0.069	-0.030	0.114	-0.068
24	18.28	21	0.6310	0.080	0.029	-0.012	-0.048	-0.072	-0.196

It appears that the pre-intervention level is about $75.95+10.79$ and the ultimate level to which sales will return is 75.95, according to this model. All estimates, including the estimated 10.79 permanent loss in sales, are significant. The geometric rate of approach to the new level is 0.56565, indicating a faster approach to the new level than that from the first model. Of course, at this point it is clear that the old model was misspecified, as it did not include LEVEL1. The AR1,1 coefficient 0.29 is quite a bit smaller than 0.53 from the first model. That is consistent with the idea that the autoregressive structure there was in part compensating for the poor fit of the mean function. You can add and subtract 1.96(2.2268) from 10.79 to get an approximate 95% confidence interval for the permanent component of the sales loss due to the contamination scare.

Other models can be tried. Seasonal dummy variables might be tried in place of the seasonal AR factor. Liu et al. suggest that some sort of trend might be added to account for a decline in consumer preference for milk. A simple linear trend gives a mild negative slope, but it is not statistically significant. The estimated permanent level shift is about the same and still significant in its presence.

4.4.4 Terrorist Attack

On September 11, 2001, terrorists used commercial airliners as weapons to attack targets in the United States, resulting in the collapse of the World Trade Center in New York City. American Airlines flights were among those involved. The stock market was closed following this incident and reopened September 17. In a second incident, an American Airlines jet crashed on November 12, 2001, in Queens, New York. An intervention analysis of American Airlines stock trading volume (in millions) is now done incorporating a pulse and level shift intervention for each of these events, defined similarly to those of the milk example in **Section 4.4.3**. Data through November 19 are used here, so there is not a lot of information about the nature of the response to the second incident. A model that seems to fit the data reasonably well, with parameters estimated from PROC ARIMA, is

$$\log(\text{Volume}) = 0.05 + (2.58–2.48B)/(1–.76B)X_t + 1.49/(1–.80B)\ P_t + (1–.52B)/(1–.84B)\ e_t$$

where X_t is a level shift variable that is 1 after September 11 and 0 before, while P_t is a pulse variable that is 1 only on the day of the second incident. The p-values for all estimates except the intercept were less than 0.0005 and those for the chi-square check of residuals were all larger than 0.35, indicating an excellent fit for the 275 log transformed volume values in the data set.

This model allows for a permanent effect of the terrorist attack of September 11 but forces the effect of the second incident to decline exponentially to 0 over time. The second incident sparked a log(volume) increase 1.49 on the day it happened, but j days later, log(volume) is $(0.80)^j(1.49)$ above what it would have otherwise been, according to the model. The permanent effect of the events of September 11 on log volume would be (2.59–2.48)/(1–.76) = 0.46 according to the model. The numerator lag for X allows a single arbitrary change from the initial shock (followed by an exponential approach at rate 0.76 to the eventual new level). In that sense, the inclusion of this lag acts like a pulse variable and likely explains why the pulse variable for September 11 was not needed in the model. The level shift variable for the second incident did not seem to be needed either, but with so little data after November 12, the existence of a permanent effect remains in question.

Output 4.37 shows a graph of the data and a forecast from this model.

Output 4.37
American Airlines Stock Volume

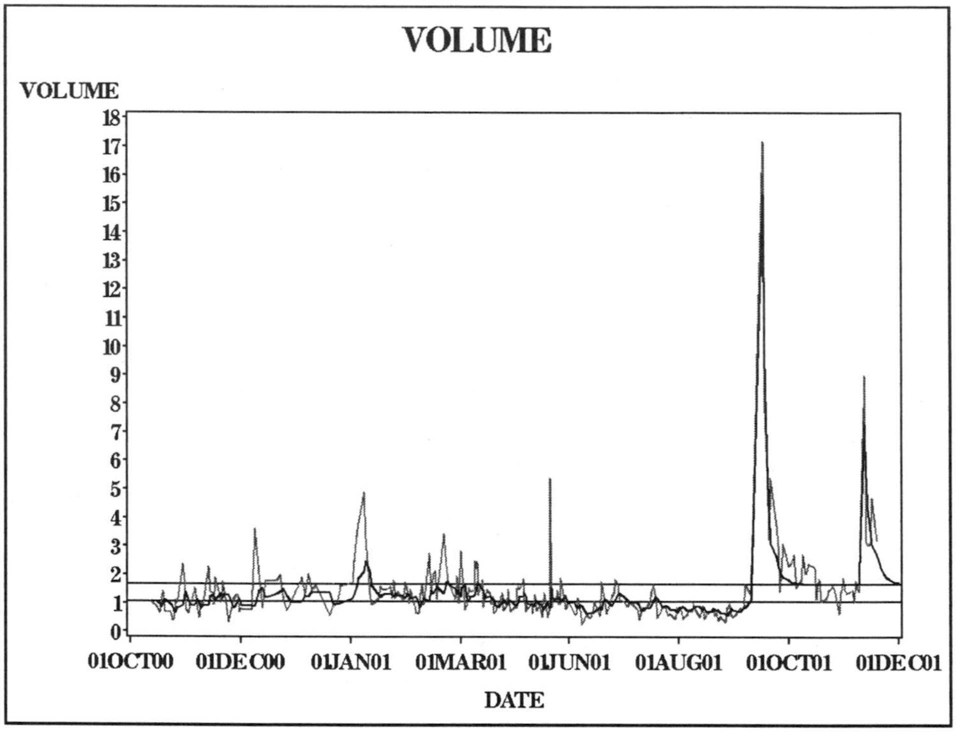

Calculations from the log model were exponentiated to produce the graph. The model was fit to the full data set, but the option BACK=42 was used in the FORECAST statement so that the data following September 11 were not used to adjust the forecasts; that is, only the X and P parts of the model are used in the post-September 11 forecasts. With that in mind—that is, with no adjustments based on recent residuals—it is striking how closely these forecasts mimic the behavior of the data after this incident. It is also interesting how similar the decay rates (denominator terms) are for the two incidents. Two horizontal lines, one at the pre-intervention level exp(0.05) = 1.05 and one at the ultimate level exp(0.05 + (2.59–2.48)/(1–.76)) = exp(0.51) = 1.66, are drawn.

The permanent effect of the event of September 11 is an increase of (2.59–2.48)/(1–.76) in log transformed volume, according to the model. That becomes a multiplicative increase of exp((2.59–2.48)/(1–.76)) = 1.58, a 58% increase in volume.

Chapter 5 The ARIMA Model: Special Applications

5.1 Regression with Time Series Errors and Unequal Variances

5.1.1 Autoregressive Errors

SAS PROC AUTOREG provides a tool to fit a regression model with autoregressive time series errors. Such a model can be written in two steps. With a response Y_t related to a single input X_t and with an AR(1) error, you can write

$$Y_t = \beta_0 + \beta_1 X_t + Z_t$$

and

$$Z_t = \alpha Z_{t-1} + e_t$$

where $|\alpha|<1$ for stationarity and $e_t \sim N(0, \sigma^2)$ with obvious extensions to multiple regression and AR(p) series. The variance of Z_t is $\sigma^2/(1-\alpha^2)$, from which the normal density function of

$Z_1 = Y_1 - \beta_0 - \beta_1 X_1$ can be derived. Furthermore, substitution of $Z_t = Y_t - \beta_0 - \beta_1 X_t$ and its lag into $e_t = Z_t - \alpha Z_{t-1}$ shows that

$$e_t = (Y_t - \beta_0 - \beta_1 X_t) - \alpha(Y_{t-1} - \beta_0 - \beta_1 X_{t-1})$$

and because $e_t \sim N(0, \sigma^2)$, the normal density of this expression can also be written down for t=2,3,…,n. Because Z_1, e_2, e_3, …, e_n are independent of each other, the product of these n normal densities constitutes the so-called joint density function of the Y values. If the observed data values Y and X are plugged into this function, the only unknowns remaining in the function are the parameters α, β_0, β_1, and σ^2, and this resulting function $L(\alpha, \beta_0, \beta_1, \sigma^2)$ is called the "likelihood function," with the values of α, β_0, β_1, and σ^2 that maximize it being referred to as "maximum likelihood estimates." This is the best way to estimate the model parameters. Other methods described below have evolved as less computationally burdensome approximations.

From the expression for e_t you can see that

$$Y_t = \alpha Y_{t-1} + \beta_0(1-\alpha) + \beta_1(X_t - \alpha X_{t-1}) + e_t$$

This suggests that you could use some form of nonlinear least squares to estimate the coefficients.

A third, less used approach to estimation of the parameters is much less computer intensive, but it does not make quite as efficient use of the data as maximum likelihood. It is called the Cochrane-Orcutt method and consists of (1) running a least squares regression of Y on X, (2) fitting an autoregressive model to the residuals, and (3) using that model to "filter" the data. Writing, as above,

$$Y_t = \alpha Y_{t-1} + \beta_0(1-\alpha) + \beta_1(X_t - \alpha X_{t-1}) + e_t$$

you observe the equation for a regression of the transformed, or filtered, variable $Y_t - \alpha Y_{t-1}$ on transformed variables $(1-\alpha)$ and $(X_t - \alpha X_{t-1})$. Because e_t satisfies the usual regression properties, this regression, done with ordinary least squares (OLS), satisfies all of the usual conditions for inference, and the resulting estimates of the parameters would be unbiased, with proper standard errors and valid t tests being given by the ordinary least squares formulas applied to these transformed variables. When α is replaced by an estimate from a model for the residuals, the statements above are approximately true. The Cochrane-Orcutt method can be modified to include an equation for the first observation as well. The method can be iterated, using the new regression estimates to produce new estimates of Z_t and hence new estimates of α, etc.; however, the simultaneous iteration on all parameters done by maximum likelihood would generally be preferred.

If the error autocorrelation is ignored and a regression of Y on X is done, the estimated slope and intercept will be unbiased and will, under rather general assumptions on X, be consistent—that is, they will converge to their true values as the sample size increases. However, the standard errors reported in this regression, unlike those for the filtered variable regression, will be wrong, as will the p-values and any inference you do with them. Thus ordinary least squares residuals can be used to estimate the error autocorrelation structure, but the OLS t tests and associated p-values for the intercept and slope(s) cannot be trusted. In PROC AUTOREG, the user sees the initial OLS regression, the estimated autocorrelation function computed from the residuals, the autoregressive parameter estimates (with insignificant ones being omitted if BACKSTEP is specified), and then the final estimation of parameters including standard errors and tests that are valid based on large sample theory.

5.1.2 Example: Energy Demand at a University

Output 5.1 shows energy demand plotted against temperature and against date. Data were collected at North Carolina State University during the 1979–1980 academic year. Three plot symbols are used to indicate non-workdays (*), workdays with no classes (dots), and teaching days (+). The goal is to relate demand for energy to temperature and type of day. The coefficient of variable WORK will be seen to be 2919. The variable WORK is 0 for non-workdays and 1 for workdays, indicating that 1(2919) is to be added to every prediction for a workday. A similar 0-1 variable called TEACH has coefficient 1011, which indicates that 1(1011) should be added to teaching days. Since all teaching days are workdays, teaching day demand is 2919+1011 = 3930 higher than non-workdays for any given temperature. Workdays that are not teaching days have demand 2919 higher than non-workdays for a given temperature. As temperatures rise, demand increases at an increasing rate. The three curves on the graph come from a model to be discussed. The plot of demand against date shows that there were a couple of workdays during class break periods, e.g., December 31, where demand was more like that for non-workdays, as you might expect. You might want to group these with the non-workdays. Also, day-of-week dummy variables can be added. A model without these modifications will be used.

Output 5.1
NCSU Energy Demand

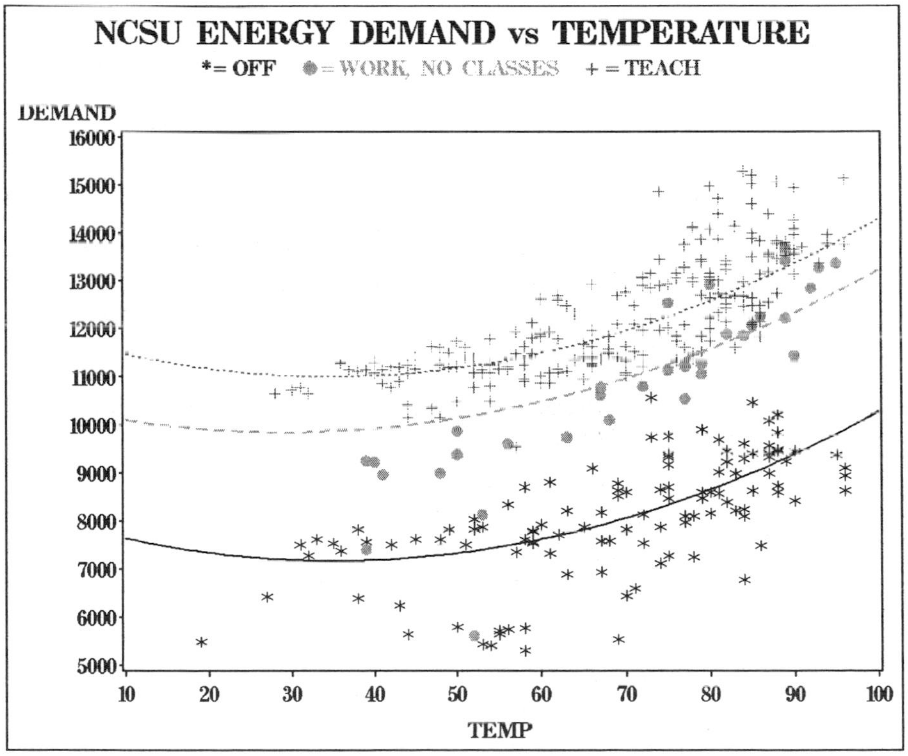

Output 5.1
NCSU Energy Demand (continued)

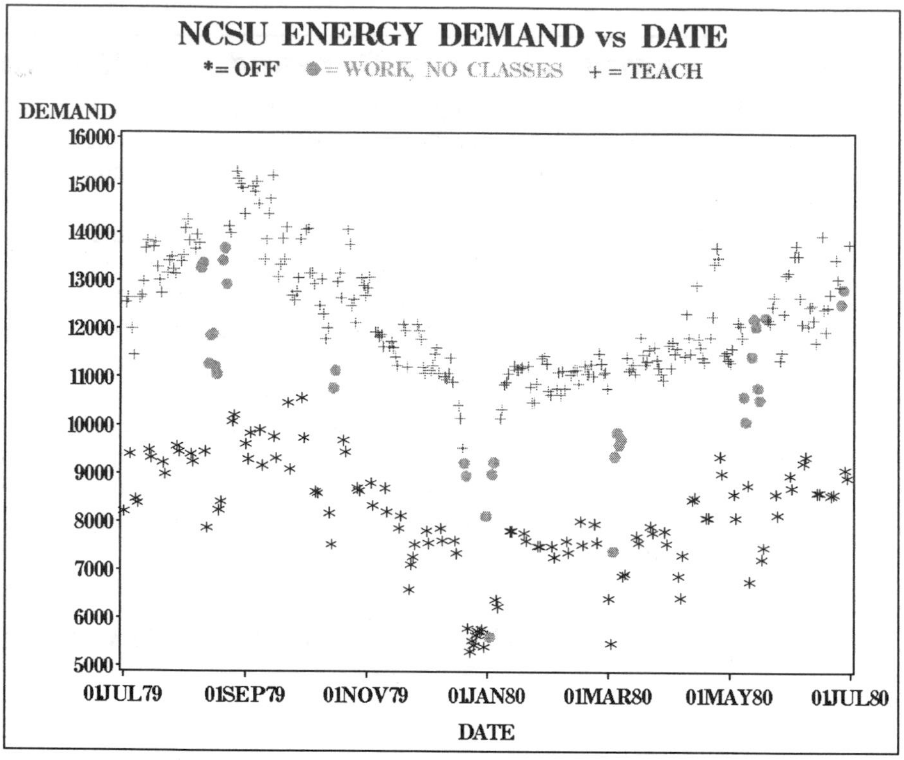

The model illustrated here has today's temperature TEMP, its square TEMPSQ, yesterday's temperature TEMP1, teaching day indicator TEACH, and workday indicator WORK as explanatory variables. Future values of TEACH and WORK would be known, but future values of the temperature variables would have to be estimated in order to forecast energy demand into the future. Future values of such inputs need to be provided (along with missing values for the response) in the data set in order to forecast. No accounting for forecast inaccuracy in future values of the inputs is done by PROC AUTOREG.

To fit the model issue this code.

```
PROC AUTOREG DATA=ENERGY;
    MODEL DEMAND = TEMP TEMPSQ TEACH WORK TEMP1
    / NLAG= 15 BACKSTEP METHOD=ML DWPROB;
RUN;
```

Output 5.2 contains the ordinary least squares regression portion of the PROC AUTOREG output.

Output 5.2 *OLS Regression*

```
                        The AUTOREG Procedure

                    Dependent Variable    DEMAND

                   Ordinary Least Squares Estimates

   SSE                      231077196    DFE                       359
   MSE                         643669    Root MSE            802.28989
   SBC                     5947.02775    AIC                5923.62837
   Regress R-Square            0.8794    Total R-Square         0.8794
   Durbin-Watson               0.5331    Pr < DW                <.0001
   Pr > DW                     1.0000

NOTE: Pr<DW is the p-value for testing positive autocorrelation,
and Pr>DW is the p-value for testing negative autocorrelation.


                                         Standard                    Approx
 Variable        DF      Estimate          Error    t Value      Pr > |t|

 Intercept        1          3593       202.8272      17.71        <.0001
 TEMP             1       27.3512         5.7938       4.72        <.0001
 TEMPSQ           1        0.7988         0.1458       5.48        <.0001
 TEACH            1          1533       150.2740      10.20        <.0001
 WORK             1          2685       158.8292      16.91        <.0001
 TEMP1            1       34.0833         5.7923       5.88        <.0001

                     Estimates of Autocorrelations

Lag  Covariance   Correlation  -1 9 8 7 6 5 4 3 2 1 0 1 2 3 4 5 6 7 8 9

  0     633088       1.000000  |                    |********************|
  1     463306       0.731819  |                    |***************     |
  2     366344       0.578662  |                    |************        |
  3     312670       0.493881  |                    |**********          |
  4     291068       0.459760  |                    |*********           |
  5     311581       0.492161  |                    |**********          |
  6     335580       0.530069  |                    |***********         |
  7     352673       0.557068  |                    |***********         |
  8     302677       0.478096  |                    |**********          |
  9     273294       0.431684  |                    |*********           |
 10     232510       0.367262  |                    |*******             |
 11     203572       0.321555  |                    |******              |
 12     183762       0.290263  |                    |******              |
 13     205684       0.324889  |                    |******              |
 14     244380       0.386012  |                    |********            |
 15     223499       0.353031  |                    |*******             |
```

This regression displays strongly autocorrelated residuals r_t. The Durbin-Watson statistic, $\text{DW}=\sum_{2}^{n}(r_t-r_{t-1})^2/\sum_{1}^{n}r_t^2 = 0.5331$, is significantly less than 2 (p<0.0001) indicating nonzero lag 1 autocorrelation in the errors. If r_t and r_{t-1} were alike (strong positive correlation), then $r_t - r_{t-1}$ would be near 0, thus showing that the DW statistic tends to be less than 2 under positive autocorrelation. The DW is expected to be near 2 for uncorrelated data. Extensions of DW to lags of more than 1 are available in PROC AUTOREG. The autocorrelation plot shows strong autocorrelations. The correction for autocorrelation reveals that lags 7 and 14 are present, indicating some sort of weekly effect. Lag 1 is also sensible, but lags 5 and 12 are a little harder to justify with intuition. All others of the 15 lags you started with are eliminated automatically by the BACKSTEP option.

Output 5.3
Autoregressive Parameter Estimates

Estimates of Autoregressive Parameters

Lag	Coefficient	Standard Error	t Value
1	-0.580119	0.040821	-14.21
5	-0.154947	0.045799	-3.38
7	-0.157783	0.048440	-3.26
12	0.127772	0.045687	2.80
14	-0.116690	0.044817	-2.60

In PROC AUTOREG, the model for the error Z_t is written with plus rather than minus signs—that is, $(1+\alpha_1 B+\alpha_2 B^2+...+\alpha_p B^p)Z_t = e_t$. Therefore the AR(14) error model in **Output 5.3** is

$$Z_t = 0.58\, Z_{t-1} + 0.15\, Z_{t-5} + 0.16\, Z_{t-7} - 0.13\, Z_{t-12} + 0.12\, Z_{t-14} + e_t$$

Using this AR(14) structure and these estimates as initial values, the likelihood function is computed and maximized, producing the final estimates with correct (justified by large sample theory) standard errors in **Output 5.4**.

Output 5.4
Final Estimates

Variable	DF	Estimate	Standard Error	t Value	Approx Pr > |t|
Intercept	1	4638	407.4300	11.38	<.0001
TEMP	1	23.4711	3.5389	6.63	<.0001
TEMPSQ	1	0.7405	0.1162	6.37	<.0001
TEACH	1	1011	114.2874	8.84	<.0001
WORK	1	2919	115.5118	25.27	<.0001
TEMP1	1	25.0550	3.5831	6.99	<.0001
AR1	1	-0.6490	0.0401	-16.18	<.0001
AR5	1	-0.1418	0.0478	-2.97	0.0032
AR7	1	-0.1318	0.0504	-2.62	0.0093
AR12	1	0.1420	0.0481	2.95	0.0033
AR14	1	-0.1145	0.0469	-2.44	0.0151

Using OLS from **Output 5.2** or from PROC REG, a 95% confidence interval for the effect, on energy demand, of teaching classes would have been incorrectly computed as 1533 $\pm$ 1.96(150), whereas the correct interval is 1011 $\pm$ 1.96(114).

Using the same regression inputs in PROC ARIMA, a model with p=(1), q=(1,7,14) showed no lack of fit in **Output 5.5**. This error model is a bit more aesthetically pleasing than that of AUTOREG, as it does not include the unusual lags 5 and 12. Note that AUTOREG cannot fit moving average terms.

***Output 5.5** PROC ARIMA for Energy Data*

The ARIMA Procedure

Maximum Likelihood Estimation

Parameter	Estimate	Standard Error	t Value	Approx Pr > |t|	Lag	Variable	Shift
MU	4468.1	363.30832	12.30	<.0001	0	DEMAND	0
MA1,1	0.25723	0.06257	4.11	<.0001	1	DEMAND	0
MA1,2	-0.19657	0.05391	-3.65	0.0003	7	DEMAND	0
MA1,3	-0.18440	0.05289	-3.49	0.0005	14	DEMAND	0
AR1,1	0.84729	0.03622	23.39	<.0001	1	DEMAND	0
NUM1	25.49771	3.45642	7.38	<.0001	0	TEMP	0
NUM2	0.74724	0.11173	6.69	<.0001	0	TEMPSQ	0
NUM3	838.08859	111.69438	7.50	<.0001	0	TEACH	0
NUM4	3085.4	114.90204	26.85	<.0001	0	WORK	0
NUM5	25.32913	3.44973	7.34	<.0001	0	TEMP1	0

Autocorrelation Check of Residuals

To Lag	Chi-Square	DF	Pr > ChiSq	--------------Autocorrelations--------------					
6	3.85	2	0.1461	0.018	-0.007	-0.042	-0.052	0.022	0.071
12	11.04	8	0.1995	-0.000	0.008	0.077	0.040	-0.036	-0.101
18	14.65	14	0.4022	-0.007	0.010	-0.019	0.084	-0.014	-0.042
24	15.71	20	0.7347	-0.003	0.011	0.049	-0.002	0.013	-0.005
30	20.87	26	0.7487	0.002	0.074	-0.073	0.043	0.007	0.012
36	28.24	32	0.6575	0.021	-0.017	0.076	0.002	0.106	-0.025
42	36.98	38	0.5163	0.052	0.015	0.050	0.018	0.042	0.117
48	38.90	44	0.6895	-0.014	0.046	-0.033	-0.021	-0.022	-0.016

The effect on energy demand of teaching classes is estimated from PROC ARIMA as 838 with standard error 112, somewhat different from PROC AUTOREG and quite different from the OLS estimates. The purely autoregressive model from PROC AUTOREG and the mixed ARMA error model can both be estimated in PROC ARIMA. Doing so (not shown here) will show the AIC and SBC criteria to be smaller (better) for the model with the mixed ARMA error. The chi-square white noise tests, while acceptable in both, have higher (better) p-values for the mixed ARMA error structure.

5.1.3 Unequal Variances

The models discussed thus far involve white noise innovations, or shocks, that are assumed to have constant variance. For long data sets, it can be quite apparent just from a graph that this constant variance assumption is unreasonable. PROC AUTOREG provides methods for handling such situations. In **Output 5.6** you see graphs of 8892 daily values (from January 1, 1920 with $Y_1 =$ 108.76 to December 31, 1949 with $Y_{8892} = 200.13$) of Y_t = the Dow Jones Industrial Average, $L_t = \log(Y_t)$, and $D_t = \log(Y_t) - \log(Y_{t-1})$. Clearly the log transformation improves the statistical properties and gives a clearer idea of the long-term increase than does the untransformed series. Many macroeconomic time series are better understood on the logarithmic scale over long periods of time. By the properties of logarithms, note that $D_t = \log(Y_t / Y_{t-1})$, and if Y_t / Y_{t-1} is near 1 then D_t is approximately $(Y_t - Y_{t-1}) / Y_{t-1}$; that is, 100 D_t represents the daily percentage change in the Dow Jones average.

To demonstrate how this works, let $\Delta = ((Y_t - Y_{t-1})/Y_{t-1})$ so $1 + \Delta = Y_t/Y_{t-1}$. Using a Taylor series expansion of log(X) at X=1, you can represent $\log(1+\Delta) = \log(1) + 1^{-1}\Delta - (1^{-2}/2)\Delta^2 + \ldots$ since $(\partial/\partial X)\log(X) = 1/X$, $(\partial/\partial X)(1/X) = (-1/(X*X))$, and so on. Since $\log(1) = 0$, the $\log(1+\Delta) = \log(Y_t/Y_{t-1})$ can be approximated by $0 + \Delta = (Y_t - Y_{t-1})/Y_{t-1}$. This also shows that $\log(Y_t/Y_{t-1})$ is essentially the overnight return on a $1 investment.

The graph of D_t shows some periods of high volatility. The five vertical graph lines represent, from left to right, Black Thursday (October 24, 1929, when the stock market crashed), the inauguration of President Franklin D. Roosevelt (FDR), the start of World War II, the bombing of Pearl Harbor, and the end of World War II. Note especially the era from Black Thursday until a bit after FDR assumed office, known as the Great Depression.

The mean of the D_t values is $\overline{D} = n^{-1}\sum_{t=2}^{n}(\log(Y_t) - \log(Y_{t-1})) = n^{-1}(\log(Y_n) - \log(Y_1))$ so that $e^{n\overline{D}} = Y_n/Y_1$, the increase in the series over the entire time period. For the data at hand, the ratio of the last to first data point is 200.13/108.76 = 1.84, so the series did not quite double over this 30-year period. You might argue that subperiods like the depression in which extreme volatility is present are not typical and should be ignored or at least downweighted in computing a rate of return that has some relevance for future periods. You decide to take a look at the variability of the D_t series.

Because there is so much data, the reduction of each month's D_t numbers to a standard deviation still leaves a relatively long time series of 360 monthly numbers. These standard deviations have a histogram with a long tail to the right. Again a logarithmic transform is used to produce a monthly series S_t =log(standard deviation) that has a more symmetric distribution. Thus S_t measures the volatility in the series, and a plot of S_t versus time is the fourth graph in **Output 5.6**.

Output 5.6 *Dow Jones Industrial Average on Several Scales*

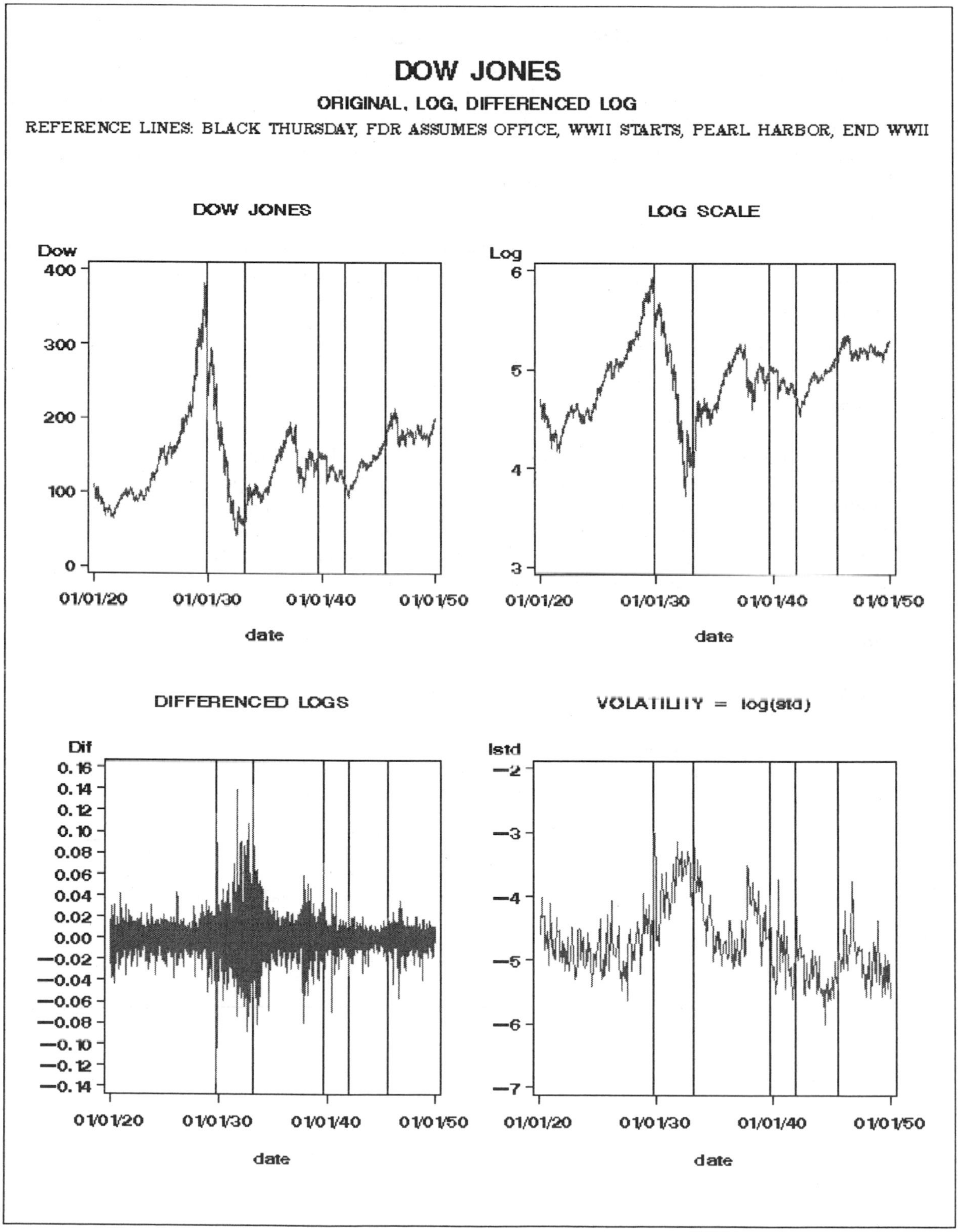

Now apply a time series model to the S_t series. The tau test for stationarity suggests a unit root process when six augmenting lags are used. The reason for choosing six lags is that the partial autocorrelation function for S_t is near 0 after lag 6 and, furthermore, a regression of $S_t - S_{t-1}$ on S_{t-1} and 20 lagged differences ($S_{t-j} - S_{t-j-1}$ for *j*=1 to 20) in PROC REG gave an insignificant F test for lags 7 through 20. A similar regression using six lagged differences showed all six to be significant according to their *t* tests. Dickey and Fuller show that such *t* tests on lagged differences are valid in large samples—only the test for the coefficient on the lagged level S_{t-1} has a nonstandard distribution. That test cannot reject the unit root hypothesis, and so a model in first differences is suggested for the log transformed standard deviation series S_t. The above results are not displayed. At this point you are ready to model S_t. You have seen that a lag 6 autoregressive model for $S_t - S_{t-1}$ seems to provide an adequate fit. Perhaps this long autoregression is an approximation of a mixed model. The following code, using LSTD as the variable name for S_t, seems to provide a reasonable ARMA(1,1) model:

```
PROC ARIMA DATA=OUT1;
   I VAR=LSTD(1) STATIONARITY=(ADF=(6));
   E  P=1 Q=1 ML NOCONSTANT;
RUN;
```

The constant was suppressed (NOCONSTANT) after an initial check showed it to be insignificant. The tau test for unit roots suggests stationarity of the differenced series (p=0.0001) when six lagged differences are used. That is, no further differencing seems to be needed. Said and Dickey (1984) show that even for mixed models, these stationarity tests are valid as long as sufficient lagged differences are included in the model. In summary, the S series appears to be well modeled as an ARIMA(1,1,1) series with parameters as shown in **Output 5.7**.

Output 5.7
ARIMA Model for S

Augmented Dickey-Fuller Unit Root Tests

Type	Lags	Rho	Pr < Rho	Tau	Pr < Tau	F	Pr > F
Zero Mean	6	322.9722	0.9999	-11.07	<.0001		
Single Mean	6	322.2521	0.9999	-11.06	<.0001	61.14	0.0010
Trend	6	322.0197	0.9999	-11.05	<.0001	61.05	0.0010

Maximum Likelihood Estimation

Parameter	Estimate	Standard Error	t Value	Approx Pr > |t|	Lag
MA1,1	0.82365	0.04413	18.66	<.0001	1
AR1,1	0.32328	0.07338	4.41	<.0001	1

Output 5.7
ARIMA Model for S (continued)

Autocorrelation Check of Residuals

To Lag	Chi-Square	DF	Pr > ChiSq	Autocorrelations					
6	2.07	4	0.7230	0.010	-0.016	-0.045	-0.053	0.015	0.016
12	7.95	10	0.6341	0.072	0.060	-0.018	0.044	-0.005	0.070
18	19.92	16	0.2236	-0.013	-0.014	-0.134	0.065	0.078	0.055
24	23.96	22	0.3494	0.053	-0.022	-0.080	-0.018	0.015	0.017
30	34.37	28	0.1890	-0.045	0.016	-0.131	-0.081	-0.002	-0.024
36	37.38	34	0.3164	0.014	0.027	0.056	-0.010	-0.057	0.010
42	38.83	40	0.5227	-0.018	0.005	-0.050	-0.016	-0.007	-0.022
48	41.69	46	0.6533	-0.015	-0.010	-0.028	-0.070	-0.004	-0.030

The model suggests the predicting equation

$$\hat{S}_t = S_{t-1} + 0.3233(S_{t-1} - S_{t-2}) - .82365\hat{e}_{t-1}$$

where $\hat{e}_{t-1}$ would be replaced by the residual $S_{t-1} - \hat{S}_{t-1}$. Exponentiation of $\hat{S}_t$ gives a *conditional* standard deviation for month *t*. Notice that because $\hat{S}_t$ is a logarithm, the resulting standard deviations will all be positive regardless of the sign of $\hat{S}_t$. This allows the variance to change over time in a way that can be predicted from the most recent few variances. The theory underlying ARIMA models is based on large sample arguments and does not require normality, so the use of log transformed standard deviations as data does not necessarily invalidate this approach. However, there are at least two major problems with approaching heterogeneous variation in the manner just used with the Dow Jones series. First, you will not often have so much data to start with, and second, the use of a month as a period for computing a standard deviation is quite arbitrary. A more statistically rigorous approach is now presented. The discussion thus far has been presented as a review of unit root test methodology as well as a motivation for fitting a nonconstant variance model that might involve a unit root. An analyst likely would use the more sophisticated approach shown in the next section.

5.1.4 ARCH, GARCH, and IGARCH for Unequal Variances

The series D_t whose variability is measured by S_t has nonconstant conditional variance. Engle (1982) introduced a model in which the variance at time *t* is modeled as a linear combination of past squared residuals and called it an ARCH (autoregressive conditionally heteroscedastic) process. Bolerslev (1986) introduced a more general structure in which the variance model looks more like an ARMA than an AR and called this a GARCH (generalized ARCH) process. Thus the usual approach to modeling ARCH or GARCH processes improves on the method just shown in substantial ways. The purpose of the monthly standard deviation approach was to illustrate the idea of an ARMA type of structure for standard deviations or variances.

The usual approach to GARCH(p,q) models is to model an error term ε_t in terms of a standard white noise $e_t \sim N(0,1)$ as $\varepsilon_t = \sqrt{h_t} e_t$ where h_t satisfies the type of recursion used in an ARMA model:

$$h_t = \omega + \sum_{i=1}^{q} \alpha_i \varepsilon_{t-i}^{\;2} + \sum_{j=1}^{p} \gamma_j h_{t-j}$$

In this way, the error term has a conditional variance that is a function of the magnitudes of past errors. Engle's original ARCH structure has γ_j=0. Because h_t is the variance rather than its logarithm, certain restrictions must be placed on the α_i, γ_j, and ω to ensure positive variances. For example, if these are all restricted to be positive, then positive initial values of h_t will ensure all h_t are positive. For this reason, Nelson (1991) suggested replacing h_t with $\log(h_t)$ and an additional modification; he called the resulting process EGARCH. These approaches allow the standard deviation to change with each observation. Nelson and Cao (1992) give constraints on the α and γ values that ensure nonnegative estimates of h_t. These are the default in PROC AUTOREG. More details are given in the PROC AUTOREG documentation and in Hamilton (1994), which is a quite detailed reference for time series.

Recall that PROC AUTOREG will fit a regression model with autoregressive errors using the maximum likelihood method based on a normal distribution. In place of the white noise shocks in the autoregressive error model you can specify a GARCH(p,q) process. If it appears, as suggested by your analysis of the Dow Jones standard deviations, that the process describing the error variances is a unit root process, then the resulting model is referred to as integrated GARCH or IGARCH. If the usual stationarity conditions are satisfied, then for a GARCH process, forecasts of h_t will revert to a long-run mean. In an IGARCH model, mean reversion is no longer a property of h_t, so forecasts of h_t will tend to reflect the most recent variation rather than the average historical variation. You would expect the variation during the Great Depression to have little effect on future h_t values in an IGARCH model of the Dow Jones data.

To investigate models of the daily percentage change in the Dow Jones Industrial Average Y_t, you will use $D_t = \log(Y_t) - \log(Y_{t-1})$. Calling this variable DDOW, you issue this code:

```
PROC AUTOREG DATA=MORE;
   MODEL DDOW = / NLAG=2
   GARCH=(P=2,Q=1,TYPE=INTEG,NOINT);
   OUTPUT OUT=OUT2 HT=HT P=F LCLI=L UCLI=U;
RUN;
```

PROC AUTOREG allows the use of regression inputs; however, here there is no apparent time trend or seasonality and no other regressors are readily available. The model statement DDOW = (with no inputs) specifies that the regression part of your model is only a mean. Note the way in which the h_t sequence, predicted values, and default upper and lower forecast limits have been requested in the data set called OUT2.

In **Output 5.8**, the estimate of the mean is seen to be 0.000363. Since DDOW is a difference, a mean is interpreted as a drift in the data, and since the data are log differences, the number $e^{0.000363}$ = 1.0003631 is an estimate of the long-run daily growth over this time period. With 8892 days in the study period, the number $e^{(0.000363)(8891)}$ = 25 represents a 25-fold increase, roughly an 11.3% yearly growth rate! This is not remotely like the rate of growth seen, except in certain portions of the graph. PROC AUTOREG starts with OLS estimates so that the average DDOW over the period is the OLS intercept 0.0000702 from **Output 5.8**. This gives $e^{(0.0000702)(8891)}$ = 1.87, indicating 87% growth for the full 30-year period. This has to be more in line with the graph because, as you saw earlier, except for rounding error it is Y_n/Y_1.

Note also the strong rejection of normality. The normality test used here is the that of Jarque and Bera (1980). This is a general test of normality based on a measurement of skewness b_1 and one of kurtosis $b_2 - 3$ using residuals r_t, where

$$b_1 = \frac{\sum_{t=1}^{n} r_t^3 / n}{(\sum_{t=1}^{n} r_t^2 / n)^{3/2}} \quad \text{and} \quad b_2 - 3 = \frac{\sum_{t=1}^{n} r_t^4 / n}{(\sum_{t=1}^{n} r_t^2 / n)^2} - 3$$

The expression $\sum_{t=1}^{n} r_t^j / n$ is sometimes called the (raw) jth moment of r. The fractions involve third and fourth moments scaled by the sample variance. The numerators are sums of approximately independent terms and thus satisfy a central limit theorem. Both have, approximately, mean 0 when the true errors are normally distributed. Approximate variances of the skewness and kurtosis are $6/n$ and $24/n$. Odd and even powers of normal errors are uncorrelated, so squaring each of these approximately normal variates and dividing by its variance produces a pair of squares of approximately independent N(0,1) variates. The sum of these squared variates, therefore, follows a chi-square distribution with two degrees of freedom under the normality null hypothesis. The Jarque-Bera test

$$\text{JB} = n(b_1^2 / 6 + (b_2 - 3)^2 / 24)$$

has (approximately) a chi-square distribution with two degrees of freedom under the null hypothesis.

Why is the IGARCH model giving a 25-fold increase? It seems unreasonable. The model indicates, and the data display, large variability during periods when there were steep drops in the Dow Jones average. A method that accounts for different variances tends to downweight observations with high variability. In fact there are some periods in which the 11.3% annual rate required for a 25-fold increase (1.113^{20}=25) was actually exceeded, such as in the periods leading up to the Great Depression, after FDR assumed office, and toward the end of WWII. The extremely large variances associated with periods of decrease or slow growth give them low weight, and that would tend to increase the estimated growth rate, but it is still not quite enough to explain the results.

Output 5.8
IGARCH Model for Dow Jones

The AUTOREG Procedure

Dependent Variable ddow

Ordinary Least Squares Estimates

SSE	1.542981	DFE	8891
MSE	0.0001735	Root MSE	0.01317
SBC	-51754.031	AIC	-51761.124
Regress R-Square	0.0000	Total R-Square	0.0000
Durbin-Watson	1.9427		

Variable	DF	Estimate	Standard Error	t Value	Approx Pr > |t|
Intercept	1	0.0000702	0.000140	0.50	0.6155

Estimates of Autoregressive Parameters

Lag	Coefficient	Standard Error	t Value
1	-0.029621	0.010599	-2.79
2	0.037124	0.010599	3.50

Algorithm converged.

Integrated GARCH Estimates

SSE	1.54537859	Observations	8892
MSE	0.0001738	Uncond Var	.
Log Likelihood	28466.2335	Total R-Square	.
SBC	-56887.003	AIC	-56922.467
Normality Test	3886.0299	Pr > ChiSq	<.0001

Variable	DF	Estimate	Standard Error	t Value	Approx Pr > |t|
Intercept	1	0.000363	0.0000748	4.85	<.0001
AR1	1	-0.0868	0.009731	-8.92	<.0001
AR2	1	0.0323	0.009576	3.37	0.0008
ARCH1	1	0.0698	0.003963	17.60	<.0001
GARCH1	1	0.7078	0.0609	11.63	<.0001
GARCH2	1	0.2224	0.0573	3.88	0.0001

Perhaps more importantly, the rejection of normality by the Jarque-Bera test introduces the possibility of bias in the estimated mean. In an ordinary least squares (OLS) regression of a column Y of responses in a matrix X of explanatory variables, the model is $\mathbf{Y}=\mathbf{X}\beta+\mathbf{e}$ and the estimated parameter vector $\hat{\beta}=(\mathbf{X}'\mathbf{X})^{-1}(\mathbf{X}'\mathbf{Y})=\beta+(\mathbf{X}'\mathbf{X})^{-1}(\mathbf{X}'\mathbf{e})$ is unbiased whenever the random vector **e** has mean 0. In regard to bias, it does not matter if the variances are unequal or even if there is correlation among the errors. These features only affect the variance of the estimates, causing biases in the standard errors for $\hat{\beta}$ but not in the estimates of β themselves. In contrast to OLS, GARCH and IGARCH models are fit by maximum likelihood assuming a normal distribution. Failure to meet this assumption could produce bias in parameter estimates such as the estimated mean.

As a check to see if bias can be induced by nonnormal errors, data from a model having the same h_t sequence as that estimated for the Dow Jones log differences data were generated for innovations $\mathbf{e}_t \sim N(0,1)$ and again for innovations $(\mathbf{e}_t^2-1)/\sqrt{2}$, so this second set of innovations used the same normal variables in a way that gave a skewed distribution still having mean 0 and variance 1. The mean was set at 0.00007 for the simulation and 50 such data sets were created. For each data set, IGARCH models were fit for each of the two generated series and the estimated means were output to a data set. The overall mean and standard deviation of each set of 50 means were as follows:

	Mean	Standard Deviation
Normal Errors	0.000071	0.0000907
Skewed Errors	0.000358	0.0001496

Thus it seems that finding a factor of 5 bias in the estimate of the mean (of the differenced logs) could be simply a result of error skewness; in fact the factor of 5 is almost exactly what the simulation shows. The simulation results also show that if the errors had been normal, good estimates of the true value, known in the simulation to be 0.00007, would have resulted.

Using type=integ specifies an IGARCH model for h_t, which, like any unit root model, will have a linearly changing forecast if an intercept is present. You thus use NOINT to suppress the intercept. Using p=2 and q=1, your h_t model has the form

$$h_t = h_{t-1} + 0.7078(h_{t-1} - h_{t-2}) + 0.2224(h_{t-2} - h_{t-3}) + 0.0698\varepsilon_{t-1}^2$$

You can look at h_t as a smoothed local estimate of the variance, computed by adding to the previous smoothed value (h_{t-1}) a weighted average of the two most recent changes in these smoothed values and the square of the most recent shock.

By default, PROC AUTOREG uses a constant variance to compute prediction limits; however, you can output the h_t values in a data set as shown and then, recalling that h_t is a local variance, add and subtract $t_{0.975}\sqrt{h_t}$ from your forecast to produce forecast intervals that incorporate the changing variance. Both kinds of prediction intervals are shown in **Output 5.9**, where the more or less horizontal bands are the AUTOREG defaults and the bands based on h_t form what looks like a border to the data. The data set MORE used for **Output 5.8** and **Output 5.9** has the historical data and 500 additional days with dates but no values of D_t. PROC AUTOREG will produce h_t values and default prediction limits for these. In general, future values of all inputs need to be included for this to work, but here the only input is the intercept.

Output 5.9
Default and h_t-Based Intervals

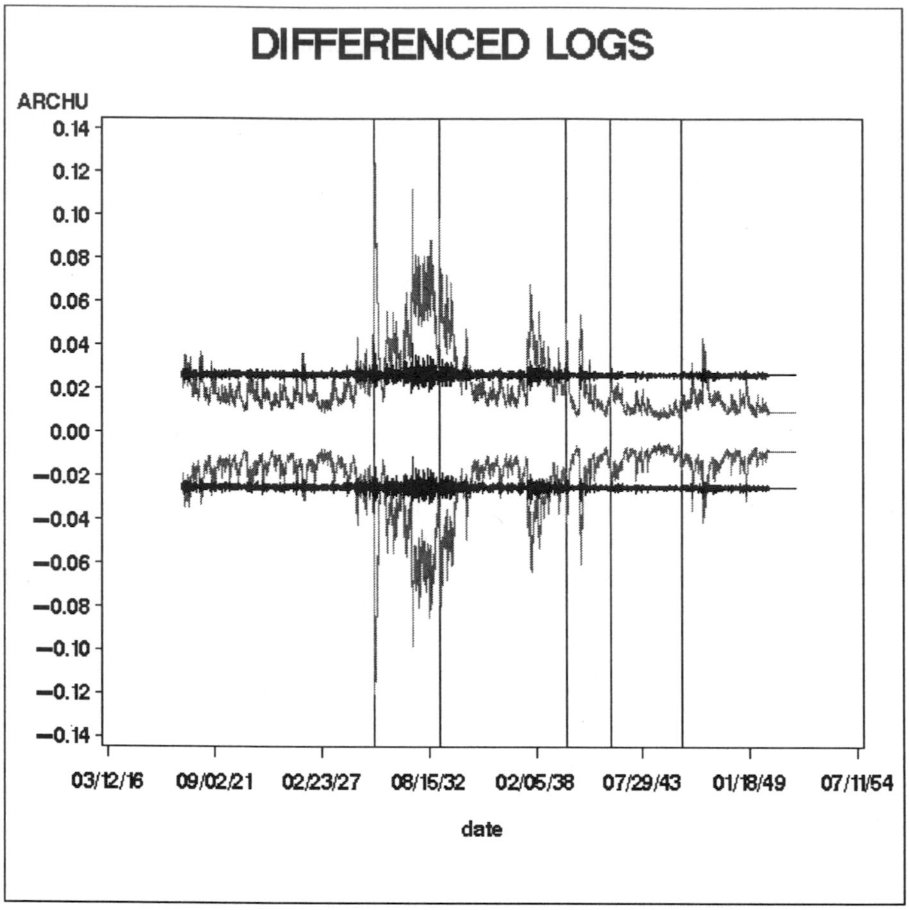

The default prediction intervals completely miss the local features of the data and come off the end of the data with a fairly wide spread. Since the last few data periods were relatively stable, the h_t-based intervals are appropriately narrower. It appears that $(h_t - h_{t-1})$, $(h_{t-1} - h_{t-2})$, and ε_t^2 were fairly small at the end of the series, contributing very little to h_{t+1} so that h_{n+1} is approximately h_n, as are all h_{n+j} for j>0. The forecast intervals coming off the end of the series thus have about the same width as the last forecast interval in the historical data. They are almost, but not exactly, two horizontal lines.

The autoregressive error model is seen to be

$$Z_t = 0.0868Z_{t-1} - 0.0323Z_{t-2} + \varepsilon_t^2$$

where $\varepsilon_t = \sqrt{h_t}e_t$. Although the lag Z coefficients are statistically significant, they are small, so their contribution to forecasts and to the width of prediction intervals into the future is imperceptible in the graph.

Clearly the IGARCH estimated mean 0.000363 is unacceptable in light of the nonnormality, the resulting danger of bias, and its failure to represent the observed growth over the period. The ordinary mean 0.00007 is an unbiased estimate and exactly reproduces the observed growth. The usual conditions leading to the (OLS) formula for the standard error of a mean do not hold here, but more will be said about this shortly. The problem is not with IGARCH versus GARCH; in fact a GARCH(2,1) model also fits the series quite nicely but still gives an unacceptable estimate of the mean of D_t. Note that the average of n independent values of ε_t has variance $n^{-2}\sum_{t=1}^{n} h_t$ if e_t has mean 0 and variance 1. The AR(2) error series

$$Z_t = \alpha_1 Z_{t-1} + \alpha_2 Z_{t-2} + \varepsilon_t$$

can be summed from 1 to n on both sides and divided by n to get $(1-\alpha_1-\alpha_2)\overline{Z}$ approximately equal to $\overline{\varepsilon}$. From $\varepsilon_t = \sqrt{h_t}e_t$ it follows that the (approximate) variance of $(1-\alpha_1-\alpha_2)\overline{Z}$ is $n^{-2}\sum_{t=1}^{n} h_t$ and that of $\overline{Z}$ is thus $(1-\alpha_1-\alpha_2)^{-2}n^{-2}\sum_{t=1}^{n} h_t$. Hamilton (1994, p. 663) indicates that maximum likelihood estimates of h_t are reasonable under rather mild assumptions for ARCH models even when the errors are not normal. Also the graphical evidence indicates that the estimated h_t series has captured the variability in the data nicely. Proceeding on that basis, you sum the estimated h_t series and use estimated autoregressive coefficients to estimate the standard deviation of the mean

$$n^{-1}\sqrt{(1-\alpha_1-\alpha_2)^2\sum_{t=1}^{n} h_t} \text{ as } 8892^{-1}\sqrt{(1-.0868+0.0323)^{-2}1.55846} = 0.0001485.$$

In this way you get

$$t = \frac{0.00007}{0.0001485} = 0.5$$

which is not significant at any reasonable level.

Interestingly, and despite the comments above, a simple t test on the D_t data, ignoring all of the variance structure, gives about the same t. A little thought shows that this could be anticipated for the special case of this model. The summing of h_t and division by n yields what might be thought of as an average variance over the period. Because the α s are small here, the average of h_t divided by n is a reasonable approximation of the variance of $\overline{Z}$ and thus of $\overline{D}$. To the extent that the squared residuals $(D_t - \overline{D})^2$ provide approximate estimates of the corresponding conditional variances h_t, the usual OLS formula, $\sqrt{n^{-1}\sum_{t=1}^{n}(D_t-\overline{D})^2/(n-1)}$, gives an estimate of the standard error of the mean. Additional care would be required, such as consideration of the assumed unit root structure for h_t and the error introduced by ignoring the α s, to make this into a rigorous argument. However, this line of reasoning does suggest that the naive t test, produced by PROC MEANS, for example, might be reasonable for these particular data. There is no reason to expect the naive approach to work well in general.

This example serves to illustrate several important points. One is that careful checking of model implications against what happens in the data is a crucial component of proper analysis. This would typically involve some graphics. Another is that failure to meet assumptions is sometimes not so important but at other times can render estimates meaningless. Careful thinking and a knowledge of statistical principles are crucial here. The naive use of statistical methods without understanding the underlying assumptions and limitations can lead to ridiculous claims. Computational software is not a replacement for knowledge.

5.2 Cointegration

5.2.1 Introduction

In this section you study a dimension *k* vector $\mathbf{V}_t$ of time series. The model $\mathbf{V}_t = \mathbf{A}_1\mathbf{V}_{t-1} + \mathbf{A}_2\mathbf{V}_{t-2} + \mathbf{e}_t$ is called a vector autoregression, a "VAR," of dimension *k* and order $p = 2$ (2 lags). It is assumed that $\mathbf{e}_t$ has a multivariate normal distribution with *k* dimensional mean vector **0**, a vector of 0s, and $k \times k$ variance matrix Σ. The *i*th element of $\mathbf{V}_t$ is the time series $Y_{it} - \mu_i$, so the deviation of each series from its mean is expressed by the model as a linear function of previous deviations of all series from their means. For example, the upper-left panel of **Output 5.10** shows the logarithms of some high and low prices for stock of the electronic retailer Amazon.com, extracted by the Internet search engine Yahoo!

Output 5.10 *Amazon.com Data with Cointegrating Plane*

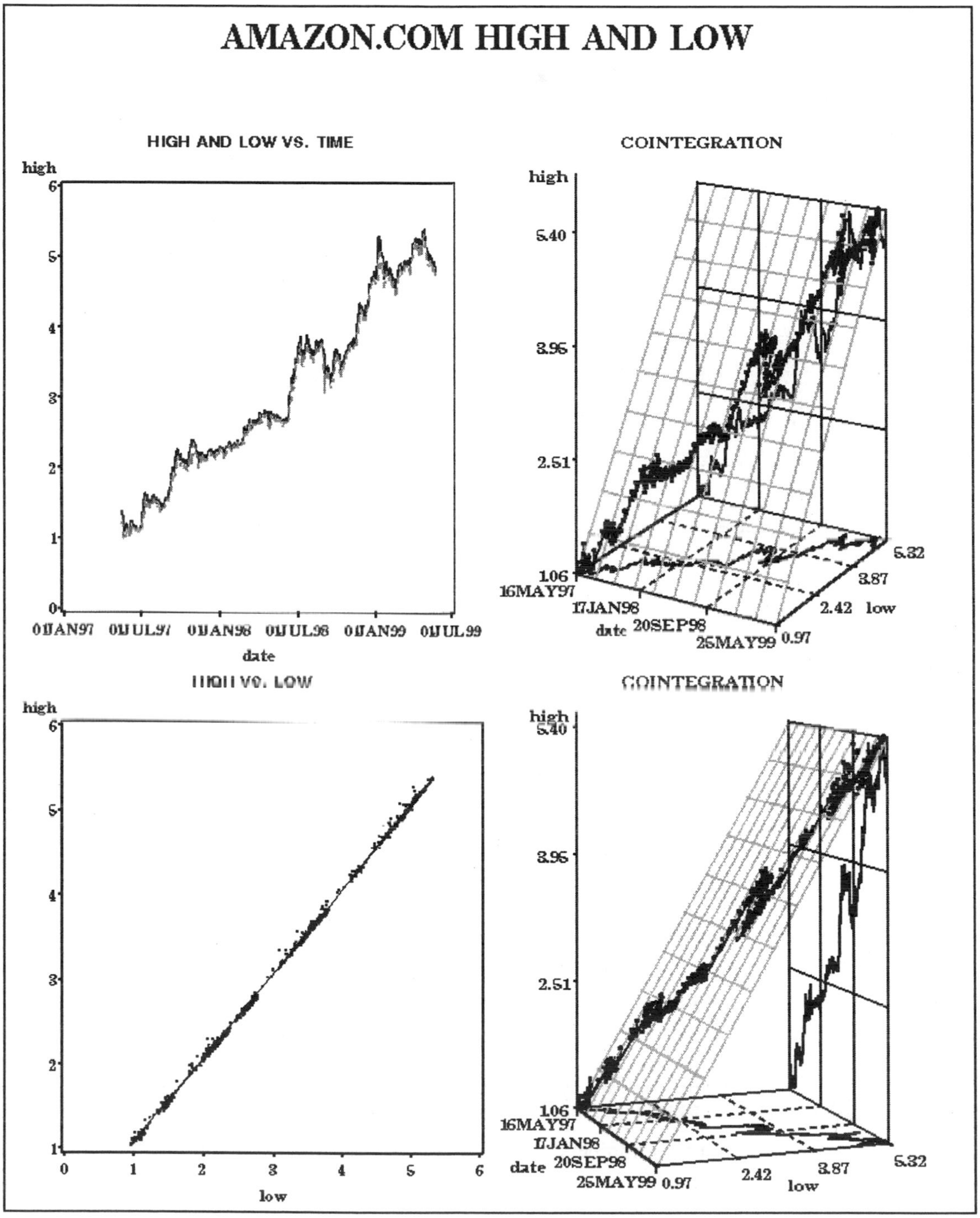

One way of fitting a vector model is to simply regress each Y_{it} on lags of itself and the other Ys thereby getting estimates of row *i* of the **A** coefficient matrices. Using just one lag you specify

```
PROC REG DATA=AMAZON;
   MODEL HIGH LOW = HIGH1 LOW1;
RUN;
```

where high1 and low1 are lagged values of the log transformed high and low prices. The partial output **Output 5.11** shows the estimates.

Output 5.11 PROC REG on Amazon.com Data

Dependent Variable: high

Variable	DF	Parameter Estimate	Standard Error	t Value	Pr > |t|
Intercept	1	0.01573	0.00730	2.15	0.0317
high1	1	0.88411	0.05922	14.93	<.0001
low1	1	0.11583	0.05979	1.94	0.0533

Dependent Variable: low

Variable	DF	Parameter Estimate	Standard Error	t Value	Pr > |t|
Intercept	1	-0.01042	0.00739	-1.41	0.1590
high1	1	0.45231	0.05990	7.55	<.0001
low1	1	0.54209	0.06047	8.96	<.0001

The estimated model becomes

$$\begin{pmatrix} Y_{1t}-\mu_1 \\ Y_{2t}-\mu_2 \end{pmatrix} = \begin{pmatrix} .8841 & .1158 \\ .4523 & .5421 \end{pmatrix} \begin{pmatrix} Y_{1,t-1}-\mu_1 \\ Y_{2,t-1}-\mu_2 \end{pmatrix} + \begin{pmatrix} e_{1t} \\ e_{2t} \end{pmatrix}$$

Recall that in a univariate AR(1) process, $Y_t = \alpha Y_{t-1} + e_t$, the requirement $|\alpha| < 1$ was imposed so that the expression $Y_t = \sum_{j=0}^{\infty} \alpha^j e_{t-j}$ for Y_t in terms of past shocks e_t would "converge"—that is, it would have weights on past shocks that decay exponentially as you move further into the past. What is the analogous requirement for the vector process $\mathbf{V}_t = \mathbf{A}\mathbf{V}_{t-1} + \mathbf{e}_t$? The answer lies in the "eigenvalues" of the coefficient matrix **A**.

5.2.2 Cointegration and Eigenvalues

Any $k \times k$ matrix has *k* complex numbers called eigenvalues or roots that determine certain properties of the matrix. The eigenvalues of **A** are defined to be the roots of the polynomial $|m\mathbf{I} - \mathbf{A}|$, where **A** is the $k \times k$ coefficient matrix, **I** is a $k \times k$ identity, and $|\ |$ denotes a determinant. For the fitted 2×2 matrix above, you find

$$\left| m\begin{pmatrix} 1 & 0 \\ 0 & 1 \end{pmatrix} - \begin{pmatrix} .8841 & .1158 \\ .4523 & .5421 \end{pmatrix} \right| = (m-.8841)(m-.5421)-(.4523)(.1158)$$

which becomes $m^2-(.8841+.5421)m+.4269=(m-.9988)(m-.4274)$, so the roots of this matrix are the real numbers 0.9988 and 0.42740. A matrix with unique eigenvalues can be expressed as $\mathbf{A}=\mathbf{ZDZ}^{-1}$, where **D** is a matrix with the eigenvalues of **A** on the main diagonal and 0 everywhere else and **Z** is the matrix of eigenvectors of **A**. Note that
$\mathbf{A}^{\mathrm{L}}=(\mathbf{ZDZ}^{-1})(\mathbf{ZDZ}^{-1})\cdots(\mathbf{ZDZ}^{-1})=\mathbf{ZD}^{\mathrm{L}}\mathbf{Z}^{-1}$. By the same reasoning as in the univariate case, the predicted deviations from the means L steps into the future are $\hat{\mathbf{V}}_{n+\mathrm{L}}=\mathbf{A}^{\mathrm{L}}\mathbf{V}_n$, where $\mathbf{V}_n$ is the last observed vector of deviations. For the 2×2 matrix currently under study you have

$$\mathbf{A}^{\mathrm{L}}=\begin{pmatrix} .8841 & .1158 \\ .4523 & .5421 \end{pmatrix}^{\mathrm{L}}=\mathbf{ZD}^{\mathrm{L}}\mathbf{Z}^{-1}$$

$$=\mathbf{Z}\begin{pmatrix} 0.9988 & 0 \\ 0 & 0.4274 \end{pmatrix}^{\mathrm{L}}\mathbf{Z}^{-1}=\mathbf{Z}\begin{pmatrix} 0.9988^{\mathrm{L}} & 0 \\ 0 & 0.4274^{\mathrm{L}} \end{pmatrix}\mathbf{Z}^{-1}$$

so that the elements of $\mathbf{A}^{\mathrm{L}}$ all converge to 0.

Output 5.12 Impulse Response, Lag 1 Model

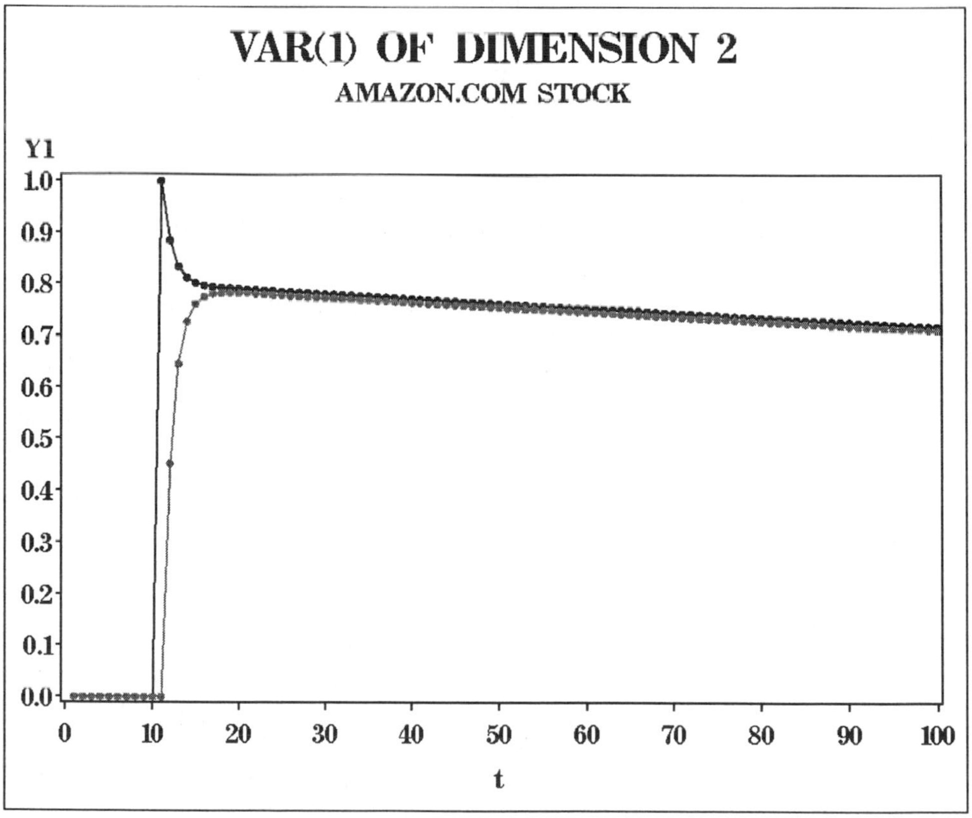

5.2.3 Impulse Response Function

To illustrate, **Output 5.12** shows a bivariate series with both Y_{1t} and Y_{2t} being 0 up to time $t = 11$, mimicking constant high and low stock price (log transformed and mean corrected). At time $t = 11$, Y_{1t} is shifted to 1 with Y_{2t} remaining at 0, thus representing a shock to the high price; that is, $\mathbf{V}_{11} = (1,0)'$. From then on, $\hat{\mathbf{V}}_{11+L} = \mathbf{A}^L \mathbf{V}_{11} = \mathbf{A}^L (1,0)'$; in other words $\hat{\mathbf{V}}_{11+L}$ traces out the path that would be followed with increasing lead L in absence of further shocks. The sequence so computed is called an *impulse response function*. It is seen that at time $t = 12$, $Y_{2,t}$ responded to the jump in $Y_{1,t}$ and increased to about 0.45 while $Y_{1,t}$ decreased, following the initial jump, to about 0.88. Continuing through time, the two series come close together then descend very slowly toward 0. This demonstrates the effect of a unit shock to the log of high price. The equilibrium, 0 deviations of both series from their mean, is approached slowly due to the eigenvalue 0.9988 being so close to 1. Clearly, if it were exactly 1.000, then 1.000^L would not decrease at all and the forecasts would not converge to the mean (0). Similarly, any attempt to represent the vector of deviations from the mean in terms of an infinite weighted sum of past error vectors will fail (i.e., not converge) if the eigenvalues or roots of the coefficient matrix **A** are one—that is, if **A** has any unit roots.

When all the eigenvalues of **A** are less than 1, we say that the vector autoregressive process of order 1, or VAR(1), is stationary, following the terminology from univariate processes. When the true **A** has unit roots, nonstandard distributions of estimates will arise just as in the univariate case. Note that the largest eigenvalue of the estimated matrix here, $\hat{\rho} = 0.9988$, is uncomfortably close to 1, and it would not be at all surprising to find that the true **A** matrix has a unit root. The roots here are analogous to the reciprocals of the roots you found for univariate series, hence the requirement that these roots be less than 1, not greater in magnitude.

5.2.4 Roots in Higher-Order Models

The requirement that the roots are all less than 1 in magnitude is called the stationarity condition. Series satisfying this requirement are said to be stationary, although technically, certain conditions on the initial observations are also required to ensure constant mean and covariances that depend only on the time separation of observations (this being the mathematical definition of stationarity).

In higher-order vector processes, it is still the roots of a determinantal equation that determine stationarity. In an order 2 VAR, $\mathbf{V}_t = \mathbf{A}_1\mathbf{V}_{t-1} + \mathbf{A}_2\mathbf{V}_{t-2} + \mathbf{e}_t$, the characteristic polynomial is $|m^2\mathbf{I} - \mathbf{A}_1 m - \mathbf{A}_2|$, and if all values of m that make this determinant 0 satisfy $|m| < 1$ then the vector process satisfies the stationarity condition.

Regressing as above on lag 1 and 2 terms in the Amazon.com high and low price series, this estimated model

$$\hat{\mathbf{V}}_t = \begin{pmatrix} 0.98486 & 0.23545 \\ 0.63107 & 0.65258 \end{pmatrix} \mathbf{V}_{t-1} + \begin{pmatrix} -0.08173 & -0.13927 \\ -0.22514 & -0.06414 \end{pmatrix} \mathbf{V}_{t-2}$$

is found, where the matrix entries are estimates coming from two PROC REG outputs:

Output 5.13
Process of Order 2

Dependent Variable: high

Variable	DF	Parameter Estimate	Standard Error	t Value	Pr > |t|
Intercept	1	0.01487	0.00733	2.03	0.0430
high1	1	0.98486	0.06896	14.28	<.0001
low1	1	0.23545	0.06814	3.46	0.0006
high2	1	-0.08173	0.07138	-1.14	0.2528
low2	1	-0.13927	0.06614	-2.11	0.0357

Dependent Variable: low

Variable	DF	Parameter Estimate	Standard Error	t Value	Pr > |t|
Intercept	1	-0.00862	0.00731	-1.18	0.2386
high1	1	0.63107	0.06877	9.18	<.0001
low1	1	0.65258	0.06795	9.60	<.0001
high2	1	-0.22514	0.07118	-3.16	0.0017
low2	1	-0.06414	0.06595	-0.97	0.3312

Inclusion of lag 3 terms seems to improve the model even further, but for simplicity of exposition, the lag 2 model will be discussed here. Keeping all the coefficient estimates, the characteristic equation, whose roots determine stationarity, is

$$\left| m^2\begin{pmatrix}1 & 0\\ 0 & 1\end{pmatrix} - m\begin{pmatrix}0.98486 & 0.23545\\ 0.63107 & 0.65258\end{pmatrix} - \begin{pmatrix}-0.08173 & -0.13927\\ -0.22514 & -0.06414\end{pmatrix}\right|$$
$$= (m-0.99787)(m-0.54974)(m-0.26767)(m+.17784)$$

Note that again, the largest eigenvalue, $\hat{\rho} = 0.99787$, is very close to 1, and it would not be at all surprising to find that the characteristic equation $\left|m^2\mathbf{I} - m\mathbf{A}_1 - \mathbf{A}_2\right| = 0$ using the true coefficient matrices has a unit root. Fountis and Dickey (1989) show that if a vector AR process has a single unit root, then the largest estimated root, normalized as $n(\hat{\rho}-1)$, has the same limit distribution as in the univariate AR(1) case. Comparing $n(\hat{\rho}-1) = 509(0.99787-1) = -1.08$ to the 5% critical value -11.3, the unit root hypothesis is not rejected. This provides a test for one versus no unit roots and hence is not as general as tests to be discussed later. Also, no diagnostics have been performed to check the model adequacy, a prerequisite for validity of any statistical test.

Using this vector AR(2) model, a bivariate vector of 0 deviations up to time $t = 11$ is generated, then a unit shock is imposed on the first component, the one corresponding to the high price, and the AR(2) used to extrapolate into the future. The code is as follows:

```
DATA SHOCK;
   Y12=0; Y22=0; Y11=0; Y21=0;
   DO T=1 TO 100;
      Y1 = .98486*Y11 + .23545*Y21 - .08173*Y12 - .13927*Y22;
      Y2 = .63107*Y11 + .65258*Y21 - .22514*Y12 - .06414*Y22;
      IF T=11 THEN Y1=1;
      OUTPUT;
      Y22=Y21; Y21=Y2; Y12=Y11; Y11=Y1;
   END;
RUN;
PROC GPLOT DATA=SHOCK; PLOT (Y1 Y2)*T/OVERLAY HREF=11;
   SYMBOL1 V=DOT I=JOIN C=RED;
   SYMBOL2 V=DOT I=JOIN C=GREEN;
RUN;
QUIT;
```

The graph of this impulse response function is shown in **Output 5.14**.

Output 5.14 *Impulse Response, Lag 2 Model*

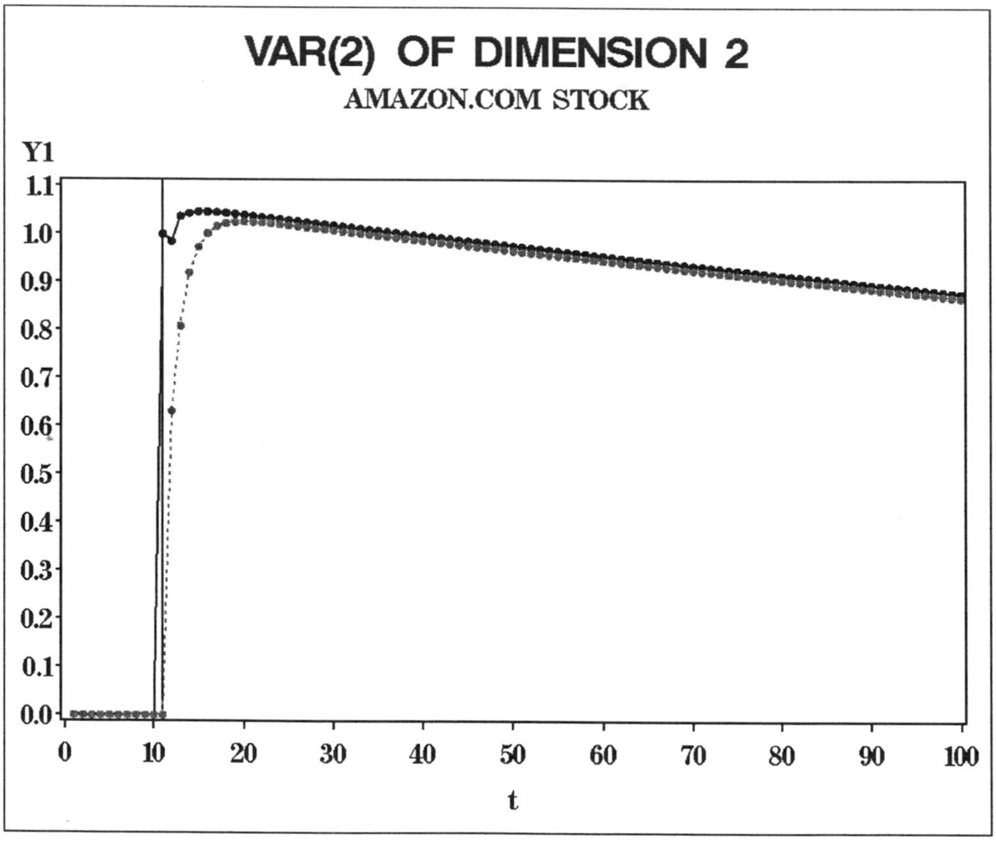

The addition of the second lag produces a more interesting pattern immediately following the shock to the high price logarithm series, but in the long run the series again approach each other and descend in tandem to the $(0,0)$ equilibrium deviation from the mean.

The forecasts might not have returned to the $(0,0)$ equilibrium point if the true coefficient matrices rather than estimates had been used. The behavior in the estimated model could simply be the result of the highest estimated root 0.99787 being a slight underestimate of a root that is really 1. Notice that a number even slightly smaller than 1 will reduce to nearly 0 when raised to a large exponent, as happens when the impulse response is extrapolated into the future. Models that allow exact unit roots in vector processes will be discussed next.

5.2.5 Cointegration and Unit Roots

An interesting class of models with exact unit roots is the class of cointegrated vector processes that can be represented in a type of model called the error correction model. Cointegration refers to a case in which a vector process, like the one with logarithms of high and low prices currently under discussion, has individually nonstationary components but there is some linear combination of them that is stationary. To make things a little clearer, suppose it is hypothesized that the ratio of high to low prices is stable; specifically, the daily price ratio series log(high/low) = log(high) – log(low) is stationary even though the log(high) and log(low) series each have unit roots. In this case, a shock to the high price series will result in an impulse response in which both series move as before, but they will not move back toward any historical mean values. Rather they will move toward some equilibrium pair of values for which log(high) – log(low) equals its long-term mean.

You can check spread = log(high) – log(low) for stationarity with no new tools—simply create the daily spread series and perform a unit root test on it. Here is some code to do the test and to check to see if 3 autoregressive lags (and hence 2 lagged differences) are sufficient to reduce the errors to white noise.

```
PROC ARIMA DATA=AMAZON;
  I VAR=SPREAD STATIONARITY = (ADF=(2));
  E P=3;
RUN;
```

As shown in **Output 5.15**, the tests strongly reject the unit root null hypothesis and thus indicate stationarity. The zero mean test would be useful only if one is willing to assume a zero mean for log(high) – log(low), and since high > low always, such an assumption is untenable for these data. Also shown are the chi-square tests for a lag 3 autoregression. They indicate that lagged differences beyond the second, $Y_{t-2} - Y_{t-3}$, are unnecessary and the fit appears to be excellent. This also suggests that an increase in the bivariate system to 3 lags might be helpful, as has previously been mentioned.

Output 5.15 *Stationary Test for High-Low Spread*

Augmented Dickey-Fuller Unit Root Tests

Type	Lags	Rho	Pr < Rho	Tau	Pr < Tau	F	Pr > F
Zero Mean	2	-18.3544	0.0026	-3.00	0.0028		
Single Mean	2	-133.290	0.0001	-7.65	<.0001	29.24	0.0010
Trend	2	-149.588	0.0001	-8.05	<.0001	32.41	0.0010

Conditional Least Squares Estimation

Parameter	Estimate	Standard Error	t Value	Approx Pr > |t|	Lag
MU	0.07652	0.0043870	17.44	<.0001	0
AR1,1	0.38917	0.04370	8.91	<.0001	1
AR1,2	0.04592	0.04702	0.98	0.3293	2
AR1,3	0.18888	0.04378	4.31	<.0001	3

Autocorrelation Check of Residuals

To Lag	Chi-Square	DF	Pr > ChiSq	--------------Autocorrelations--------------					
6	4.63	3	0.2013	-0.001	-0.018	-0.009	-0.047	0.072	-0.033
12	9.43	9	0.3988	0.037	0.041	0.035	0.025	0.029	0.058
18	12.71	15	0.6248	-0.018	-0.046	0.025	-0.014	0.016	0.052
24	21.14	21	0.4506	0.017	0.023	-0.067	-0.074	0.059	0.038
30	25.13	27	0.5669	0.026	0.049	0.014	-0.012	-0.006	0.063
36	28.86	33	0.6734	0.013	0.038	0.049	-0.023	0.016	-0.045
42	33.05	39	0.7372	0.049	0.055	0.023	0.010	0.039	0.003
48	36.51	45	0.8125	0.030	-0.035	-0.050	-0.038	0.006	-0.004

It appears that $\text{spread} = \log(\text{high}) - \log(\text{low})$ is stationary according to the unit roots tests. That means standard distribution theory should provide accurate tests since the sample size $n = 502$ is not too small. In that light, notice that the mean estimate 0.07652 for spread is significantly different from 0. An estimate of the number toward which the ratio of high to low prices tends to return is $e^{0.07652} = 1.08$ with a 95% confidence interval extending from $e^{0.07652-(1.96)(0.004387)} = 1.07$ to $e^{0.07652+(1.96)(0.004387)} = 1.09$. You conclude that the high tends to be 7% to 9% higher than the low in the long run.

You see that testing for cointegration is easy if you can prespecify the linear combination—e.g., S_t = spread = log(high) – log(low). Often one only suspects that *some* linear combination $Y_{1t} - \beta Y_{2t}$ is stationary, where (Y_{1t}, Y_{2t}) is a bivariate time series, so the problem involves estimating β as well as testing the resulting linear combination for stationarity. Engle and Granger (1987) argue that if you use regression to estimate β, your method is somewhat like sorting through all linear combinations of log(high) and log(low) to find the most stationary-looking linear combination. Therefore if you use the standard critical values for this test as though you *knew* β from some external source, your nominal level 0.05 would understate the true probability of falsely rejecting the unit root null hypothesis. Their solution was to compute residuals $r_t = Y_{1t} - \hat{\beta} Y_{2t}$ from a least squares regression of Y_{1t} on Y_{2t} and run a unit root test on these residuals, but then to compare the test statistic to special critical values that they supplied. This is a relatively easy and intuitively pleasing approach; however, it is not clear which of two or more series to use as the dependent variable in such a regression.

More symmetric approaches were suggested by Stock and Watson (1988) and Johansen (1988, 1991). Stock and Watson base their approach on a principal components decomposition of the vector time series, and Johansen's method involves calculating standard quantities, canonical correlations, from a multivariate multiple regression and then figuring out what distributions these would have in the vector time series case with multiple unit roots. Both strategies allow testing for multiple unit roots. For further comparisons among these approaches and an application to a macroeconomic vector series, see Dickey, Janssen, and Thornton (1991).

5.2.6 An Illustrative Example

To get a little better feeling for cointegration, consider this system with known coefficients:

$$\begin{pmatrix} Y_{1t} - 15 \\ Y_{2t} - 10 \end{pmatrix} = \begin{pmatrix} 1.84 & -0.24 \\ -0.06 & 1.66 \end{pmatrix} \begin{pmatrix} Y_{1,t-1} - 15 \\ Y_{2,t-1} - 10 \end{pmatrix} + \begin{pmatrix} -0.88 & 0.28 \\ 0.07 & -0.67 \end{pmatrix} \begin{pmatrix} Y_{1,t-2} - 15 \\ Y_{2,t-2} - 10 \end{pmatrix} + \begin{pmatrix} e_{1,t} \\ e_{2,t} \end{pmatrix}$$

Suppose $Y_{1t} = 15$ and $Y_{2t} = 10$ up to time 11, where a shock takes place. What happens after time 11 if no further shocks come along? That is, what does the impulse response function look like?

Output 5.16 *Impulse Responses*

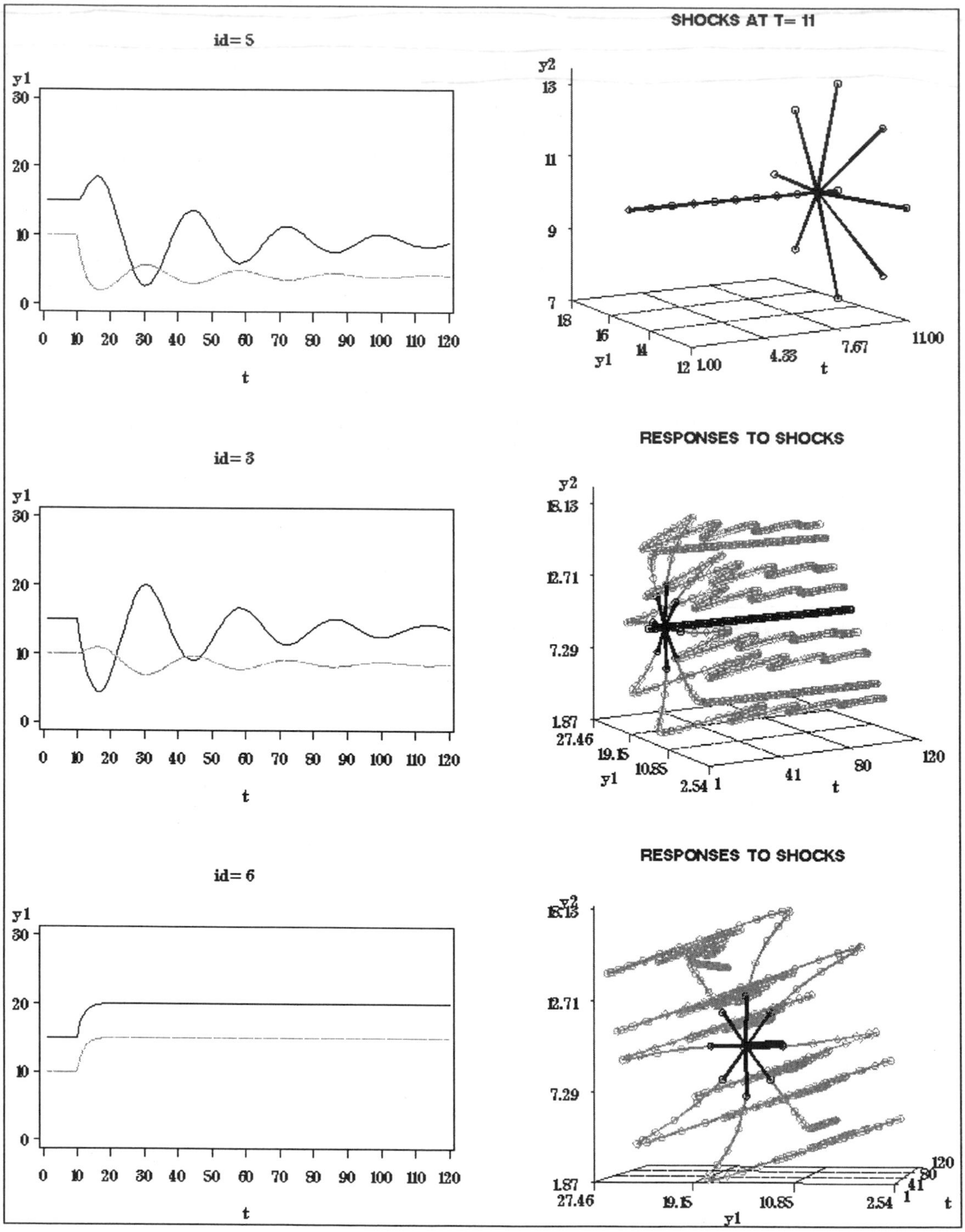

The left panels of **Output 5.16** show the results of setting the pair (Y_1, Y_2) to $(15,7)$, $(15,13)$, and $(17,12)$. It is seen that a change in either coordinate at time 11 results in the ultimate shifting of both coordinates. Also it is seen that there can be a lot of wiggling as the new levels are approached or there can be a relatively monotone approach of each coordinate to its new level. An insight into this behavior is given by the plots of these three impulse response functions and several others in the three-dimensional plots in the right column of the graph.

The axes represent Y_1, Y_2, and time *t*. All series set $(Y_1, Y_2) = (15,10)$ up to time $t = 11$, thus forming a "means axis." The top-right panel shows eight possible shocks at time $t = 11$, fanning out in an asterisk-shaped pattern. The middle plot on the right adds in the eight resulting impulse response curves, and the bottom-right plot is just a rotated view of the middle plot, with time measured by depth into the plot. In the first and second plots, time increases with movement to the right, the height of a point is Y_1, and its depth back into the plot is Y_2. The plots include a 0 shock case that forms a continuation of the means axis. For a while after the shock at time 11, there can be substantial wiggling or relatively smooth movement. What is striking is that as time passes, the points all seem to align in a plane. This plane is interpreted as a long-term relationship that will be approached over time after a shock bumps the point off of it (the plane). This gives rise to the term "error correction," meaning that movement off the plane is an "error," and in the long run in the absence of shocks, the points will move back to the equilibrium represented by the plane—an error "correction." A single shock can send the system into fairly wild fluctuations that, depending on what the series represent, might frighten investors, but these are temporary and the vector ultimately will settle near the plane of equilibrium. This equilibrium plane is interpreted as a relationship that cannot be dramatically violated for long periods of time by the system. Envision the plane as an "attractor," exerting a force like gravity on the points to settle them down after a shock.

Further insights are given by a bit of mathematics. Note that a vector VAR(2) model of dimension *k*, $\mathbf{V}_t = \mathbf{A}_1\mathbf{V}_{t-1} + \mathbf{A}_2\mathbf{V}_{t-2} + \mathbf{e}_t$, can be algebraically written in terms of differenced vectors $\nabla\mathbf{V}_t = \mathbf{V}_t - \mathbf{V}_{t-1}$ and a lagged vector $\mathbf{V}_{t-1}$ as $\nabla\mathbf{V}_t = -(\mathbf{I} - \mathbf{A}_1 - \mathbf{A}_2)\mathbf{V}_{t-1} - \mathbf{A}_2\nabla\mathbf{V}_{t-1} + \mathbf{e}_t$, where $(\mathbf{I} - \mathbf{A}_1 - \mathbf{A}_2)$ is $\mathbf{I}m^2 - \mathbf{A}_1 m - \mathbf{A}_2$ evaluated at $m = 1$. So if $|\mathbf{I} - \mathbf{A}_1 - \mathbf{A}_2| = 0$ (that is, if this matrix is less than full rank) then the time series has a unit root $m = 1$. Any $k \times k$ matrix $\mathbf{\Pi}$ that has rank $r < k$ can be written as $\mathbf{\Pi} = \boldsymbol{\alpha}\boldsymbol{\beta}'$, where $\boldsymbol{\alpha}$ and $\boldsymbol{\beta}$ are full-rank $k \times r$ matrices. Using the **A** matrices currently under discussion, the model

$$\begin{pmatrix} Y_{1t} - 15 \\ Y_{2t} - 10 \end{pmatrix} = \begin{pmatrix} 1.84 & -0.24 \\ -0.06 & 1.66 \end{pmatrix} \begin{pmatrix} Y_{1,t-1} - 15 \\ Y_{2,t-1} - 10 \end{pmatrix} + \begin{pmatrix} -0.88 & 0.28 \\ 0.07 & -0.67 \end{pmatrix} \begin{pmatrix} Y_{1,t-2} - 15 \\ Y_{2,t-2} - 10 \end{pmatrix} + \begin{pmatrix} e_{1,t} \\ e_{2,t} \end{pmatrix}$$

becomes

$$\begin{aligned}
\begin{pmatrix} \nabla Y_{1t} \\ \nabla Y_{2t} \end{pmatrix} &= -\begin{pmatrix} 0.04 & -0.04 \\ -0.01 & 0.01 \end{pmatrix} \begin{pmatrix} Y_{1,t-1} - 15 \\ Y_{2,t-1} - 10 \end{pmatrix} + \begin{pmatrix} 0.88 & -0.28 \\ -0.07 & 0.67 \end{pmatrix} \begin{pmatrix} \nabla Y_{1,t-1} \\ \nabla Y_{2,t-1} \end{pmatrix} + \begin{pmatrix} e_{1,t} \\ e_{2,t} \end{pmatrix} \\
&= \begin{pmatrix} -0.04 \\ 0.01 \end{pmatrix} \begin{pmatrix} 1 & -1 \end{pmatrix} \begin{pmatrix} Y_{1,t-1} - 15 \\ Y_{2,t-1} - 10 \end{pmatrix} + \begin{pmatrix} 0.88 & -0.28 \\ -0.07 & 0.67 \end{pmatrix} \begin{pmatrix} \nabla Y_{1,t-1} \\ \nabla Y_{2,t-1} \end{pmatrix} + \begin{pmatrix} e_{1,t} \\ e_{2,t} \end{pmatrix} \\
&= \begin{pmatrix} -0.04 \\ 0.01 \end{pmatrix} \left(Y_{1,t-1} - Y_{2,t-1} - 5 \right) + \begin{pmatrix} 0.88 & -0.28 \\ -0.07 & 0.67 \end{pmatrix} \begin{pmatrix} \nabla Y_{1,t-1} \\ \nabla Y_{2,t-1} \end{pmatrix} + \begin{pmatrix} e_{1,t} \\ e_{2,t} \end{pmatrix}
\end{aligned}$$

so that $\mathbf{\Pi} = -\begin{pmatrix} 0.04 & -0.04 \\ -0.01 & 0.01 \end{pmatrix} = \begin{pmatrix} -0.04 \\ 0.01 \end{pmatrix} \begin{pmatrix} 1 & -1 \end{pmatrix} = \boldsymbol{\alpha}\boldsymbol{\beta}'$.

The interpretation here is that $S_t = Y_{1t} - Y_{2t} - 5$ is stationary—that is, it tends to be near 0 so that the difference $Y_{1t} - Y_{2t}$ tends to be near 5. This algebraic form of the model is known as the "error correction model," or ECM. The plane satisfying $Y_{1t} = Y_{2t} + 5$ at every *t* is the attractor toward which all the impulse response functions are moving in the three-dimensional plots. A vector (a,b) such that $(a,b)(Y_{1t}, Y_{2t})'$ is stationary is called a *cointegrating vector,* so in this case $\boldsymbol{\beta}' = (-1,1)$ is such a vector, as are $(-2,2)$, $(1,-1)$, and any nonzero vector of the form $(-\phi,\phi) = \phi(-1,1)$. The set of all linear combinations of the rows of $\boldsymbol{\beta}'$ constitutes the set of all possible cointegrating vectors in the general case.

Next consider $N_t = Y_{1t} + 4Y_{2t} = (1,4)(Y_{1t}, Y_{2t})'$ and note that $\nabla N_t = (1,4)(\nabla Y_{1t}, \nabla Y_{2t})'$. Multiplying the vector equation on both sides by the row vector $(1,4)$, it is seen that ∇N_t involves lagged levels only through the term

$$(1,4)\begin{pmatrix} -0.04 \\ 0.01 \end{pmatrix}(Y_{1,t-1} - Y_{2,t-1} - 5) = 0$$

That is, ∇N_t in fact does not involve the lag levels of the variables at all. It is strictly expressible in terms of differences, so N_t is a unit root process. Also, because the only constant in the model, −5, is captured in the term $Y_{1,t-1} - Y_{2,t-1} - 5$ and is thus annihilated in the ∇N_t equation, it follows that N_t has no drift. N_t is a stochastic *common trend* shared by Y_{1t} and Y_{2t}. The two interesting linear combinations, the nonstationary $N_t = Y_{1t} + 4Y_{2t}$ and the stationary $S_t = Y_{1t} - Y_{2t} - 5$, can be written as

$$\begin{pmatrix} N_t \\ S_t \end{pmatrix} = \begin{pmatrix} 1 & 4 \\ 1 & -1 \end{pmatrix}\begin{pmatrix} Y_{1t} \\ Y_{2t} \end{pmatrix} + \begin{pmatrix} 0 \\ -5 \end{pmatrix} = \mathbf{T}\begin{pmatrix} Y_{1t} \\ Y_{2t} \end{pmatrix} + \begin{pmatrix} 0 \\ -5 \end{pmatrix}$$

from which you see that

$$\begin{pmatrix} Y_{1t} \\ Y_{2t} \end{pmatrix} = \mathbf{T}^{-1}\begin{pmatrix} N_t \\ S_t + 5 \end{pmatrix} = \begin{pmatrix} 0.2 & 0.8 \\ 0.2 & -.2 \end{pmatrix}\begin{pmatrix} N_t \\ S_t + 5 \end{pmatrix}$$

Thus it becomes clear exactly how the nonstationary common trend N_t is part of both Y series. For $\boldsymbol{\Pi} = \boldsymbol{\alpha}\boldsymbol{\beta}'$, with $\boldsymbol{\alpha}$ and $\boldsymbol{\beta}$ both $k \times r$ matrices, the matrix $\mathbf{T}$ can always be constructed by stacking $\boldsymbol{\alpha}'_p$ above $\boldsymbol{\beta}'$, where $\boldsymbol{\alpha}'_p$ is a $(k-r) \times k$ matrix such that $\boldsymbol{\alpha}'_p\boldsymbol{\alpha} = \mathbf{0}$.

As a final insight, multiply both sides of the VAR in error correction form by the transformation matrix **T** to get

$$\mathbf{T}\begin{pmatrix}\nabla Y_{1t}\\ \nabla Y_{2t}\end{pmatrix}=\begin{pmatrix}\nabla N_t\\ \nabla S_t\end{pmatrix}$$

$$=\begin{pmatrix}0\\ -0.05\end{pmatrix}(Y_{1,t-1}-Y_{2,t-1}-5)+\mathbf{T}\begin{pmatrix}0.88 & -0.28\\ -0.07 & 0.67\end{pmatrix}\mathbf{T}^{-1}\mathbf{T}\begin{pmatrix}\nabla Y_{1,t-1}\\ \nabla Y_{2,t-1}\end{pmatrix}+\mathbf{T}\begin{pmatrix}e_{1,t}\\ e_{2,t}\end{pmatrix}$$

$$=\begin{pmatrix}0\\ -0.05\end{pmatrix}S_{t-1}+\mathbf{T}\begin{pmatrix}0.88 & -0.28\\ -0.07 & 0.67\end{pmatrix}\mathbf{T}^{-1}\begin{pmatrix}\nabla N_{t-1}\\ \nabla S_{t-1}\end{pmatrix}+\mathbf{T}\begin{pmatrix}e_{1,t}\\ e_{2,t}\end{pmatrix}$$

$$=\begin{pmatrix}0\\ -0.05\end{pmatrix}S_{t-1}+\begin{pmatrix}0.60 & 0\\ 0 & 0.95\end{pmatrix}\begin{pmatrix}\nabla N_{t-1}\\ \nabla S_{t-1}\end{pmatrix}+\begin{pmatrix}z_{1,t}\\ z_{2,t}\end{pmatrix}$$

where the z white noise errors are linear combinations of the e errors. The coefficient matrix for the lagged differences of N and S is diagonal, which would not be the case in general. Nevertheless there does always exist a transformation matrix **T** such that the vector $\mathbf{TV}_t$ contains

1. as many unit root processes as the series has unit roots, followed by
2. stationary processes (provided none of the original Y series requires second differencing to achieve stationarity).

The period of the sinusoidal waves follows from the mathematical model. With the diagonal coefficient matrix in this example, it is easy to describe the stationary component as $\nabla S_t=-.05S_{t-1}+.95\nabla S_{t-1}+z_{2,t}$ or $S_t=1.9S_{t-1}-.95S_{t-2}+z_{2,t}$ with characteristic polynomial $m^2-1.9m+.95=(m-\sqrt{.95}e^{i\theta})(m-\sqrt{.95}e^{-i\theta})$. Here the representation of a complex number as $re^{i\theta}=r[\cos(\theta)+i\sin(\theta)]$ can be used with the fact that $\sin(-\theta)=-\sin(\theta)$ to show that $\sqrt{.95}(e^{i\theta}+e^{-i\theta})=2\sqrt{.95}\cos(\theta)$ must equal 1.9 so that the angle θ is $\theta=\arccos\left(1.90/2\sqrt{.95}\right)=12.92$ degrees. In the graphs one expects $110(12.92/360)=4$ cycles in the 110 observations after the shock. This is precisely what the graphs in **Output 5.16** show, with $\sqrt{.95}^{L}$ giving the amplitude damping factor L periods after the shock.

The relationship of S_t to Y_{1t} and Y_{2t} determines the orientation of this sinusoidal fluctuation in the three-dimensional plots. For the cases with equal shocks to Y_{1t} and Y_{2t}, no fluctuations were seen. That is because for these cases $S_t=Y_{1t}-Y_{2t}-5$ is no different after the shocks than before, so the shocked points are still in the cointegrating plane. With $\nabla N_t=.60\nabla N_{t-1}$ describing the component of motion in the cointegrating plane, one expects an exponential increase, in the equal shock cases, to a new horizontal line contained in the cointegrating plane. That indeed is what happens. The cases with unequal shocks to the two Y components force the point off the cointegrating plane, initiating a ripple-like fluctuation about the plane as the new levels are approached. In the bottom-right plot of **Output 5.16**, where passing time moves you toward the back of the plot, it is seen that the cointegrating plane slopes from the upper left to the lower right while the sinusoidal fluctuations seem to move from lower left to upper right and back again repeatedly as time passes.

You have learned some of the terminology and seen some geometric implications of cointegration in a hypothetical model with known parameters. You have seen from the graphs, or in more detail from the mathematical analysis, that the error correction model defines a simple linear attractor, a line, plane, or hyperplane, toward which forecasts gravitate. It can capture some fairly complicated short-term dynamics. You now look for cointegrating relationships like the S_t formula and common trends like N_t in the Amazon.com data.

For the Amazon.com stocks it appeared that the relationship $\log(\text{high}) - \log(\text{low})$ was stationary with average value about 0.0765. In a three-dimensional plot of $H_t = \log(\text{high})$, $L_t = \log(\text{low})$, and time *t*, one would expect the points to stay close to a plane having $\log(\text{high}) = \log(\text{low}) + 0.0765$ over time. That plane and the data were seen in **Output 5.10**. In the upper-left panel both series are plotted against time and it is seen that they almost overlay each other. The upper-right panel plots L_t versus *t* in the floor and H_t versus *t* in the back wall. These are projections into the floor and back wall of the points (t, L_t, H_t), which are seen moving from the lower left to upper right while staying quite close to a sloping plane. This is the cointegrating plane. The lower-right panel shows this same output rotated so that points move out toward the observer as time passes, and to its left the rotation continues so you are now looking directly down the edge of the cointegrating plane. This is also the graph of H_t versus L_t and motivates the estimation of the cointegrating plane by regression, as suggested by Engle and Granger (1987).

Using ordinary least squares regression you estimate an error correction model of the form

$$\begin{pmatrix} \nabla L_t \\ \nabla H_t \end{pmatrix} = \begin{pmatrix} -.4185 \\ 0.1799 \end{pmatrix} S_{t-1} + \begin{pmatrix} -.0429 & 0.2787 \\ 0.4170 & -.1344 \end{pmatrix} \begin{pmatrix} \nabla L_{t-1} \\ \nabla H_{t-1} \end{pmatrix}$$

where H_t and L_t are log transformed high and low prices, $S_t = H_t - L_t - .0765$, and .0765 is the estimated mean of $H_t - L_t$. Thus $0.18L_t + 0.42H_t$ is the common trend unit root process. It can be divided by 0.6 to make the weights sum to 1, in which case the graph will, of course, look like those of the original series that were so similar to each other in this example. It is a weighted average of things that are almost the same as each other.

5.2.7 Estimating the Cointegrating Vector

In the Amazon.com data, it was easy to guess that log(high/low) would be stationary and hence that $\log(\text{high}) - \log(\text{low})$ is the cointegrating relationship between these two series. This made the analysis pretty straightforward. A simple unit root test on $\log(\text{high}) - \log(\text{low})$ sufficed as a cointegration test. The high and low prices are so tightly cointegrated that it is clear from the outset the data will produce a nice example. In other cases, the data may not be so nice and the nature of the cointegrating plane might not be easily anticipated as it was here. The complete cointegration machinery includes tests that several series are cointegrated and methods for estimating the cointegrating relationships. If you estimate that $H_t - bL_t$ is stationary in the Amazon.com example, you might want to test to see if *b* is an estimate of 1.00 to justify the coefficient of L_t in $S_t = H_t - L_t - .0765$. The techniques include tests of such hypotheses about the cointegrating parameters.

The number of cointegrating relations in a process with known parameters is the rank of the coefficient matrix, **Π**, on the lagged levels in the error correction representation. In the previous hypothetical known parameter example you saw that this matrix was

$$\Pi = -\begin{pmatrix} 0.04 & -0.04 \\ -0.01 & 0.01 \end{pmatrix} = \begin{pmatrix} -0.04 \\ 0.01 \end{pmatrix} \begin{pmatrix} 1 & -1 \end{pmatrix} = \boldsymbol{\alpha\beta}'$$

which is clearly a rank-one matrix. This factoring of the matrix not only shows that there is one cointegrating relationship, but also reveals its nature: from the vector $(1 \quad -1)$ it is seen that the difference in the bivariate vector's elements is the linear combination that is stable—that is, it stays close to a constant. This happens to be the same cointegrating relationship that seemed to apply to the Amazon.com case and was displayed in the lower-left corner of **Output 5.10**. A vector time series of dimension 3 could move around anywhere in three-dimensional space as time passes. However, if its lag level coefficient matrix is

$$\boldsymbol{\Pi} = \begin{pmatrix} .1 \\ -.3 \\ -.2 \end{pmatrix} \begin{pmatrix} 1 & -.5 & -.5 \end{pmatrix}$$

then the points (Y_{1t}, Y_{2t}, Y_{3t}) will stay near the plane $Y_{1t} - .5Y_{2t} - .5Y_{3t} = C$ for some constant C as time passes. This is a plane running obliquely through three-dimensional (Y_{1t}, Y_{2t}, Y_{3t}) space just as the line in the lower-left corner of **Output 5.10** runs obliquely through two-dimensional space. In this case there is one cointegrating vector (1, –.5, –.5) and thus two common trends. You can think of these as two dimensions in which the series is free to float without experiencing a "gravitational pull" back toward the plane, just as our bivariate series was free to float up and down along the diagonal line in the lower-left corner of **Output 5.10**. Because time added to Y_{1t}, Y_{2t}, Y_{3t} introduces a fourth dimension, no graph analogous to the plane in the Amazon.com example is possible.

As a second example, if

$$\boldsymbol{\Pi} = \begin{pmatrix} 0.2 & 0.2 \\ 0.1 & -.5 \\ 0.8 & 0.1 \end{pmatrix} \begin{pmatrix} 1 & -.5 & -.5 \\ 1 & .2 & -.6 \end{pmatrix}$$

then the points (Y_{1t}, Y_{2t}, Y_{3t}) will stay near the line formed by the intersection of two planes: $Y_{1t} - .5Y_{2t} - .5Y_{3t} - C_1$ and $Y_{1t} + .2Y_{2t} - .6Y_{3t} - C_2$. In this last example, there are two cointegrating vectors and one common trend. That is, there is one dimension, the line of intersection of the planes, along which points are free to float.

SAS/ETS software provides PROC VARMAX to do this kind of modeling as well as allowing exogenous variables and moving average terms (hence the X and MA in VARMAX). Note that a lag 1 and a lag 2 bivariate autoregression have been fit to the Amazon.com data, but no check has yet been provided as to whether 2 lags are sufficient. In fact, a regression of the log transformed high and low stock prices on their lags indicates that 3 lags may in fact be needed. A popular method by

Johansen will be described next. It involves squared canonical correlations. Let **W** and **Y** be two random vectors. Pick a linear combination of elements of **W** and one of **Y** in such a way as to maximize the correlation. That correlation is the highest canonical correlation. Using only the linear combinations of **W** that are not correlated with the first, and similarly for **Y**, pick the linear combination from each set that produces the most highly correlated pair. That's the second highest canonical correlation, etc.

Let **W** and **Y** be two random mean 0 vectors related by $\mathbf{Y} = \mathbf{\Pi W} + \mathbf{e}$, where $\mathbf{\Pi}$ is a $k \times k$ matrix of rank r. Let $\mathbf{\Sigma}_{\text{YY}}$, $\mathbf{\Sigma}_{\text{WW}}$, and $\mathbf{\Sigma}$ denote the variance matrices of **Y**, **W**, and **e**, and assume **W** and **e** are uncorrelated. Let $\mathbf{\Sigma}_{\text{YW}} = \text{E}\{\mathbf{YW}'\} = \mathbf{\Pi\Sigma}_{\text{WW}}$. The problem of finding vectors $\boldsymbol{\gamma}_j$ and scalars λ_j for which

$$(\mathbf{\Sigma}_{\text{YW}}\mathbf{\Sigma}_{\text{WW}}^{-1}\mathbf{\Sigma}_{\text{YW}}' - \lambda_j\mathbf{\Sigma}_{\text{YY}})\boldsymbol{\gamma}_j = \mathbf{0}$$

or equivalently

$$(\mathbf{\Sigma}_{\text{YW}}\mathbf{\Sigma}_{\text{WW}}^{-1}\mathbf{\Sigma}_{\text{YW}}'\mathbf{\Sigma}_{\text{YY}}^{-1} - \lambda_j\mathbf{I})\boldsymbol{\gamma}_j = \mathbf{0}$$

is an eigenvalue problem. The solutions λ_j are the squared canonical correlations between **Y** and **W**, and since the rank of $\mathbf{\Pi}$ is r, there must be $k - r$ linearly independent vectors γ_j such that $\mathbf{\Sigma}_{\text{YW}}'\boldsymbol{\gamma}_j = \mathbf{\Sigma}_{\text{WW}}'\mathbf{\Pi}'\boldsymbol{\gamma}_j = 0$. For these you can solve the eigenvalue equation using $\lambda_j = 0$; that is, there are $k - r$ eigenvalues equal to 0. It is seen that finding the number of cointegrating vectors r is equivalent to finding the number of nonzero eigenvalues for the matrix $\mathbf{\Sigma}_{\text{YW}}\mathbf{\Sigma}_{\text{WW}}^{-1}\mathbf{\Sigma}_{\text{YW}}'\mathbf{\Sigma}_{\text{YY}}^{-1}$. Johansen's test involves estimating these variance and covariance matrices and testing the resulting estimated eigenvalues.

Begin with a lag 1 model $\mathbf{V}_t = \mathbf{AV}_{t-1} + \mathbf{e}_t$ or $\nabla\mathbf{V}_t = -(\mathbf{I} - \mathbf{A})\mathbf{V}_{t-1} + \mathbf{e}_t$. Johansen's method (1988, 1991) consists of a regression of $\nabla\mathbf{V}_t$ on $\mathbf{V}_{t-1}$; that is, each element of $\nabla\mathbf{V}_t$ is regressed on all the elements in $\mathbf{V}_{t-1}$ to produce the rows of the estimated $-(\mathbf{I} - \mathbf{A})$ coefficient matrix. For a lag 1 model, $(\mathbf{I} - \mathbf{A}) = \mathbf{\Pi} = \boldsymbol{\alpha\beta}'$, where the rows of $\boldsymbol{\beta}'$ are the cointegrating vectors and the following three numbers are all the same:

r = the rank of $\mathbf{I} - \mathbf{A}$

r = the number of cointegrating vectors, or rows of $\boldsymbol{\beta}'$

r = the number of nonzero squared canonical correlations between the elements of $\nabla\mathbf{V}_t$ and those of $\mathbf{V}_{t-1}$.

Johansen suggested studying the estimated squared canonical correlation coefficients to decide how many of them are significantly different from 0 and thereby estimate *r*. Standard procedures such as PROC CANCORR will deliver the desired estimates, just as an ordinary regression program will deliver the test statistics for a univariate unit root test but not the right p-values. As with the univariate unit root tests, the distributions of tests based on the squared canonical correlation coefficients are nonstandard for unit root processes, such as those found in the error correction model. Johansen tabulated the required distributions, thus enabling a test for *r*, the number of cointegrating vectors.

In the Amazon.com data, it appeared that $\boldsymbol{\beta}'$ could be taken as any multiple of the vector $\mathbf{H}' = (1, -1)$. Johansen also provides a test of the null hypothesis that $\boldsymbol{\beta} = \mathbf{H}\boldsymbol{\phi}$, where **H** is, as in the case of the Amazon.com data, a matrix of known constants. The test essentially compares the squared canonical correlations between $\nabla\mathbf{V}_t$ and $\mathbf{V}_{t-1}$. to those between $\nabla\mathbf{V}_t$ and $\mathbf{H}'\mathbf{V}_{t-1}$. If $\nabla\mathbf{V}_t = \boldsymbol{\alpha\beta}'\mathbf{V}_{t-1} + \mathbf{e}_t$ and $\boldsymbol{\beta} = \mathbf{H}\boldsymbol{\phi}$, you can easily see that $\nabla\mathbf{V}_t = \boldsymbol{\alpha\phi}'(\mathbf{H}'\mathbf{V}_{t-1}) + \mathbf{e}_t$, which motivates the test. In the Amazon.com data, if $\boldsymbol{\beta}'$ is some multiple of $\mathbf{H}' = (1, -1)$, one would expect the two squared canonical correlations between $\nabla\mathbf{V}_t$ and $\mathbf{V}_{t-1}$ to consist of one number near 0 and another number nearly equal to the squared canonical correlation between $\nabla\mathbf{V}_t$ and $S_t = \text{spread} = \mathbf{H}'\mathbf{V}_{t-1} = (1, -1)\mathbf{V}_{t-1} = \log(\text{high}) - \log(\text{low})$. The test that the first number is near 0 is a test for the cointegrating rank and involves nonstandard distributions. Given that there is one cointegrating vector, the test that its form is $(1, -1)$ is the one involving comparison of two eigenvalues and, interestingly, is shown by Johansen to have a standard chi-square distribution under the null hypothesis in large samples.

5.2.8 Intercepts and More Lags

PROC VARMAX gives these tests and a lot of additional information for this type of model. Before using PROC VARMAX on the Amazon.com data, some comments about higher-order processes and the role of the intercept are needed. Up to now, vector $\mathbf{V}_t$ was assumed to have been centered so that no intercept was needed in the model. Suppose now that

$$\mathbf{V}_t - \boldsymbol{\mu} = \mathbf{A}(\mathbf{V}_{t-1} - \boldsymbol{\mu}) + \mathbf{e}_t \qquad \text{and} \qquad \mathbf{V}_t = \boldsymbol{\lambda} + \mathbf{A}\mathbf{V}_{t-1} + \mathbf{e}_t$$

where $\boldsymbol{\mu}$ is a vector of means. The left-hand equation will be referred to as the "deviations form" for the model. In order for these two equations to be equivalent, the "intercept restriction" $\boldsymbol{\lambda} = (\mathbf{I} - \mathbf{A})\boldsymbol{\mu}$ must hold. Subtracting $(\mathbf{V}_{t-1} - \boldsymbol{\mu})$ from both sides of the first equation and subtracting $\mathbf{V}_{t-1}$ from both sides of the second gives

$$\nabla\mathbf{V}_t = (\mathbf{A} - \mathbf{I})(\mathbf{V}_{t-1} - \boldsymbol{\mu}) + \mathbf{e}_t \qquad \text{and} \qquad \nabla\mathbf{V}_t = \boldsymbol{\lambda} + (\mathbf{A} - \mathbf{I})\mathbf{V}_{t-1} + \mathbf{e}_t$$

In the cointegration case, recall that $\mathbf{A} - \mathbf{I} = \boldsymbol{\alpha\beta}'$, with $\boldsymbol{\alpha}_p$ representing a matrix of the same dimensions as $\boldsymbol{\alpha}$ such that $\boldsymbol{\alpha}_p'\boldsymbol{\alpha} = \mathbf{0}$ and so $\boldsymbol{\alpha}_p'(\mathbf{A} - \mathbf{I}) = \boldsymbol{\alpha}_p'\boldsymbol{\alpha\beta}' = 0$. Multiplying by $\boldsymbol{\alpha}_p'$ displays the "common trends" in the vector process. The equations become

$$\boldsymbol{\alpha}_p'\nabla\mathbf{V}_t = 0 + \boldsymbol{\alpha}_p'\mathbf{e}_t \qquad \text{and} \qquad \boldsymbol{\alpha}_p'\nabla\mathbf{V}_t = \boldsymbol{\alpha}_p'\boldsymbol{\lambda} + 0 + \boldsymbol{\alpha}_p'\mathbf{e}_t$$

The elements of vector $\boldsymbol{\alpha}'_p\mathbf{V}_t$ are seen to be driftless random walks in the left-hand equation since their first differences are white noise processes. The right-hand equation appears to describe random walks with drift terms given by the elements of vector $\boldsymbol{\alpha}'_p\boldsymbol{\lambda}$. Of course, once you remember the "intercept restriction" $\boldsymbol{\lambda} = (\mathbf{I}-\mathbf{A})\boldsymbol{\mu}$ you see that $\boldsymbol{\alpha}'_p\boldsymbol{\lambda} = \boldsymbol{\alpha}'_p(\mathbf{I}-\mathbf{A})\boldsymbol{\mu} = \mathbf{0}$. Nevertheless, some practitioners are interested in the possibility of an unrestricted (nonzero) drift in such data. Such data will display rather regular upward or downward trends. As in the case of univariate unit root tests, you might prefer to associate the unrestricted drift case with a deviations form that allows for such trends. Subtracting $\mathbf{V}_{t-1} - \boldsymbol{\mu} - \boldsymbol{\lambda}(t-1)$ from both sides of

$$\mathbf{V}_t - \boldsymbol{\mu} - \boldsymbol{\lambda}t = \mathbf{A}(\mathbf{V}_{t-1} - \boldsymbol{\mu} - \boldsymbol{\lambda}(t-1)) + \mathbf{e}_t$$

gives

$$\nabla\mathbf{V}_t - \boldsymbol{\lambda} = -(\mathbf{I}-\mathbf{A})(\mathbf{V}_{t-1} - \boldsymbol{\mu} - \boldsymbol{\lambda}(t-1)) + \mathbf{e}_t$$

Multiplying by $\boldsymbol{\alpha}'_p$ on both sides and remembering that $\boldsymbol{\alpha}'_p\ (\mathbf{I}-\mathbf{A})\boldsymbol{\mu} = \mathbf{0}$, the common trends for this model are given by

$$\boldsymbol{\alpha}'_p\nabla\mathbf{V}_t = \boldsymbol{\alpha}'_p\boldsymbol{\lambda} + 0 + \boldsymbol{\alpha}'_p\mathbf{e}_t$$

Further discussion about the role of the intercept in cointegration can be found in Johansen (1994).

In the case of higher-order models such as $\mathbf{V}_t = \mathbf{A}_1\mathbf{V}_{t-1} + \mathbf{A}_2\mathbf{V}_{t-2} + \mathbf{e}_t$ or $\nabla\mathbf{V}_t = (\mathbf{A}_1 + \mathbf{A}_2 - \mathbf{I})\mathbf{V}_{t-1} - \mathbf{A}_2\nabla\mathbf{V}_{t-1} + \mathbf{e}_t$, the estimate of $(\mathbf{A}_1 + \mathbf{A}_2 - \mathbf{I})$ that would be obtained by mutivariate multiple regression can be obtained in three stages as follows:

1. Regress $\nabla\mathbf{V}_t$ on $\nabla\mathbf{V}_{t-1}$ getting residual matrix $\mathbf{R}_{1t}$.
2. Regress $\mathbf{V}_{t-1}$ on $\nabla\mathbf{V}_{t-1}$ getting residuals $\mathbf{R}_{2,t-1}$.
3. Regress $\mathbf{R}_{1t}$ on $\mathbf{R}_{2,t-1}$.

In higher-order models, then, you can simply replace $\nabla\mathbf{V}_t$ and $\mathbf{V}_{t-1}$ with $\mathbf{R}_{1t}$ and $\mathbf{R}_{2,t-1}$ and follow the same steps as described earlier for a lag 1 model. In a lag *p* model, steps 1 and 2 would have regressors $\nabla\mathbf{V}_{t-1},\ldots,\nabla\mathbf{V}_{t-p+1}$, and furthermore, Johansen shows that seasonal dummy variables can be added as regressors without altering the limit distributions of his tests. The procedure has been described here in a manner that emphasizes its similarity to univariate unit root testing. The reader familiar with Johansen's method may note that he uses a slightly different parameterization that places the lag levels at the furthest lag rather than lag 1. For example, $\mathbf{V}_t = \mathbf{A}_1\mathbf{V}_{t-1} + \mathbf{A}_2\mathbf{V}_{t-2} + \mathbf{e}_t$ becomes $\nabla\mathbf{V}_t = (\mathbf{A}_1 - \mathbf{I})\nabla\mathbf{V}_{t-1} + (\mathbf{A}_1 + \mathbf{A}_2 - \mathbf{I})\mathbf{V}_{t-2} + \mathbf{e}_t$. The same "impact matrix," as it is called, $\boldsymbol{\Pi} = \mathbf{A}_1 + \mathbf{A}_2 - \mathbf{I}$, appears in either format, and inferences about its rank are the same either way.

5.2.9 PROC VARMAX

Returning to the Amazon.com data, PROC VARMAX is used to produce some of the cointegration computations that have just been discussed.

```
PROC VARMAX DATA=AMAZON;
   MODEL HIGH LOW/P=3 LAGMAX=5 ECM=(RANK=1 NORMALIZE=HIGH)
   COINTTEST;
   COINTEG RANK=1 H=(1 , -1 );
   OUTPUT OUT=OUT1 LEAD=50;
   ID T INTERVAL=DAY;
RUN;
```

This requests a vector autoregressive model of order 3, VAR(3), on variables high and low. They are the log transformed high and low prices for Amazon.com stock. Diagnostics of fit will be given up to lagmax = 5. The error correction model, ECM, is assigned a rank 1, meaning that the impact matrix $\mathbf{A}_1 + \mathbf{A}_2 + \mathbf{A}_3 - \mathbf{I} = \mathbf{\Pi} = \boldsymbol{\alpha\beta}'$ is such that $\boldsymbol{\alpha}$ and $\boldsymbol{\beta}$ are column vectors (rank 1). Here $\mathbf{A}_1$, $\mathbf{A}_2$, and $\mathbf{A}_3$ represent the VAR coefficient matrices. The normalize option asks PROC VARMAX to report the multiple of $\boldsymbol{\beta}'$ that has 1 as the coefficient of high. Recall that if $\boldsymbol{\beta}'\mathbf{V}_t$ is a stationary linear combination of elements of the random vector $\mathbf{V}_t$, then so is any multiple of it. The COINTTEST option asks for a test of the cointegrating rank, while the COINTEG statement tests the hypothesis that the cointegrating vector $\boldsymbol{\beta}'$ can be expressed as a multiple of $\mathbf{H}' = (1,\ \ 1)$. Only a few of the many items produced by PROC VARMAX are shown in **Output 5.17**.

Output 5.17 VARMAX on Amazon.com Data, Part 1

The VARMAX Procedure

Number of Observations 509
Number of Pairwise Missing 0

Variable	Type	NoMissN	Mean	StdDev	Min	Max
high	DEP	509	3.12665	1.23624	1.06326	5.39929
low	DEP	509	3.05067	1.22461	0.96508	5.31812

Cointegration Rank Test

H_0: Rank=r	H_1: Rank>r	Eigenvalue	Trace	Critical Value	Drift InECM	DriftIn Process
0	0	0.1203	65.50 ❶	15.34	Constant	Linear
1	1	0.0013	0.66	3.84		

Cointegration Rank Test under the Restriction

H_0: Rank=r	H_1: Rank>r	Eigenvalue	Trace	Critical Value	Drift InECM	DriftIn Process
0	0	0.1204	71.15	19.99	Constant	Constant
1	1	0.0123	6.26	9.13		

Whether or not the intercept restriction (that $\boldsymbol{\alpha}'_p$ anihilates the intercept) is imposed, the hypothesis of $r=0$ cointegrating vectors is rejected. For example, in the unrestricted case, Johansen's "trace test" has value 65.50, exceeding the critical value 15.34 ❶, so $r=0$ is rejected. The test for $r=1$ versus $r>1$ does not reject the $r=1$ null hypothesis. Thus Johansen's test indicates a single $(r=1)$ cointegrating vector, and hence a single ($k-r=2-1=1$) common trend. Note that the tests are based on eigenvalues, as might be anticipated from the earlier discussion linking squared canonical correlations to eigenvalues. From the graph, it would seem that a drift or linear trend term would be appropriate here, so the test without the restriction seems appropriate, though both tests agree that $r=1$ anyway. Assuming a rank $r=1$, the null hypothesis that $\boldsymbol{\alpha}'_p$ anihilates the intercept is tested by comparing eigenvalues of certain matrices with and without this intercept restriction. The null hypothesis that the intercept restriction holds is rejected using the chi-square test 5.60 with 1 degree of freedom. In light of the plot, **Output 5.10**, it is not surprising to find a drift in the common trend.

Output 5.17a *VARMAX on Amazon.com Data, Part 2*

Test of the Restriction when Rank=r

Rank	Eigenvalue On Restrict	Eigenvalue	Chi-Square	DF	Prob> ChiSq
0	0.1204	0.1203	5.65	2	0.0593
1	0.0123	0.0013	5.60	1	0.0179

Long-Run Parameter BETA Estimates

Variable	Dummy 1	Dummy 2
high	1.00000	1.00000
low	-1.01036	-0.24344

Adjustment Coefficient ALPHA Estimates

Variable	Dummy 1	Dummy 2
high	-0.06411	-0.00209
low	0.35013	-0.00174

The long-run parameter estimates in **Output 5.17a** allow the user to estimate impact matrices $\Pi = \boldsymbol{\alpha\beta}'$ of various ranks. For this data the rank 1 and rank 2 versions of Π are

$$\begin{pmatrix} -0.06411 \\ 0.35013 \end{pmatrix} \begin{pmatrix} 1.00000 & -1.01036 \end{pmatrix} = \begin{pmatrix} -0.064 & 0.0648 \\ 0.350 & -0.3548 \end{pmatrix}$$

and

$$\begin{pmatrix} -0.06411 & -0.00209 \\ 0.35013 & -0.00174 \end{pmatrix} \begin{pmatrix} 1.00000 & -1.01036 \\ 1.00000 & -0.24344 \end{pmatrix} = \begin{pmatrix} -0.0662 & 0.065283 \\ 0.34839 & -0.353334 \end{pmatrix}$$

These are almost the same, as might be expected since there was very little evidence from the test that the rank is greater than 1. In this computation, no restriction is made on the intercept.

Now suppose $\mathbf{W}_{t-1}$ is an augmented version of $\mathbf{V}_{t-1}$, namely a vector whose last entry is 1 and whose first entries are the same as those of $\mathbf{V}_{t-1}$. For simplicity consider the lag 1 model. Write the model as $\nabla\mathbf{V}_t = \boldsymbol{\alpha}\boldsymbol{\beta}_+'\mathbf{W}_{t-1} + \mathbf{e}_t$, where $\boldsymbol{\beta}_+'$ is the same as $\boldsymbol{\beta}'$ except for its last column. Recall the previously mentioned transformation matrix $\mathbf{T}$ constructed by stacking $\boldsymbol{\alpha}_p'$ above $\boldsymbol{\beta}'$, where $\boldsymbol{\alpha}_p'$ is a $(k-r)\times k$ matrix such that $\boldsymbol{\alpha}_p'\boldsymbol{\alpha} = \mathbf{0}$. Because $\nabla\boldsymbol{\alpha}_p'\mathbf{V}_t = \boldsymbol{\alpha}_p'\boldsymbol{\alpha}\boldsymbol{\beta}_+'\mathbf{W}_{t-1} + \boldsymbol{\alpha}_p'\mathbf{e}_t = \boldsymbol{\alpha}_p'\mathbf{e}_t$, it follows that the elements of $\boldsymbol{\alpha}_p'\mathbf{V}_t$ are driftless unit root processes. These are the first $k-r$ elements of $\mathbf{TV}_t$. The last r elements are the stationary linear combinations. They satisfy $\nabla\boldsymbol{\beta}'\mathbf{V}_t = \boldsymbol{\beta}'\boldsymbol{\alpha}\boldsymbol{\beta}_+'\mathbf{W}_{t-1} + \boldsymbol{\beta}'\mathbf{e}_t$. The elements in the last column of $\boldsymbol{\beta}'\boldsymbol{\alpha}\boldsymbol{\beta}_+'$ get multiplied by 1, the last entry of $\mathbf{W}_{t-1}$. In other words, they represent the intercepts for the stationary linear combinations. This shows how the addition of an extra element, a 1, to $\mathbf{V}_{t-1}$ forces a model in which the unit root components do not drift. The result is the same in higher-order models. PROC VARMAX gives "dummy variables" for this case as well. Because the last column of ALPHA is 0, the "Dummy 3" columns could be omitted as you might expect from the preceding discussion. Having previously rejected the restriction of no drift in the common trends, you are not really interested in these results that assume the restriction. In another data set, they might be of interest and hence are shown for completeness.

Output 5.17b *VARMAX on Amazon.com Data, Part 3*

```
            Long-Run Coefficient BETA based
                on the Restricted Trend

Variable        Dummy 1        Dummy 2        Dummy 3

high            1.00000        1.00000        1.00000
low            -1.01039       -0.79433        0.03974
1              -0.04276       -1.48420       -2.80333

              Adjustment Coefficient ALPHA
               based on the Restricted Trend

Variable        Dummy 1        Dummy 2        Dummy 3

high           -0.05877       -0.00744    1.49463E-16
low             0.35453       -0.00614    -9.0254E-16
```

5.2.10 Interpreting the Estimates

A list of estimates ❶❷follows (**Output 5.17c**) that shows that the fitted rank 1 model for the log transformed high and low prices, H_t and L_t, is

$$\begin{pmatrix}\nabla \mathrm{H}_t\\ \nabla \mathrm{L}_t\end{pmatrix} = \begin{pmatrix}0.00857\\ -0.01019\end{pmatrix} + \begin{pmatrix}-0.064 & 0.065\\ 0.350 & -0.354\end{pmatrix}\begin{pmatrix}\mathrm{H}_{t-1}\\ \mathrm{L}_{t-1}\end{pmatrix}$$
$$+\begin{pmatrix}0.044 & 0.191\\ 0.294 & 0.013\end{pmatrix}\begin{pmatrix}\nabla \mathrm{H}_{t-1}\\ \nabla \mathrm{L}_{t-1}\end{pmatrix} + \begin{pmatrix}-0.097 & 0.039\\ 0.068 & -0.132\end{pmatrix}\begin{pmatrix}\nabla \mathrm{H}_{t-2}\\ \nabla \mathrm{L}_{t-2}\end{pmatrix} + \begin{pmatrix}e_{1t}\\ e_{2t}\end{pmatrix}$$

and error variance matrix ❸

$$\Sigma = \frac{1}{1000}\begin{pmatrix} 2.98 & 2.29 \\ 2.29 & 2.95 \end{pmatrix}$$

indicating a correlation 0.77 between the errors.

Output 5.17c *VARMAX on Amazon.com Data, Part 4*

```
            Long-Run Parameter
              BETA Estimates
              given RANK = 1

          Variable        Dummy 1

          high            1.00000
          low            -1.01036

          Adjustment Coefficient
              ALPHA Estimates
              given RANK = 1

          Variable        Dummy 1

          high           -0.06411
          low             0.35013

            Constant Estimates

          Variable       Constant

          high            0.00857   ❶
          low            -0.01019

       Parameter ALPHA * BETA' Estimates

   Variable            high            low

   high            -0.06411        0.06478
   low              0.35013       -0.35376

          AR Coefficient Estimates ❷

DIF_Lag     Variable           high            low

      1     high            0.04391        0.19078
            low             0.29375        0.01272
      2     high           -0.09703        0.03941
            low             0.06793       -0.13209

    Covariance Matrix for the Innovation
   Variable            high            low
   high             0.00298        0.00229 ❸
   low              0.00229        0.00295
```

5.2.11 Diagnostics and Forecasts

There follows a series of diagnostics. The regression of ∇H_t on the lagged levels and two lagged differences of both H and L is seen to have a model F test 4.91 ❶ and R square 0.0558 ❷, and a similar line describing the ∇L_t regression is found just below this. (See **Output 5.17d.**)The residuals from these models are checked for normality and unequal variance of the autoregressive conditional heteroscedastic, or ARCH, type. Both of these departures from assumptions are found. The Durbin-Watson DW(1) statistics are near 2 for both residual series, and autoregressive models fit to these residuals up to 4 lags show no significance. These tests indicate uncorrelated residuals.

Output 5.17d
VARMAX on Amazon.com Data, Part 5

Univariate Model Diagnostic Checks

Variable	R-square	StdDev	F Value	Prob>F
high	0.0558 ❷	0.0546	4.91 ❶	<.0001
low	0.1800	0.0543	18.25	<.0001

Univariate Model Diagnostic Checks

Variable	DW(1)	Normality ChiSq	Prob> ChiSq	ARCH1 F Value	Prob>F
high	1.98	82.93	<.0001	19.06	<.0001
low	1.98	469.45	<.0001	144.47	<.0001

Univariate Model Diagnostic Checks

Variable	AR1 F Value	AR1 Prob>F	AR1-2 F Value	AR1-2 Prob>F	AR1-3 F Value	AR1-3 Prob>F	AR1-4 F Value	AR1-4 Prob>F
high	0.02	0.8749	0.07	0.9294	0.97	0.4057	1.31	0.2666
low	0.00	0.9871	0.33	0.7165	0.65	0.5847	0.81	0.5182

Recall that the spread $H_t - L_t$ was found to be stationary using a standard unit root test, and that the estimated cointegrating relationship was $H_t - 1.01L_t$. Given these findings, it is a bit surprising that the test that $\boldsymbol{\beta}' = \phi(1, -1)$ rejects that hypothesis. However, the sample size $n = 509$ is somewhat large, so rather small and practically insignificant departures from the null hypotheses might still be statistically significant. In a similar vein, one might look at plots of residual histograms to see if they are approximately bell shaped before worrying too much about the rejection of normality. The test that $\boldsymbol{\beta}' = \phi(1, -1)$ is referred to as a test of the restriction matrix **H**. The test compares eigenvalues 0.1038 and 0.1203 by comparing

$$(n-3)[\log(1-0.1038) - \log(1-0.1203)] = 506(0.01858) = 9.40$$

to a chi square with 1 degree of freedom ❶. (See **Output 5.17e.**)

Output 5.17e
VARMAX on Amazon.com Data, Part 6

```
                Restriction Matrix H
                with respect to BETA

               Variable        Dummy 1

               high            1.00000
               low            -1.00000

                Long-Run Coefficient
                BETA with respect to
                 Hypothesis on BETA

               Variable        Dummy 1

               high            1.00000
               low            -1.00000

               Adjustment Coefficient
                ALPHA with respect to
                 Hypothesis on BETA

               Variable        Dummy 1

               high           -0.07746
               low             0.28786

     Test for Restricted Long-Run Coefficient BETA

          Eigenvalue                       Chi-              Prob>
 Index    OnRestrict     Eigenvalue      Square      DF     ChiSq

     1        0.1038         0.1203        9.40 ❶     1     0.0022
```

The fitted model implies one common trend that is a unit root with drift process and one cointegrating vector. The last bit of code requests forecasts using the VAR(3) in rank 1 error correction form. These are put into an output data set, a few observations from which are shown in **Output 5.17f**. An additional complication with these data is that the market is closed on the weekends, so the use of the actual dates as ID variables causes a missing data message to be produced. An easy fix here is to use t = observation number as an ID variable, thus making the implicit assumption that the correlation between a Monday and the previous Friday is the same as between adjacent days. A portion of these data, including standard errors and upper and lower 95% confidence limits, is shown.

Output 5.17f
VARMAX on Amazon.com Data, Last Part

Obs	t	high	FOR1	RES1	STD1	LCI1	UCI1
508	508	4.86272	4.88268	-0.019970	0.05461	4.77565	4.98972
509	509	4.79682	4.85402	-0.057193	0.05461	4.74698	4.96105
510	510	.	4.79125	.	0.05461	4.68422	4.89829
511	511	.	4.80030	.	0.08475	4.63420	4.96640
512	512	.	4.80704	.	0.10715	4.59704	5.01704

Obs	low	FOR2	RES2	STD2	LCI2	UCI2
508	4.75359	4.81222	-0.058634	0.05430	4.70580	4.91865
509	4.71290	4.75931	-0.046406	0.05430	4.65288	4.86574
510	.	4.70442	.	0.05430	4.59799	4.81084
511	.	4.70671	.	0.08605	4.53806	4.87537
512	.	4.71564	.	0.10855	4.50288	4.92840

You can observe the quick spreading of confidence intervals, typical of data whose logarithms contain a unit root. The fact that the unit root is in some sense shared between the two series does not do much to narrow the intervals. The drift in the underlying unit root process, or common trend, is apparent in the forecasts. The short-term dynamics do not seem to contribute much to the forecasts, suggesting that the last few observations were quite near the cointegrating plane. (See **Output 5.18.**)

Output 5.18 *Forecasts Using Cointegration*

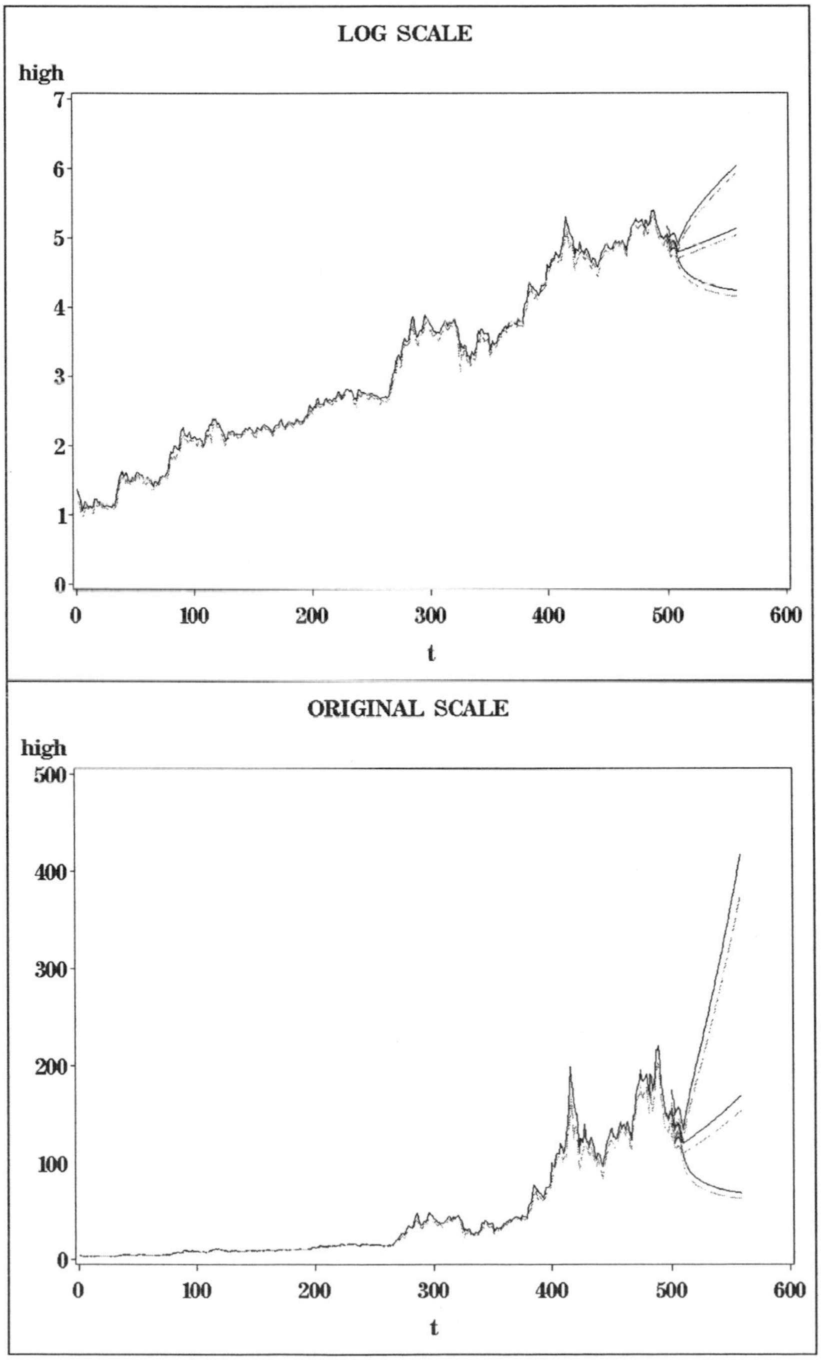

Chapter 6 State Space Modeling

6.1 Introduction

In ARIMA modeling, one of the difficult tasks is to select a model. Also, if you have several related time series, they must satisfy some restrictive conditions in order to justify the kind of transfer function modeling that is available in PROC ARIMA. There must be no feedback, and, for proper identification and forecast intervals, multiple inputs must be independent of each other and enough differencing must be specified to render the series stationary. PROC STATESPACE allows estimation under less restrictive conditions and provides some automatic model specification ability, although the user is still responsible for making the series stationary.

In Chapter 5, another procedure, PROC VARMAX, was discussed. This procedure also handles multiple series and, unlike STATESPACE, can perform cointegration analysis, which is appropriate when your series display unit root nonstationarity but some linear combination of the series is stationary. In other words, the transformation to stationarity is not just differencing.

The basic idea in state space modeling is to discover the "state vector." The state vector consists of the current values of all series under investigation plus enough forecasts into the future so that all forecasts, no matter how far away, are linear combinations of these.

6.1.1 Some Simple Univariate Examples

To get started, here are some models, all with mean 0, and their forecasting equations. As is customary in discussing state space models, the symbol $\hat{Y}_{t+L|t}$ denotes a forecast of Y_{t+L} using information available at time t. In model discussions in this section, the default assumption is that the mean has already been subtracted.

Table 6.1 *One-, Two-, and Three-Step-Ahead Prediction for Different Models*

Name	Formula	$\hat{Y}_{t+1\|t}$	$\hat{Y}_{t+2\|t}$	$\hat{Y}_{t+3\|t}$
AR(1)	$Y_t = \alpha Y_{t-1} + e_t$	αY_t	$\alpha^2 Y_t$	$\alpha^3 Y_t$
AR(2)	$Y_t = \alpha_1 Y_{t-1} + \alpha_2 Y_{t-2} + e_t$	$\hat{Y}_{t+1\|t} = \alpha_1 Y_t + \alpha_2 Y_{t-1}$	$\alpha_1 \hat{Y}_{t+1\|t} + \alpha_2 Y_t$	$\alpha_1 \hat{Y}_{t+2\|t} + \alpha_2 \hat{Y}_{t+1\|t}$
MA(2)	$Y_t = e_t + \beta_1 e_{t-1} + \beta_2 e_{t-2}$	$\beta_1 e_t + \beta_2 e_{t-1}$	$\beta_2 e_t$	0
ARMA(1,1)	$Y_t = \alpha Y_{t-1} + e_t + \beta e_{t-1}$	$\alpha Y_t + \beta e_t$	$\alpha(\alpha Y_t + \beta e_t)$	$\alpha^2(\alpha Y_t + \beta e_t)$

Numerical examples and further discussion of models like these appear in **Section 6.2**.

A "linear combination" of a set of variables is a sum of constant coefficients times variables. For example, $2X + 3Y$ and $5X - 2Y$ are linear combinations of X and Y. Notice that $6(2X + 3Y) - 4(5X - 2Y)$ is automatically also a linear combination of X and Y; that is, linear combinations of linear combinations are themselves also linear combinations. Note that $0X + 0Y = 0$ is also a valid linear combination of X and Y. Considering Y_t and $\hat{Y}_{t+L|t}$ for $L = 1, 2, \ldots$ to be the variables and considering functions of model parameters, like α^L, to be constants, the state vector is defined to be $(Y_t, \hat{Y}_{t+1|t}, \hat{Y}_{t+2|t}, \ldots, \hat{Y}_{t+k|t})$, where *k* is the smallest value such that all remaining forecasts $\hat{Y}_{t+L|t}$ with $L > k$ are linear combinations of the state vector elements. For the AR(1) model all forecasts are linear combinations (multiples) of Y_t, so the state vector is just (Y_t). For the AR(2) the state vector is $(Y_t, \hat{Y}_{t+1|t})$. It can't be just Y_t because $\hat{Y}_{t+1|t}$ involves Y_{t-1} whose value cannot be determined from Y_t. However, $\hat{Y}_{t+2|t}$ is a linear combination, $\alpha_1 \hat{Y}_{t+1|t} + \alpha_2 Y_t$, of state vector entries, and $\hat{Y}_{t+3|t}$ is a linear combination, $\alpha_1(\alpha_1 \hat{Y}_{t+1|t} + \alpha_2 Y_t) + \alpha_2 \hat{Y}_{t+1|t}$, of them too. The expressions get more complicated, but by the "linear combination of linear combinations" argument it is clear that *all* forecasts are linear combinations of Y_t and $\hat{Y}_{t+1|t}$. You can see that for an AR(*p*) the state vector will have *p* elements. For moving averages it is assumed that current and past e_t s can be well approximated from the observed data—that is, MA models need to be invertible. Acting as though e_ts that have already occurred are known, it is clear from the MA(2) example that for an MA(*q*) model, forecasts more than *q* steps ahead are trivial linear combinations (0) of state vector elements. Finally, for mixed models the forecasts are eventually determined through autoregressive type recursions and, by the linear combination of linear combinations argument, must be linear combinations of state vector elements from that point on.

The state vector contains all the information needed to forecast into the infinite future. During an early space shuttle mission in which the landing was broadcast, the mission control engineers were heard to say, "Your state vector is looking good." What did that mean? Numerical measurements of height, velocity, deceleration, and so forth were being taken, and from them, forecasts of the flight path into the future were being computed. Of course these state vector entries were being updated quickly and state space forecasting is based on this updating idea. At time $t + 1$, the state vector will be updated to $(Y_{t+1}, \hat{Y}_{t+2|t+1}, \hat{Y}_{t+3|t+1}, \ldots, \hat{Y}_{t+k+1|t+1})$. The updating equation is the model in PROC STATESPACE and it is the thing that you are trying to estimate from the data. In the space shuttle example, if the elements of the state vector included height, deceleration, and location information, then a state vector that "looked good" would be one whose projections forecast a landing on the runway.

6.1.2 A Simple Multivariate Example

Now suppose a vector process is of interest. An easy case to consider is

$$\begin{pmatrix} Y_{1t} \\ Y_{2t} \end{pmatrix} = \begin{pmatrix} 1.3 & -0.9 \\ 0.1 & 0.7 \end{pmatrix} \begin{pmatrix} Y_{1,t-1} \\ Y_{2,t-1} \end{pmatrix} + \begin{pmatrix} -0.4 & 0 \\ 0 & 0 \end{pmatrix} \begin{pmatrix} Y_{1,t-2} \\ Y_{2,t-2} \end{pmatrix} + \begin{pmatrix} e_{1t} \\ e_{2t} \end{pmatrix}$$

This is a vector autoregressive model, VAR, of dimension 2 (2 elements) and of order 2 (maximum lag is 2). The state vector is

$$\mathbf{Z}_t = \begin{pmatrix} Y_{1t} \\ Y_{2t} \\ \hat{Y}_{1,t+1|t} \end{pmatrix}$$

To see why this is the case, first note from the bottom row in the model equation that the one-step-ahead predictor $\hat{Y}_{2,t+1|t} = 0.1Y_{1t} + 0.7Y_{2t}$ is clearly a linear combination of state vector elements and thus does not need to be included in the state vector. Next, note that if the best predictor is used and the coefficients are known as is assumed here, then the forecast $\hat{Y}_{1,t+1|t}$ will differ from $Y_{1,t+1}$ only by $e_{1,t+1}$, which is the error term that, at time t, has yet to be realized. The same is true for Y_{2t}, and using $\hat{Y}_{2,t+1|t} = 0.1Y_{1t} + 0.7Y_{2t}$ you thus have

$$\begin{aligned} Y_{1,t+1} &= \hat{Y}_{1,t+1|t} + e_{1,t+1} \\ Y_{2,t+1} &= 0.1Y_{1t} + 0.7Y_{2t} + e_{2,t+1} \end{aligned}$$

Noting from the top row of the model equation that $Y_{1,t+2} = 1.3Y_{1,t+1} - 0.9Y_{2,t+1} - 0.4Y_{1t} + e_{1,t+2}$, it is seen that forecasting one step ahead using information available up through time $t+1$ would produce

$$\begin{aligned} \hat{Y}_{1,t+2|t+1} &= 1.3Y_{1,t+1} - 0.9Y_{2,t+1} - 0.4Y_{1t} \\ &= 1.3(\hat{Y}_{1,t+1|t} + e_{1,t+1}) - 0.9(0.1Y_{1t} + 0.7Y_{2t} + e_{2,t+1}) - 0.4Y_{1t} \\ &= -0.49Y_{1t} - 0.63Y_{2t} + 1.3\hat{Y}_{1,t+1|t} + 1.3e_{1,t+1} - 0.9e_{2,t+1} \end{aligned}$$

These three equations show how to update from $\mathbf{Z}_t$ to $\mathbf{Z}_{t+1}$. You have

$$\begin{pmatrix} Y_{1,t+1} \\ Y_{2,t+1} \\ \hat{Y}_{1,t+2|t+1} \end{pmatrix} = \begin{pmatrix} 0 & 0 & 1 \\ 0.1 & 0.7 & 0 \\ -0.49 & -0.63 & 1.3 \end{pmatrix} \begin{pmatrix} Y_{1,t} \\ Y_{2,t} \\ \hat{Y}_{1,t+1|t} \end{pmatrix} + \begin{pmatrix} 1 & 0 \\ 0 & 1 \\ 1.3 & -0.9 \end{pmatrix} \begin{pmatrix} e_{1,t+1} \\ e_{2,t+1} \end{pmatrix}$$

which has the form

$$\mathbf{Z}_{t+1} = \mathbf{F}\mathbf{Z}_t + \mathbf{G}\mathbf{E}_{t+1}$$

This looks quite a bit like a vector autoregressive model, and you might think of the state space approach as an attempt to put all vector ARMA processes in a canonical form that looks like an AR(1), because it happens that every possible vector ARMA process of any dimension can be cast into the state space form $\mathbf{Z}_{t+1} = \mathbf{F}\mathbf{Z}_t + \mathbf{G}\mathbf{E}_{t+1}$. While this eliminates the problem of identifying the autoregressive and moving average orders, it introduces a new problem—namely, deciding from the observed data what elements are needed to construct the state vector.

Prior to discussing this new problem, a simulation of 2000 values from this bivariate VAR model is used to produce some state space output. The data set TEST contains the variables Y and X, corresponding to Y_{2t} and Y_{1t}, respectively. The code

```
PROC STATESPACE DATA=TEST;
   VAR Y X;
RUN;
```

is all that is needed to produce the results in **Output 6.1**.

Output 6.1 PROC STATESPACE on Generated Data

```
                 The STATESPACE Procedure
         Selected Statespace Form and Fitted Model

                       State Vector

         Y(T;T)            X(T;T)            Y(T+1;T)

               Estimate of Transition Matrix

                 0                 0                 1
          0.112888          0.683036           -0.0182
          -0.52575           -0.5764           1.34468


                Input Matrix for Innovation

                         1                 0
                         0                 1
                  1.322741          -0.85804

              Variance Matrix for Innovation

                  1.038418          0.018465
                  0.018465            0.9717

                    Parameter Estimates

                                      Standard
       Parameter      Estimate           Error      t Value

       F(2,1)         0.112888        0.027968         4.04
       F(2,2)         0.683036        0.040743        16.76
       F(2,3)         -0.01820        0.028964        -0.63
       F(3,1)         -0.52575        0.037233       -14.12
       F(3,2)         -0.57640        0.054350       -10.61
       F(3,3)         1.344680        0.038579        34.86
       G(3,1)         1.322741        0.021809        60.65
       G(3,2)         -0.85804        0.022550       -38.05
```

The state vector has been correctly identified as containing Y_{1t}, Y_{2t}, and $\hat{Y}_{1,t+1|t}$, as is seen in the notation `Y(T;T) X(T;T) Y(T+1;T)` using the X Y variable names. Had this not been the case, the user could specify

```
FORM Y 2 X 1
```

to force a one-step-ahead Y predictor and no predictions of future X to enter the state vector. Of course this assumes the unlikely scenario that the user has some prior knowledge of the state vector's true form. The matrix **F** is referred to as the *transition matrix* and **G** as the *input matrix* in the output. Comparing the true and estimated **F** matrices you see that

$$\mathbf{F} = \begin{pmatrix} 0 & 0 & 1 \\ 0.1 & 0.7 & 0 \\ -0.49 & -0.63 & 1.3 \end{pmatrix} \text{ and } \hat{\mathbf{F}} = \begin{pmatrix} 0 & 0 & 1 \\ 0.11 & 0.68 & -0.02 \\ -0.53 & -0.58 & 1.34 \end{pmatrix}$$

and for the input matrix **G**

$$\mathbf{G} = \begin{pmatrix} 1 & 0 \\ 0 & 1 \\ 1.3 & -0.9 \end{pmatrix} \text{ and } \hat{\mathbf{G}} = \begin{pmatrix} 1 & 0 \\ 0 & 1 \\ 1.32 & -0.86 \end{pmatrix}$$

Entries of 0 or 1 are known once the state vector has been determined. They are structural parameters that do not require estimation. No elements of **F** or **G** are more than 2 standard errors away from the true values, and all estimates are quite close to the true values both numerically and statistically. Knowing that the estimate –0.02 is in fact an estimate of 0, you would expect its *t* statistic to be smaller than 2 in magnitude, which it is $(t = -0.63)$. You might want to drop that term from your model by forcing its coefficient to 0, using the statement

```
RESTRICT F(2,3)=0;
```

in the PROC STATESPACE step to restrict that row 2, column 3 element to 0. Doing so produces the results in **Output 6.2**.

Output 6.2
RESTRICT Statement in PROC STATESPACE

```
                  The STATESPACE Procedure
            Selected Statespace Form and Fitted Model

                        State Vector

     Y(T;T)              X(T;T)              Y(T+1;T)


               Estimate of Transition Matrix

             0                  0                  1
      0.095449           0.707423                  0
      -0.51028           -0.59826           1.328525


                Input Matrix for Innovation

                        1                  0
                        0                  1
                 1.322604           -0.85752


              Variance Matrix for Innovation

                 1.038417           0.018442
                 0.018442           0.971882

                    Parameter Estimates

                                    Standard
   Parameter        Estimate          Error       t Value

   F(2,1)           0.095449       0.003469         27.51
   F(2,2)           0.707423       0.012395         57.07
   F(3,1)           -0.51028       0.028997        -17.60
   F(3,2)           -0.59826       0.043322        -13.81
   F(3,3)           1.328525       0.029876         44.47
   G(3,1)           1.322604       0.021816         60.62
   G(3,2)           -0.85752       0.022552        -38.03
```

The estimated elements of **F** and **G** are again close to their true values. Plots of both series and their forecasts are seen in **Output 6.3**.

Output 6.3 *Forecasts for Generated Data*

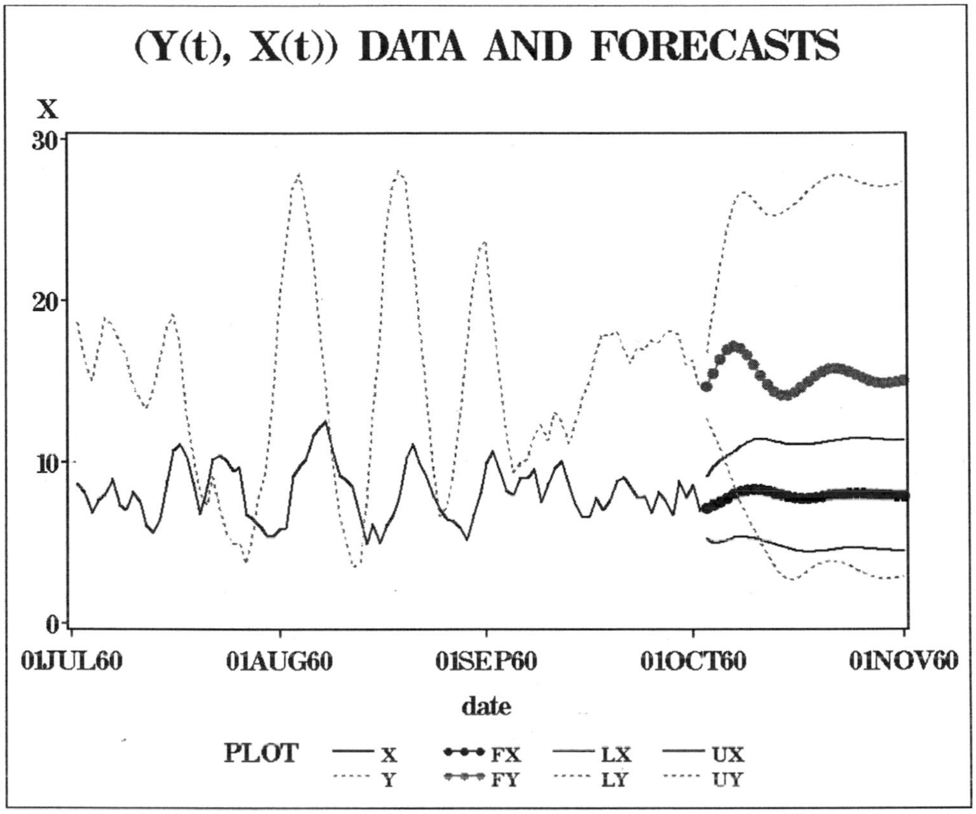

The forecasts seem to have a little more interesting structure than some you have previously seen. This has to do with the nature of the roots of the characteristic equation.

As in a univariate series, the behavior of a VAR of the form

$$\mathbf{Y}_t = \mathbf{A}_1\mathbf{Y}_{t-1} + \mathbf{A}_2\mathbf{Y}_{t-2} + \cdots + \mathbf{A}_p\mathbf{Y}_{t-p} + \mathbf{E}_t$$

is determined by the roots of a "characteristic equation," and the same is true for a vector ARMA. Here $\mathbf{E}_t$ is a vector of random normal variables that can be contemporaneously correlated in any arbitrary way, but must be uncorrelated across time. $\mathbf{Y}_t$ is a dimension k vector of deviations from means. For $k = 3$ these might be the time t deviations of GDP, unemployment, and interest rates from their long-term means, and each $\mathbf{A}_j$ is a $k \times k$ matrix of parameters to be estimated. The characteristic equation involves the determinant

$$\left| m^p\mathbf{I} - m^{p-1}\mathbf{A}_1 - m^{p-2}\mathbf{A}_2 - \cdots - \mathbf{A}_p \right|$$

and the values of m that make this 0 are the roots. In the VAR example currently under study,

$$\left| m^2\begin{pmatrix} 1 & 0 \\ 0 & 1 \end{pmatrix} - m\begin{pmatrix} 1.3 & -0.9 \\ 0.1 & 0.7 \end{pmatrix} - \begin{pmatrix} -0.4 & 0 \\ 0 & 0 \end{pmatrix} \right| = \left| \begin{pmatrix} m^2 - 1.3m + 0.4 & 0.9m \\ -0.1m & m^2 - 0.7m \end{pmatrix} \right|$$
$$= \left(m^2 - 1.3m + .4\right)\left(m^2 - 0.7m\right) + 0.09m^2$$
$$= m\left(m^3 - 2m^2 + 1.4m - 0.28\right)$$
$$= m\left(m - 0.32966342378782\right)\left(m^2 - 1.6703366m + 0.84935\right)$$

whose roots are 0, 0.32966 and the complex pair $0.83517 \pm 0.3987i$. The complex pair of roots has representation $\sqrt{0.84935}\left(\cos\theta \pm i\sin\theta\right)$, so $\theta = \text{Atan}\left(.3987/.83517\right) = 25.5$ degrees, implying a damped sinusoidal component with damping rate $\sqrt{0.84935}^{\,L}$ and period $360/25.5 = 14.1$ time periods as you forecast L periods ahead. This seems consistent with the graph. At a lead of around $\text{L} = 14$, the forecasts hit local low points as they did at the end of the data. Each low is about $\sqrt{0.84935}^{\,14} = 0.32$, or about 1/3 of what it was then. All of these roots are less than 1 in magnitude, this being the stationarity condition. Some sources write the characteristic equation in ascending powers of m, namely,

$$\left|\mathbf{I} - m\mathbf{A}_1 - m^2\mathbf{A}_2 - \cdots - m^p\mathbf{A}_p\right| = 0$$

whose roots they then require to all be greater than 1 in magnitude for stationarity.

In a VAR of order 2 you have

$$\mathbf{Y}_t = \mathbf{A}_1\mathbf{Y}_{t-1} + \mathbf{A}_2\mathbf{Y}_{t-2} + \mathbf{E}_t$$

which is sometimes written in a matrix form like this:

$$\begin{pmatrix} \mathbf{Y}_{t-1} \\ \mathbf{Y}_t \end{pmatrix} = \begin{pmatrix} \mathbf{0} & \mathbf{I} \\ \mathbf{A}_2 & \mathbf{A}_1 \end{pmatrix}\begin{pmatrix} \mathbf{Y}_{t-2} \\ \mathbf{Y}_{t-1} \end{pmatrix} + \begin{pmatrix} \mathbf{0} \\ \mathbf{I} \end{pmatrix}\mathbf{E}_t$$

This simply says that $\mathbf{Y}_{t-1} = \mathbf{Y}_{t-1}$ and $\mathbf{Y}_t = \mathbf{A}_2\mathbf{Y}_{t-2} + \mathbf{A}_1\mathbf{Y}_{t-1} + \mathbf{E}_t$, so it consists of a trivial identity and the original AR(2). If you substitute the **A** matrices of the current example, you have

$$\begin{pmatrix} Y_{1,t-1} \\ Y_{2,t-1} \\ Y_{1t} \\ Y_{2t} \end{pmatrix} = \begin{pmatrix} 0 & 0 & 1 & 0 \\ 0 & 0 & 0 & 1 \\ -0.4 & 0 & 1.3 & -0.9 \\ 0 & 0 & 0.1 & 0.7 \end{pmatrix}\begin{pmatrix} Y_{1,t-2} \\ Y_{2,t-2} \\ Y_{1,t-1} \\ Y_{2,t-1} \end{pmatrix} + \begin{pmatrix} 0 \\ 0 \\ e_{1t} \\ e_{2t} \end{pmatrix}$$

which looks somewhat similar to the state space representation; however, this system has dimension 4, not 3 as you had previously. Had the matrix $\mathbf{A}_2$ been of full rank, the system would have had full rank, the state vector would have had dimension 4, and the above representation would be another type of state space representation of the vector process. It would differ from the process used in PROC STATESPACE in that current and lagged Ys, rather than current Ys and predictions, constitute the state vector elements. As it stands, the system is not full rank and can be reduced by simply eliminating the second row and second column of the coefficient matrix. That second row produces the trivial identity $Y_{2,t-1} = Y_{2,t-1}$, which, of course, is true whether you put it in the system or not. The second column is all 0s, so leaving out the second row and column makes no real change in the system. The resulting reduction gives

$$\begin{pmatrix} Y_{1,t-1} \\ Y_{1t} \\ Y_{2t} \end{pmatrix} = \begin{pmatrix} 0 & 1 & 0 \\ -0.4 & 1.3 & -0.9 \\ 0 & 0.1 & 0.7 \end{pmatrix} \begin{pmatrix} Y_{1,t-2} \\ Y_{1,t-1} \\ Y_{2,t-1} \end{pmatrix} + \begin{pmatrix} 0 \\ e_{1t} \\ e_{2t} \end{pmatrix}$$

again having a familiar form $\mathbf{Z}_t = \mathbf{F}\mathbf{Z}_{t-1} + \mathbf{G}\mathbf{E}_t$. The first row gives a trivial identity, but it is needed to make the system square. This 3×3 system is observationally equivalent to the 4×4 system in that, for any *e* sequence and any given initial values, the two systems will produce exactly the same sequence of Ys. In the theoretical research on STATESPACE methods there are several ways to formulate the state vector, as has been demonstrated. The size of the state vector **Z** and the general form of the updating recursion is the same for all ways of writing a state vector. The entries of **Z** and of the matrices **F** and **G** depend on the particular formulation.

Every state vector $\mathbf{Z}_t$ that arises from a vector ARMA satisfies a recursive relationship of the form $\mathbf{Z}_{t+1} = \mathbf{F}\mathbf{Z}_t + \mathbf{G}\mathbf{E}_{t+1}$. In PROC STATESPACE the state vector $\mathbf{Z}_t$ always consists of the current observations—say, Y_{1t}, Y_{2t}, and Y_{3t} if you have 3 observed series at each time *t*—along with predictions into the future. For example $\mathbf{Z}_t$ might contain Y_{1t}, Y_{2t}, Y_{3t}, $Y_{1,t+1|t}$, $Y_{2,t+1|t}$, and $Y_{1,t+2|t}$. There will be no "gaps"; that is, if the two-step-ahead predictor $Y_{1,t+2|t}$ is included, so must be the Y_1 predictors *up to* two steps ahead.

How do you decide what to put in the state vector? Returning to the bivariate VAR of order 2 that is being used as an example, it is possible from the model to compute the autocorrelation between any Y_{it} and Y_{js} for the same $(i - j)$ or different vector elements at the same $(t = s)$ or different times. The data were generated using the innovations variance matrix Σ defined as

$$\Sigma = \text{Var}\begin{pmatrix} e_{1t} \\ e_{2t} \end{pmatrix} = \begin{pmatrix} 1 & 0 \\ 0 & 1 \end{pmatrix}$$

The covariance matrix between column vector $\mathbf{Y}_t$ and row vector $\mathbf{Y}'_{t+j}$, symbolized as $\mathbf{\Gamma}(j)$, is defined as an expected value, namely, you define $\mathbf{\Gamma}(j) = \text{E}\{\mathbf{Y}_t \quad \mathbf{Y}'_{t+j}\}$ (assuming $\mathbf{Y}_t$ has mean 0). Multiplying the AR(2) on both sides by $\mathbf{Y}'_{t-j}$ and taking expected values, you see that

$$\mathbf{\Gamma}(-j) = \text{E}\left\{\mathbf{Y}_t \quad \mathbf{Y}'_{t-j}\right\} = \mathbf{A}_1\text{E}\left\{\mathbf{Y}_{t-1} \quad \mathbf{Y}'_{t-j}\right\} + \mathbf{A}_2\text{E}\left\{\mathbf{Y}_{t-2} \quad \mathbf{Y}'_{t-j}\right\} = \mathbf{A}_1\mathbf{\Gamma}(-j+1) + \mathbf{A}_2\mathbf{\Gamma}(-j+2)$$

for $j > 0$. For $j = 0$ you find

$$\mathbf{\Gamma}(0) = \mathbf{A}_1\mathbf{\Gamma}(1) + \mathbf{A}_2\mathbf{\Gamma}(2) + \Sigma$$

Now $\Gamma(-j) = \Gamma'(j)$, so for $j > 0$ you have

$$\Gamma(j) = \Gamma(j-1)\mathbf{A}_1' + \Gamma(j-2)\mathbf{A}_2'$$

These, then, constitute the multivariate Yule-Walker equations that can be solved to give all the covariances from the known **A** coefficient matrices and the innovations variance matrix Σ. Thus it would be possible to compute, say, the 6×6 covariance matrix **M** between the vector $(Y_{1t}\ \ Y_{2t}\ \ Y_{1,t-1}\ \ Y_{2,t-1}\ \ Y_{1,t-2}\ \ Y_{2,t-2})$, these identifying the columns of **M**, and the vector $(Y_{1t}\ \ Y_{2t}\ \ Y_{1,t+1}\ \ Y_{2,t+1}\ \ Y_{1,t+2}\ \ Y_{2,t+2})$, these identifying the rows. That matrix would have what is known as a *block Hankel* form:

$$\mathbf{M} = \begin{pmatrix} \Gamma(0) & \Gamma(1) & \Gamma(2) \\ \Gamma(1) & \Gamma(2) & \Gamma(3) \\ \Gamma(2) & \Gamma(3) & \Gamma(4) \end{pmatrix}$$

State space researchers describe **M** as the covariance matrix between a set of current and lagged Ys and a set of current and future Ys. For such a matrix **M**, the following numbers are all the same:

1. The size of the state vector
2. The rank of the covariance matrix **M**
3. The number of nonzero canonical correlations between the set of current and lagged Ys and the set of current and future Ys.

Items 2 and 3 are always the same for any covariance matrix. (See the PROC CANCOR documentation for more information on canonical correlations.) Thus the size of the state vector and the nature of the corresponding state space equations can be deduced by studying the covariance matrix **M**.

With only data, rather than a model with known coefficients, the covariances must be estimated. The strategy used is to fit a long vector autoregression whose length is determined by some information criterion, and then to use the fitted model as though it were the true structure to construct an estimate of **M**. The initial autoregressive approximation provides an upper bound for the size of **M**. Returning to the order 2 VAR with known coefficient matrices, by substitution, you can see that these $\Gamma(j)$ matrices satisfy the multivariate Yule-Walker equations.

$$\Gamma(0) = \begin{pmatrix} 39.29 & 1.58 \\ 1.58 & 3.17 \end{pmatrix} \qquad \Gamma(1) = \begin{pmatrix} 35.47 & 5.04 \\ -2.81 & 2.37 \end{pmatrix} \qquad \Gamma(2) = \begin{pmatrix} 25.86 & 7.07 \\ -6.42 & 1.38 \end{pmatrix}$$

$$\Gamma(3) = \begin{pmatrix} 13.06 & 7.54 \\ -8.46 & 0.32 \end{pmatrix} \qquad \Gamma(4) = \begin{pmatrix} -0.14 & 6.58 \\ -8.73 & -0.62 \end{pmatrix}$$

These in turn lead to a matrix **M** formed by stacking together the Γ matrices in the block Hankel form previously suggested, namely,

$$\mathbf{M} = \begin{pmatrix} \Gamma(0) & \Gamma(1) & \Gamma(2) \\ \Gamma(1) & \Gamma(2) & \Gamma(3) \\ \Gamma(2) & \Gamma(3) & \Gamma(4) \end{pmatrix} = \begin{pmatrix} 39.29 & 1.58 & 35.47 & 5.04 & 25.86 & 7.07 \\ 1.58 & 3.17 & -2.81 & 2.37 & -6.42 & 1.38 \\ 35.47 & 5.04 & 25.86 & 7.07 & 13.06 & 7.54 \\ -2.81 & 2.37 & -6.42 & 1.38 & -8.46 & 0.32 \\ 25.86 & 7.07 & 13.06 & 7.54 & -0.14 & 6.58 \\ -6.42 & 1.38 & -8.46 & 0.32 & -8.73 & -0.62 \end{pmatrix}$$

You can diagnose the column dependencies and rank of matrix **M** using a clever trick. If any column of a matrix is a linear combination of some others, then a regression of that first column on those others (no intercept) will fit perfectly. Regressing column 2 of **M** on column 1 and column 3 on columns 1 and 2, you find nonzero error sums of squares, indicating that columns 1, 2, and 3 form a linearly independent set. Regressing any other column on columns 1, 2, and 3 gives a perfect fit and so shows that the rank of matrix **M** is 3. For example, you find by regression that $(\text{column } 4) = 0.1(\text{column } 1) + .7(\text{column } 2) + 0(\text{column } 3)$, so that the covariance between the column 4 variable, $Y_{2,t+1}$, and any future Y is the same as that between $0.1Y_{1t} + 0.7Y_{2t}$ and that same future Y. Linear forecasts such as we are considering are functions only of the covariances. Therefore, adding $Y_{2,t+1|t}$ to a set of predictors that already contains Y_{1t} and Y_{2t} does not add any more prediction accuracy.

Note that the second row of the state space transition matrix **F** is 0.1, 0.7, 0, so the same regression that displayed the dependency gives the corresponding row of **F**. On the other hand, column 3 is not a perfect linear combination of columns 1 and 2. You get a positive error mean square when regressing column 3 on columns 1 and 2. Regression reveals that (column 5)= –0.49(column 1)–0.63(column 2)+1.3(column 3) with 0 error sum of squares. Again note that the coefficients give a row of **F**. Column 5 is associated with $Y_{1,t+2}$, so even though you needed $Y_{1,t+1|t}$ in the state vector, there is nothing to be gained by including $Y_{1,t+2|t}$. The dependent columns, 4 and 5, thus far considered are the first columns associated with series Y_2 and with series Y_1 that show dependencies. These dependencies reveal the number of forecasts of each series that appear in the state vector (one less than the lag number associated with the dependent column) and the row of the **F** matrix associated with the last occurrence of that series in the state vector. Once the first dependency in each variable has been discovered, the state vector has been completely determined and no further investigation is needed. Column 6 is automatically a linear combination of columns 1,2, and 3 at this point.

A perfectly fitting regression corresponds to a canonical correlation 0 in matrix **M**. In particular, you can build a sequence of matrices by sequentially appending columns of **M**. When you use the first four columns of **M** you will get a 0 canonical correlation, but not before. That tells you the fourth column, and hence $Y_{2,t+1|t}$, is redundant information. Leave out that redundant fourth column and consider a matrix consisting of column 1, 2, 3, and 5 of **M**. If that matrix had no 0 canonical correlations, then $Y_{1,t+2|t}$ (associated with column 5) would have been included in the state vector, but in this example, the addition of column 5 also produces a 0 canonical correlation. Since dependencies for both series have been discovered, you need not look any further.

When estimated covariances are used to get an estimated **M** matrix, that matrix, $\hat{\mathbf{M}}$, will almost certainly be of full rank, possibly with some small but nonzero canonical correlations. What is needed is a statistic to decide if a small *estimated* canonical correlation in $\hat{\mathbf{M}}$ is consistent with the hypothesis that **M** has corresponding *true* canonical correlation 0. A criterion DIC to do so has been proposed by Akaike. If you build matrices as described above by appending columns of $\hat{\mathbf{M}}$, then the DIC criterion is expected to be negative when the column just added introduces an approximate dependency. That column would then be omitted from the matrix being built, as would all columns to its right in $\hat{\mathbf{M}}$ that correspond to lagged values of that series. Then the appending would continue, using only columns of the other series, until dependencies have been discovered in each of the series. Like any statistical criterion, the DIC is not infallible and other tests, such as Bartlett's test for canonical correlations, could also be used to test the hypothesis that the newly added column has introduced a dependency in the system.

Now, if there are moving average components in the series, things become a little more complicated and, of course, estimates of the elements of **G** are also needed. But if you have followed the example, you have the idea of how the STATESPACE procedure starts. The long autoregression is run, the estimated **M** matrix is computed from it, the rank is diagnosed, and initial elements of **F** and **G** are computed by treating estimated covariances as though they were the true ones. Thus the initial estimates of **F** and **G** fall into the "method of moments" category of estimates. Such estimates are approximate and are often used, as is the case here, as starting values for more accurate methods such as maximum likelihood. Another nice feature of the maximum-likelihood method is that large sample approximate standard errors, based on the derivatives of the likelihood function, can be computed. Examples of these standard errors and *t* tests were seen in **Output 6.1** and **Output 6.2.**

Additional numerical examples and discussion are given in **Section 6.2**. Some ideas are reiterated there and some details filled in. The reader who feels that **Section 6.1** has provided enough background may wish to move directly to **Section 6.3**. The following section is for those interested in a more general theoretical discussion.

6.1.3 Equivalence of State Space and Vector ARMA Models

A general discussion of the state space model, under the name "Markovian representation," is given by Akaike (1974). The following summarizes a main idea from that paper.

Let $\mathbf{Y}_t$ represent a dimension *k* vector ARMA(p,q) process with mean vector **0**, and let $\mathbf{E}_t$ be an uncorrelated sequence of multivariate, mean 0 normal variables with variance matrix $\mathbf{\Sigma}$. The ARMA(p,q) process is

$$\mathbf{Y}_t = \mathbf{A}_1\mathbf{Y}_{t-1} + \cdots + \mathbf{A}_p\mathbf{Y}_{t-p} + \mathbf{E}_t - \mathbf{B}_1\mathbf{E}_{t-1} - \cdots - \mathbf{B}_q\mathbf{E}_{t-q}$$

At time $t-1$,

$$\mathbf{Y}_{t-1} = \mathbf{A}_1\mathbf{Y}_{t-2} + \cdots + \mathbf{A}_p\mathbf{Y}_{t-1-p} + \mathbf{E}_{t-1} - \mathbf{B}_1\mathbf{E}_{t-2} - \cdots - \mathbf{B}_q\mathbf{E}_{t-1-q}$$

so substituting in the original expression gives

$$\mathbf{Y}_t = \mathbf{A}_1\left(\mathbf{A}_1\mathbf{Y}_{t-2} + \cdots + \mathbf{A}_p\mathbf{Y}_{t-1-p} + \mathbf{E}_{t-1} - \mathbf{B}_1\mathbf{E}_{t-2} - \cdots - \mathbf{B}_q\mathbf{E}_{t-1-q}\right)$$
$$+ \mathbf{A}_2\mathbf{Y}_{t-2} + \cdots + \mathbf{A}_p\mathbf{Y}_{t-p} + \mathbf{E}_t - \mathbf{B}_1\mathbf{E}_{t-1} - \cdots - \mathbf{B}_q\mathbf{E}_{t-q}$$

which involves current and lagged **E** vectors and **Y** vectors prior to time $t-1$. Repeated back substitution produces a convergent expression only in terms of the **E** vectors, say, $\mathbf{Y}_t = \sum_{j=0}^{\infty}\psi_j\mathbf{E}_{t-j}$, provided the series is stationary. The forecast of any $\mathbf{Y}_{t+L} = \sum_{j=0}^{\infty} \psi_j \mathbf{E}_{t+L-j}$ using information up to time t would just be the part of this sum that is known at time t, namely, $\mathbf{Y}_{t+L|t} = \sum_{j=L}^{\infty} \psi_j \mathbf{E}_{t+L-j}$, and the forecast of the same thing at time $t+1$ just adds one more term so that $\mathbf{Y}_{t+L|t+1} = \mathbf{Y}_{t+L|t} + \psi_{L-1}\mathbf{E}_{t+1}$. So if the state vector $\mathbf{Z}_t$ contains $\mathbf{Y}_{t+L|t}$, then it will also contain the predecessor $\mathbf{Y}_{t+L-1|t}$. Thus $\mathbf{Z}_{t+1}$ will contain $\mathbf{Y}_{t+L|t+1}$, and the relationship between these, $\mathbf{Y}_{t+L|t+1} = \mathbf{Y}_{t+L|t} + \psi_{L-1}\mathbf{E}_{t+1}$, will provide k (the dimension of Y) rows of the state space equation $\mathbf{Z}_{t+1} = \mathbf{F}\mathbf{Z}_{t+1} + \mathbf{G}\mathbf{E}_{t+1}$. In particular, using a question mark (?) for items not yet discussed, you have

$$\begin{pmatrix} \mathbf{Y}_{t+1} \\ \mathbf{Y}_{t+2|t+1} \\ \vdots \\ \mathbf{Y}_{t+M-1|t+1} \\ \mathbf{Y}_{t+M|t+1} \end{pmatrix} = \begin{pmatrix} \mathbf{0} & \mathbf{I} & \mathbf{0} & \cdots & \mathbf{0} \\ \mathbf{0} & \mathbf{0} & \mathbf{I} & \cdots & \mathbf{0} \\ & & & \ddots & \\ \mathbf{0} & \mathbf{0} & \mathbf{0} & \cdots & \mathbf{I} \\ ? & ? & ? & & ? \end{pmatrix} \begin{pmatrix} \mathbf{Y}_{t} \\ \mathbf{Y}_{t+1|t} \\ \vdots \\ \mathbf{Y}_{t+M-2|t} \\ \mathbf{Y}_{t+M-1|t} \end{pmatrix} + \begin{pmatrix} \mathbf{I} \\ \psi_1 \\ \vdots \\ \psi_{M-2} \\ ? \end{pmatrix} \mathbf{E}_{t+1}$$

What should be the size of the state vector—that is, what should you use as the subscript M? To answer that, you look for dependencies. At time $t+L$ the model becomes

$$\mathbf{Y}_{t+L} = \mathbf{A}_1\mathbf{Y}_{t+L-1} + \cdots + \mathbf{A}_p\mathbf{Y}_{t+L-p} + \mathbf{E}_{t+L} - \mathbf{B}_1\mathbf{E}_{t+L-1} - \cdots - \mathbf{B}_q\mathbf{E}_{t+L-q}$$

If there were only information up to time t, the forecast of $\mathbf{Y}_{t+L}$ would be

$$\mathbf{Y}_{t+L|t} = \mathbf{A}_1\mathbf{Y}_{t+L-1|t} + \cdots + \mathbf{A}_p\mathbf{Y}_{t+L-p|t} + \mathbf{E}_{t+L|t} - \mathbf{B}_1\mathbf{E}_{t+L-1|t} - \cdots - \mathbf{B}_q\mathbf{E}_{t+L-q|t}$$

where $\mathbf{E}_{t+j|t}$ is 0 for $j > 0$ and is the vector of one-step-ahead forecast errors at time $t+j$ if $j < 0$. For leads L exceeding the moving average length q, this becomes

$$\mathbf{Y}_{t+L|t} = \mathbf{A}_1\mathbf{Y}_{t+L-1|t} + \cdots + \mathbf{A}_p\mathbf{Y}_{t+L-p|t} \qquad (\text{for } L > q)$$

and so if the state vector $\mathbf{Z}_t$ contains $\mathbf{Y}_t$, $\mathbf{Y}_{t+1|t}, \ldots, \mathbf{Y}_{t+M|t}$, where M is $\max(p, q+1)$, then every forecast $\mathbf{Y}_{t+L|t}$ with $L > M$ will be a linear combination of these. This establishes M.

Finally, to replace the "?" in **F** and **G,** note that $\mathbf{Y}_{t+M|t} = \mathbf{A}_1\mathbf{Y}_{t+M-1|t} + \cdots + \mathbf{A}_p\mathbf{Y}_{t+M-p|t}$, which combined with $\mathbf{Y}_{t+M|t+1} = \mathbf{Y}_{t+M|t} + \psi_{M-1}\mathbf{E}_{t+1}$ gives the full set of equations. Expanding the set of autoregressive coefficient matrices with $\mathbf{A}_j = \mathbf{0}$ when $j > p$, you have the complete set of equations

$$\begin{pmatrix} \mathbf{Y}_{t+1} \\ \mathbf{Y}_{t+2|t+1} \\ \vdots \\ \mathbf{Y}_{t+M-1|t+1} \\ \mathbf{Y}_{t+M|t+1} \end{pmatrix} = \begin{pmatrix} \mathbf{0} & \mathbf{I} & \mathbf{0} & \cdots & \mathbf{0} \\ \mathbf{0} & \mathbf{0} & \mathbf{I} & \cdots & \mathbf{0} \\ & & & \ddots & \\ \mathbf{0} & \mathbf{0} & \mathbf{0} & \cdots & \mathbf{I} \\ \mathbf{A}_M & \mathbf{A}_{M-1} & \mathbf{A}_{M-2} & \cdots & \mathbf{A}_1 \end{pmatrix} \begin{pmatrix} \mathbf{Y}_{t+1} \\ \mathbf{Y}_{t+1|t} \\ \vdots \\ \mathbf{Y}_{t+M-2|t} \\ \mathbf{Y}_{t+M-1|t} \end{pmatrix} + \begin{pmatrix} \mathbf{I} \\ \psi_1 \\ \vdots \\ \psi_{M-2} \\ \psi_{M-1} \end{pmatrix} \mathbf{E}_{t+1}$$

This would be the final state space form *if* the system were of full rank. Such a full-rank system is called "block identifiable." In this case the link between the ARIMA representation and the state space representation is relatively easy to see, unlike the **Section 6.2.2** example

$$\begin{pmatrix} Y_t \\ X_t \end{pmatrix} = \begin{pmatrix} 1.3 & -0.9 \\ 0.1 & 0.7 \end{pmatrix} \begin{pmatrix} Y_{t-1} \\ X_{t-1} \end{pmatrix} + \begin{pmatrix} -0.4 & 0 \\ 0 & 0 \end{pmatrix} \begin{pmatrix} Y_{t-2} \\ X_{t-2} \end{pmatrix} + \begin{pmatrix} e_{1t} \\ e_{2t} \end{pmatrix}$$

where the 4×4 system arising from the current discussion was not block identifiable. It had a linear dependency, and a reduction in the size of the system was called for, ultimately producing a state space representation with dimension 3.

Consider the state space representation

$$\begin{pmatrix} Y_{t+1} \\ X_{t+1} \\ Y_{t+2|t+1} \end{pmatrix} = \begin{pmatrix} 0 & 0 & 1 \\ -.44 & .58 & .8 \\ .044 & -.058 & .72 \end{pmatrix} \begin{pmatrix} Y_t \\ X_t \\ Y_{t+1|t} \end{pmatrix} + \begin{pmatrix} 1 & 0 \\ 0 & 1 \\ .8 & -.6 \end{pmatrix} \begin{pmatrix} e_{1,t+1} \\ e_{2,t+1} \end{pmatrix}$$

Suppose you want to find the equivalent bivariate ARMA representation for $(X_t \;\; Y_t)'$. From row 2 you have $X_{t+1|t} = -.44Y_t + .58X_t + .8Y_{t+1|t}$ so that

$$X_{t+2|t+1} = (-.44 \quad .58 \quad .8) \begin{pmatrix} Y_{t+1} \\ X_{t+1} \\ Y_{t+2|t+1} \end{pmatrix}$$

$$= (-.44 \quad .58 \quad .8) \begin{pmatrix} 0 & 0 & 1 \\ -.44 & .58 & .8 \\ .044 & -.058 & .72 \end{pmatrix} \begin{pmatrix} Y_t \\ X_t \\ Y_{t+1|t} \end{pmatrix} + (-.44 \quad .58 \quad .8) \begin{pmatrix} 1 & 0 \\ 0 & 1 \\ .8 & -.6 \end{pmatrix} \begin{pmatrix} e_{1,t+1} \\ e_{2,t+1} \end{pmatrix}$$

$$= (-.22 \quad 0.29 \quad 0.60) \begin{pmatrix} Y_t \\ X_t \\ Y_{t+1|t} \end{pmatrix} + (.2 \quad .1) \begin{pmatrix} e_{1,t+1} \\ e_{2,t+1} \end{pmatrix}$$

Inserting this extra line into the state space equations you get

$$\begin{pmatrix} Y_{t+1} \\ X_{t+1} \\ Y_{t+2|t+1} \\ X_{t+2|t+1} \end{pmatrix} = \begin{pmatrix} 0 & 0 & 1 & 0 \\ -.44 & .58 & .8 & 0 \\ .044 & -.058 & .72 & 0 \\ -.22 & .29 & .6 & 0 \end{pmatrix} \begin{pmatrix} Y_t \\ X_t \\ Y_{t+1|t} \\ X_{t+1|t} \end{pmatrix} + \begin{pmatrix} 1 & 0 \\ 0 & 1 \\ .8 & -.6 \\ .2 & .1 \end{pmatrix} \begin{pmatrix} e_{1,t+1} \\ e_{2,t+1} \end{pmatrix}$$

Now anytime you see $-.44\mathbf{Y}_t + .58\mathbf{X}_t + .8\mathbf{Y}_{t+1|t}$ you can replace it with $\mathbf{X}_{t+1|t}$, so row 2 of **F** can be replaced by $0\,0\,0\,1$. The result is

$$\begin{pmatrix} Y_{t+1} \\ X_{t+1} \\ Y_{t+2|t+1} \\ X_{t+2|t+1} \end{pmatrix} = \begin{pmatrix} 0 & 0 & 1 & 0 \\ 0 & 0 & 0 & 1 \\ .044 & -.058 & .72 & 0 \\ -.22 & .29 & .6 & 0 \end{pmatrix} \begin{pmatrix} Y_t \\ X_t \\ Y_{t+1|t} \\ X_{t+1|t} \end{pmatrix} + \begin{pmatrix} 1 & 0 \\ 0 & 1 \\ .8 & -.6 \\ .2 & .1 \end{pmatrix} \begin{pmatrix} e_{1,t+1} \\ e_{2,t+1} \end{pmatrix}$$

Anytime you see a multiple of $-.44\mathbf{Y}_t + .58\mathbf{X}_t$ as the leading term in a row, you can re-express that row using $\mathbf{X}_{t+1|t}$. Row 3 gives

$$\begin{aligned} Y_{t+2|t+1} &= -0.1(-.44Y_t + .58X_t + .8Y_{t+1|t}) + 0.8Y_{t+1|t} \\ &= -0.1X_{t+1|t} + 0.8Y_{t+1|t} \end{aligned}$$

and row 4 is $X_{t+2|t+1} = 0.5X_{t+1|t} + 0.2Y_{t+1|t}$. This system results:

$$\begin{pmatrix} Y_{t+1} \\ X_{t+1} \\ Y_{t+2|t+1} \\ X_{t+2|t+1} \end{pmatrix} = \begin{pmatrix} 0 & 0 & 1 & 0 \\ 0 & 0 & 0 & 1 \\ 0 & 0 & .8 & -.1 \\ 0 & 0 & .2 & .5 \end{pmatrix} \begin{pmatrix} Y_t \\ X_t \\ Y_{t+1|t} \\ X_{t+1|t} \end{pmatrix} + \begin{pmatrix} 1 & 0 \\ 0 & 1 \\ .8 & -.6 \\ .2 & .1 \end{pmatrix} \begin{pmatrix} e_{1,t+1} \\ e_{2,t+1} \end{pmatrix}$$

It is seen to be an ARMA(2,1) with $\mathbf{A}_2 = \mathbf{0}$; in other words it is the vector ARMA(1,1) $\mathbf{Y}_t = \mathbf{A}_1\mathbf{Y}_{t-1} + \mathbf{E}_t - \mathbf{B}_1\mathbf{E}_{t-1}$ with

$$\mathbf{A}_1 = \begin{pmatrix} 0.8 & -1 \\ .2 & 0.5 \end{pmatrix} \quad \text{and} \quad \psi_1 = \begin{pmatrix} .8 & -.6 \\ .2 & .1 \end{pmatrix}$$

and $\mathbf{Y}_t = (Y_t \quad X_t)$. It can be expressed as $\mathbf{Y}_t = \mathbf{E}_t + (\mathbf{A}_1 - \mathbf{B}_1)\mathbf{E}_{t-1} + \psi_2\mathbf{E}_{t-2} + \psi_3\mathbf{E}_{t-3} + \cdots$, and setting

$$\psi_1 = \begin{pmatrix} .8 & -.6 \\ .2 & .1 \end{pmatrix} = \mathbf{A}_1 - \mathbf{B}_1 = \begin{pmatrix} 0.8 & -.1 \\ .2 & 0.5 \end{pmatrix} - \mathbf{B}_1$$

you find that

$$\mathbf{B}_1 = \begin{pmatrix} 0.8 & -.1 \\ .2 & 0.5 \end{pmatrix} - \begin{pmatrix} .8 & -.6 \\ .2 & .1 \end{pmatrix} = \begin{pmatrix} 0 & .5 \\ 0 & .4 \end{pmatrix}$$

So you have recovered the original ARMA(1,1) model. A useful feature of such a structure is that it sometimes gives a nice interpretation; for example, it is clear from the ARMA model that lagged shocks to Y do not have any effect on Y or X, while that is not so clear from the state space representation.

6.2 More Examples

6.2.1 Some Univariate Examples

Univariate models can also be expressed in state space form, and doing so provides some insights.

Consider the AR(1) model where

$$Y_t - 100 = .6(Y_{t-1} - 100) + e_t$$

Suppose you are given Y_1, Y_2, Y_3, . . . , Y_{100}. You can forecast Y_{101}, Y_{102}, Y_{103}, ... In fact, the only value you need to know is Y_{100}. If Y_{100}=150, then $\hat{Y}_{101} = 130$, $\hat{Y}_{102} = 118$, $\hat{Y}_{103} = 110.8$, ... If you observe Y_{101}=140 at time 101, the forecasts of Y_{102}, Y_{103}, Y_{104}, . . . change to $\hat{Y}_{102} = 124$, $\hat{Y}_{103} = 114.4$, ... The point is that given the model, you need to know only the last Y, Y_n to forecast as far into the future as you like. The forecasts are updated as new information is obtained.

Consider the AR(2) model

$$Y_t - 100 = 1.2(Y_{t-1} - 100) - .36(Y_{t-2} - 100) + e_t$$

Again, suppose you know Y_1, Y_2, . . . , Y_{100} with Y_{100}=150. Knowing Y_{100} is not enough to forecast Y_{101}. You need more information. If you know Y_{99}=110, then

$$\hat{Y}_{101} = 100 + 1.2(50) - .36(10) = 156.4$$

and

$$\hat{Y}_{102} = 100 + 1.2(56.4) - .36(50) = 149.68, \ldots$$

In this example, you need to know two pieces of information: Y_{99}=110 and Y_{100}=150, or Y_{100}=150 and $\hat{Y}_{101} = 156.4$. Either pair of numbers allows you to forecast the series as far into the future as you like. The vector with the information you need is the state vector $\mathbf{Z}_t$.

For the AR(2) model, the state vector is

$$\mathbf{Z}_t = (Y_t - 100, Y_{t+1|t} - 100)'$$

where the prime symbol (′) indicates the transpose of the row vector. Recall that $Y_{t+k|t}$ denotes the forecast of Y_{t+k} given the data Y_1, Y_2, . . . , Y_t.

Now

$$Y_{t+1|t} - 100 = 1.2(Y_t - 100) - .36(Y_{t-1} - 100)$$

for the AR(2) model. When the data at time t+1 become available, the state vector changes to

$$\mathbf{Z}_{t+1} = (Y_{t+1} - 100,\ Y_{t+2|t+1} - 100)'$$
$$= (Y_{t+1} - 100,\ 1.2(Y_{t+1} - 100) - .36(Y_t - 100))'$$

Because

$$Y_{t+1} - 100 = (Y_{t+1|t} - 100) + e_{t+1}$$

you can write

$$\mathbf{Z}_{t+1} = \begin{bmatrix} 0 & 1 \\ -.36 & 1.2 \end{bmatrix} \mathbf{Z}_t + \begin{bmatrix} 1 \\ 1.2 \end{bmatrix} e_{t+1}$$

The first line of the matrix equation is simply

$$Y_{t+1} - 100 = Y_{t+1|t} - 100 + e_{t+1}$$

or

$$Y_{t+1} - 100 = 1.2(Y_t - 100) - .36(Y_{t-1} - 100) + e_{t+1}$$

The last line of the matrix equation becomes

$$Y_{t+2|t+1} - 100 = -.36(Y_t - 100) + 1.2(Y_{t+1|t} - 100) + 1.2e_{t+1}$$
$$= -.36(Y_t - 100) + 1.2(Y_{t+1} - 100)$$

because

$$Y_{t+1|t} - 100 + e_{t+1} = Y_{t+1} - 100$$

Two examples have been discussed thus far. In the first, an AR(1) model, the state vector was

$$\mathbf{Z}_t = Y_t - 100$$

and

$$\mathbf{Z}_{t+1} = .6\mathbf{Z}_t + e_{t+1}$$

In the second example, an AR(2) model, the state vector was

$$\mathbf{Z}_{t+1} = (Y_t - 100,\ Y_{t+1|t} - 100)'$$

and

$$\mathbf{Z}_{t+1} = \mathbf{F}\mathbf{Z}_t + \mathbf{G}\mathbf{E}_{t+1}$$

where

$$\mathbf{F} = \begin{bmatrix} 0 & 1 \\ -.36 & 1.2 \end{bmatrix} \text{ and } \mathbf{G} = \begin{bmatrix} 1 \\ 1.2 \end{bmatrix}$$

Every ARMA model has an associated state vector $\mathbf{Z}_t$ and an updating equation of the form

$$\mathbf{Z}_{t+1} = \mathbf{F}\mathbf{Z}_t + \mathbf{G}\mathbf{E}_{t+1}$$

To determine the state vector for a univariate ARMA time series Y_t, consider the sequence Y_t, $Y_{t+1|t}$, $Y_{t+2|t}$, ... At some point, such as k+1, you find that $Y_{t+k+1|t}$ is a linear combination of the previous sequence of elements Y_t, $Y_{t+1|t}$, . . . $Y_{t+k|t}$; that is,

$$Y_{t+k+1|t} = \alpha_0 Y_t + \alpha_1 Y_{t+1|t} + \ldots + \alpha_k Y_{t+k|t}$$

In the AR(2) example,

$$Y_{t+2|t} - 100 = 1.2\left(Y_{t+1|t} - 100\right) - .36\left(Y_t - 100\right)$$

so k=1. This determines the state vector as

$$\mathbf{Z}_t = \left(Y_t, Y_{t+1|t}, \ldots, Y_{t+k|t}\right)'$$

Furthermore, any prediction of $Y_{t+R|t}$ with R>k is also a linear combination of state vector elements. Think of constructing the state vector by sequentially including forecasts $Y_{t+j|t}$ into $\mathbf{Z}_t$ until you reach the first forecast that is linearly dependent on forecasts already in $\mathbf{Z}_t$. At that point, stop expanding $\mathbf{Z}_t$. **Section 6.1.2** shows how this can be accomplished using canonical correlations. One more univariate example follows.

Suppose

$$Y_t = e_t + .8e_{t-1}$$

Then, because

$$Y_{t+1|t} = .8e_t$$

and, for j>1,

$$Y_{t+j|t} = 0$$

(which is $\alpha_0 Y_t + \alpha_1 Y_{t+1|t}$ with $\alpha_0 = \alpha_1 = 0$), the state vector has the following form:

$$\mathbf{Z}_t = \begin{bmatrix} Y_t \\ Y_{t+1|t} \end{bmatrix} = \begin{bmatrix} Y_t \\ .8e_t \end{bmatrix}$$

and

$$\mathbf{Z}_{t+1} = \begin{bmatrix} Y_{t+1} \\ .8e_{t+1} \end{bmatrix}$$

Note that

$$\begin{bmatrix} Y_{t+1} \\ .8e_{t+1} \end{bmatrix} = \begin{bmatrix} 0 & 1 \\ 0 & 0 \end{bmatrix} \begin{bmatrix} Y_t \\ .8e_t \end{bmatrix} + \begin{bmatrix} 1 \\ .8 \end{bmatrix} e_{t+1}$$

which is equivalent to the equation

$$Y_{t+1} = e_{t+1} + .8e_t$$

along with the identity

$$.8e_{t+1} = .8e_{t+1}$$

Thus,

$$\mathbf{F} = \begin{bmatrix} 0 & 1 \\ 0 & 0 \end{bmatrix} \text{and } \mathbf{G} = \begin{bmatrix} 1 \\ .8 \end{bmatrix}$$

for the moving average (MA) model.

The truly useful fact is that all multivariate ARMA models have state space equations. To construct the state vector for a multivariate process, you forecast each element of the process. For the bivariate process (X_t, Y_t), for example, consider the sequence X_t, Y_t, $X_{t+1|t}$, $Y_{t+1|t}$, $X_{t+2|t}$, $Y_{t+2|t}$, ... When you first reach a forecast of X (or Y) that is a linear combination of elements currently in the state vector, do not include that forecast or any future forecasts of that variable in the state vector. Continue including forecasts of the other variable in $\mathbf{Z}_t$ until you reach a point of linear dependence in that variable. You have seen an AR(2) of dimension 2 in **Section 6.1.2.** A bivariate ARMA(1,1) is shown next.

6.2.2 ARMA(1,1) of Dimension 2

Consider the model

$$\begin{bmatrix} X_t \\ Y_t \end{bmatrix} = \begin{bmatrix} .5 & .3 \\ .3 & .5 \end{bmatrix} \begin{bmatrix} X_{t-1} \\ Y_{t-1} \end{bmatrix} + \begin{bmatrix} \varepsilon_{1,t} \\ \varepsilon_{2,t} \end{bmatrix} - \begin{bmatrix} .2 & .1 \\ 0 & 0 \end{bmatrix} \begin{bmatrix} \varepsilon_{1,t-1} \\ \varepsilon_{2,t-1} \end{bmatrix}$$

or

$$X_t = .5X_{t-1} + .3Y_{t-1} + \varepsilon_{1,t} - .2\varepsilon_{1,t-1} - .1\varepsilon_{2,t-1}$$

and

$$Y_t = .3X_{t-1} + .5Y_{t-1} + \varepsilon_{2,t}$$

from which

$$X_{t+1|t} = .5X_t + .3Y_t - .2\varepsilon_{1,t} - .1\varepsilon_{2,t}$$

$$Y_{t+1|t} = .3X_t + .5Y_t$$

and

$$X_{t+2|t} = .5X_{t+1|t} + .3Y_{t+1|t} = .5X_{t+1|t} + .09X_t + .15Y_t$$

so

$$\mathbf{Z}_t = (X_t, Y_t, X_{t+1|t})'$$

Finally, the state space form is

$$\begin{bmatrix} X_{t+1} \\ Y_{t+1} \\ X_{t+2|t+1} \end{bmatrix} = \begin{bmatrix} 0 & 0 & 1 \\ .3 & .5 & 0 \\ .09 & .15 & .5 \end{bmatrix} \begin{bmatrix} X_t \\ Y_t \\ X_{t+1|t} \end{bmatrix} + \begin{bmatrix} 1 & 0 \\ 0 & 1 \\ .3 & .2 \end{bmatrix} \begin{bmatrix} \varepsilon_{1,t+1} \\ \varepsilon_{2,t+1} \end{bmatrix}$$

6.3 PROC STATESPACE

The general outline of PROC STATESPACE is as follows:

1. For a multivariate or vector series, for example,
   $\mathbf{X}_t = (X_{1,t}, X_{2,t}, X_{3,t})'$

   fit a multivariate AR model

   $$X_t = A_1X_{t-1} + A_2X_{t-2} + \ldots + A_kX_{t-k} + E_t$$

   You can do this row by row. That is, the regression of $X_{1,t}$, on $X_{1,t-1}$, $X_{2,t-1}$, $X_{3,t-1}$, $X_{1,t-2}$, $X_{2,t-2}$, $X_{3,t-2}, \ldots, X_{3,t-k}$ produces the top rows of matrices $\mathbf{A}_1, \mathbf{A}_2, \mathbf{A}_3, \ldots, \mathbf{A}_k$. Using $X_{2,t}$ and $X_{3,t}$ as dependent variables in the regression produces the second and third rows of the **A** matrices. This is essentially what is done in PROC STATESPACE. To decide on *k*, you use a version of Akaike's information criterion (AIC). This criterion is

   AIC = –2LOG(maximized likelihood) + 2(number of parameters in the model)

   Note that AIC is made smaller by a decrease in the number of model parameters or an increase in the likelihood function. Thus, it trades off precision of fit against the number of parameters used to obtain that fit. Select *k* to minimize AIC.

2. The model is now called a vector autoregression of order *k*, and you have a measure, AIC, of its fit. The question becomes whether the fit can be improved by allowing MA terms and setting to 0 some of the elements of the A matrices. In other words, search the class of all vector ARMA models that can be reasonably approximated by a vector autoregression of order *k*. The smallest canonical correlation is used to assess the fit of each model against the fit of the preliminary vector autoregression.

For example, a vector ARMA(1,1) of dimension 3 can be

$$\begin{bmatrix} X_{1,t} \\ X_{2,t} \\ X_{3,t} \end{bmatrix} = \begin{bmatrix} .72 & 0 & 0 \\ 0 & .6 & .4 \\ 0 & .4 & .8 \end{bmatrix} \begin{bmatrix} X_{1,t-1} \\ X_{2,t-1} \\ X_{3,t-1} \end{bmatrix} + \begin{bmatrix} e_{1,t} \\ e_{2,t} \\ e_{3,t} \end{bmatrix} - \begin{bmatrix} .4 & .3 & .1 \\ 0 & .2 & .1 \\ 0 & 0 & .4 \end{bmatrix} \begin{bmatrix} e_{1,t-1} \\ e_{2,t-1} \\ e_{3,t-1} \end{bmatrix}$$

or

$$\mathbf{X}_t = \mathbf{A}\mathbf{X}_{t-1} + \mathbf{E}_t - \mathbf{B}\mathbf{E}_{t-1}$$

Check to see if this model fits as well as the original vector autoregression. If k=4, for example, the original autoregression contains four **A** matrices, each with nine parameters. If the vector ARMA(1,1) fits about as well as the vector AR(4) in likelihood, the inherent penalty in the information criterion for a large number of parameters can make the information criterion for the vector ARMA(1,1) smaller than for the vector AR(4), and thus the difference will be negative. An information criterion for model selection based on this idea is called DIC in PROC STATESPACE.

3. The comparison in step 2 is easier than it first appears. All vector ARMA models can be expressed in state space form. Thus, comparing state space models and determining the best model is equivalent to finding the dimension of the best model's state vector Z_t, because all state space models have the same basic form,

$$\mathbf{Z}_t = \mathbf{F}\mathbf{Z}_{t-1} + \mathbf{G}\mathbf{E}_t$$

The key to this decision is an organized sequential formulation of the state vector. Start by including $X_{1,t}$, $X_{2,t}$, and $X_{3,t}$. Next, check $X_{1,t+1|t}$ to see if it is a linear combination of $X_{1,t}$, $X_{2,t}$, and $X_{3,t}$. If it is, it provides no new information and is not added to the state vector. Otherwise, the state vector is augmented to ($X_{1,t}$, $X_{2,t}$, $X_{3,t}$, $X_{1,t+1|t}$).

The next question is whether $X_{2,t+1|t}$ should be included in the state vector. Include it only if it cannot be written as a linear combination of elements already in the state vector. The state vector is formulated sequentially in this fashion. Suppose $X_{1,t+1|t}$ is included and both $X_{2,t+1|t}$ and $X_{3,t+1|t}$ have been tested. Next, consider testing $X_{1,t+2|t}$ for inclusion in the state vector. If $X_{1,t+2|t}$ is not included in the state vector, pioneering work by Akaike shows that $X_{1,t+j|t}$ is not included for any j>2. That is, if a forecast $X_{1,t+k|t}$ is a linear combination of elements already in the state vector, $X_{1,t+j|t}$ also is such a linear combination for any j>k. At this point, stop considering the forecast of X_1, but continue to consider forecasts of X_2 and X_3 (unless $X_{2,t+1|t}$ or $X_{3,t+1|t}$ was found earlier to be a linear combination of elements already in the state vector) and continue in this fashion.

For this example, the state vector may be

$$\mathbf{Z}_t = (X_{1,t}, X_{2,t}, X_{3,t}, X_{1,t+1|t}, X_{3,t+1|t}, X_{1,t+2|t})'$$

In this case, $X_{1,t+3|t}$, $X_{2,t+1|t}$, and $X_{3,t+2|t}$ are linear combinations of elements already present in the state vector.

4. PROC STATESPACE uses the initial vector AR(k) approximation to estimate covariances between the elements of Z_{t+1} and Z_t. It uses these in a manner similar to the Yule-Walker equations to compute initial estimates of the elements of **F** and **G**. Recall that the model, in state space form, is

$$\mathbf{Z}_{t+1} = \mathbf{F}\mathbf{Z}_t + \mathbf{G}\mathbf{E}_{t+1}$$

for any underlying multivariate vector ARMA model. Assuming $\mathbf{E}_t$ is a sequence of independent normal random vectors with mean 0 and variance-covariance matrix Σ, you can write out the likelihood function. Start with the initial estimates, and use a nonlinear search routine to find the maximum-likelihood (ML) estimates of the parameters and their asymptotically valid standard errors.

You can obtain an estimate, $\hat{\Sigma}$, of Σ from the sums of squares and crossproducts of the one-step-ahead forecast errors of each series. Such a multivariate setting has several error variances (these are the diagonal elements of Σ). A general measure of the size of $\hat{\Sigma}$ is its determinant $|\hat{\Sigma}|$, which can be printed out in PROC STATESPACE. This determinant should be minimized because it is a general measure of prediction error variance.

5. Because the first few elements of the state vector make up the multivariate series to be forecast, use the state space equation to forecast future values Z_{t+k} and then extract the first elements. These are the forecasts of the multivariate series. In addition, the state space equation yields the prediction error variances. Consider, for example, forecasting three periods ahead. Now

$$\begin{aligned}\mathbf{Z}_{t+3} &= \mathbf{F}\mathbf{Z}_{t+2} + \mathbf{G}\mathbf{E}_{t+3} = \mathbf{F}^2\mathbf{Z}_{t+1} + \mathbf{F}\mathbf{G}\mathbf{E}_{t+2} + \mathbf{G}\mathbf{E}_{t+3}\\ &= \mathbf{F}^3\mathbf{Z}_t + (\mathbf{F}^2\mathbf{G}\mathbf{E}_{t+1} + \mathbf{F}\mathbf{G}\mathbf{E}_{t+2} + \mathbf{G}\mathbf{E}_{t+3})\end{aligned}$$

If the original vector process has the three elements $X_{1,t}$, $X_{2,t}$, and $X_{3,t}$ considered before, the forecasts $X_{1,t+3}$, $X_{2,t+3}$, and $X_{3,t+3}$ are the first three elements of $\mathbf{F}^3\mathbf{Z}_t$. The variance-covariance matrix of the quantity in parentheses contains the prediction-error, variance-covariance matrix. It can be estimated from estimates of **F**, **G**, and Σ.

6. In PROC STATESPACE, an output data set is created that contains forecasts, forecast standard errors, and other information useful for displaying forecast intervals. The theoretical result that allows this process to be practical is the use of canonical correlations to accomplish step 3. Akaike (1976) showed how this can be done. Note that up to this point, state vectors have been derived only for known models. In practice, you do not know the vector ARMA model form, and you must use the canonical correlation approach (see **Section 6.3.2**) to compute the state vector.

6.3.1 State Vectors Determined from Covariances

PROC STATESPACE computes the sequence of information criterion values for

$k = 1, 2, \ldots, 10$ (or ARMAX)

and selects the model that gives the minimum.

This vector AR model for the time series is used to compute a variance-covariance matrix **M** between the set of current and past values and between the set of current and future values.

These two facts are relevant:

1. All predictions are linear combinations of the observations of $Y_t, Y_{t-1}, Y_{t-2}, \ldots$ where for practical purposes this list can be truncated at Y_{t-k}, as determined by the initial vector autoregression.

2. The covariance between a prediction $Y_{t+j|t}$ and a current or past value Y_{t-i} is the same as that between Y_{t+j} and Y_{t-i}.

Akaike uses these facts to show that analyzing the covariances in the matrix **M** is equivalent to determining the form of the state vector. Canonical correlations are used in this case, and the elements of **M** are replaced by their sample values.

6.3.2 Canonical Correlations

Suppose the covariance matrix between the set of current and past values Y_t, Y_{t-1}, Y_{t-2} and the set of current and future values Y_t, Y_{t+1}, Y_{t+2} for a univariate series is given by

$$\mathbf{M} = \begin{bmatrix} 8 & 4 & 2 \\ 4 & 2 & 1 \\ 2 & 1 & .5 \end{bmatrix}$$

Note that there are no zero correlations. (You will find, however, that some canonical correlations are zero.)

Canonical correlation analysis proceeds as follows:

1. Find the linear combination of elements in the first vector (Y_t, Y_{t-1}, Y_{t-2}) and second vector (Y_t, Y_{t+1}, Y_{t+2}), with maximum cross-correlation. This canonical correlation is the largest, and the linear combinations are called canonical variables. In the example, Y_t and Y_t are perfectly correlated, so $(1, 0, 0)(Y_t, Y_{t-1}, Y_{t-2})'$ and $(1, 0, 0)(Y_t, Y_{t+1}, Y_{t+2})'$ are the canonical variables.

2. Now consider all linear combinations of elements in the original vectors that are not correlated with the first canonical variables. Of these, the two most highly correlated give the second-highest canonical correlation. In this case, you can show that the next-highest canonical correlation is, in fact, zero.

3. At each stage, consider linear combinations that are uncorrelated with the canonical variables found thus far. Pick the two (one for each vector being analyzed) with the highest cross-correlation.

Akaike establishes that the following numbers are all the same for general vector ARMA models:

- the rank of an appropriate **M**
- the number of nonzero canonical correlations
- the dimension of the state vector.

When you look at the example matrix **M**, you see that the covariance between Y_{t-k} and Y_{t+j} is always $8(.5^{j+k})$ for $j,k \geq 0$. Thus, **M** is the covariance matrix of an AR(1) process. All rows of **M** are direct multiples of the first, so **M** has rank 1. Finally, the canonical correlations computed from **M** are 1,0,0.

When the general sample covariance matrix **M** is used, the analysis proceeds as follows (illustration for bivariate series $(X_t, Y_t)'$):

1. Determine the number of lags into the past $(X_t, Y_t, X_{t-1}, Y_{t-1}, \ldots, Y_{t-k})$.

2. Do a canonical correlation of (X_t, Y_t) with current and past values. This produces correlations 1,1.

3. Next, do a canonical correlation analysis of $(X_t, Y_t, X_{t+1|t})$ with current and past values of X_t, Y_t from step 1.

4. a) If the smallest canonical correlation is not close to zero, include $X_{t+1|t}$ in the state vector and analyze $(X_t, Y_t, X_{t+1|t}, Y_{t+1|t})$.
   b) If the smallest canonical correlation is close to zero, exclude from consideration $X_{t+1|t}$ and all $X_{t+j|t}$ for j>1. In this case, the analysis of $(X_t, Y_t, Y_{t+1|t})$ is next.

5. Continue until you have determined the first predictions, $X_{t+j|t}$ and $Y_{t+s|t}$, that introduce zero canonical correlations. Then $X_{t+j-1|t}$ and $Y_{t+s-1|t}$ are the last predictions of X and Y to be included in the state vector.

PROC STATESPACE executes this procedure automatically. The sample canonical correlations are judged by the aforementioned DIC. A chi-square test statistic attributed to Bartlett (1947) is computed. The significance of Bartlett's statistic indicates a nonzero canonical correlation. Robinson (1973) suggests potential problems with Bartlett's test for MA models. Thus, DIC is used as the default criterion.

PROC STATESPACE uses the estimated covariance matrix and the identified state vector to compute initial estimates of matrices **F** and **G** in the state space representation. The advantage of PROC STATESPACE is its automatic identification of a model and preliminary parameter estimation, but the user is responsible for any transformation necessary to produce stationarity and approximate normality. Also note that the STATESPACE theory does not include deterministic components like polynomials in time.

Use the NOEST option to view the preliminary model before fine-tuning the parameter estimates through nonlinear iterative least squares (LS) or ML estimation.

You may want to use the RESTRICT statement to set certain elements of **F** and **G** to zero. (You have seen several cases where **F** and **G** contain zeros.)

6.3.3 Simulated Example

To see how PROC STATESPACE works with a known univariate model, consider 100 observations from an MA(1) model

$$Y_t = e_t + .8e_{t-1}$$

Note that the model can be re-expressed as

$$Y_t = (.8Y_{t-1} - .64Y_{t-2} + (.8)^3Y_{t-3} - (.8)^4Y_{t-4} \ldots) + e_t$$

Thus, the initial AR approximation should have coefficients near .8, –.64, .512, –.4096, ... Use the following SAS statements for the analysis:

```
PROC STATESPACE CANCORR ITPRINT DATA=TEST2;
   VAR Y;
RUN;
```

As shown in **Output 6.4**, the CANCORR option displays the sequential construction of the state vector. The ITPRINT option shows the iterative steps of the likelihood maximization.

In Output 6.4, observe the sample mean, $\overline{Y}$ ❶, and standard deviation and the sequence of AICs ❷ for up to ten AR lags. The smallest AIC in the list is 9.994428 ❸, which occurs at lag 4. Thus, the initial AR approximation involves four lags ❹ and is given by

$$Y_t = .79Y_{t-1} - .58Y_{t-2} + .31Y_{t-3} - .24Y_{t-4} + e_t$$

This corresponds reasonably well with the theoretical results. Schematic representations of the autocorrelation function (ACF) ❺ and partial autocorrelation function (PACF) ❻ are also given. A plus sign (+) indicates a value more than two standard errors above 0, a period (.) indicates a value within two standard errors of 0, and a minus sign (-) indicates a value more than two standard errors below 0. Based on results from Chapter 3, "The General ARIMA Model," you would expect the following sequences of + and - signs in the theoretical ACF and PACF plots.

LAG	0	1	2	3	4	5	6	7	8	9	10
ACF	+	+	.	.	.	.	.	.	.	.	.
PACF	+	–	+	–	+	–	+	–	+	–	+

You also would expect the estimated PACF to drop within two standard errors of 0 after a few lags. The estimated functions correspond fairly well with the theoretical functions.

Note the canonical correlation analysis ❼. Initially, consideration is given to adding $Y_{t+1|t}$ to the state vector containing Y_t. The canonical correlation, 0.454239 ❽, is an estimate of the second-largest canonical correlation between the set of variables (Y_t, Y_{t+1}) and the set of variables (Y_t, Y_{t-1}, Y_{t-2}, Y_{t-3}, Y_{t-4}). The first canonical correlation is always 1 because both sets of variables contain Y_t. The question is whether 0.4542 is an estimate of 0. PROC STATESPACE concludes that a correlation is 0 if DIC<0. In this case, DIC=15.10916 ❾, so 0.4542 is not an estimate of 0. This implies that the portion of Y_{t+1} that cannot be predicted from Y_t is correlated with the past of the time series and, thus, that $Y_{t+1|t}$ should be included in the state vector. Another test statistic, Bartlett's test, is calculated as 22.64698 ❿. The null hypothesis is that the second-highest canonical correlation is 0 and the test statistic is to be compared to a chi-squared table with four degrees of freedom ⓫. The hypothesis of zero correlation is rejected, and Bartlett's test agrees with DIC to include $Y_{t+1|t}$ in the state vector.

Output 6.4 *Modeling Simulated Data in PROC STATESPACE with the CANCORR and ITPRINT Options*

```
                    The STATESPACE Procedure

               Number of Observations    100

                                       Standard
           ❶ Variable         Mean        Error

             Y             -0.40499    1.322236


        ❷ Information Criterion for Autoregressive Models

 Lag=0     Lag=1     Lag=2     Lag=3    Lag=4     Lag=5     Lag=6     Lag=7     Lag=8     Lag=9
                                       ❸
55.86487 27.05937 13.52724  13.8878 9.994428 11.95021  12.7582 10.71094 12.64703  14.5518

                          Information
                         Criterion for
                         Autoregressive
                            Models
                            Lag=10

                           16.48957


            Schematic Representation of Correlations
  ❺
  Name/Lag    0    1    2    3    4    5    6    7    8    9    10

  Y           +    +    .    .    .    .    .    .    +    .    .

      + is > 2*std error,  - is < -2*std error,  . is between


     Schematic Representation of Partial Autocorrelations
   ❻
   Name/Lag    1    2    3    4    5    6    7    8    9    10

   Y           +    -    .    -    .    .    .    .    .    .

      + is > 2*std error,  - is < -2*std error,  . is between


            Yule-Walker Estimates for Minimum AIC

           ❹  --Lag=1-     --Lag=2-     --Lag=3-     --Lag=4-
                    Y            Y            Y            Y

     Y         0.789072     -0.58225     0.308989     -0.23923
```

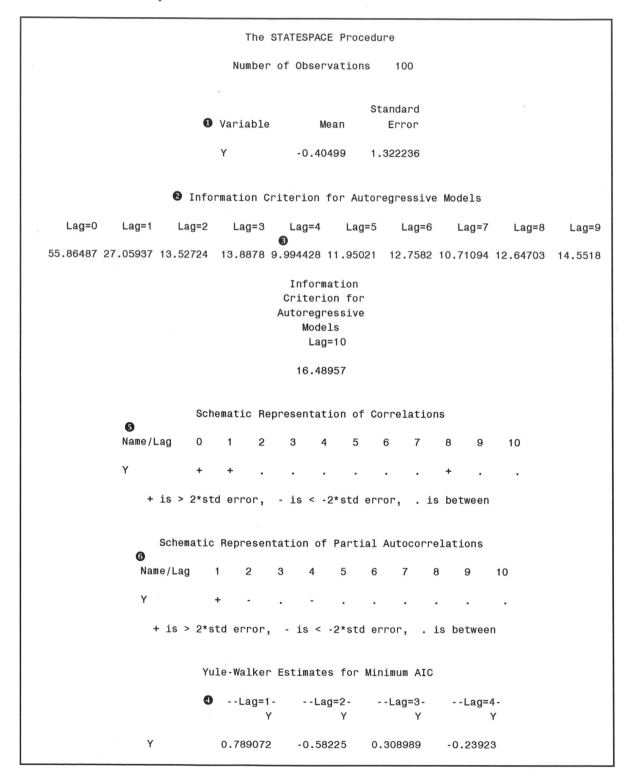

Output 6.4 *Modeling Simulated Data in PROC STATESPACE with the CANCORR and ITPRINT Options (continued)*

❼ Canonical Correlations Analysis

Y(T;T)	Y(T+1;T)	Information Criterion	Chi Square	DF
1	❽ 0.454239	❾ 15.10916	❿ 22.64698	⓫ 4

Y(T;T)	Y(T+1;T)	Y(T+2;T)	Information Criterion	Chi Square	DF
1	0.489796	⓬ 0.208571	⓭ -1.55238	⓮ 4.38091	⓯ 3

Selected Statespace Form and Preliminary Estimates

State Vector

⓰ Y(T;T) Y(T+1;T)

Estimate of Transition Matrix

0	1
-0.16297	0.290093

⓱ Input Matrix for Innovation

1
0.789072

Variance Matrix for Innovation

1.020144

Iterative Fitting: Maximum Likelihood Estimation

Iter	Half	Determinant	Lambda	F(2,1)	F(2,2)	G(2,1)	Sigma(1,1)
0	0	1.040261	0.1	-0.1629742	0.29009252	0.78907196	1.04026133
1	0	1.025047	0.01	-0.1157232	0.20149505	0.83329356	1.02504685
2	0	1.020305	0.001	-0.0570093	0.12653818	0.84215994	1.02030521
3	0	1.020245	0.0001	-0.051494	0.11987985	0.8409064	1.02024527
4	2	1.020245 ㉒	0.001 ㉑	-0.0515121	0.1198994	0.84093914	1.02024507

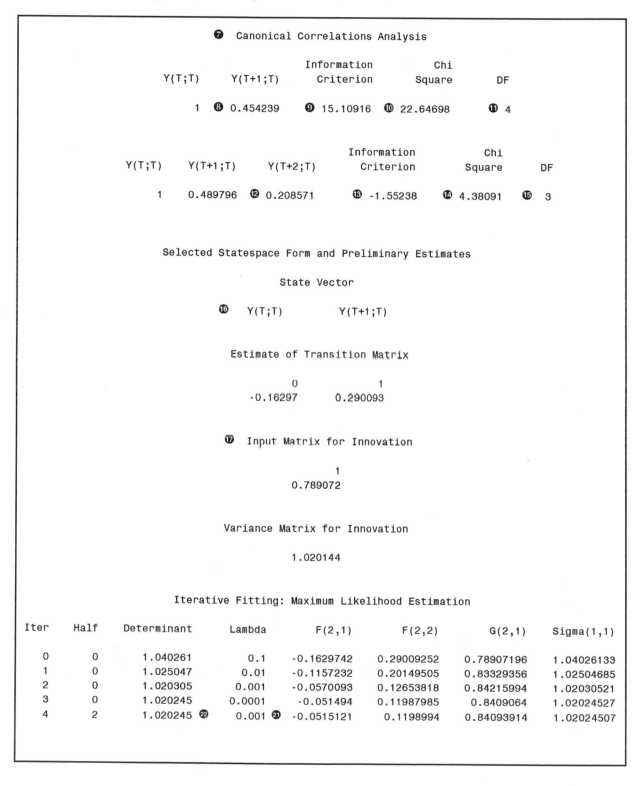

Output 6.4 *Modeling Simulated Data in PROC STATESPACE with the CANCORR and ITPRINT Options (continued)*

```
                Maximum likelihood estimation has converged.

                      Selected Statespace Form and Fitted Model

                                  State Vector

                      Y(T;T)              Y(T+1;T)

                   ⓲  Estimate of Transition Matrix

                                0              1
                         -0.05151       0.119899

                    ⓳  Input Matrix for Innovation

                                       1
                                0.840939

                      Variance Matrix for Innovation

                                1.020245

                           Parameter Estimates

                                            Standard         ⓴
           Parameter      Estimate            Error     t Value

           F(2,1)         -0.05151         0.132797       -0.39
           F(2,2)         0.119899         0.151234        0.79
           G(2,1)         0.840939         0.099856        8.42
```

Now consider the portion of $Y_{t+2|t}$ that you cannot predict from Y_t and $Y_{t+1|t}$. If this portion is correlated with the past of the series, you can produce a better predictor of the future than one that uses only Y_t and $Y_{t+1|t}$. Add $Y_{t+2|t}$ to the state vector unless the third-highest canonical correlation between the set (Y_t, Y_{t+1}, Y_{t+2}) and the set (Y_t, Y_{t-1}, Y_{t-2}, . . . , Y_{t-4}) is 0. The estimate of the third-highest canonical correlation is 0.208571 ⓬. PROC STATESPACE assumes that 0.208571 is just an estimate of 0 because DIC is negative (–1.55238) ⓭. This means that once you have predicted Y_{t+2} from Y_t and $Y_{t+1|t}$, you have the best predictor available.

The past data do not improve the forecast. Thus, $Y_{t+2|t}$ is not added to the state vector. Bartlett's test statistic, 4.38091 ⓮, is not significant compared to a chi-squared table with three degrees of freedom ⓯ (with a critical value of 7.81).

Again, the two tests agree that $Y_{t+2|t}$ is a linear combination of Y_t and $Y_{t+1|t}$. Thus, the only information you need to predict arbitrarily far into the future is in

$$\mathbf{Z}_t = (Y_t, Y_{t+1|t})'$$

When you compare this to the theoretical analysis of an MA(1), you see that PROC STATESPACE has correctly identified the state vector as having two elements. The theoretical analysis gives the state space representation as

$$\mathbf{Z}_{t+1} = \begin{bmatrix} 0 & 1 \\ 0 & 0 \end{bmatrix} \mathbf{Z}_t + \begin{bmatrix} 1 \\ .8 \end{bmatrix} e_{t+1}$$

PROC STATESPACE estimates these matrices to be

$$\mathbf{F} = \begin{bmatrix} 0 & 1 \\ -.163 & .290 \end{bmatrix} \text{ and } \mathbf{G} = \begin{bmatrix} 1 \\ .79 \end{bmatrix}$$

initially ⓰⓱ and

$$\mathbf{F} = \begin{bmatrix} 0 & 1 \\ -.051 & .12 \end{bmatrix} \text{ and } \mathbf{G} = \begin{bmatrix} 1 \\ .841 \end{bmatrix}$$

finally ⓲⓳.

Note that the *t* statistics ⓴ on **F** (2,1) and **F** (2,2) are, as expected, not significant. The true entries of **F** are zeros in those positions. Finally, observe the nonlinear search ㉑ beginning with the initial values in **Output 6.4** ⓰⓱ and then moving to the final values in **Output 6.4** ⓲⓳. Note that $|\hat{\Sigma}|$ decreases at each step ㉒.

To force the correct form on the matrix **F**, use the RESTRICT statement:

```
PROC STATESPACE ITPRINT COVB DATA=TEST2;
   RESTRICT F(2,1)=0 F(2,2)=0;
   VAR Y;
RUN;
```

The RESTRICT statement may also include restrictions on the entries of **G**. (See **Output 6.5.**)

As requested, the bottom row of **F** has been set to 0 0 ❶. The initial **G** matrix ❷and the final **G** matrix ❸ are close to the theoretical matrix, namely G=(1 .8)′. The COVB option requests the variance-covariance matrix of the parameter estimates ❹, which is a scalar in the case of a single parameter estimate.

Output 6.5 *Modeling Simulated Data in PROC STATESPACE with the RESTRICT Statement*

```
                     The STATESPACE Procedure

                 Number of Observations     100

                                         Standard
               Variable          Mean       Error

               Y             -0.40499    1.322236



          Information Criterion for Autoregressive Models

   Lag=0     Lag=1     Lag=2     Lag=3     Lag=4     Lag=5     Lag=6     Lag=7     Lag=8     Lag=9

55.86487 27.05937 13.52724  13.8878 9.994428 11.95021  12.7582 10.71094 12.64703  14.5518

                              Information
                             Criterion for
                            Autoregressive
                                Models

                                Lag=10

                               16.48957

             Schematic Representation of Correlations

        Name/Lag     0    1    2    3    4    5    6    7    8    9    10

        Y            +    +    .    .    .    .    .    .    +    .    .

            + is > 2*std error,  - is < -2*std error,  . is between

          Schematic Representation of Partial Autocorrelations

           Name/Lag     1    2    3    4    5    6    7    8    9    10

           Y            +    -    .    -    .    .    .    .    .    .

             + is > 2*std error,  - is < -2*std error,  . is between


                Yule-Walker Estimates for Minimum AIC

                   --Lag=1-    --Lag=2-    --Lag=3-    --Lag=4-
                          Y           Y           Y           Y

           Y       0.789072    -0.58225    0.308989    -0.23923
```

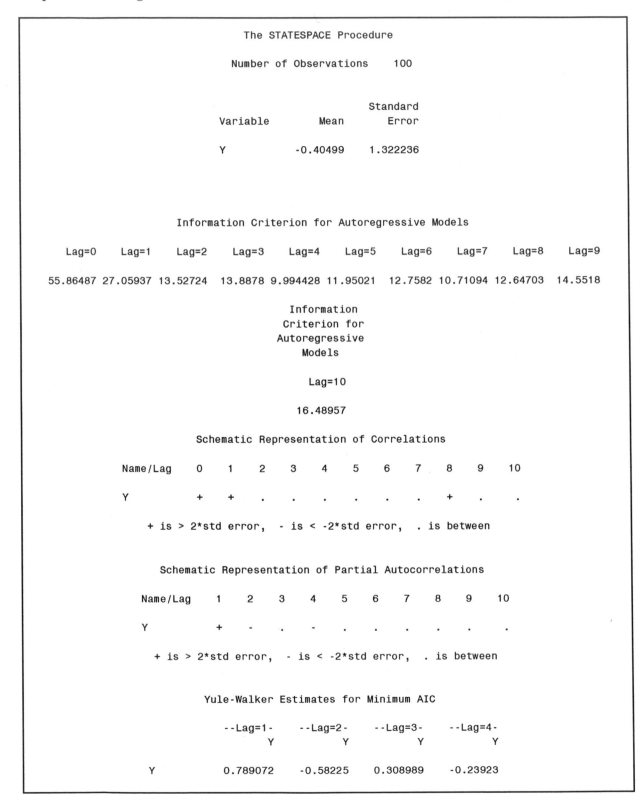

Output 6.5 *Modeling Simulated Data in PROC STATESPACE with the RESTRICT Statement (continued)*

```
                 Selected Statespace Form and Preliminary Estimates

                                    State Vector
                              Y(T;T)          Y(T+1;T)

                           Estimate of Transition Matrix

                                    0               1
                                  ❶ 0               0

                            Input Matrix for Innovation

                                            1
                                  ❷ 0.789072

                           Variance Matrix for Innovation

                                     1.020144

                    Iterative Fitting: Maximum Likelihood Estimation

       Iter     Half    Determinant       Lambda          G(2,1)       Sigma(1,1)

          0        0        1.02648          0.1      0.78907196        1.0264803
          1        0       1.026474         0.01      0.77926993       1.02647358
          2        0       1.026445        0.001      0.78027074       1.02644522

        WARNING: No improvement after 10 step halvings. Convergence has
        been assumed.

                     Selected Statespace Form and Fitted Model

                                    State Vector

                              Y(T;T)          Y(T+1;T)

                           Estimate of Transition Matrix

                                    0               1
                                    0               0

                            Input Matrix for Innovation

                                              1
                                   ❸  0.780271

                           Variance Matrix for Innovation

                                     1.026445
```

Output 6.5 *Modeling Simulated Data in PROC STATESPACE with the RESTRICT Statement (continued)*

```
                    Parameter Estimates

                                   Standard
         Parameter     Estimate       Error    t Value

         G(2,1)        0.780271    0.062645      12.46


              Covariance of Parameter Estimates

                                      G(2,1)

                   ❹  G(2,1)       0.0039244


             Correlation of Parameter Estimates

                                   G(2,1)

                    G(2,1)        1.00000
```

You can find other options for PROC STATESPACE in the *SAS/ETS User's Guide.*

It is dangerous to ignore the autocorrelations. The theory behind PROC STATESPACE assumes the input series are stationary. You have no guarantee of a reasonable result if you put nonstationary series into PROC STATESPACE. Often, you see almost all plus signs in the ACF diagram, which indicates a very slow decay and, consequently, possible nonstationarity. Differencing is specified exactly as in PROC ARIMA. For example, the following SAS statements specify a first and span 12 difference to be applied to Y:

```
PROC STATESPACE;
   VAR Y(1,12);
RUN;
```

The FORM statement is used to specify a form for the state vector. This statement can be helpful if you want to specify a state vector different from what DIC automatically chooses (for example, Bartlett's test may give a different result than DIC, and you may prefer Bartlett's test). For example, the statements

```
PROC STATESPACE;
   VAR X Y;
   FORM X 2 Y 1;
RUN;
```

specify the state vector as

$$\mathbf{Z}_t = (X_t, Y_t, X_{t+1|t})'$$

Now consider an interesting data set that cannot be modeled correctly as a transfer function because of feedback. The data are counts of mink and muskrat pelts shipped to Europe from Canada by the Hudson's Bay Company. The logarithms are analyzed, and both the logarithms and the original data are plotted in **Output 6.6**.

Output 6.6
Plotting Original and Logged Data

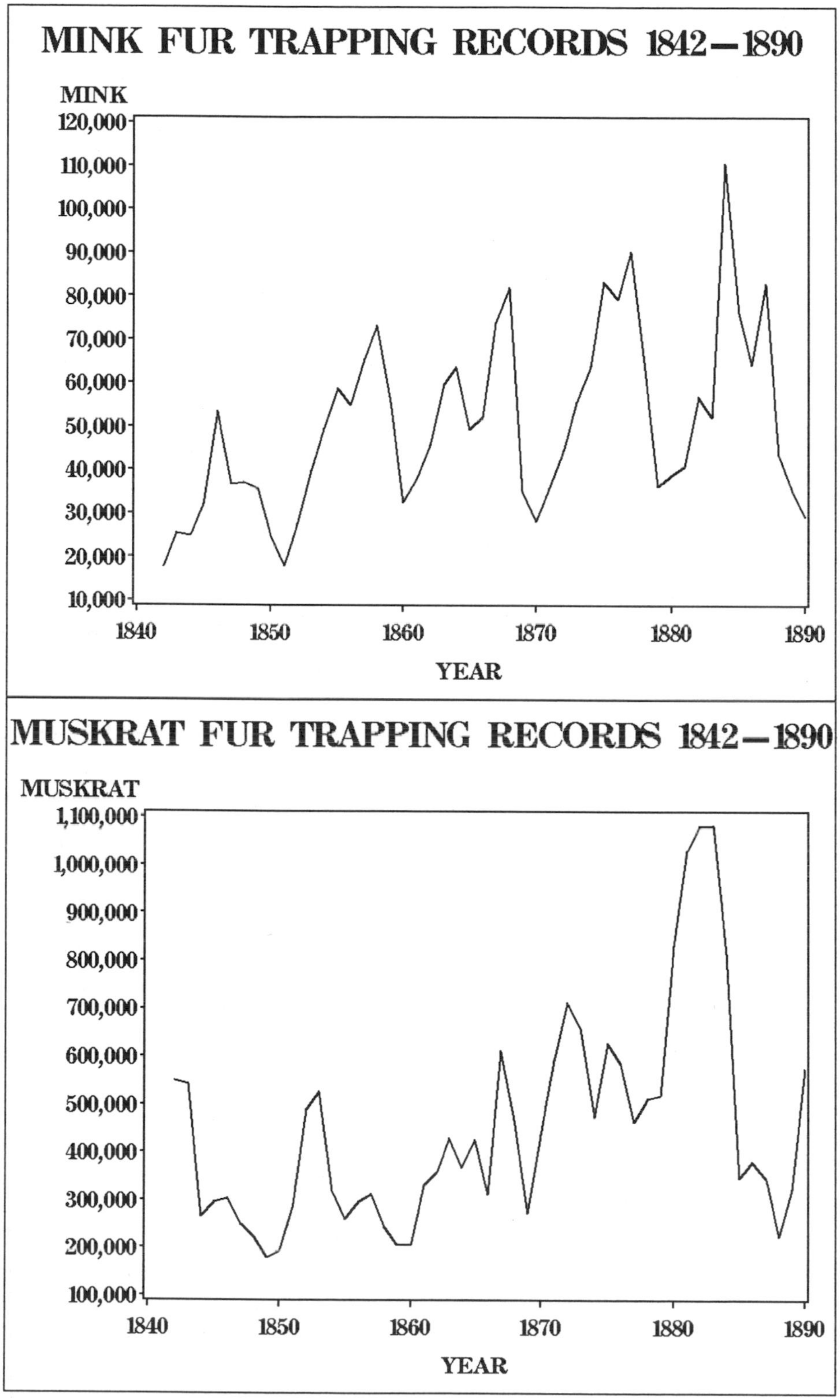

Output 6.6 *Plotting Original and Logged Data (continued)*

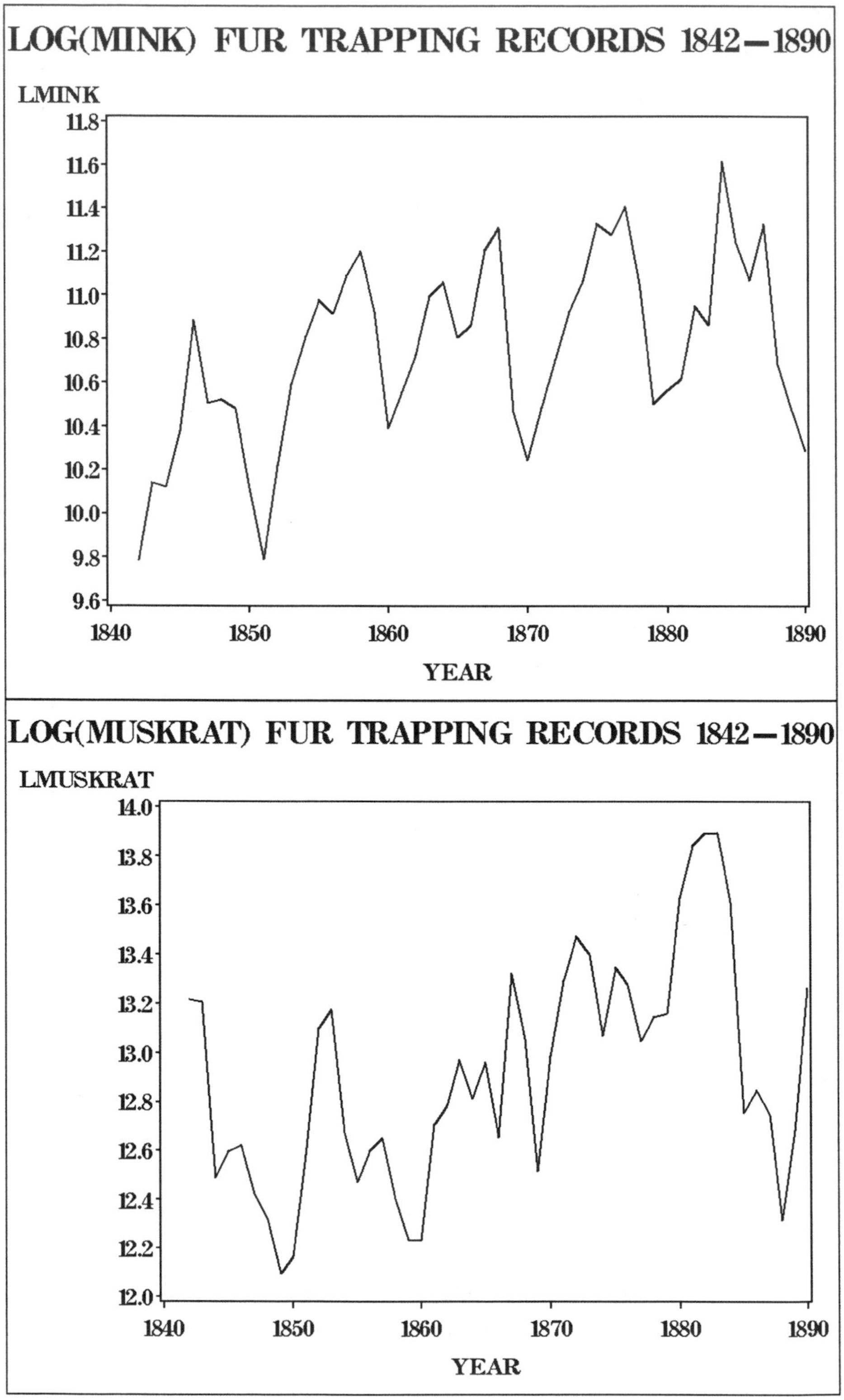

You have an increasing, seemingly linear trend in the data plots. PROC REG is appropriate for detrending the data if the trend is due to increased trapping and does not reveal anything about the relationship between these two species. In that case, the dynamic relationship between mink (predator) and muskrat (prey) is best displayed in residuals from the trend. Another approach is to difference the two data series and analyze the resulting changes in pelt numbers. The approach you choose depends on the true nature of the series. The question becomes whether it is a unit root process or a time trend plus stationary error process. The regression detrending approach is used here simply to display the technique and is not necessarily recommended over differencing. The following SAS code detrends the logged data and submits the detrended data (residuals) to PROC STATESPACE for analysis:

```
DATA DETREND;
   SET MINKMUSK;
   T+1;
RUN;
PROC REG DATA=DETREND NOPRINT;
   MODEL LMINK LMUSKRAT=T;
   OUTPUT OUT=RESID R=RMINK RMUSKRAT;
RUN;
PROC STATESPACE NOCENTER DATA=RESID;
   VAR RMINK RMUSKRAT;
   TITLE 'HUDSON"S BAY FUR TRAPPING RECORDS 1842-1890';
      TITLE2 'RESIDUALS FROM LINEAR TREND';
RUN;
```

The results are shown in **Output 6.7**.

Because the data are detrended, you do not need to subtract the mean. Thus, you can specify NOCENTER. Note that the ACF ❶ schematic plot shows several plus and minus signs but not enough to indicate nonstationarity. (However, it is notoriously difficult to detect nonstationarity visually in a series that has been detrended.)

Note that the ACF at lag 1 is represented by a matrix of plus and minus signs because you have a bivariate series. If you consider a bivariate series in general as (X_t, Y_t) and the lag 1 matrix

$$\begin{array}{c} \\ X_t \\ Y_t \end{array} \begin{array}{c} \begin{array}{cc} X_{t-1} & Y_{t-1} \end{array} \\ \begin{bmatrix} + & + \\ - & + \end{bmatrix} \end{array}$$

then the + in the upper-left corner indicates a positive covariance between X_t and X_{t-1}. The + in the upper-right corner indicates a positive covariance between X_t and Y_{t-1}. The – in the lower-left corner indicates a negative covariance between Y_t and X_{t-1} and, finally, the + in the lower-right corner indicates a positive covariance between Y_t and Y_{t-1} In terms of the current example, X_t represents RMINK, and Y_t represents RMUSKRAT, so the signs make sense with respect to the predator–prey relationship.

The PACF ❷ looks like that of a vector AR of dimension 2 and order 1 (one lag). Thus, you expect the initial AR approximation ❸ to have only one lag and to be very close to the final model ❹ chosen by PROC STATESPACE. This is, in fact, the case here.

The state vector ❺ is simply the vector of inputs, so the vector ARMA model is easily derived from the state space model ❹. When X_t=RMINK (mink residuals at time *t*) and Y_t=RMUSKRAT (muskrat residuals at time *t*) are used, the state vector is simply

$$\mathbf{Z}_t = (X_t, Y_t)'$$

and the model is

$$\begin{matrix}\text{RMINK} \\ \text{RMUSKRAT}\end{matrix} \begin{bmatrix} X_{t+1} \\ Y_{t+1} \end{bmatrix} = \begin{bmatrix} .569 & .298 \\ -.468 & .627 \end{bmatrix} \begin{bmatrix} X_t \\ Y_t \end{bmatrix} + \begin{bmatrix} e_{1,t+1} \\ e_{2,t+1} \end{bmatrix}$$

Here, for example, the number –.468 indicates that large mink values (predator) at time *t* are associated with small muskrat values (prey) at time *t*+1. X_t and Y_t are not related by a transfer function because you can use the *t* statistic ❻ to reject the hypothesis that –.468 is an estimate of 0. That is, each series is predicted by using lagged values of the other series. The transfer function methodology in PROC ARIMA is not appropriate.

Output 6.7 *Using PROC REG to Detrend the Data and PROC STATESPACE to Analyze the Residuals*

```
                    HUDSON'S BAY FUR TRAPPING RECORDS 1842-1890
                           RESIDUALS FROM LINEAR TREND
                            The STATESPACE Procedure

                          Number of Observations    49

                                         Standard
                             Variable       Error

                             RMINK       0.367741
                             RMUSKRAT    0.397009

                      HUDSON'S BAY FUR TRAPPING RECORDS 1842-1890
                            RESIDUALS FROM LINEAR TREND
                             The STATESPACE Procedure

                     Information Criterion for Autoregressive Models

   Lag=0     Lag=1     Lag=2     Lag=3     Lag=4     Lag=5     Lag=6     Lag=7     Lag=8     Lag=9

-188.624  -251.778  -245.705  -242.548  -236.654   -230.49   -229.25  -227.127  -225.203  -221.891

                                   Information
                                  Criterion for
                                 Autoregressive
                                     Models

                                     Lag=10

                                    -221.112
```

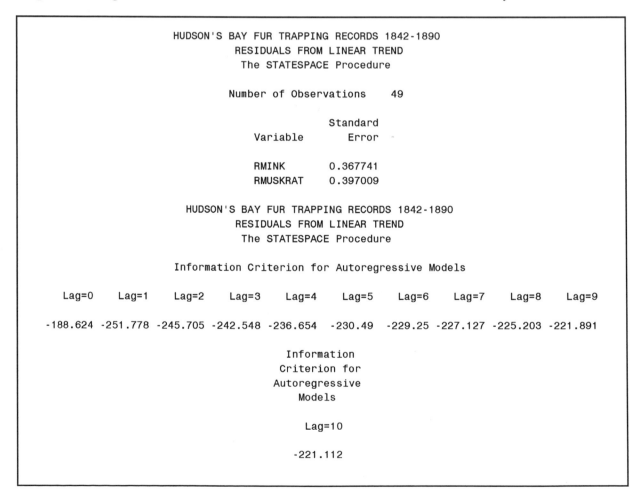

Output 6.7 *Using PROC REG to Detrend the Data and PROC STATESPACE to Analyze the Residuals (continued)*

```
                    Schematic Representation of Correlations

 Name/Lag     0     1     2     3     4     5     6     7     8     9    10
 RMINK        +.    ++    .+    .+    -.    -.    -.    .-    .-    +-    +.
 RMUSKRAT     .+    -+    -.    -.    ..    ..    +.    +.    +.    ..    ..  ❶

      + is > 2*std error,  - is < -2*std error,  . is between


              Schematic Representation of Partial Autocorrelations

   Name/Lag      1     2     3     4     5     6     7     8     9    10

   RMINK         +.    ..    ..    ..    ..    ..    ..    ..    ..    ..
   RMUSKRAT      -+    ..    ..    ..    ..    ..    ..    ..    ..    ..  ❷

      + is > 2*std error,  - is < -2*std error,  . is between


                          Yule-Walker Estimates
                             for Minimum AIC

                            --------Lag=1-------
                              RMINK     RMUSKRAT

                RMINK        0.568749    0.298262  ❸
                RMUSKRAT    -0.46839     0.627158

              HUDSON'S BAY FUR TRAPPING RECORDS 1842-1890
                      RESIDUALS FROM LINEAR TREND
                       The STATESPACE Procedure
              Selected Statespace Form and Preliminary Estimates

                          ❺  State Vector

              RMINK(T;T)               RMUSKRAT(T;T)


                       Estimate of Transition Matrix

                         0.568749       0.298262
                  ❹     -0.46839        0.627158


                        Input Matrix for Innovation

                              1              0
                              0              1
```

Output 6.7 *Using PROC REG to Detrend the Data and PROC STATESPACE to Analyze the Residuals (continued)*

```
                    Variance Matrix for Innovation

                        0.079131      0.002703
                        0.002703      0.063072


        Maximum likelihood estimation has converged.

                HUDSON'S BAY FUR TRAPPING RECORDS 1842-1890
                        RESIDUALS FROM LINEAR TREND
                         The STATESPACE Procedure
                 Selected Statespace Form and Fitted Model

                              State Vector

                 RMINK(T;T)             RMUSKRAT(T;T)


                    Estimate of Transition Matrix

                        0.568749      0.298262
                        -0.46839      0.627158


                     Input Matrix for Innovation

                               1             0
                               0             1


                    Variance Matrix for Innovation

                        0.079131      0.002703
                        0.002703      0.063072


                         Parameter Estimates

                                      Standard        ❻
                Parameter   Estimate      Error    t Value

                F(1,1)      0.568749   0.109253       5.21
                F(1,2)      0.298262   0.101203       2.95
                F(2,1)      -0.46839   0.097544      -4.80
                F(2,2)      0.627158   0.090356       6.94
```

If you had (mistakenly) decided to fit a transfer function, you could have fit an AR(2) to the mink series and computed the prewhitened cross-correlations. You observe a somewhat subtle warning in the cross-correlations plot—namely, that there are nonzero correlations at both positive and negative lags as shown in **Output 6.8**.

Output 6.8 *Cross-Correlations*

```
                      HUDSON'S BAY FUR TRAPPING RECORDS 1842-1890

                                The ARIMA Procedure

                                Cross-Correlations

Lag   Covariance    Correlation   -1 9 8 7 6 5 4 3 2 1 0 1 2 3 4 5 6 7 8 9 1

-10    0.0023812      0.02848     |                .    |*     .               |
 -9    -0.015289      -.18288     |                . ****|      .               |
 -8    -0.033853      -.40494     |              ********|      .               |
 -7   -0.0089179      -.10668     |                .   **|      .               |
 -6   -0.0012433      -.01487     |                .     |      .               |
 -5   -0.0010694      -.01279     |                .     |      .               |
 -4    0.0089324      0.10685     |                .     |**    .               |
 -3     0.018517      0.22149     |                .     |****  .               |
 -2   -0.0007918      -.00947     |                .     |      .               |
 -1     0.022822      0.27299     |                .     |*****.               |
  0     0.010153      0.12145     |                .     |**    .               |
  1    -0.040174      -.48055     |            **********|      .               |
  2   -0.0095530      -.11427     |                .   **|      .               |
  3   -0.0090755      -.10856     |                .   **|      .               |
  4   -0.0001946      -.00233     |                .     |      .               |
  5    0.0048362      0.05785     |                .     |*     .               |
  6     0.012202      0.14596     |                .     |***   .               |
  7     0.013651      0.16329     |                .     |***   .               |
  8     0.010270      0.12284     |                .     |**    .               |
  9     0.015405      0.18427     |                .     |****  .               |
 10   -0.0067477      -.08071     |                .   **|      .               |

                        "." marks two standard errors


          Both variables have been prewhitened by the following filter:

                                Prewhitening Filter

                               Autoregressive Factors

                  Factor 1:  1 - 0.78452 B**(1) + 0.29134 B**(2)
```

Chapter 7 Spectral Analysis

7.1 Periodic Data: Introduction

The modeling of time series data using sinusoidal components is called *spectral analysis*. The main tool here is the *periodogram*. A very simple model appropriate for spectral analysis is a mean plus a sinusoidal wave plus white noise:

$$Y_t = \mu + \alpha(\sin(\omega t + \delta)) + e_t = \mu + \alpha(\sin(\delta)\cos(\omega t) + \cos(\delta)\sin(\omega t)) + e_t$$

where the formula, $\sin(A+B) = \cos(A)\sin(B) + \sin(A)\cos(B)$ for the sine of the sum of two angles, has been applied. The function $\mu + \alpha(\sin(\omega t + \delta))$ oscillates between $\mu - \alpha$ and $\mu + \alpha$ in a smooth and exactly periodic fashion. The number α is called the *amplitude*. The number δ, in radians, is called the *phase shift* or *phase angle*. The number ω is called the *frequency* and is also measured in radians. If an arc of length *r* is measured along the circumference of a circle whose radius is *r*, then the angle obtained by connecting the arc's ends to the circle center is one radian. There are 2π radians in a full 360-degree circle, and one radian is thus $360/(2\pi) = 360/6.2832 = 57.3$ degrees. A plot of $\mu + \alpha(\sin(\omega t + \delta))$ versus *t* is a sine wave that repeats every $2\pi/\omega$ time units; that is, the *period* is $2\pi/\omega$. A sinusoid of period 12 would "go through" $\omega = 2\pi/12 = 0.52$ radians per observation.

Letting $A = \alpha\sin(\delta)$ and $B = \alpha\cos(\delta)$, we see that

$$Y_t = \mu + A\cos(\omega t) + B\sin(\omega t) + e_t$$

This is a very nice expression in that, if ω is known, variables $\sin(\omega t)$ and $\cos(\omega t)$ can be constructed in a DATA step and the parameters μ, A, and B can be estimated by ordinary least squares as in PROC REG. From the expressions for A and B it is seen that $B/A = \tan(\delta)$ and $\sqrt{A^2 + B^2} = \alpha\sqrt{\cos^2(\delta) + \sin^2(\delta)} = \alpha$, so phase angle and amplitude estimates can be constructed from estimates of A and B.

7.2 Example: Plant Enzyme Activity

As an example, Chiu-Yueh Hung, in the Department of Genetics at North Carolina State University, collected observations on leaf enzyme activity Y every 4 hours over 5 days. There are 6 observations per day and 30 observations in all. Each observation is an average of several harvested leaves. The researcher anticipated a 12-hour enzyme cycle, which corresponds to 3 observations. To focus this discussion on periodic components, the original data have been detrended using linear regression.

First read in the data, creating the sine and cosine variables for a period 3 (frequency $2\pi/3$ cycles per observation), and then regress Y on these two variables.

```
DATA PLANTS;
   TITLE "ENZYME ACTIVITY";
   TITLE2 "(DETRENDED)";
   DO T=1 TO 30;   INPUT Y @@; PI=3.1415926;
      S1=SIN(2*PI*T/3); C1=COS(2*PI*T/3);
      OUTPUT;
   END;
CARDS;
265.945    290.385    251.099    285.870    379.370    301.173
283.096    306.199    341.696    246.352    310.648    276.348
234.870    314.744    261.363    321.780    313.289    253.460
307.988    303.909    284.128    252.886    317.432    287.160
213.168    308.458    296.351    283.666    333.544    316.998
;
RUN;
PROC REG DATA=PLANTS;
   MODEL Y = S1 C1/SS1;
   OUTPUT OUT=OUT1 PREDICTED=P RESIDUAL=R;
RUN;
```

The analysis of variance table is shown in **Output 7.1**.

Output 7.1
Plant Enzyme Sinusoidal Model

The REG Procedure
Model: MODEL1
Dependent Variable: Y

Analysis of Variance

Source	DF	Sum of Squares	Mean Square	F Value	Pr > F
Model	2	11933	5966.44591	7.09	0.0034
Error	27	22724	841.62128		
Corrected Total	29	34657			

Root MSE	29.01071	R-Square	0.3443
Dependent Mean	291.44583	Adj R-Sq	0.2957
Coeff Var	9.95407		

Parameter Estimates

Variable	DF	Parameter Estimate	Standard Error	t Value	Pr > |t|	Type I SS
Intercept	1	291.44583	5.29661	55.03	<.0001	2548220
s1	1	-27.84889	7.49053	-3.72	0.0009	11633
c1	1	-4.46825	7.49053	-0.60	0.5558	299.47884

The sum of squares for the intercept is $n\overline{Y}^2 = 30(291.44583^2) = 2548220$, and the sum of squares for the model, which is the sum of squares associated with frequency $\omega = 2\pi/3$, is 11933 and has 2 degrees of freedom. It is seen to be statistically significant based on the F test, $F = 7.09$ ($P = .0034$). It appears that the sine term is significant but not the cosine term; however, such a splitting of the two degree of freedom sum of squares is not meaningful in that, if $t = 0$ had been used as the first time index rather than $t = 1$, both would have been significant. The sum of squares 11933 would *not* change with any such time shift. The sum of squares 11933 associated with frequency $\omega = 2\pi/3$ is called the *periodogram ordinate* at that frequency. A given set of data may have important fluctuations at several frequencies. **Output 7.2** shows the actual and fitted values for the plant enzyme data.

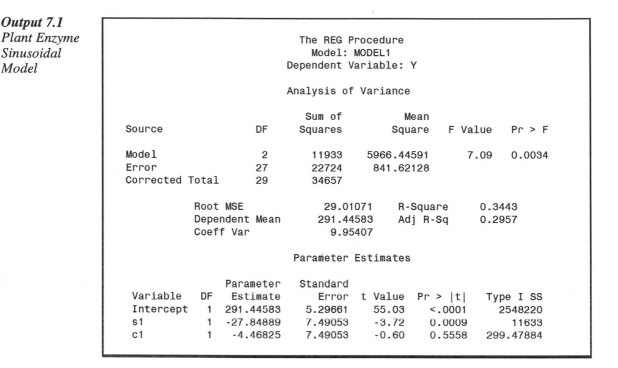

Output 7.2 *Data and Sinusoidal Predictions*

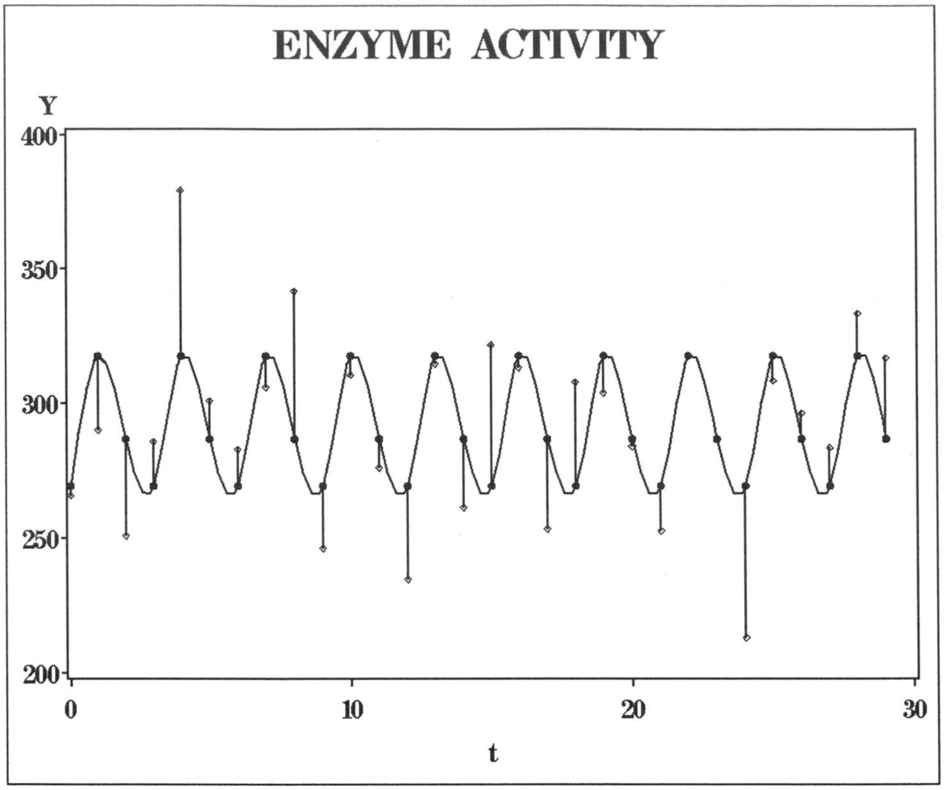

7.3 PROC SPECTRA Introduced

Periodogram ordinates are calculated for a collection of frequencies known as the *Fourier frequencies*. With $2m+1$ observations Y, there are m of these, each with 2 degrees of freedom, so that a multiple regression of Y on the $2m$ sine and cosine columns fits the data perfectly; that is, there are no degrees of freedom for error. The Fourier frequencies are $(2\pi j/n)$, where j runs from 1 to m. For each j, two columns $\sin(2\pi jt/n)$ and $\cos(2\pi jt/n)$, $t=1,2,\ldots n$, are created. The model sum of squares, when the data are regressed on these two columns, is the jth *periodogram ordinate*. At the jth Fourier frequency, the sine and cosine run through j cycles in the time period covered by the data. If $n=2m$, an even number, there are still m periodogram ordinates and j still runs from 1 to m, but when $j=m$ the frequency $2\pi j/n$ becomes $2\pi m/(2m)=\pi$ and $\sin(\pi t)=0$. Thus for even n, the last Fourier frequency has only one degree of freedom associated with it, arising from the cosine term, $\cos(\pi t)=(-1)^t$, only. It does not matter whether a multiple regression using all the Fourier sine and cosine columns or m bivariate regressions, one for each j, are run. The columns are all orthogonal to each other and the sums of squares (periodogram ordinates) are the same either way.

PROC SPECTRA calculates periodogram ordinates at all the Fourier frequencies. With the 30 plant enzyme measurements there are 15 periodogram ordinates, the last having 1 degree of freedom and the others 2 each. Since $2\pi 10/30=2\pi/3$, the Fourier frequency for $j=10$ should have periodogram ordinate equal to the previously computed model sum of squares, 11933. You might expect the other periodogram ordinates to add to 22724, the error sum of squares. However, PROC SPECTRA associates twice the correction term, $2n\overline{Y}^2 = 5096440$, with frequency 0 and twice the sum of squares at frequency π (when n is even) with that frequency, so one must divide the frequency π ordinate by 2 to get its contribution to the error sum of squares from regression. This doubling is done here because, after doubling some, division of all ordinates by 2 becomes the same as dividing unadjusted numbers by their degrees of freedom. The frequency 0 ordinate is replaced with 0 when the option ADJMEAN is used in PROC SPECTRA. These ideas are illustrated here for the plant enzyme data:

```
PROC SPECTRA DATA=PLANTS OUT=OUT2 COEFF;
   VAR Y;
RUN;
DATA OUT2; SET OUT2; SSE = P_01;
   TITLE J=CENTER "ENZYME DATA";
   IF PERIOD=3 OR PERIOD=. THEN SSE=0;
   IF ROUND (FREQ, .0001) = 3.1416 THEN SSE = .5*P_01;
RUN;
PROC PRINT DATA=OUT2;
   SUM SSE;
RUN;
```

The option COEFF in PROC SPECTRA adds the regression coefficients (cos_01 and sin_01) to the data. Looking at the period 3 line of **Output 7.3**, you see the regression sum of squares $11933 = \text{P\_01}$, which matches the regression output. The coefficients $A = -21.88$ and $B = 17.79$ are those that would have been obtained if time t had been labeled as $0,1,\ldots,29$ (as PROC SPECTRA does) instead of $1,2,\ldots,30$. Any periodogram ordinate with 2 degrees of freedom can be computed as $(n/2)(A^2 + B^2)$, where A and B are its Fourier coefficients. You see that $(30/2)((-21.884)^2 + (17.7941)^2) = 11933$. (See **Output 7.3**.)

Output 7.3 OUT Data Set from PROC SPECTRA

Enzyme activity
(detrended)

Obs	FREQ	PERIOD	COS_01	SIN_01	P_01	SSE
1	0.00000	.	582.892	0.0000	5096440.43	0.00
2	0.20944	30.0000	5.086	6.3073	984.78	984.78
3	0.41888	15.0000	1.249	6.6450	685.74	685.74
4	0.62832	10.0000	-6.792	-11.9721	2841.87	2841.87
5	0.83776	7.5000	-1.380	-9.6111	1414.15	1414.15
6	1.04720	6.0000	-10.035	-7.5685	2369.85	2369.85
7	1.25664	5.0000	0.483	-9.3467	1313.91	1313.91
8	1.46608	4.2857	4.288	-1.0588	292.58	292.58
9	1.67552	3.7500	17.210	-2.7958	4560.07	4560.07
10	1.88496	3.3333	-6.477	-0.1780	629.66	629.66
11	2.09440	3.0000	-21.884	17.7941	11932.89	0.00
12	2.30383	2.7273	-5.644	-2.7635	592.35	592.35
13	2.51327	2.5000	6.760	13.8749	3573.13	3573.13
14	2.72271	2.3077	-9.830	-10.0278	2957.94	2957.94
15	2.93215	2.1429	-0.022	-5.4249	441.45	441.45
16	3.14159	2.0000	2.973	0.0000	132.60	66.30
						========
						22723.78

PROC SPECTRA automatically creates the column FREQ of Fourier frequencies equally spaced in the interval 0 to π and the column PERIOD of corresponding periods. It is customary to plot the periodogram versus frequency or period, omitting frequency 0.

Output 7.4 *Periodogram with a Single Important Frequency*

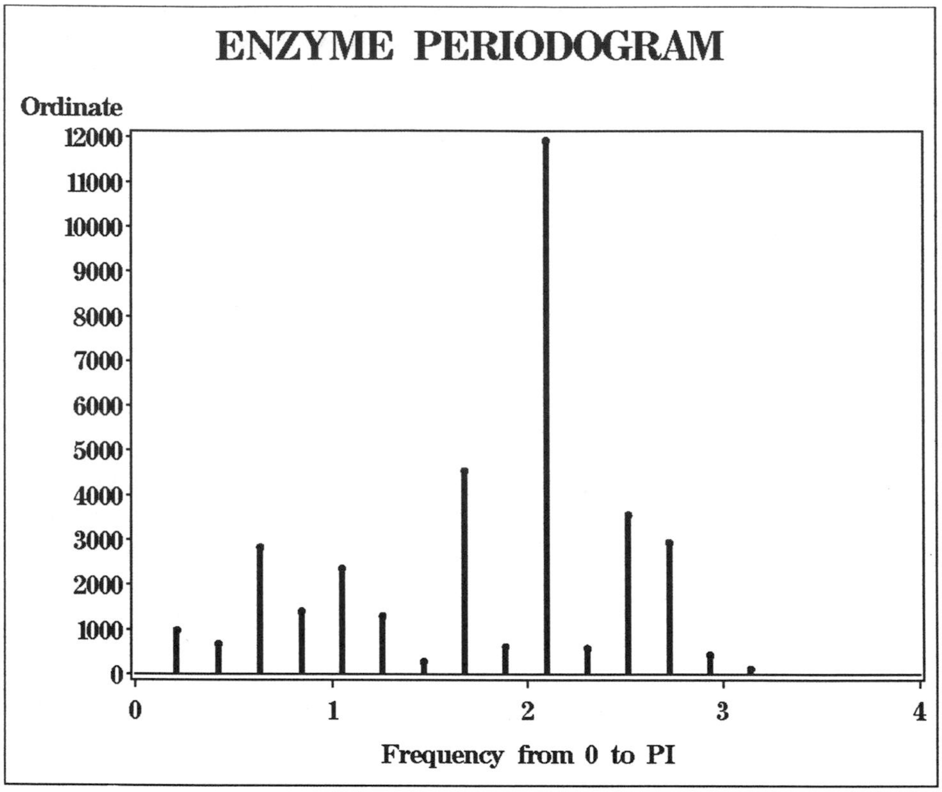

Output 7.4 shows the unusually large ordinate 11933 at the anticipated frequency of one cycle per 12 hours—that is, one cycle per 3 observations. The researcher was specifically looking for such a cycle and took sufficient observations to make the frequency of interest a Fourier frequency. If the important frequency is not a Fourier frequency, the periodogram ordinates with frequencies *near* the important one will be large. Of course, by creating their own sine and cosine columns, researchers can always investigate any frequency using regression. The beauty of the Fourier frequencies is the orthogonality of the resulting collection of regression columns (sine and cosine functions).

7.4 Testing for White Noise

For a normal white noise series with variance σ^2, the periodogram ordinates are independent and, when divided by σ^2, have chi-square distributions with 2 degrees of freedom (df). These properties lead to tests of the white noise null hypothesis.

You are justified in using an F test for the single sinusoid plus white noise model when the appropriate ω is known in advance, as in **Section 7.2**. You would *not* be justified in testing the largest observed ordinate (just because it is the largest) with F. If you test for a *period 3* component in multiple sets of white noise data (your null hypothesis), the F test statistic will have an F distribution. However, if you always test the *largest* ordinate whether or not it occurs at period 3, then this new F statistic will never be less than the F for period 3 and will usually be larger. Clearly this new "F" statistic cannot have the same F distribution.

Fisher computed the distribution for the largest periodogram ordinate divided by the mean of all the 2 df ordinates under the white noise null hypothesis. In the plant enzyme data, omission of the 1 df ordinate 132.6 gives *Fisher's kappa test* statistic $11933/[(22723.8-132.6/2+11933)/14]=4.83$. Fuller (1996) discusses this test along with the *cumulative periodogram test*. The latter uses C_k, which is the ratio of the sum of the first k periodogram ordinates to the sum of all the ordinates (again dropping any 1 df ordinate). The set of these C_k should behave like an ordered sample from a uniform distribution if the data are white noise. Therefore a standard distributional test, like those in PROC UNIVARIATE, can be applied to these cumulative C_k ratios, resulting in a test of the white noise null hypothesis. Traditionally the Kolmogorov-Smirnov test is applied. (See Fuller, page 363, for more details.)

Interpolating in Fuller's table of critical values for Fisher's kappa with 14 ordinates gives 4.385 as the 10% and 4.877 as the 5% critical value. Our value 4.83 is significant at 10% but not quite at 5%. Therefore, if you were just searching for a large ordinate rather than focusing from the start on a 12-hour cycle, your evidence for a 12-hour cycle would be nowhere near as impressive. This illustrates the increase in statistical power that can be obtained when you know something about your subject matter. You obtain both white noise tests using the WHITETEST option, as shown in **Output 7.5**.

```
PROC SPECTRA DATA=PLANTS WHITETEST;
   VAR Y;
RUN;
```

Output 7.5 Periodogram-Based White Noise Tests

```
                    The SPECTRA Procedure

           Test for White Noise for Variable Y

                M-1                    14
                Max(P(*))        11932.89
                Sum(P(*))        34590.36

        Fisher's Kappa: (M-1)*Max(P(*))/Sum(P(*))

                    Kappa     4.829682

         Bartlett's Kolmogorov-Smirnov Statistic:
      Maximum absolute difference of the standardized
      partial sums of the periodogram and the CDF of a
              uniform(0,1) random variable.

     Test Statistic                             0.255984
```

For 14 periodogram ordinates, tables of the Kolmogorov-Smirnov (K-S) statistic indicate that a value larger than about 0.36 would be needed for significance at the 5% level so that 0.256 is not big enough. Fisher's test is designed to detect a single sinusoid buried in white noise and so would be expected to be more powerful under the model proposed here than the K-S test, which is designed to have some power against any departure from white noise.

7.5 Harmonic Frequencies

Just because a function is periodic does not necessarily mean it is a pure sinusoid. For example, the sum of a sinusoid of period *k* and another of period $k/2$ is a periodic function of period *k* but is not expressible as a single sinusoid. On the other hand, any periodic function of period *k* defined on the integers can be represented as the sum of sinusoids of period *k*, $k/2$, $k/3$, etc. For a fundamental period *k*, periods k/j for $j = 2,3,\ldots$ are called "harmonics." Harmonics affect the wave shape but not the period. A period of 2 is the shortest period detectable in a periodogram, and its associated frequency, π, is sometimes called the *Nyquist frequency*. Thus the plant enzyme measurements were not taken frequently enough to investigate harmonics of the fundamental frequency $2\pi/3$ (period 3). Even the first harmonic has period $3/2<2$ and frequency $4\pi/3$, which exceeds the Nyquist frequency π.

To further illustrate the idea of harmonics, imagine $n=36$ monthly observations where there is a fundamental frequency $2\pi/12$ and possibly contributions from the harmonic frequencies $2(2\pi)/12$ and $3(2\pi)/12$ plus white noise. To fit the model you create three sine and three cosine columns. The sine column for the fundamental frequency would have tth entry $\sin(2\pi t/12)$ and would go through 3 cycles in 36 observations. Now look at **Output 7.6**.

Output 7.6
Fundamental and Harmonic Sinusoids

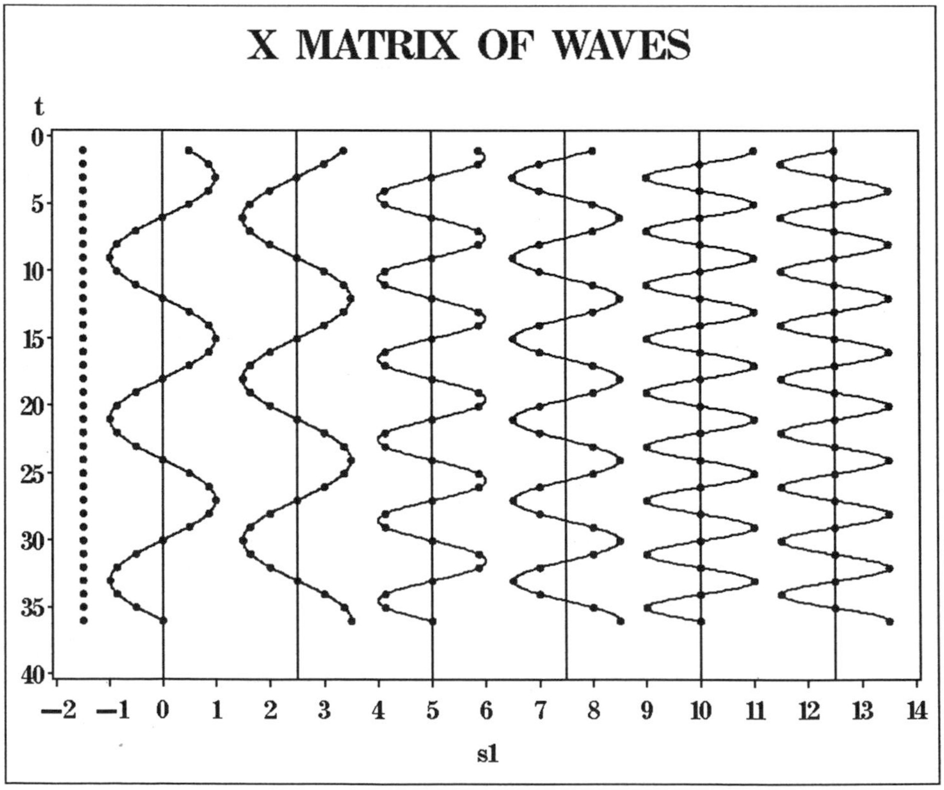

Output 7.6 is a schematic representation of the regression X matrix just described and is interpreted as follows. On the left, a vertical column of dots represents the intercept column, a column of 1s. Just to its right is a wave that represents $\cos(2\pi t/12)$, and to its right is another wave representing $\sin(2\pi t/12)$. Run your finger down one of these two waves. Your finger cycles between one unit left and one unit right of the wave center line. There are three cycles in each of these two columns. Writing the deviations of dots from centers as numbers supplies the entries of the corresponding column of the X matrix. The two waves, or columns of X, currently under discussion will have regression coefficients A_1, B_1. By proper choice of these, the regression will exactly fit any sinusoid of frequency $2\pi t/12$ regardless of its amplitude and phase.

Similar comments apply to the other two pairs of waves, but note that, as you run your finger down any of these, the left-to-right oscillation is faster and thus there are more cycles: $36/6=6$ for the middle pair and $36/4=9$ for the rightmost pair, where $12/2=6$ and $12/3=4$ are the periods corresponding to the two harmonic frequencies. Three more pairs of columns, with periodicities $12/4=3$, $12/5$, and $12/6=2$, fill out a full set of harmonics for a period 12 function measured at integer time points. They would add 6 more columns for a total of 12 waves, seeming to contradict the fact that a period 12 function has 11, not 12, degrees of freedom. However, at period $12/6=2$, the sine column becomes $\sin(2\pi t/2)=\sin(\pi t)=0$ for all *t*. Such a column of 0s would, of course, be omitted, leaving 11 columns (11 degrees of freedom) plus an intercept column associated with the period 12 function. If 36 consecutive observations from any period 12 function were regressed on this 12 column X matrix, the fit would be perfect at the observed points but would not necessarily interpolate well between them. A perfect fit *at the observation times* would result even if the sequence Y_t were repeated sets of six 1s followed by six -1 s. The fitted values would exactly match the observed -1,1 pattern at integer values of *t*, but interpolated values, say, at time $t=5.9$, would not be restricted to -1 or 1. One might envision the harmonics as fine-tuning the wave shape as you move up through the higher harmonic frequencies (shorter period fluctuations). This motivates the statistical problem of separating the frequencies that contribute to the true process from those that are fitting just random noise so that a good picture of the wave shape results. Periodograms and associated tests are useful here.

The following outputs are generated from a sinusoid of period $k=12$ plus another at the first harmonic, period $12/2=6$. Each sinusoid is the sum of a sine and cosine component, thus allowing an arbitrary phase angle. For interpolation purposes, sine and cosine terms are generated for *t* in increments of 0.1, but Y exists only at integer *t*.

Output 7.7
Increased Resolution Using Harmonics

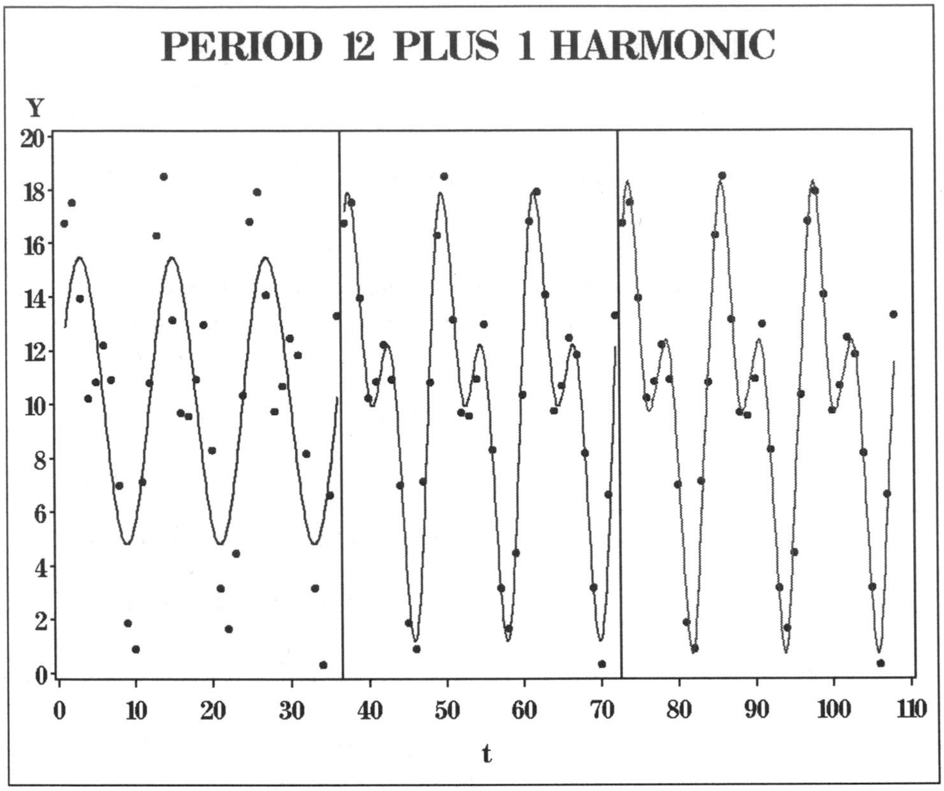

Output 7.7 shows three sets of fitted values. The sine and cosine at the fundamental frequency $\omega_1 = 2\pi/12$ are used to produce the fitted values on the left side. These fitted values do not capture the double peak in each interval of 12 time points, and they miss the low and high extremes. Including the first harmonic $\omega_2 = 2(2\pi/12)$ gives a better fit and gives an idea of what the data-generating function looks like. The fitted values on the right side are those coming from the fundamental and *all* harmonic frequencies $j(2\pi/12)$ for $j = 1,2,\ldots,6$, omitting the sine at $j = 6$. The minor wiggles there are due to the frequencies with $j > 2$. Adding all those extra parameters does not seem to have produced any useful new features in the fitted values. From PROC REG (see **Output 7.9**), the F test 1.53 ❶ for frequencies with $j = 3,4,\ldots,6$ is not significant, and the Type I sums of squares for $j = 1$ and 2 are large enough that neither the $j = 1$ nor $j = 2$ frequencies can be omitted. Recall that you would not eliminate just a sine or cosine; they are treated as pairs. Rearrangement of terms or deletion of some terms would not affect the sums of squares here because the sine and cosine columns correspond to Fourier frequencies, so they are orthogonal to each other.

The following PROC SPECTRA code is used to generate **Output 7.8** and **Output 7.10**.

```
PROC SPECTRA DATA=COMPRESS P S ADJMEAN OUT=OUTSPECTRA;
   VAR Y;
   WEIGHTS 1 2 3 4 3 2 1;
RUN;
```

Output 7.8
Periodogram with Two Independent Frequencies

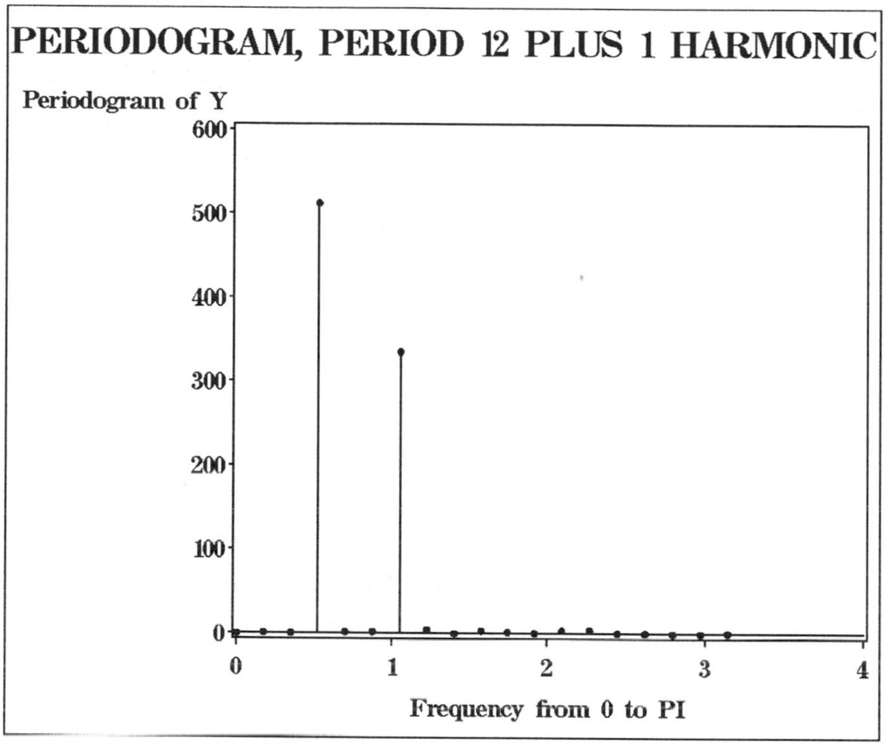

The periodogram, shown in **Output 7.8**, makes it quite clear that there are two dominant frequencies, $2\pi/12$ and its first harmonic, $4\pi/12$. The last few lines of the program deliver a smoothed version of the periodogram, shown in **Output 7.10**, that will be discussed in **Section 7.9**. Smoothing is not helpful in this particular example.

Output 7.9
Regression Estimates and F Test

Parameter Estimates

Variable	DF	Parameter Estimate	Standard Error	t Value	Pr > |t|	Type I SS
Intercept	1	10.13786	0.14369	70.55	<.0001	3699.94503
s1	1	5.33380	0.20321	26.25	<.0001	512.09042
c1	1	0.10205	0.20321	0.50	0.6201	0.18745
s2	1	3.90502	0.20321	19.22	<.0001	274.48586
c2	1	1.84027	0.20321	9.06	<.0001	60.95867
s3	1	-0.33017	0.20321	-1.62	0.1173	1.96220
s4	1	-0.41438	0.20321	-2.04	0.0526	3.09082
s5	1	-0.17587	0.20321	-0.87	0.3954	0.55672
c3	1	-0.27112	0.20321	-1.33	0.1947	1.32314
c4	1	-0.16725	0.20321	-0.82	0.4186	0.50351
c5	1	-0.02326	0.20321	-0.11	0.9098	0.00974
c6	1	0.11841	0.14369	-0.82	0.4180	0.50473

Test Harmonics Results for Dependent Variable Y

Source	DF	Mean Square	F Value	Pr > F
Numerator	7	1.13584	❶ 1.53	0.2054
Denominator	24	0.74330		

Output 7.10
Smoothed Periodogram

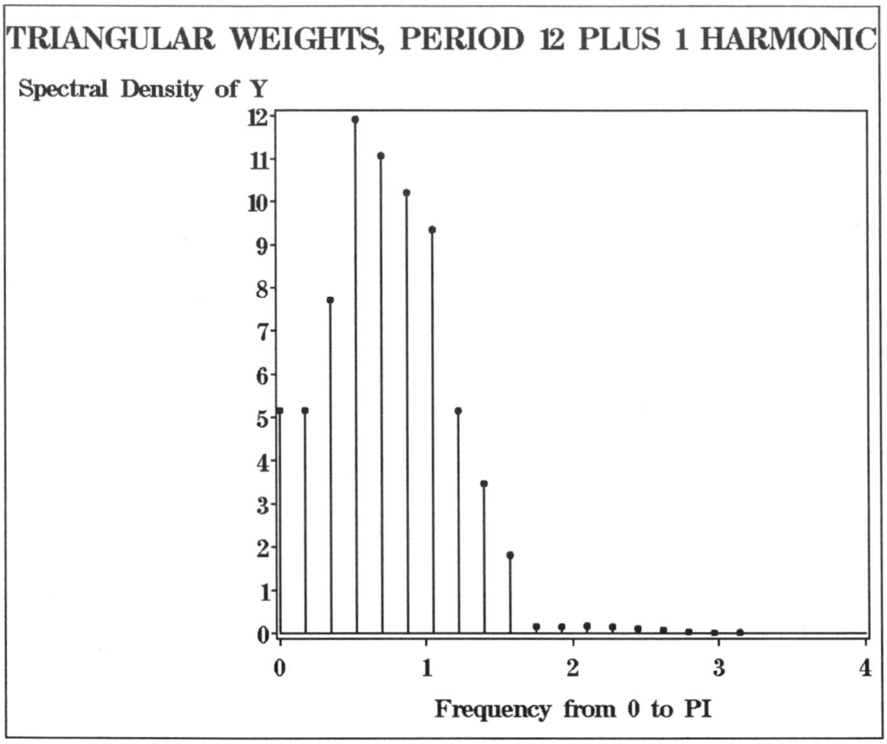

7.6 Extremely Fast Fluctuations and Aliasing

Suppose a series actually has a frequency larger (faster fluctuations) than the Nyquist frequency π radians per observation—for example, $4\pi/3 > \pi$. Imagine a wheel with a dot on its edge, and an observer who looks at the wheel each second. If the wheel rotates clockwise $4\pi/3$ radians per second, at the first observation, the dot will now be $2\pi/3$ radians *counterclockwise*—i.e., $-2\pi/3$ radians—from its previous position, and similarly for subsequent observations. Based on the dot's position, the observer only knows that the frequency of rotation is $-2\pi/3 + 2\pi j$ for some integer j. These frequencies are all said to be *aliased* with $-2\pi/3$ where this frequency was selected because it is in the interval $[-\pi, \pi]$. Another alias will be seen to be $2\pi/3$, as though the observer had moved to the other side of the wheel.

Because $A\cos(\omega t)+B\sin(\omega t)=A\cos(-\omega t)-B\sin(-\omega t)$, it is not possible to distinguish a cycle of frequency ω from one of $-\omega$ using the periodogram. Thus it is sufficient and customary to compute periodogram ordinates at the Fourier frequencies $2\pi j/n$ with $j=1,2,\ldots,m$ so that $0\le 2\pi j/n\le\pi$. Recall that the number of periodogram ordinates m is either $(n-1)/2$ if n is odd or $n/2$ if n is even.

Imagine a number line with reference points at πj for all integers j, positive, negative, and zero. Folding that line back and forth in accordion fashion at these reference points maps the whole line into the interval $[0,\pi]$. The set of points that map into any ω are its aliases. For that reason, the Nyquist frequency π is also referred to as the *folding frequency*. The reason that this frequency has names instead of always being called π is that some people prefer radians or cycles per second, per hour, per day, etc., rather than radians per observations as a unit of measure. If observations are taken every 15 minutes, the Nyquist frequency π radians per observation would convert to 4π radians per hour, or 2 cycles per hour. In this book, radians per observation and the Nyquist frequency π will be the standard.

When the periodogram is plotted over $[0,\pi]$ and there appears to be a cycle at a bizarre frequency in $[0,\pi]$, ask yourself if this might be coming from a cycle beyond the Nyquist frequency.

7.7 The Spectral Density

Consider three processes: $W_t=10+(5/3)e_t$, $Y_t=10+.8(Y_{t-1}-10)+e_t$, and $Z_t=10-.8(Z_{t-1}-10)+e_t$, where $e_t\sim N(0,0.36)$ is white noise. Each process has mean 10 and variance 1.

The spectral density function of a process is defined as $f(\omega)=\frac{1}{2\pi}\sum_{h=-\infty}^{\infty}\gamma(h)\cos(\omega h)$, where $\gamma(h)$ is the autocovariance function. The function is symmetric: $f(\omega)=f(-\omega)$. For W_t the variance is $\gamma(0)=\sigma^2=1$ and $\gamma(h)=0$ if h is not 0. The spectral density for W_t becomes just

$f_W(\omega)=\frac{1}{2\pi}\sum_{h=-\infty}^{\infty}\gamma(h)\cos(\omega h)=\frac{1}{2\pi}\gamma(0)\cos(0)=\frac{1}{2\pi}$ and is $\frac{\sigma^2}{2\pi}$ for a general white noise series with variance σ^2. Sometimes the spectral density is plotted over the interval $-\pi\le\omega\le\pi$. Since for white noise, $f(\omega)$ is $\frac{\sigma^2}{2\pi}$, the plot is just a rectangle of height $\frac{\sigma^2}{2\pi}$ over an interval of width 2π. The area of the rectangle, $2\pi\frac{\sigma^2}{2\pi}=\sigma^2$, is the variance of the series and this (area = variance) will be true in general of a spectral density whether or not it is plotted as a rectangle.

Because the plot of $f_W(\omega)$ has equal height at each ω, it is said that all frequencies contribute equally to the variance of W_t. This is the same idea as white light, where all frequencies of the light spectrum are equally represented, or white noise in acoustics. In other words the time series is conceptualized as the sum of sinusoids at various frequencies with white noise having equal contributions for all frequencies. In general, then, the interpretation of the spectral density is the decomposition of the variance of a process into components at different frequencies.

An interesting mathematical fact is that if the periodogram is computed for data from any ARMA model, the periodogram ordinate at any Fourier frequency $0 < \omega < \pi$ estimates $4\pi f(\omega)$; that is, $4\pi f(\omega)$ is (approximately) the periodogram ordinate's *expected value*. Dividing the periodogram ordinate by 4π thus gives an almost unbiased estimate of $f(\omega)$. If the plot over $-\pi \leq \omega \leq \pi$ is desired (so that the area under the curve is the variance), use the symmetry of $f(\omega)$ and plot the estimate at both ω and $-\omega$. For white noise, of course, $f(\omega)$ estimates the same thing $(\sigma^2/2\pi)$ at each ω, and so averaging several $f(\omega)$ values gives an even better estimate. You will see that local averaging of estimates often, but not always, improves estimation. Often only the positive frequency half of the estimated spectral density is plotted, and it is left to the reader to remember that the variance is twice the area of such a plot.

What do the spectral densities of Y_t and Z_t look like? Using a little intuition, you would expect the positively autocorrelated series Y_t to fluctuate at a slower rate around its mean than does W_t. Likewise you would expect the negatively autocorrelated series Z_t to fluctuate faster than W_t since, for Z_t, positive deviations tend to be followed by negative and negative by positive. The slower fluctuation in Y_t should show up as longer period waves—that is, higher periodogram ordinates at low frequencies. For Z_t you'd expect the opposite—large contributions to the variance from frequencies near $-\pi$ or π.

The three graphs at the top of **Output 7.11** show the symmetrized periodograms for W, Y, and Z each computed from 1000 simulated values and having each ordinate plotted at the associated ω and its negative to show the full symmetric spectrum. The behavior is as expected—high values near $\omega = 0$ indicating low-frequency waves in Y_t, high values near the extreme ωs for Z_t indicating high-frequency fluctuations, and a flat spectrum for W_t. Two other periodograms are shown. The first, in the bottom-left corner, is for $D_t = Y_t - Y_{t-1}$. Because D_t is a moving linear combination of Y_t values, D_t is referred to as a *filtered* version of Y_t. Note that if the filter $Y_t - .8Y_{t-1}$ had been applied, the filtered series would just be white noise and the spectral density just a horizontal line. It is seen that linear filtering of this sort is a way of altering the spectral density of a process. The differencing filter has overcompensated for the autocorrelation, depressing the middle (near 0 frequency) periodogram ordinates of D_t too much so that instead of being level, the periodogram dips down to 0 at the middle.

In the wide middle panel of **Output 7.11**, a sinusoid $0.2\sin(2\pi t/25+.1)$ has been added to Y_t and the first 200 observations from both the original and altered series have been plotted. Because the amplitude of the sinusoid is so small, the plot of the altered Y_t is nearly indistinguishable from that of the original Y_t. The same is true of the autocorrelations (not shown). In contrast, the periodogram in the middle of the bottom row shows a strong spike at the Fourier frequency $2\pi/25 = 40(2\pi/1000) = 0.2513$ radians, clearly exposing the modification to Y_t. The middle graphs in the top (original Y_t) and bottom rows are identical except for the spikes at frequency ± 0.2513.

The bottom-right graph of **Output 7.11** contains plots of three smoothed spectral density estimates along with the theoretical spectral densities that they estimate. See **Section 7.8** for details about smoothing the periodogram to estimate the spectral density. The low horizontal line associated with white noise has been discussed already. For autoregressive order 1 series, AR(1) like Y and Z, the theoretical spectral density is $f(\omega) = \frac{\sigma^2}{2\pi}/(1+\rho^2 - 2\rho\cos(\omega))$, where ρ is the lag 1 autoregressive coefficient, .8 for Y and $-.8$ for Z. For a moving average (MA) such as $X_t = e_t - \theta e_{t-1}$, the spectral density is $f_X(\omega) = \frac{\sigma^2}{2\pi}(1+\theta^2 - 2\theta\cos(\omega))$. Both the AR and MA spectral densities involve the white noise spectral density $\frac{\sigma^2}{2\pi}$. It is either multiplied (MA) or divided (AR) by a trigonometric function involving the ARMA coefficients. Note that X_t is a filtered version of e_t. If X_t had been defined in terms of a more general time series V_t as $X_t = V_t - \theta V_{t-1}$, the spectral density of X_t would have been similarly related to that of V_t as $f_X(\omega) = f_V(\omega)(1+\theta^2 - 2\theta\cos(\omega))$, where $f_V(\omega)$ is the spectral density of V_t. As an example, if Y_t has spectral density

$$f_Y(\omega) = \frac{\sigma^2}{2\pi}/(1+\rho^2 - 2\rho\cos(\omega))$$

and is filtered to get $D_t = Y_t - \theta Y_{t-1}$, then the spectral density of D_t is

$$\begin{aligned} f_D(\omega) &= (1+\theta^2 - 2\theta\cos(\omega))f_Y(\omega) \\ &= \frac{\sigma^2}{2\pi}(1+\theta^2 - 2\theta\cos(\omega))/(1+\rho^2 - 2\rho\cos(\omega)) \end{aligned}$$

Output 7.11 *Spectral Graphics (see text)*

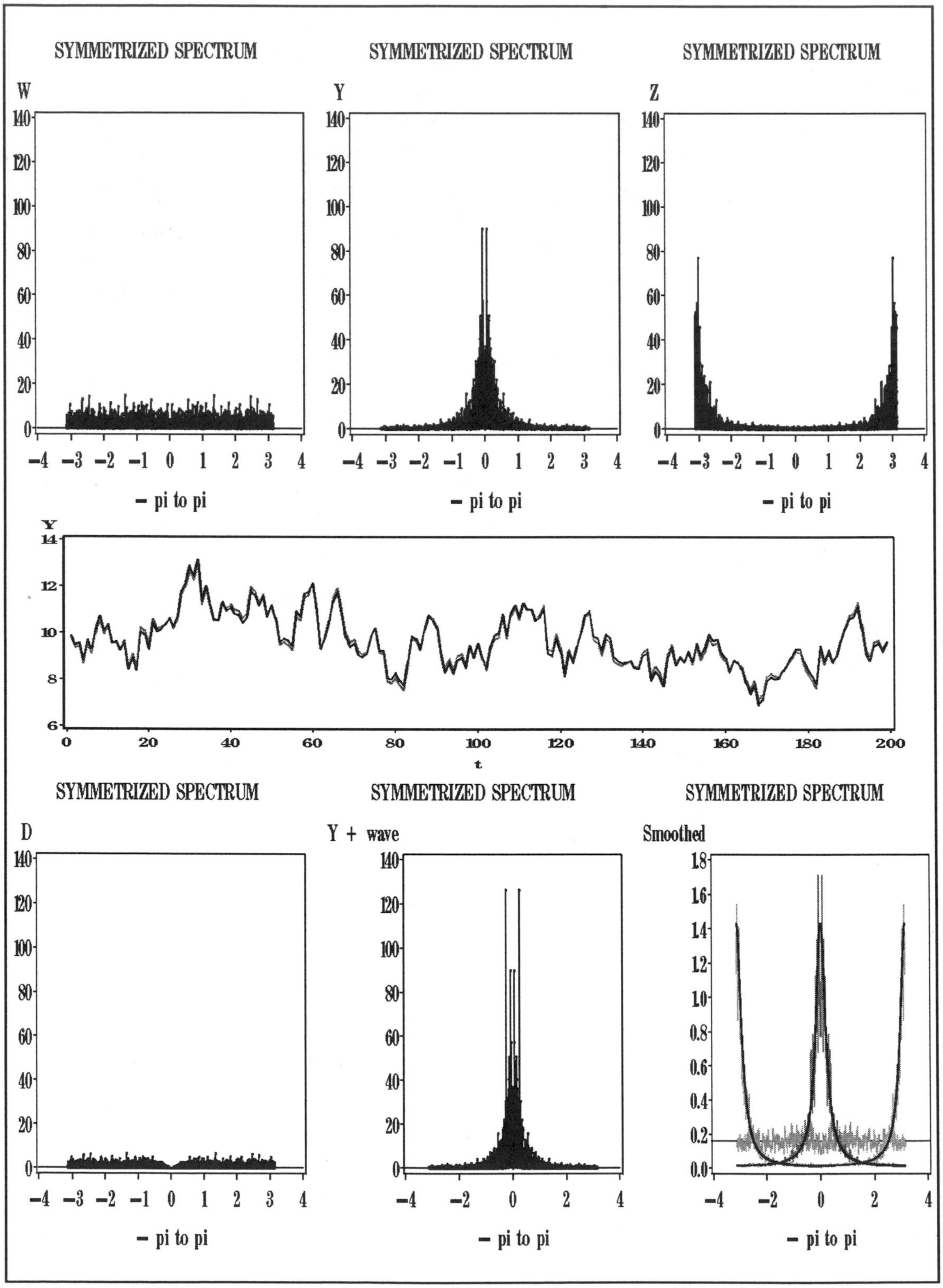

If $D_t = Y_t - Y_{t-1}$ then $\theta = 1$, so

$$f_D(\omega) = (2 - 2\cos(\omega))\frac{\sigma^2}{2\pi}/(1 + \rho^2 - 2\rho\cos(\omega))$$

which is seen to be 0 at frequency $\omega = 0$. This gives some insight into the behavior of the D_t periodogram displayed in the bottom-left corner of **Output 7.11**. Filtering affects different frequencies in different ways. The multiplier associated with the filter, such as $(1 + \theta^2 - 2\theta\cos(\omega))$ in the examples above, is sometimes called the *squared gain* of the filter in that amplitudes of some waves get increased ($\text{gain} > 1$) and some get reduced ($\text{gain} < 1$). Designing filters to amplify certain frequencies and reduce or eliminate others has been studied in some fields. The term *squared* gain is used because the spectral density decomposes the variance, not the standard deviation, into frequency components.

7.8 Some Mathematical Detail (Optional Reading)

The spectral densities for general ARMA processes can be defined in terms of complex exponentials $e^{i\omega} = \cos(\omega) + i\sin(\omega)$. Here i represents an imaginary number whose square is -1. Although that concept may be hard to grasp, calculations done with such terms often result ultimately in expressions not involving *i*, so the use of *i* along the way is a convenient mechanism for calculation of quantities that ultimately do not involve imaginary numbers.

Using the backshift operator B, the ARMA(p,q) model is expressed as

$$(1 - \alpha_1 B - \cdots - \alpha_p B^p)Y_t = (1 - \theta_1 B - \cdots - \theta_q B^q)e_t$$

You now understand that these expressions in the backshift can be correctly referred to as filters. Replace B^j with $e^{i\omega j}$, getting $A(\omega) = (1 - \alpha_1 e^{i\omega} - \alpha_2 e^{i\omega 2} - \cdots - \alpha_p e^{i\omega p})$ on the autoregressive and $M(\omega) = (1 - \theta_1 e^{i\omega} - \theta_2 e^{i\omega 2} - \cdots - \theta_q e^{i\omega q})$ on the moving average side. The complex polynomials $A(\omega)$ and $M(\omega)$ have corresponding *complex conjugate* expressions $A^*(\omega)$ and $M^*(\omega)$ obtained by replacing $e^{i\omega} = \cos(\omega) + i\sin(\omega)$ everywhere with $e^{-i\omega} = \cos(\omega) - i\sin(\omega)$. Start with the spectral density of e_t, which is $\frac{\sigma^2}{2\pi}$. The spectral density for the ARMA(p,q) process Y_t becomes $f_Y(\omega) = \frac{\sigma^2}{2\pi}\frac{M(\omega)M^*(\omega)}{A(\omega)A^*(\omega)}$. When a complex expression is multiplied by its complex conjugate, the product involves only real numbers and is positive.

If there are unit roots on the autoregressive side, the denominator, $A(\omega)A^*(\omega)$, will be zero for some ω and the theoretical expression for $f_Y(\omega)$ will be undefined there. Of course such a process does not have a covariance function $\gamma(h)$ that is a function of h only, so the spectral density $f(\omega) = \frac{1}{2\pi}\sum_{h=-\infty}^{\infty} \gamma(h)\cos(\omega h)$ cannot exist either. Despite the fact that $f(\omega)$ does not exist for unit root autoregressions, the periodogram can still be computed. Akdi and Dickey (1997) and Evans (1998) discuss normalization and distributional properties for the periodogram in this situation. Although they find the expected overall gross behavior (extremely large ordinates near frequency 0), they also find some interesting distributional departures from the stationary case. Unit roots on the moving average side are not a problem; they simply cause $f_Y(\omega)$ to be 0 at some ω values. An example of this is D_t, whose periodogram is shown in the bottom-left corner of **Output 7.11** and whose spectral density is 0 at $\omega = 0$.

7.9 Estimating the Spectrum: The Smoothed Periodogram

The graph in the bottom-right corner of **Output 7.11** contains theoretical spectral densities for W_t, Y_t, and Z_t as well as estimates derived from the periodogram plotted symmetrically over the full interval $-\pi \le \omega \le \pi$. These estimates are derived by locally smoothing the periodogram. Smoothed estimates are local weighted averages, and in that picture a simple average of 21 ordinates centered at the frequency of interest is taken and then divided by 4π. It is seen that these are good estimates of the theoretical spectral densities that are overlaid in the plot. The 4π divisor is used so that the area under the spectral density curve over $-\pi \le \omega \le \pi$ will be the series variance. Weighted averages concentrated more on the ordinates near the one of interest can also be used. In the PROC SPECTRA output data set, the spectral density estimates are named S_01, S_02, etc., and the periodogram ordinates are named P_01, P_02, etc., for variables in the order listed in the VAR statement.

Let $I_n(\omega_j)$ denote the periodogram ordinate at Fourier frequency $\omega_j = 2\pi j/n$ constructed from n observations on some time series Y_t. Suppose you issue these statements:

```
PROC SPECTRA P S ADJMEAN OUT=OUTSPEC;
   WEIGHTS 1 2 3 4 3 2 1;
   VAR X R Y;
RUN;
```

The smoothed spectral density estimate for Y will have variable name S_03 in the data set OUTSPEC and for $j > 2$ will be computed as

$\frac{1}{16}[I_n(\omega_{j-3}) + 2I_n(\omega_{j-2}) + 3I_n(\omega_{j-1}) + 4I_n(\omega_j) + 3I_n(\omega_{j+1}) + 2I_n(\omega_{j+2}) + I_n(\omega_{j+3})]/(4\pi)$, where the divisor 16 is the sum of the numbers in the WEIGHT statement. Modifications are needed for $j < 4$ and $j > m-3$, where m is the number of ordinates. **Output 7.10** shows the results of smoothing the **Output 7.8** periodogram. Note that much of the detail has been lost.

From the graphs in **Output 7.8** and **7.10**, it is seen that the sinusoids are indicated more strongly by the unsmoothed P_01 than by the smoothed spectrum S_01. That is because the smoothing spreads the effect of the sinusoid into neighboring frequencies where the periodogram concentrates it entirely on the true underlying Fourier frequency. On the other hand, when the true spectrum is fairly smooth, as with X, Y, Z, and D in **Output 7.11**, the estimator should be smoothed. This presents a dilemma for the researcher who is trying to discover the nature of the true spectrum: the best way to smooth the spectrum for inspection is not known without knowing the nature of the true spectrum, in which case inspecting its estimate is of no interest. To address this, several graphs are made using different degrees of smoothing. The less smooth ones reveal spikes and the more smooth ones reveal the shape of the smooth regions of the spectrum.

Dividing each periodogram ordinate by the corresponding spectral density $f(\omega)$ results in a set of almost independent variables, each with approximately (exactly if the data are normal white noise) a chi-square distribution with 2 degrees of freedom, a highly variable distribution. The weights applied to produce the spectral density lower the variance while usually introducing a bias. The set of weights is called a *spectral window,* and the effective number of periodogram ordinates involved in an average is called the *bandwidth* of the window. The estimated spectral density approximates a weighted average of the true spectral density in an interval surrounding the target frequency rather than just at the target frequency. The interval is larger for larger bandwidths and hence the resulting potential for bias increased, whereas the variance of the estimate is decreased by increasing the bandwidth.

7.10 Cross-Spectral Analysis

7.10.1 Interpreting Cross-Spectral Quantities

Interpreting cross-spectral quantities is closely related to the transfer function model in which an output time series, Y_t, is related to an input time series, X_t, through the equation

$$Y_t = \Sigma_{j=-\infty}^{\infty} v_j X_{t-j} + \eta_t$$

and where η_t is a time series independent of the input, X_t. For the moment, assume $\eta_t = 0$.

For example, let Y_t and X_t be related by the transfer function

$$Y_t - .8Y_{t-1} = X_t$$

Then

$$Y_t = \Sigma_{j=0}^{\infty} (.8)^j X_{t-j}$$

which is a weighted sum of current and previous inputs.

Cross-spectral quantities tell you what happens to sinusoidal inputs. In the example, suppose X_t is the sinusoid

$$X_t = \sin(\omega t)$$

where

$$\omega = 2\pi/12$$

Using trigonometric identities shows that Y_t satisfying

$$Y_t - .8Y_{t-1} = \sin(\omega t)$$

must be of the form

$$Y_t = A\sin(\omega t - B)$$

Solving

$$\begin{aligned} &A\sin(\omega t - B) - .8A\sin(\omega t - B - \omega) \\ &\quad = \left(A\cos(B) - .8A\cos(B+\omega)\right)\sin(\omega t) \\ &\qquad + \left(-A\sin(B) + .8A\sin(B+\omega)\right)\cos(\omega t) \\ &\quad = \sin(\omega t) \end{aligned}$$

you have

$$A\cos B - .8A\cos(B+\omega) = 1$$

and

$$-A\sin B + .8A\sin(B+\omega) = 0$$

The solution is

$$\tan(B) = .8\ \sin(\omega)/(1 - .8\ \cos(\omega)) = .4/\left(1-(.4)\sqrt{3}\right) = 1.3022$$

and

$$A = 1/\left\{.6091 - .8\left[.6091\left(\sqrt{3}/2\right) - .7931(1/2)\right]\right\} = 1.9828$$

The transfer function produces output with amplitude 1.9828 times that of the input; it has the same frequency and a phase shift of $\arctan(1.3022) = 52.5° = .92$ radians. These results hold only for $\omega = 2\pi/12$. The output for any noiseless linear transfer function is a sinusoid of frequency ω when the input X is such a sinusoid. Only the amplitude and phase are changed.

In cross-spectral analysis, using arbitrary input and its associated output, you simultaneously estimate the gain and phase at all Fourier frequencies. An intermediate step is the computation of quantities called cospectrum and quadrature spectrum.

The theoretical cross-spectrum, $f_{xy}(\omega)$, is the Fourier transform of the cross-covariance function, $\gamma_{xy}(h)$, where

$$\gamma_{xy}(h) = \mathrm{E}\{[\mathrm{X}_t - \mathrm{E}(X_t)][\mathrm{Y}_{t+h} - \mathrm{E}(\mathrm{Y}_t)]\}$$

The real part of $f_{xy}(\omega) = c(\omega) - iq(\omega)$ is the cospectrum, $c(\omega)$, and the imaginary part gives the quadrature spectrum, $q(\omega)$. In the example

$$\mathrm{Y}_t - .8\mathrm{Y}_{t-1} = \mathrm{X}_t$$

multiply both sides by X_{t-h} and take the expected value. You obtain

$$\gamma_{xy}(h) - .8\gamma_{xy}(h-1) = \gamma_{xx}(h)$$

where $\gamma_{xx}(h)$ is the autocorrelation function (ACF) for x.

Now when $\gamma(h)$ is absolutely summable,

$$f(\omega) = (2\pi)^{-1} \ \Sigma_{h=-\infty}^{\infty}\left(\gamma(h)e^{-i\omega h}\right)$$

From these last two equations

$$(2\pi)^{-1}\Sigma_{h=-\infty}^{\infty}\left(\gamma_{xy}(h)e^{-i\omega h} - .8\gamma_{xy}(h-1)e^{-i\omega h}\right)$$
$$= (2\pi)^{-1}\Sigma_{h=-\infty}^{\infty}\left(\gamma_{xx}(h)e^{-i\omega h}\right)$$

or

$$(2\pi)^{-1}\Sigma_{h=-\infty}^{\infty}\left(\gamma_{xy}(h)e^{\ i\omega h} - .8\gamma_{xy}(h-1)e^{-i\omega(h-1)}e^{-i\omega}\right)$$
$$= (2\pi)^{-1}\Sigma_{h=-\infty}^{\infty}\left(\gamma_{xx}(h)e^{-i\omega h}\right)$$

or

$$f_{xy}(\omega) - \ .8f_{xy}(\omega)e^{-i\omega} = f_{xx}(\omega)$$

However,

$$e^{-i\omega} = \cos(\omega) - i\sin(\omega)$$

so

$$f_{xy}(\omega)(1 - .8\cos(\omega) + .8i\sin(\omega)) = f_{xx}(\omega)$$

Multiplying and dividing the left side by the complex conjugate $(1-.8\cos(\omega)-.8i\sin(\omega))$, you obtain

$$f_{xy}(\omega) = (1 - .8\cos(\omega) - .8i\sin(\omega)) / (1.64 - 1.6\cos(\omega)) f_{xx}(\omega)$$

You then have the cospectrum of X by Y (that of Y by X is the same)

$$c(\omega) = f_{xx}(\omega)(1-.8\cos(\omega))/(1.64-1.6\cos(\omega))$$

and the quadrature spectrum of X by Y (that of Y by X is $-q(\omega)$)

$$q(\omega) = f_{xx}(\omega)\{.8\sin(\omega)/[1.64-1.6\cos(\omega)]\}$$

In **Output 7.12** (pp. 348–349) the cospectrum and quadrature spectrum of Y by X along with their estimates from PROC SPECTRA are graphed for the case

$$X_t = .5X_{t-1} + e_t$$

7.10.2 Interpreting Cross-Amplitude and Phase Spectra

The cross-amplitude spectrum is defined as

$$A_{xy}(\omega) = |f_{xy}(\omega)| = (c^2(\omega)+q^2(\omega))^{0.5}$$

In this example,

$$A_{xy}(\omega) = (1.64-1.6\cos(\omega))^{-0.5} f_{xx}(\omega)$$

The gain is defined as the amplitude divided by the spectral density of X, or

$$A_{xy}(\omega)/f_{xx}(\omega)$$

provided $f_{xx}(\omega) \neq 0$. Thus, the gain is the multiplier applied to the sinusoidal component of X at frequency ω to obtain the amplitude of the frequency ω component of Y in a noiseless transfer function, in our case $(1.64-1.6\cos(\omega))^{-0.5}$

The phase spectrum $\Psi_{xy}(\omega)$ of X by Y is defined as

$$\Psi_{xy}(\omega) = \arctan(q(\omega)/c(\omega))$$

and that of Y by X is $\arctan(-q(\omega)/c(\omega))$.

This is the phase difference between the output and input at frequency ω. In this example,

$$\Psi_{yx}(\omega) = \arctan\left\{-.8 \sin(\omega)/\left[1-.8 \cos(\omega)\right]\right\}$$

These cross-amplitude and phase spectra are graphed along with their estimates from PROC SPECTRA in **Output 7.12**. The graphs explain the effect of the transfer function on a sinusoidal input. Its amplitude is changed $(A_{xy}(\omega))$, and it undergoes a phase shift $(\Psi_{xy}(\omega))$. The graphs show how these changes are a function of frequency (ω). The cross-spectrum can be expressed as

$$f_{xy}(\omega) = A_{xy}(\omega)\exp(i\Psi_{xy}(\omega))$$

Transfer function relationships are not perfect (noiseless), so an error series is introduced into the model as

$$Y_t = \Sigma_{j=-\infty}^{\infty} v_j X_{t-j} + \eta_t$$

where η_t is uncorrelated with X_t. Now, in analogy to the correlation coefficient, the squared coherency is defined as

$$K_{xy}^2(\omega) = \left|f_{xy}(\omega)\right|^2 / \left(f_{xx}(\omega) f_{yy}(\omega)\right)$$

This measures the strength of the relationship between X and Y as a function of frequency. The spectrum $f_\eta(\omega)$ of η_t satisfies

$$f_\eta(\omega) = f_{yy}(\omega) - f_{xy}(\omega) f_{xx}^{-1}(\omega) f_{xy}(\omega) = f_{yy}(\omega)\left(1 - K_{xy}^2(\omega)\right)$$

To compute the theoretical coherency for the example, you need assumptions on X and η. Assume

$$X_t = .5X_{t-1} + e_t$$

with var(e_t)=1 and η_t is white noise with variance 1. Then

$$K_{xy}^2(\omega) = \left\{1 + \left[1.64 - 1.6 \cos(\omega)\right]/\left[1.25 - \cos(\omega)\right]\right\}^{-1}$$

The true squared coherency and its estimate from PROC SPECTRA for the example are also graphed in **Output 7.12** (p. 347).

7.10.3 PROC SPECTRA Statements

PROC SPECTRA gives these names to estimates of the cross-spectral quantities for the first two variables in the VAR list:

Cospectrum	CS_01_02
Quadrature Spectrum	QS_01_02
Cross-Amplitude Spectrum	A_01_02
Phase Spectrum	PH_01_02
Squared Coherency	K_01_02

PROC SPECTRA options for cross-spectral analysis are as follows:

```
PROC SPECTRA DATA=IN OUT=O1 COEF P S
   CROSS A K PH WHITETEST ADJMEAN;
   VAR Y1 Y2;
   WEIGHTS 1 1 1 1 1;
RUN;
```

CROSS indicates that cross-spectral analysis is to be done. It produces the cospectrum C_01_02 and the quadrature spectrum Q_01_02 when used in conjunction with S. CROSS produces the real part RP_01_02 and the imaginary part IP_01_02 of the cross-periodogram when used in conjunction with P. Thus, RP and IP are unweighted estimates, and C and Q are weighted and normalized estimates of the cospectrum and quadrature spectrum. A, K, and PH request, respectively, estimation of cross-amplitude, squared coherency, and phase spectra (CROSS must be specified also). Weighting is necessary to obtain a valid estimate of the squared coherency.

Consider the following 512 observations Y_t generated from the model

$$V_t = .8V_{t-1} + X_t \quad \text{(the noiseless transfer function)}$$

and

$$Y_t = V_t + \eta_t \quad \text{(adding a noise term)}$$

where X_t is an autoregressive (AR) process

$$X_t = .5X_{t-1} + e_t$$

with variance 1.3333 and where η_t is white noise with variance 1.

The following SAS code produces appropriate spectral estimates:

```
PROC SPECTRA DATA=A OUT=OOO P S CROSS A K PH;
   WEIGHTS 1 1 1 1 1 1 1 1 1 1 1;
   VAR Y X;
RUN;
```

Plots of estimated and true spectra are overlaid in **Output 7.12**.

Output 7.12
Plots of Estimated and True Spectra

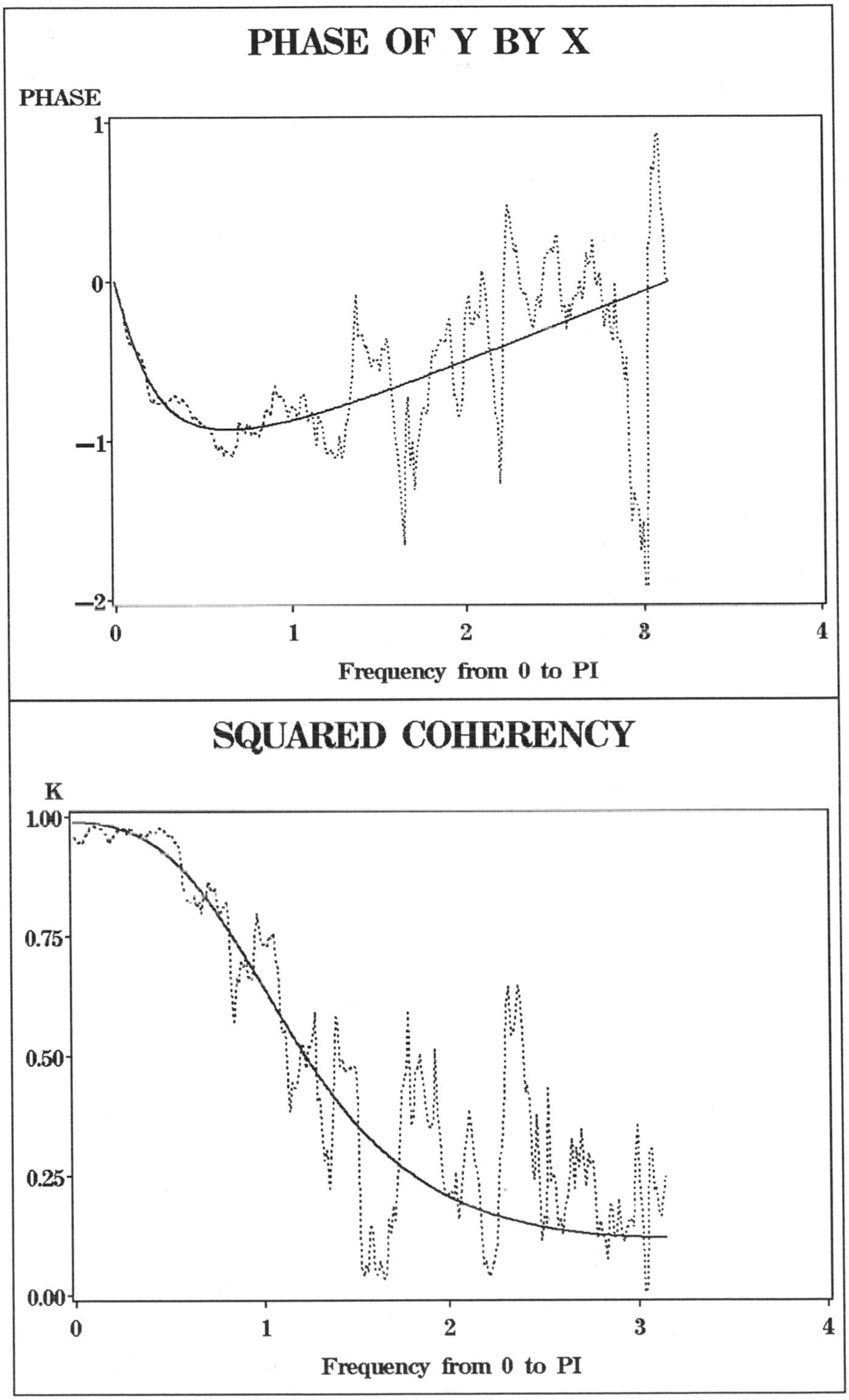

Output 7.12 *Plots of Estimated and True Spectra (continued)*

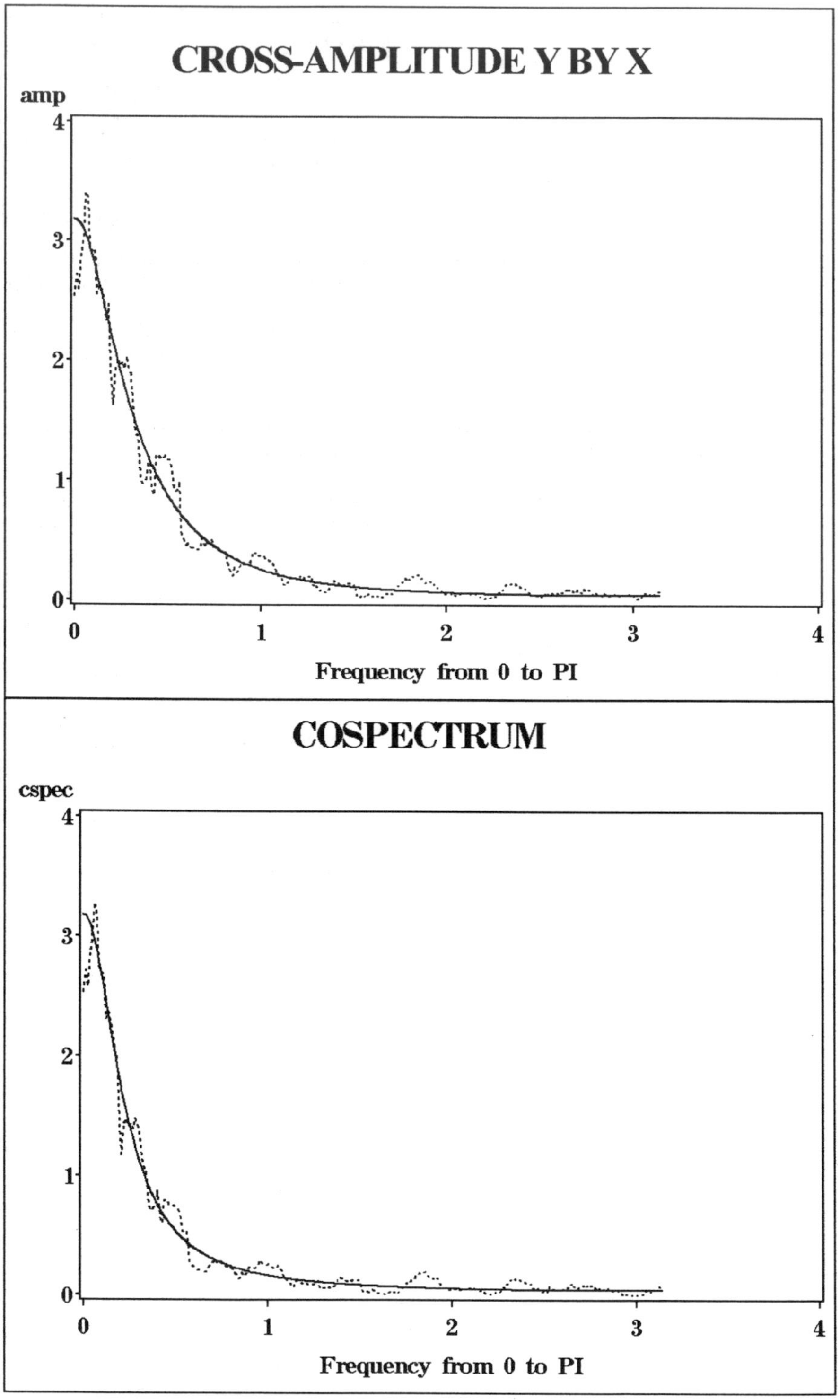

Output 7.12
Plots of Estimated and True Spectra (continued)

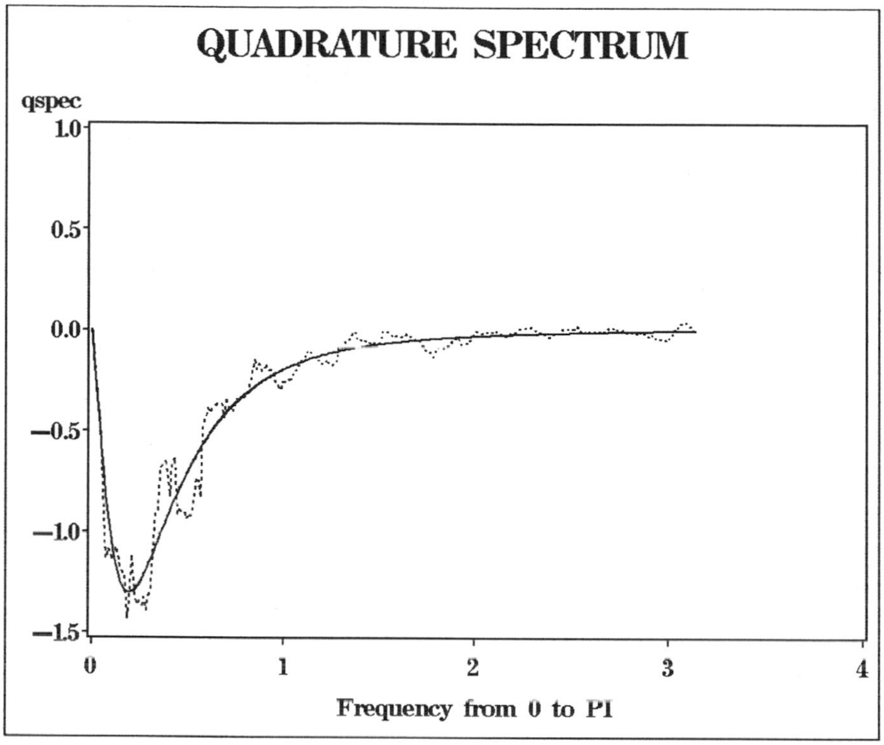

Although the data are artificial, think of X and Y as representing furnace and room temperatures in a building. The phase spectrum shows that long-term fluctuations (ω near zero) and short-term fluctuations (ω near π) for furnace and room temperatures are nearly in phase. The phase spectrum starts at zero and then decreases, indicating that X (the furnace temperature) tends to peak slightly before room temperature at intermediate frequencies. This makes sense if the furnace is connected to the room by a reasonably long pipe.

The squared coherency is near one at low frequencies, indicating a strong correlation between room temperature and furnace temperature at low frequencies. The squared coherency becomes smaller at the higher frequencies in this example. The estimated phase spectrum can vary at high frequencies as a result of this low correlation between furnace and room temperatures at high frequencies. Because of mixing as the air travels from the furnace to the room, high-frequency oscillations in furnace temperatures tend not to be strongly associated with temperature fluctuations in the room.

The gain, A_01_02/S_02, behaves like the cross-amplitude spectrum A_01_02 for this example. This behavior shows that low-frequency fluctuations in the furnace produce high-amplitude fluctuations at room temperature, while high-frequency fluctuations produce low-amplitude (small variance) fluctuations at room temperature. The transfer function tends to smooth the high-frequency fluctuations. Because of mixing in the pipe leading from the furnace to the room, it is not surprising that high-frequency (fast oscillation) temperature changes in the furnace are not transferred to the room.

7.10.4 Cross-Spectral Analysis of the Neuse River Data

In Chapter 3, "The General ARIMA Model," the differenced log flow rates of the Neuse River at Kinston (Y) and Goldsboro (X) are analyzed with the transfer function model

$$(1 - 1.241B + .291B^2 + .117B^3)X_t = (1 - .874B)e_t$$

or

$$X_t - 1.241X_{t-1} + .291X_{t-2} + .117X_{t-3} = e_t - .874e_{t-1}$$

with

$$\sigma_e^2 = .0399$$

and

$$Y_t = .495X_{t-1} + .273X_{t-2} + \varepsilon_t$$

where

$$\varepsilon_t = 1.163\varepsilon_{t-1} - .48\varepsilon_{t-2} + v_t - .888v_{t-1}$$

and v_t is a white noise series with

$$\sigma_v^2 = .0058$$

The spectral quantities discussed above are computed and plotted using the estimated model parameters. First, the model-based spectral quantities are developed. Then, the direct estimates (no model) of the spectral quantities from PROC SPECTRA are plotted.

When the models above are used, the spectrum of Goldsboro is

$$\begin{aligned} f_{XX}(\omega) &= \left(\left(1-.874e^{i\omega}\right)\left(1-.874e^{-i\omega}\right)\right)/\left(\left(1-1.241e^{i\omega}+.291e^{2i\omega}+.117e^{3i\omega}\right)\right. \\ &\quad \left.\left(1-1.241e^{-i\omega}+.291e^{-2i\omega}+.117e^{-3i\omega}\right)\right)\left(.0399/(2\pi)\right) \\ &= \left[1+.874^2-2(.874)\cos(\omega)\right]/\left\{\left[1+1.241^2+.291^2+.117^2\right]\right. \\ &\quad -2\cos(\omega)\left[1.241+1.241(.291)-.291(.117)\right]-2\cos(2\omega)\left[1.241(.117)-.291\right] \\ &\quad \left.+\ 2\cos(3\omega)[.117]\right\}\left[.0399/(2\pi)\right] \end{aligned}$$

Note that the cross-covariance of Y_t with X_{t-j} is the same as the cross-covariance of X_{t-j} with

$$.495X_{t-1} + .273X_{t-2}$$

so you obtain

$$\gamma_{XY}(j) = .495\gamma_{XX}(j-1) + .273\gamma_{XX}(j-2)$$

Thus, the cross-spectrum is

$$f_{XY}(\omega) = \left(.495e^{-i\omega} + .273e^{-2i\omega}\right)f_{XX}(\omega)$$

The real part (cospectrum) is

$$c(\omega) = (.495\cos(\omega) + .273\cos(2\omega))\, f_{XX}(\omega)$$

and the quadrature spectrum is

$$q(\omega) = (.495\sin(\omega) + .273\sin(2\omega))\, f_{XX}(\omega)$$

The phase spectrum is

$$\Psi_{XY}(\omega) = \arctan(q(\omega)/c(\omega))$$

The spectrum of Kinston (Y) is

$$f_{YY}(\omega) = (.495e^{i\omega} + .273e^{2i\omega})(.495e^{-i\omega} + .273e^{-2i\omega})\, f_{XX}(\omega) + f_{\varepsilon\varepsilon}(\omega)$$

where

$$f_{\varepsilon\varepsilon}(\omega) = \left(\frac{((1-.888e^{i\omega})(1-.888e^{-i\omega}))}{((1-1.163e^{i\omega}+.48e^{2i\omega})(1-1.163e^{-i\omega}+.48e^{-2i\omega}))}\right)\left(\frac{.0058}{2\pi}\right)$$

The squared coherency is simply

$$K^2_{XY}(\omega) = |f_{XY}(\omega)|^2 / (f_{XX}(\omega) f_{YY}(\omega))$$

Consider this pure delay transfer function model:

$$Y_t = \beta X_{t-c}$$

Using the Fourier transform, you can show the following relationship:

$$f_{XY}(\omega) = \Sigma_{h=-\infty}^{\infty}\left(\gamma_{XY}(h)e^{i\omega h}\right) = \beta\, \Sigma_{h=-\infty}^{\infty}\left(\gamma_{XX}(h-c)e^{i\omega(h-c)}e^{ic\omega}\right)$$

so

$$f_{XY}(\omega) = f_{XX}(\omega)\beta(\cos(c\omega) + i\sin(c\omega))$$

Thus, the phase spectrum is

$$\Psi_{XY}(\omega) = \arctan(\tan(c\omega)) = c\omega$$

When you use the ordinates in the plot of the phase spectrum as dependent variable values and frequency as the independent variable, a simple linear regression using a few low frequencies gives 1.34 as an estimate of c. This indicates a lag of 1.34 days between Goldsboro and Kinston. Because ARIMA models contain only integer lags, this information appears as two spikes, at lags 1 and 2, in the prewhitened cross-correlations. However, with the cross-spectral approach, you are not restricted to integer lags. In **Output 7.13** the irregular plots are the cross-spectral estimates from PROC SPECTRA. These are overlaid on the (smooth) plots computed above from the transfer function fitted by PROC ARIMA.

Output 7.13 *Overlaying the Smoothed Model-Derived Plots from PROC ARIMA and the Irregular PROC SPECTRA Plots*

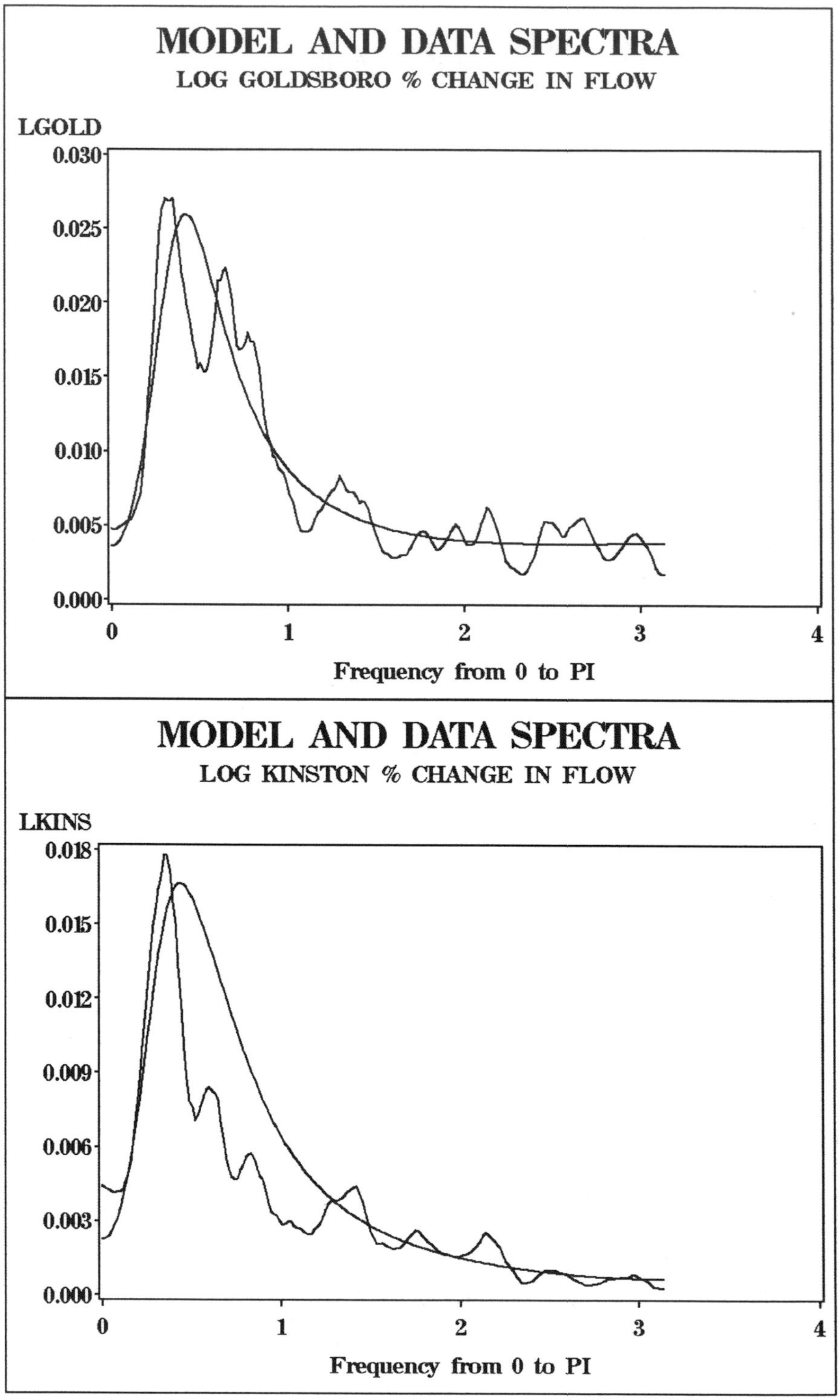

Output 7.13 *Overlaying the Smoothed Model-Derived Plots from PROC ARIMA and the Irregular PROC SPECTRA Plots (continued)*

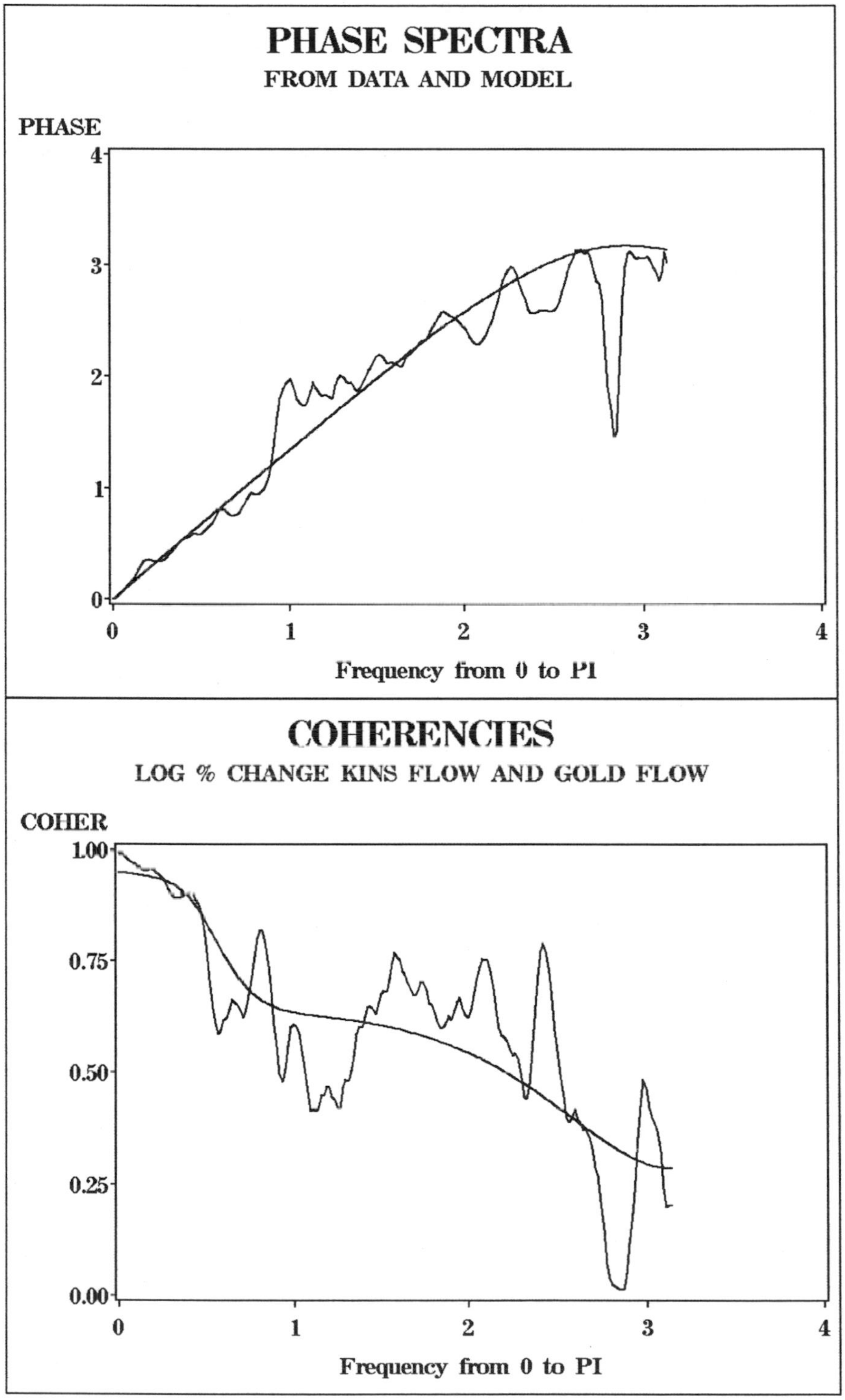

From one viewpoint, the closeness of the PROC SPECTRA plots to the model-derived plots provides a check on the ARIMA transfer function model and estimates. From another viewpoint, the model-based spectral plots provide a highly smoothed version of the PROC SPECTRA output.

7.10.5 Details on Gain, Phase, and Pure Delay

Suppose X_t is a perfect sine wave $X_t = \alpha\sin(\omega t + \delta)$. Now suppose $Y_t = 3X_{t-1} = 3\alpha\sin(\omega t - \omega + \delta)$. Y is also a perfect sine wave. The phase $-\omega + \delta$ of Y is ω radians less than the phase of X, and the amplitude of Y is 3 times that of X. You could also say that the phase of X is ω radians *more* than the phase of Y and the amplitude of X is $1/3$ that of Y. You have seen that the idea of cross-spectral analysis is to think of a general pair of series X and Y as each being composed of sinusoidal terms, then estimating how the sinusoidal components of Y are related, in terms of amplitude and phase, to those of the corresponding sinusoidal component of X.

With two series, Y and X, there is a phase of Y by X and a phase of X by Y. If $Y_t = 3X_{t-1}$ then Y is behind X by 1 time unit; that is, the value of X at time *t* is a perfect predictor of Y at time $t+1$. Similarly X is ahead of Y by 1 time unit. This program creates $X_t = e_t$ and $Y_t = 3X_{t-1}$, so it is an example of a simple noiseless transfer function. With $e_t \sim N(0,1)$, the spectrum $f_{XX}(\omega)$ of X is $f_{XX}(\omega) = 1/(2\pi) = 0.1592$ at all frequencies ω, and Y_t has spectrum $9/(2\pi) = 1.4324$.

```
DATA A;
   PI = 4*ATAN(1);
   X=0;
   DO T = 1 TO 64;
      Y = 3*X;   *Y IS 3 TIMES PREVIOUS X*;
      X=NORMAL(1827655);
      IF T=64 THEN X=0;
      OUTPUT;
   END;
RUN;
PROC SPECTRA DATA=A P S CROSS A K PH OUT=OUT1 COEFF;
   VAR X Y;
RUN;
PROC PRINT LABEL DATA=OUT1;
   WHERE PERIOD > 12;
   ID PERIOD FREQ;
RUN;
```

Since no weights were specified, no smoothing has been done. Only a few frequencies are printed out.

Output 7.14 *X and Y Series*

Period	Frequency from 0 to PI	Cosine Transform of X	Sine Transform of X	Cosine Transform of Y	Sine Transform of Y	Periodogram of X
64.0000	0.09817	0.16213	-0.09548	0.51212	-0.23739	1.13287
32.0000	0.19635	-0.24721	-0.13649	-0.64748	-0.54628	2.55166
21.3333	0.29452	0.09053	0.26364	0.03031	0.83572	2.48656
16.0000	0.39270	0.37786	-0.15790	1.22856	-0.00383	5.36666
12.8000	0.49087	-0.32669	-0.20429	-0.57543	-1.00251	4.75075

Period	Frequency from 0 to PI	Periodogram of Y	Spectral Density of X	Spectral Density of Y	Real Periodogram of X by Y	Imag Periodogram of X by Y
64.0000	0.09817	10.1959	0.09015	0.81136	3.3823	0.33312
32.0000	0.19635	22.9650	0.20305	1.82749	7.5079	1.49341
21.3333	0.29452	22.3791	0.19787	1.78087	7.1385	2.16543
16.0000	0.39270	48.2999	0.42707	3.84359	14.8744	6.16119
12.8000	0.49087	42.7567	0.37805	3.40247	12.5694	6.71846

Period	Frequency from 0 to PI	Cospectra of X by Y	Quadrature of X by Y	Coherency**2 of X by Y	Amplitude of X by Y	Phase of X by Y
64.0000	0.09817	0.26015	0.02651	1	0.27045	0.09817
32.0000	0.19635	0.59746	0.11884	1	0.60916	0.19635
21.3333	0.29452	0.56806	0.17232	1	0.59362	0.29452
16.0000	0.39270	1.18367	0.49029	1	1.28120	0.39270
12.8000	0.49087	1.00024	0.53464	1	1.13416	0.49087

It is seen that at period 64, X has a component

$0.16213\cos(2\pi t/64) - 0.09548\sin(2\pi t/64) = 0.188156\sin(2\pi t/64 + 2.10302)$ and Y has a component

$0.51212\cos(2\pi t/64) - 0.23739\sin(2\pi t/64) = 0.564465\sin(2\pi t/64 - 2.00486)$, where $0.564465/0.188156=3$ is the amplitude increase in going from X to Y. The phase shift is $2.10302 - 2.00486 = 0.09817$ radians. Each periodogram ordinate is $(n/2)$ times the sum of squares of the two coefficients, $(64/2)[(0.16213)^2 + (0.09548)^2] = 1.13287$ for X at period 64, for example.

Each Y periodogram ordinate is 3^2 times the corresponding X periodogram ordinate. This exact relationship would not hold if noise were added to Y. Within the class of ARMA models, the periodogram $I_n(\omega)$ divided by $2\pi f\omega$ (where the true spectral density of the process is $f(\omega)$) has approximately a chi-square distribution with 2 degrees of freedom, a distribution with mean 2. This motivates $I_n(\omega)/4\pi$ as an estimator of $f(\omega)$ for both Y and X. Each spectral density estimator is the corresponding periodogram ordinate divided by 4π. For example, $1.13287/(4\pi) = 0.0902$ for X at period 64.

In the VAR statement of PROC SPECTRA, the order of variables is X Y, and you see that this produces the phase of X by Y, not Y by X. The phase $-\omega + \delta$ of Y is ω radians less than the phase δ of X as was shown above. Thus the entries in the phase column are exactly the same as the frequencies. The plot of phase by frequency is a straight line with slope 1, and this slope gives the pure delay d for $Y_t = CX_{t-d}$ so d=1. Had the variables been listed in the order Y X, $-\omega$ would have appeared as the phase spectrum estimate.

The slope of the phase plot near the origin gives some idea of the lag relationship between Y and X in a transfer function model with or without added noise, as long as the coherency there is reasonably strong. The delay need not be an integer, as was illustrated with the river data earlier. The phase plot of the generated data that simulated furnace and room temperatures had a negative slope near the origin. The room temperature Y is related to lagged furnace temperature X, and with the variables listed in the order Y X, the phase of Y by X is produced, giving the negative slope. Had the order been X Y, the plot would be reflected about the $\text{phase} = 0$ horizontal line and an initial positive slope would have been seen. For the river data, you see that the sites must have been listed in the order Goldsboro Kinston in PROC SPECTRA, since the phase slope is positive and Goldsboro (X) is upstream from Kinston (Y).

If $Y_t = 3X_{t-1}$ and if X_t has an absolutely summable covariance function $\gamma_{XX}(h)$, which is the case in the current example, then Y also has a covariance function

$$\begin{aligned}\gamma_{YY}(h) &= E\{Y_t Y_{t+h}\}\\ &= 9E\{X_{t-1}X_{t-1+h}\}\\ &= 9\gamma_{XX}(h)\end{aligned}$$

By definition, the theoretical spectral density $f_{XX}(\omega)$ of X is the Fourier transform of the covariance sequence: $f_{XX}(\omega) = \frac{1}{2\pi}\sum_{h=-\infty}^{\infty} e^{-i\omega h}\gamma_{XX}(h)$ and similarly $f_{YY}(\omega) = 9f_{XX}(\omega)$. The absolute summability assumption ensures the existence of the theoretical spectral densities. The processes also have a cross-covariance function

$$\begin{aligned}\gamma_{XY}(h) &= E\{X_t Y_{t+h}\}\\ &= 3E\{X_t X_{t-1+h}\}\\ &= 3\gamma_{XX}(h-1)\end{aligned}$$

whose Fourier transform is the cross-spectral density of Y by X:

$$\begin{aligned}f_{XY}(\omega) &= \frac{1}{2\pi}\sum_{h=-\infty}^{\infty} e^{-i\omega h}\gamma_{XY}(h) = \frac{3}{2\pi}\sum_{h=-\infty}^{\infty} e^{-i\omega}e^{-i\omega(h-1)}\gamma_{XX}(h-1)\\ &= \frac{3}{2\pi}\sum_{h=-\infty}^{\infty}[\cos(\omega) - i\sin(\omega)]e^{-i\omega(h-1)}\gamma_{XX}(h-1)\\ &= 3[\cos(\omega) - i\sin(\omega)]f_{XX}(\omega)\end{aligned}$$

Writing $\frac{1}{2\pi}\sum_{h=-\infty}^{\infty} e^{-i\omega h}\gamma_{XY}(h)$ as $c(\omega) - iq(\omega)$, the real part $c(\omega)$ is the cospectrum of X by Y and the coefficient of $-i$ is the quadrature spectrum $q(\omega)$. In this example $c(\omega) = 3\cos(\omega)f_{XX}(\omega)$ and $q(\omega) = 3\sin(\omega)f_{XX}(\omega)$. For example, at period 32 you find $3\cos(2\pi/32) = 2.9424$ and $3\sin(2\pi/32) = 0.5853$. Multiplying these by the estimated X spectral density gives

$(2.9424)(0.20305) = 0.5974$, the estimated cospectrum of X by Y for period 32, and similarly $(0.5853)(.20305) = 0.1188$, the estimated quadrature spectrum of X by Y on the printout.

The phase and amplitude spectra are transformations of $q(\omega)$ and $c(\omega)$ and are often easier to interpret. The phase of X by Y is $\text{Atan}(q(\omega)/c(\omega)) = \text{Atan}(\sin(\omega)/\cos(\omega)) = \omega$ and that of Y by X is $-\omega$, as would be expected from the previous discussion of phase diagrams. The phase shows you the lag relationship between the variables, as has been mentioned several times. For $f_{XY}(\omega) = 3[\cos(\omega) - i\sin(\omega)]f_{XX}(\omega)$, the amplitude of the frequency ω component is $\sqrt{c^2(\omega)+q^2(\omega)} = \text{A}(\omega) = 3\sqrt{\cos^2(\omega)+\sin^2(\omega)}f_{XX}(\omega) = 3f_{XX}(\omega)$. This is called the amplitude of X by Y, and in the printout, each of these entries is the corresponding spectral density of X estimate multiplied by 3. The quantity $\text{A}^2(\omega)/f_{XX}(\omega)$ is the spectral density for that part of Y that is exactly related to X, without any added noise. Since Y is related to X by a noiseless transfer function, the spectral density of Y should be $\text{A}^2(\omega)/f_{XX}(\omega)$. For example, at period 32 you find $(0.60916)^2/(0.20305) = 1.82749$. Recall that the quantity $\text{A}(\omega)/f_{XX}(\omega)$ has been referred to earlier as the "gain." It represents the amplitude multiplier for the frequency ω component in going from X to Y in a model where Y is related to X without noise. In our case the gain is thus $3\sqrt{\cos^2(\omega)+\sin^2(\omega)} = 3$.

A more realistic scenario is that an observed series W_t consists of Y_t plus an added noise component N_t independent of X (and thus Y). Here the phase, amplitude, and gain using W and X as data have their same interpretation, but refer to relationships between X and Y—that is, between X and the part of W that is a direct transfer function of X. You can think of fluctuations in X over time as providing energy that is transferred into Y, such as vibrations in an airplane engine transferred to the wing or fuselage. The fluctuations in that object consist of the transferred energy plus independent fluctuations such as wind movements while flying. The spectral density $f_{WW}(\omega)$ of W will no longer be $\text{A}^2(\omega)/f_{XX}(\omega)$ but will be this plus the noise spectrum. In a system with noise, then, the quantity $\text{A}^2(\omega)/[f_{XX}(\omega)f_{WW}(\omega)]$ provides an R^2 measure as a function of frequency. Its symbol is $\kappa^2(\omega)$, and it is called the squared coherency. In a noiseless transfer function, like $\text{Y}_t = 3\text{X}_{t-1}$, the squared coherency between Y and X would be 1 at all frequencies because $[\text{A}^2(\omega)/f_{XX}(\omega)]/f_{YY}(\omega) = f_{YY}(\omega)/f_{YY}(\omega) = 1$ in that case. This appears in the output; however, in the absence of smoothing weights, the squared coherency is really meaningless, as would be an R^2 of 1 in a simple linear regression with only 2 points.

This small example without smoothing is presented to show and interpret the cross-spectral calculations. In practice, smoothing weights are usually applied so that more accurate estimates can be obtained. Another practical problem arises with the phase. The phase is usually computed as the angle in $[-\pi/2, \pi/2]$ whose tangent is $q(\omega)/c(\omega)$. If a phase angle a little less than $\pi/2$ is followed by one just a bit bigger than $\pi/2$, the interval restriction will cause this second angle to be reported as an angle just a little bigger than $-\pi/2$. The phase diagram can thus show phases jumping back and forth between $-\pi/2$ and $\pi/2$ when in fact they could be represented as not changing much at all. Some practitioners choose to add and subtract multiples of π from the phase at selected frequencies in order to avoid excessive fluctuations in the plot.

Fuller (1996) gives formulas for the cross-spectral estimates and confidence intervals for these quantities in the case that there are $2d+1$ equal smoothing weights.

Chapter 8 Data Mining and Forecasting

8.1 Introduction

This chapter deals with the process of forecasting many time series with little intervention by the user. The goal is to illustrate a modern automated interface for a collection of forecasting models, including many that have been discussed so far. Most models herein, such as damped trend exponential smoothing and Winters method, are equivalent to specific ARIMA models. Some of these were developed in the literature without using ARIMA ideas and were later recognized as being ARIMA models. The examples focus on Web traffic data that accumulate very quickly over time and require a demanding warehousing and analytics strategy to automate the process. Analysis of such large amounts of data is often referred to as "data mining."

In this chapter SAS Web Analytics are used to read the Web traffic data, summarize the information for detailed and historical analyses, and define the information into a data warehouse. The SAS Web Analytics reports provide important details about your Web traffic—who is visiting your site, how long they stay, and what material or pages they are viewing. This information can then be accumulated over time to construct a set of metrics that enables you to optimize your e-business investment. Results are displayed on the Web and accessed by an Internet browser.

In addition, the SAS/ETS software Time Series Forecasting System (TSFS) is examined. This system provides a menu-driven interface to SAS/ETS and SAS/GRAPH procedures to facilitate quick and easy analysis of time series data.

The HPF (High Performance Forecasting) procedure is used here to provide an automated way to generate forecasts for many time series in one step. All parameters associated with the forecast model are optimized based on the data.

Finally, the chapter uses a scorecard to integrate, distribute, and analyze the information enterprise-wide to help make the right decisions. This interface helps business users analyze data in new and different ways to anticipate business trends and develop hypotheses. They can receive automated alerts to early indicators of excellence or poor performance. The interface enables IT (information technology) professionals to fully automate and personalize the collection and distribution of knowledge across the organization.

The application presented here is available through the SAS IntelliVisor for Retail service. The delivery mechanism is provided through an ASP (application service provider) infrastructure.

8.2 Forecasting Data Model

Under the ASP framework, each night we receive customer Web logs after 12:00 AM local time. The Web logs are unzipped, placed in a file directory, and analyzed using SAS Web Analytics. The data examine key metrics used to describe activity during the 24 hours of e-retailing in a given day. One company using this approach is the online retailer the Vermont Country Store. (They provided a modified version of their data for illustration here. See www.vermontcountrystore.com.)

The variables and their descriptions are provided in Table 8.1, followed by a listing of some of the data.

Table 8.1 *Variables and Descriptions*

Variable	Description
date	SAS Date variable formatted in DATE9.
revenue	Revenue (TARGET)
buyer	Number of Purchasing Sessions
dollars_per_purch_session	Average Order Value
items_per_purch_session	Average Items per Purchasing Session
catalog_quick_purch_perc	%CQS Buyers
perc_abandon_carts	Abandon Carts %
num_session	Number of Sessions
requestcatalog_con	Number of Catalog Requests
productsuggestion_pages	Number of Product Suggestion Pages Viewed
new_cust_perc	New/Total Sessions × 100
purch_perc	Purchase Response Rate
new_buy_perc	New/Total Buyers × 100

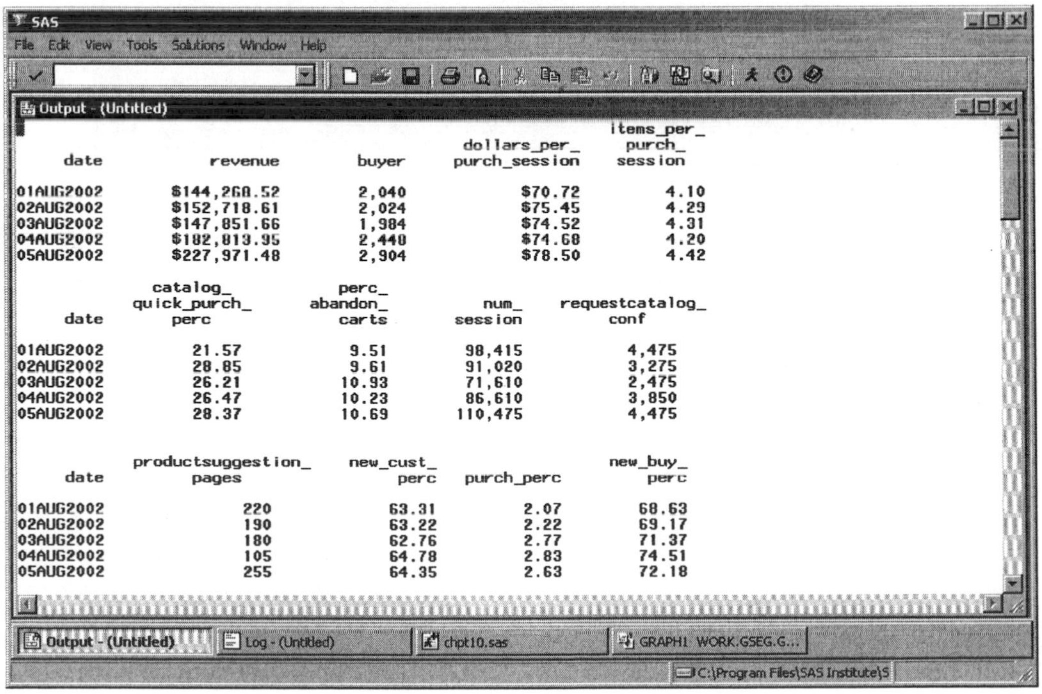

date	revenue	buyer	dollars_per_purch_session	items_per_purch_session
01AUG2002	$144,268.52	2,040	$70.72	4.10
02AUG2002	$152,718.61	2,024	$75.45	4.29
03AUG2002	$147,851.66	1,984	$74.52	4.31
04AUG2002	$182,813.95	2,440	$74.68	4.20
05AUG2002	$227,971.48	2,904	$78.50	4.42

date	catalog_quick_purch_perc	perc_abandon_carts	num_session	requestcatalog_conf
01AUG2002	21.57	9.51	98,415	4,475
02AUG2002	28.85	9.61	91,020	3,275
03AUG2002	26.21	10.93	71,610	2,475
04AUG2002	26.47	10.23	86,610	3,850
05AUG2002	28.37	10.69	110,475	4,475

date	productsuggestion_pages	new_cust_perc	purch_perc	new_buy_perc
01AUG2002	220	63.31	2.07	68.63
02AUG2002	190	63.22	2.22	69.17
03AUG2002	180	62.76	2.77	71.37
04AUG2002	105	64.78	2.83	74.51
05AUG2002	255	64.35	2.63	72.18

8.3 The Time Series Forecasting System

Open the TSFS and select the data set to be accessed. The TSFS automatically identifies the Time ID variable DATE and recognizes that the data are at daily intervals. Since revenue is the target or main response variable of interest, select the graph button to evaluate revenue behavior over time.

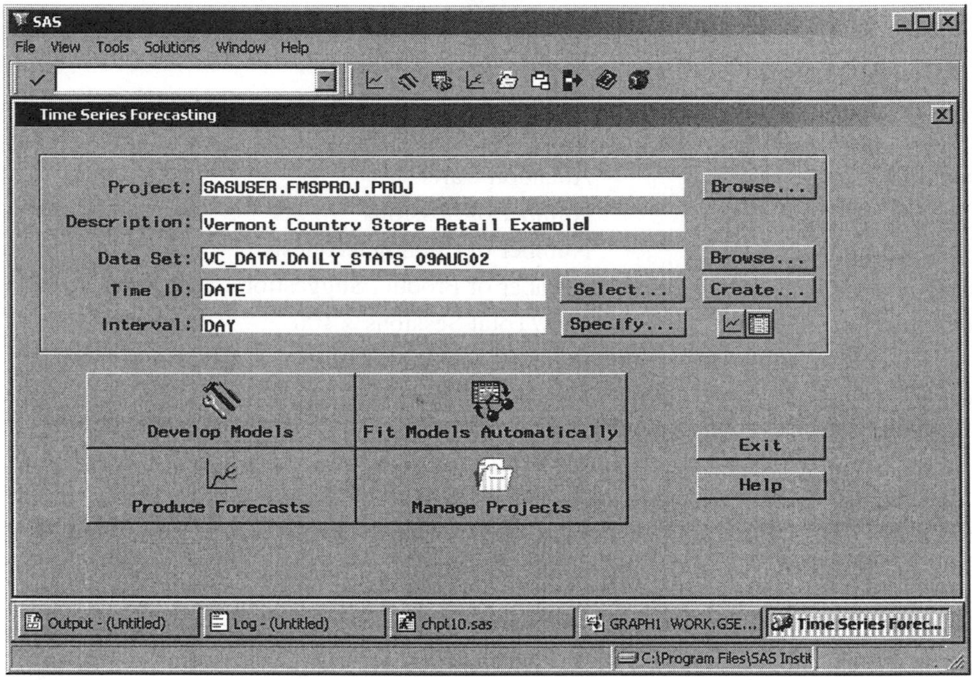

Select the Revenue variable and then select the **Graph** button.

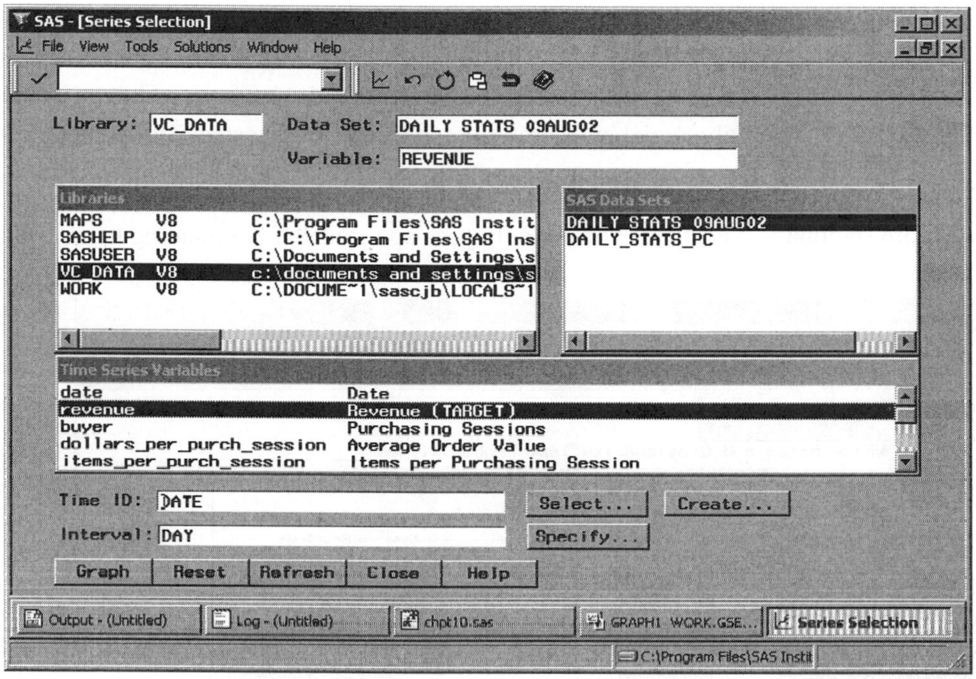

The Revenue variable shows a decrease in variability over time with some periodic tendencies. This is not unusual. Retail sales over the Web tend to show a daily cycle over time. (Again, this graph represents a display that does not reflect the true revenue at the Vermont Country Store.)

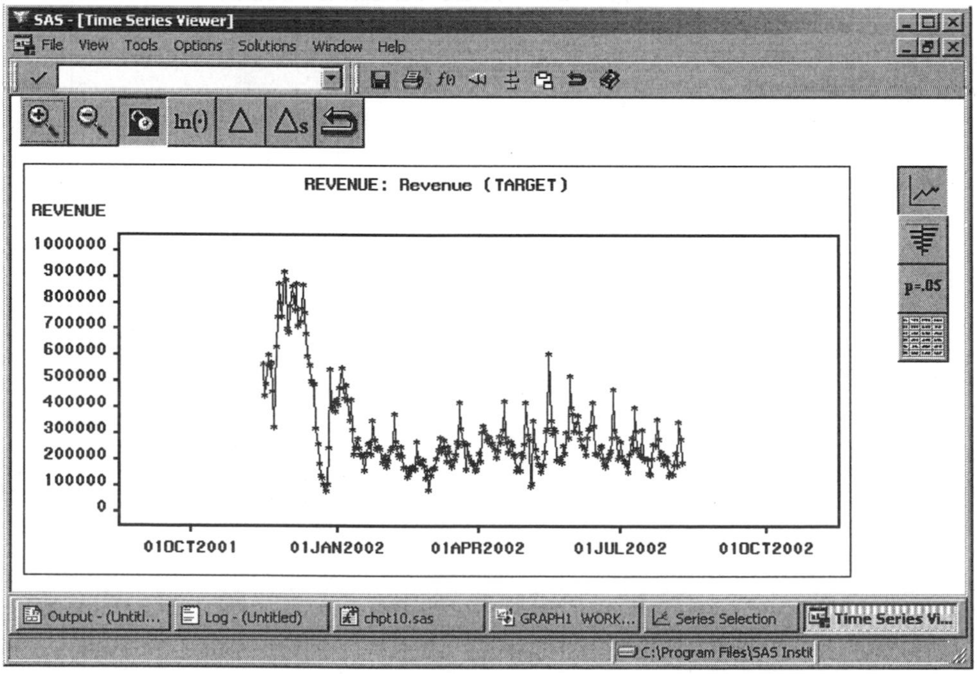

The series looks nonstationary, and examining the autocorrelation plots suggests the need to difference. By selecting the **p=.05** button you can access the Dickey-Fuller unit root test. This test is previously described and fails to reject the null hypothesis of nonstationarity only with four augmenting lags. The TSFS employs ordinary unit root tests $(1-\phi\ B)$ and unit root tests for the seasonal polynomial $(1-\phi\ B^s)$ using k lagged differences as augmenting terms. That is, these are factors in an autoregressive polynomial of order $k + 1$ and $H_0: \phi = 1$ is tested. The user should always entertain the possibility of fitting a model outside the class of models considered here. For example, had the pre-Christmas surge in sales been modeled, say, with a separate mean, the residuals might look more stationary. The display below only goes up through 5 augmenting terms.

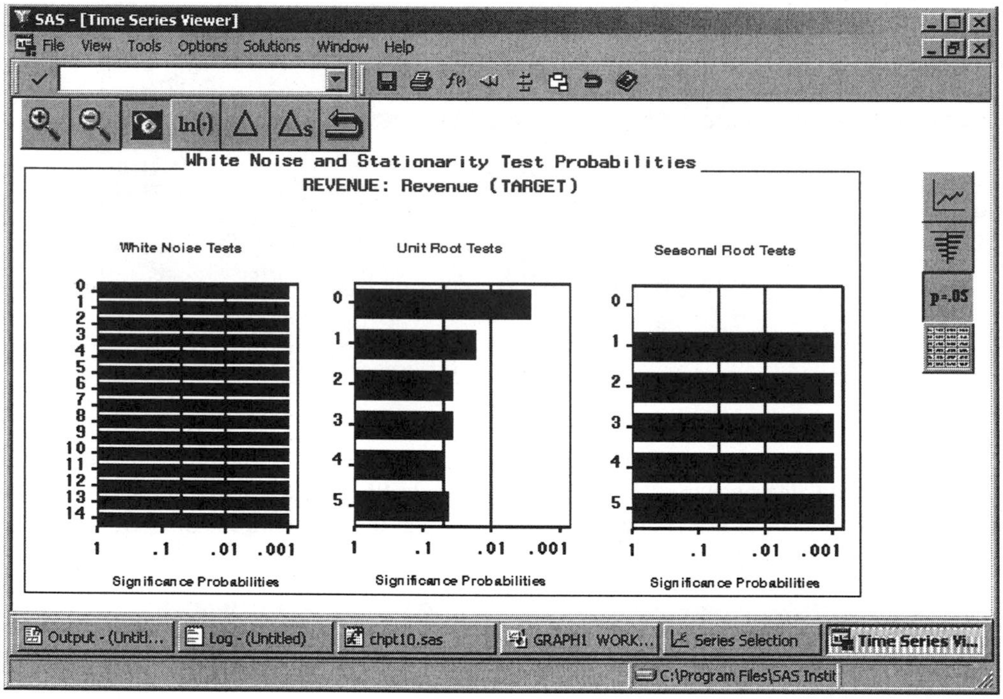

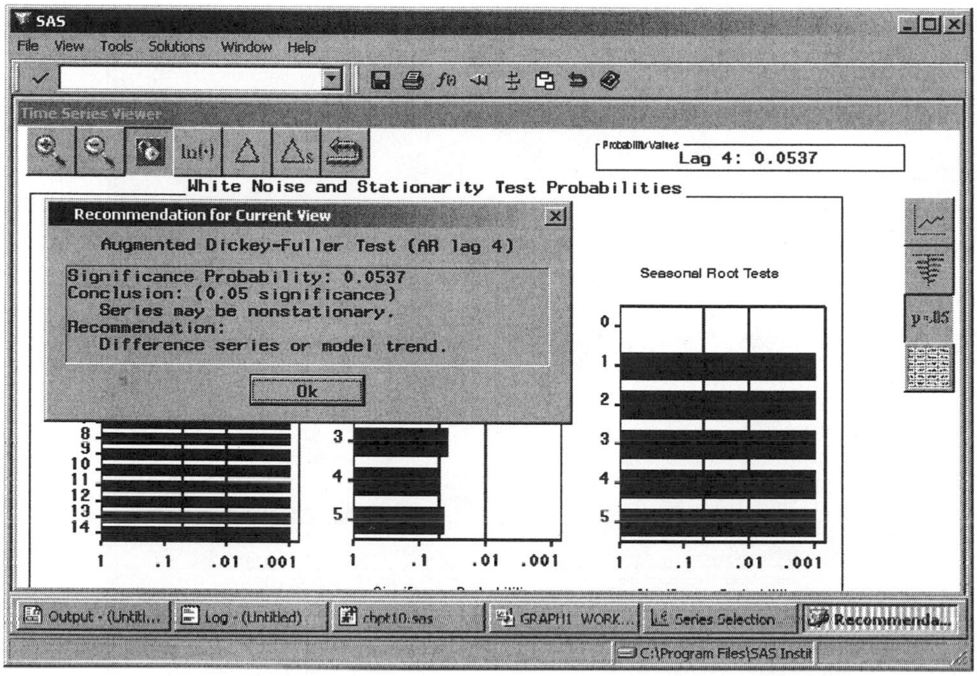

Go back to the main window and request that the TSFS automatically fit models for every series.

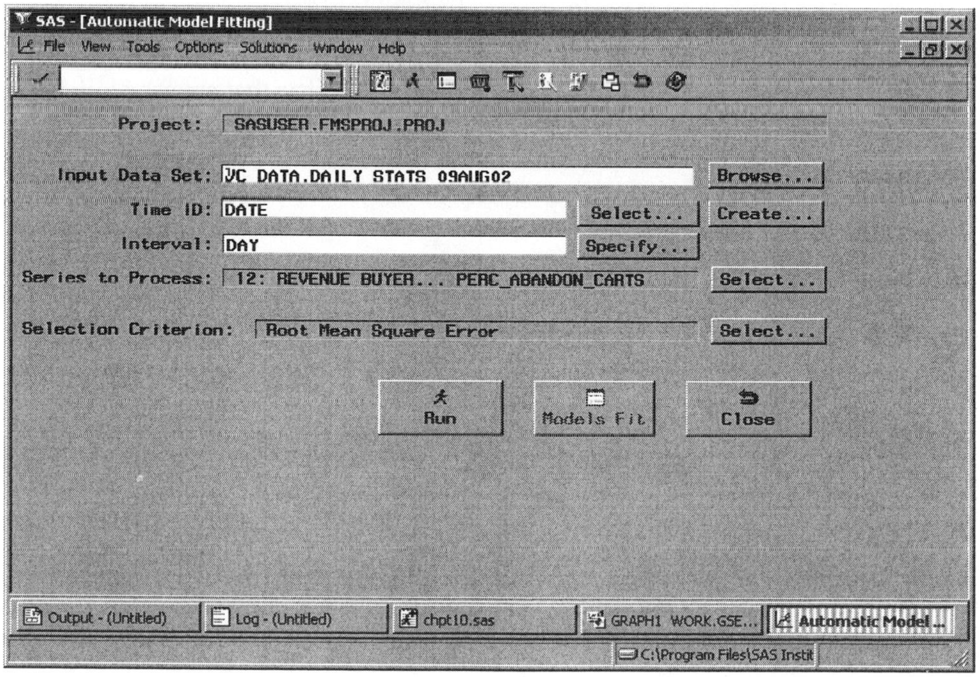

We're notified that 12 models will be fit for each series.

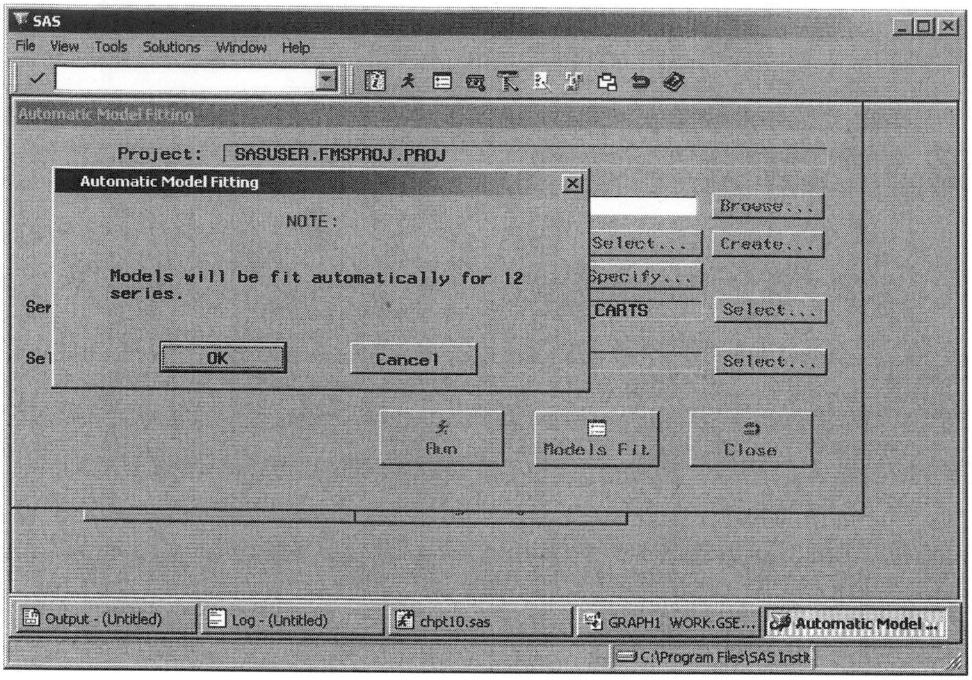

Notice the TSFS selects a seasonal exponential smoothing model for revenue. The TSFS provides an assortment of different seasonal and nonseasonal models and chooses the "best" model based on an information criterion that in this case is minimizing the root mean square error. The user has some control over the list of potential models and simple features of the data that are used initially to pare down the list, in this case, to 12 models that might fit well.

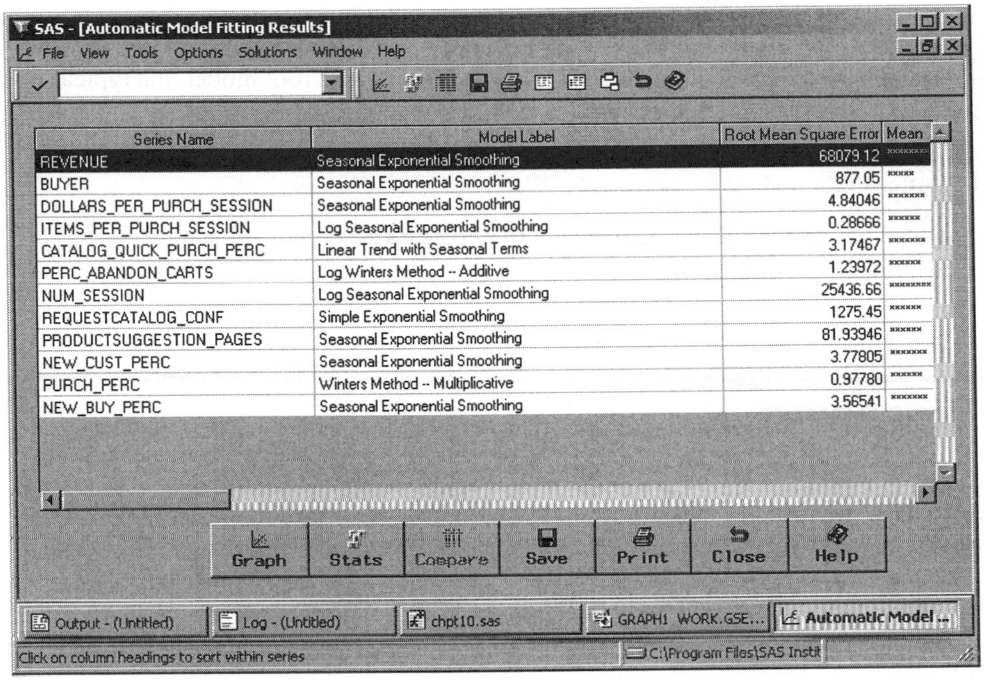

Select the **Graph** button to see the forecasts, and then select the forecast graph button to see the forecasts and confidence intervals.

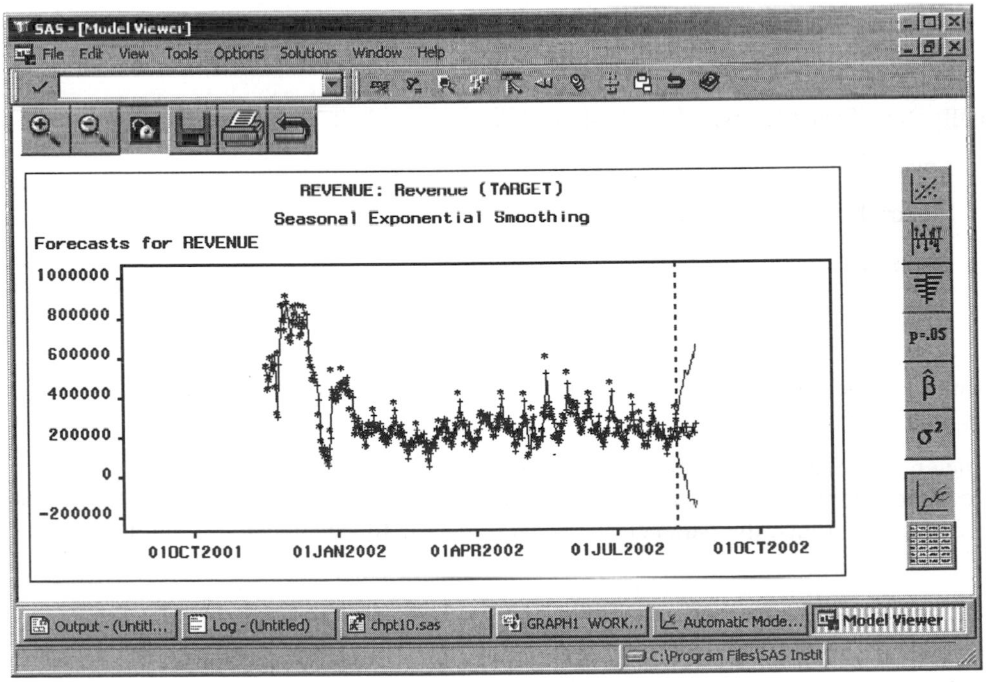

The graph and review of the data and forecasts using the data button suggest the seasonal exponential smoothing model does not fit the larger revenue spikes very well, although it does a reasonable job overall. Because exponential smoothing is analogous to fitting a unit root model, the typical fast spreading prediction intervals are seen as the forecast goes beyond one or two steps.

You can also go back to the **Automatic Fitting Results** screen to evaluate the forecasts for each series individually.

The TSFS can also be further automated by using the `forecast` command and the SAS/AF Forecast Application Command Builder.

8.4 HPF Procedure

The HPF procedure can forecast millions of time series at a time, with the series organized into separate variables or across BY groups. You can use the following forecasting models:

Smoothing Models:

- Simple
- Double
- Linear
- Damped Trend
- Seasonal
- Winters Method (additive and multiplicative)

Additionally, transformed versions of these models are provided:

- Log
- Square Root
- Logistic
- Box-Cox

For intermittent time series (series where a large number of values are zero values), you can use Croston's method (Croston 1977).

All parameters associated with the forecast model are optimized based on the data. The HPF procedure writes the time series with extrapolated forecasts, the series summary statistics, the forecast confidence limits, the parameter estimates, and the fit statistics to output data sets.

The HPF procedure step below examines the application of the automatic forecasting technique to the evaluation of seven different forecasting methods described above. The program creates a Forecasts data set that contains forecasts for seven periods beyond the end of the input data set VC_DATA.DAILY_STATS_09AUG02. The data represent daily values for Revenue, a variable describing the total number of purchasing dollars for a given day. The daily variable indicator, date, is formatted date9.

The GPLOT procedure is used to display the actual values, predicted values, and upper and lower confidence limits overlaid on the same graph.

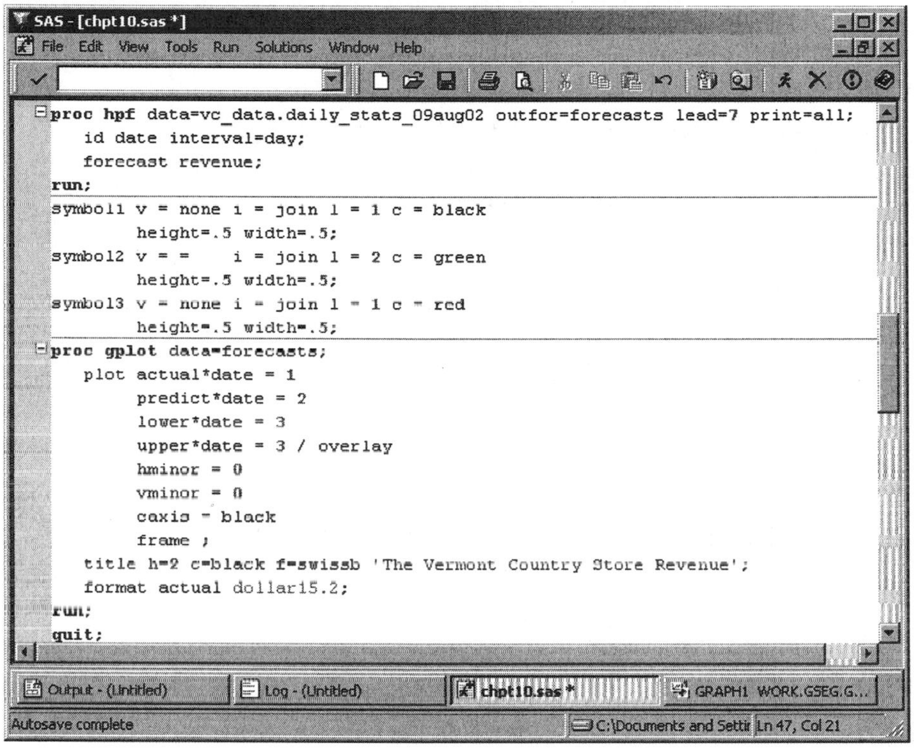
SAS - [chpt10.sas *]
File Edit View Tools Run Solutions Window Help
proc hpf data=vc_data.daily_stats_09aug02 outfor=forecasts lead=7 print=all;
id date interval=day;
forecast revenue;
run;
symbol1 v = none i = join l = 1 c = black
height=.5 width=.5;
symbol2 v = = i = join l = 2 c = green
height=.5 width=.5;
symbol3 v = none i = join l = 1 c = red
height=.5 width=.5;
proc gplot data=forecasts;
plot actual*date = 1
predict*date = 2
lower*date = 3
upper*date = 3 / overlay
hminor = 0
vminor = 0
caxis = black
frame ;
title h=2 c=black f=swissb 'The Vermont Country Store Revenue';
format actual dollar15.2;
run;
quit;
Output - (Untitled)
Log - (Untitled)
chpt10.sas *
GRAPH1 WORK.GSEG.G...
Autosave complete
C:\Documents and Settir
Ln 47, Col 21

The HPF procedure describes the input data set WORK.DAILY_STATS_09AUG02 and the Time ID variable DATE. There are 268 observations in the data set and no missing observations. The descriptive statistics are also provided.

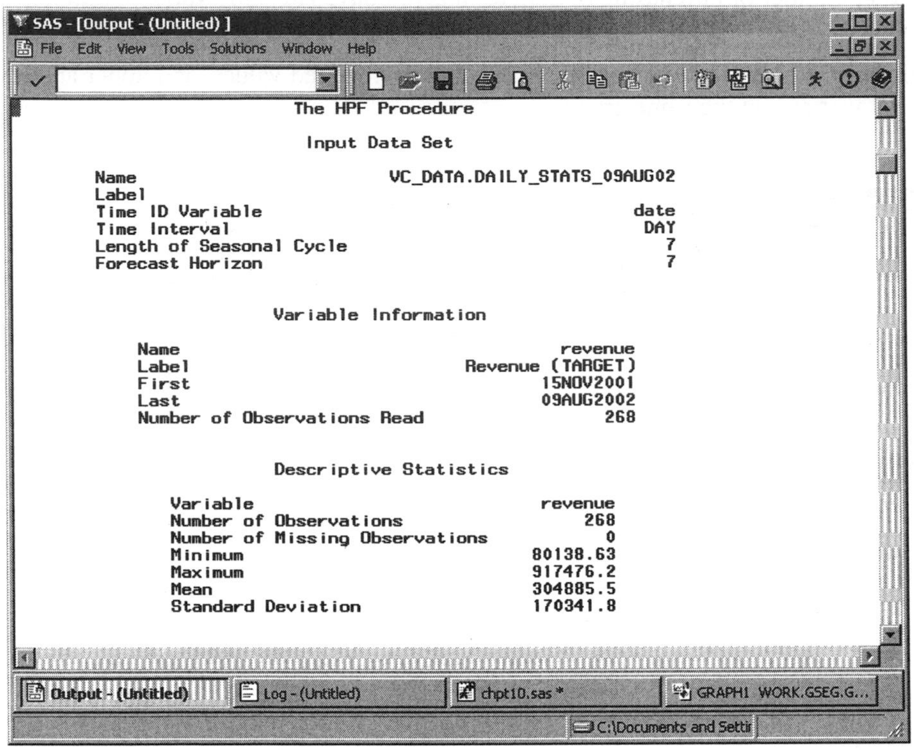

```
                    The HPF Procedure

                     Input Data Set

     Name                        VC_DATA.DAILY_STATS_09AUG02
     Label
     Time ID Variable                                   date
     Time Interval                                       DAY
     Length of Seasonal Cycle                              7
     Forecast Horizon                                      7


                  Variable Information

        Name                                     revenue
        Label                           Revenue (TARGET)
        First                                  15NOV2001
        Last                                   09AUG2002
        Number of Observations Read                  268


                  Descriptive Statistics

          Variable                             revenue
          Number of Observations                   268
          Number of Missing Observations             0
          Minimum                             80138.63
          Maximum                             917476.2
          Mean                                304885.5
          Standard Deviation                  170341.8
```

The Winters additive method seasonal exponential smoothing model fits best based on the RMSE statistic, and only the level weight is statistically different from 0. When performing these operations in an automatic fashion on many series, it is often found that the models tend to be overparameterized.

Model Selection Criterion = RMSE

Model	Statistic
Seasonal Exponential Smoothing	68079.123
Winters Method (Additive)	68075.664
Winters Method (Multiplicative)	68562.973

Winters Method (Additive) Parameter Estimates

Parameter	Estimate	Standard Error	t Value	Approx Pr > |t|
Level Weight	0.83528	0.04398	18.99	<.0001
Trend Weight	0.0010000	0.01320	0.08	0.9397
Seasonal Weight	0.0010000	0.07984	0.01	0.9900

The forecasts for the next seven days are displayed below, in addition to the standard errors and upper and lower 95% prediction intervals. The lower 95% confidence interval falls below 0 as you extend well beyond the end of the historical data.

The HPF Procedure

Forecasts for Variable revenue

Obs	Time	Forecasts	Standard Error	95% Confidence Limits	
269	10AUG2002	191993.8337	68459.913	57814.8695	326172.7979
270	11AUG2002	216322.8322	89236.741	41422.0328	391223.6315
271	12AUG2002	237213.4797	106048.02	29363.1856	445063.7739
272	13AUG2002	210125.9288	120564.19	-26175.5419	446427.3995
273	14AUG2002	260382.5872	133535.87	-1342.9095	522108.0838
274	15AUG2002	232600.2767	145377.10	-52333.6130	517534.1664
275	16AUG2002	191211.5513	156344.98	-115218.976	497642.0784

The statistics of fit for the selected model are given as a reference for model comparison. As noted, these are calculated based on the full range of data. A detailed description of these summary statistics can be found by consulting the SAS System 9 documentation.

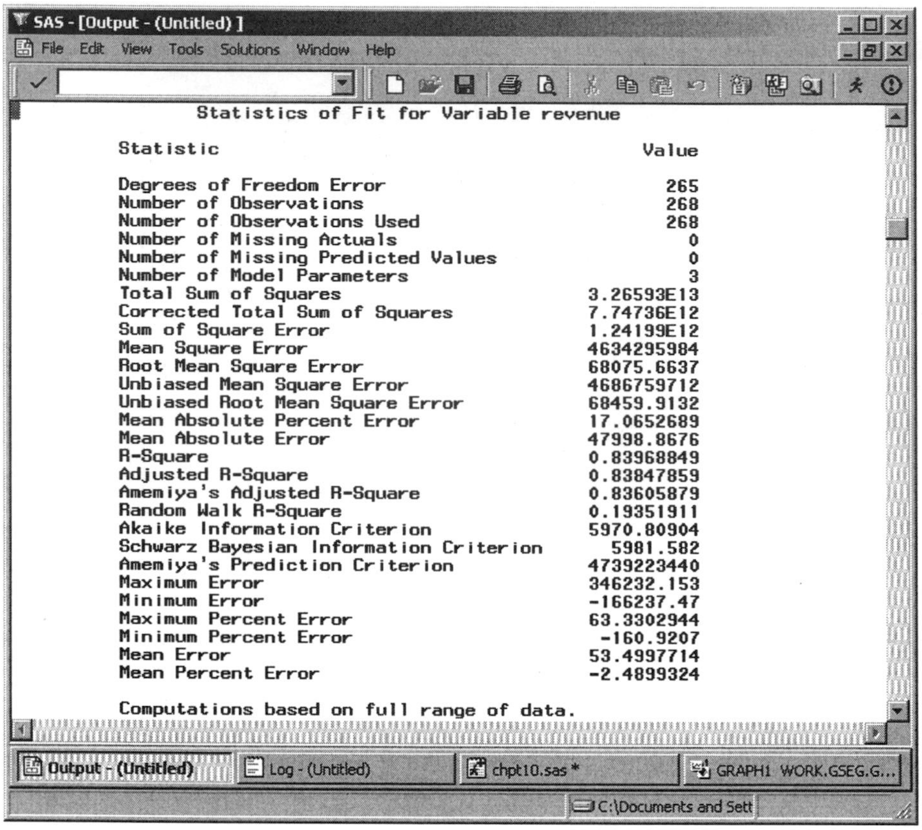

Statistics of Fit for Variable revenue

Statistic	Value
Degrees of Freedom Error	265
Number of Observations	268
Number of Observations Used	268
Number of Missing Actuals	0
Number of Missing Predicted Values	0
Number of Model Parameters	3
Total Sum of Squares	3.26593E13
Corrected Total Sum of Squares	7.74736E12
Sum of Square Error	1.24199E12
Mean Square Error	4634295984
Root Mean Square Error	68075.6637
Unbiased Mean Square Error	4686759712
Unbiased Root Mean Square Error	68459.9132
Mean Absolute Percent Error	17.0652689
Mean Absolute Error	47998.8676
R-Square	0.83968849
Adjusted R-Square	0.83847859
Amemiya's Adjusted R-Square	0.83605879
Random Walk R-Square	0.19351911
Akaike Information Criterion	5970.80904
Schwarz Bayesian Information Criterion	5981.582
Amemiya's Prediction Criterion	4739223440
Maximum Error	346232.153
Minimum Error	-166237.47
Maximum Percent Error	63.3302944
Minimum Percent Error	-160.9207
Mean Error	53.4997714
Mean Percent Error	-2.4899324

Computations based on full range of data.

A forecast summary shows values for the next seven days, and a sum forecast for the seven-day total is displayed at the bottom.

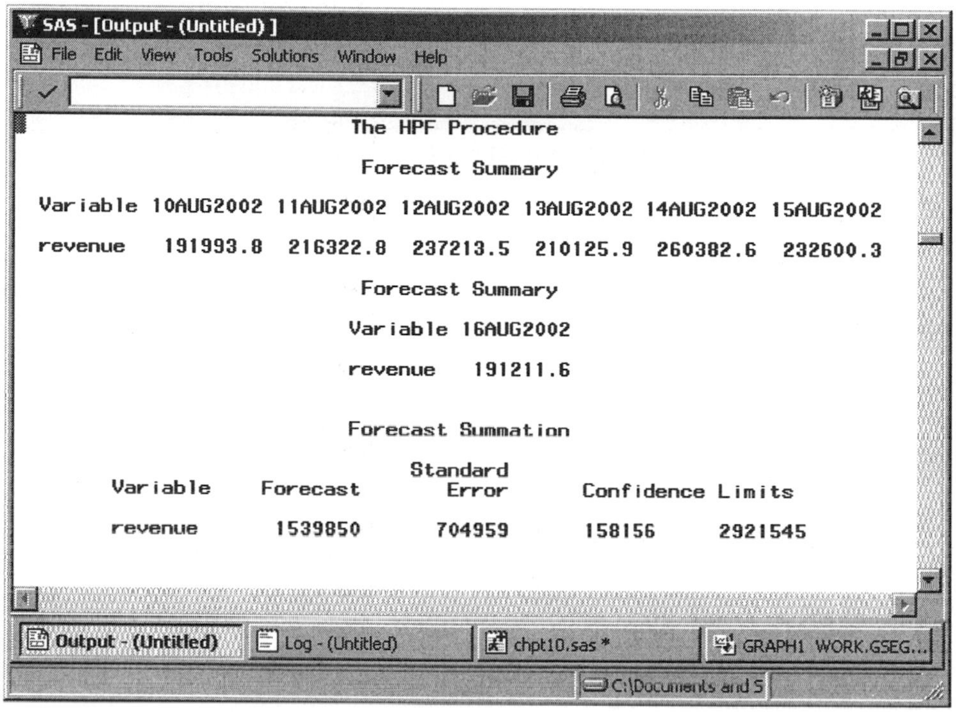

The graph below suggests a drop in purchasing sessions in early January. The Winters additive method of seasonal exponential smoothing does a nice job of tracking the historical data shown by the heavy middle graph line.

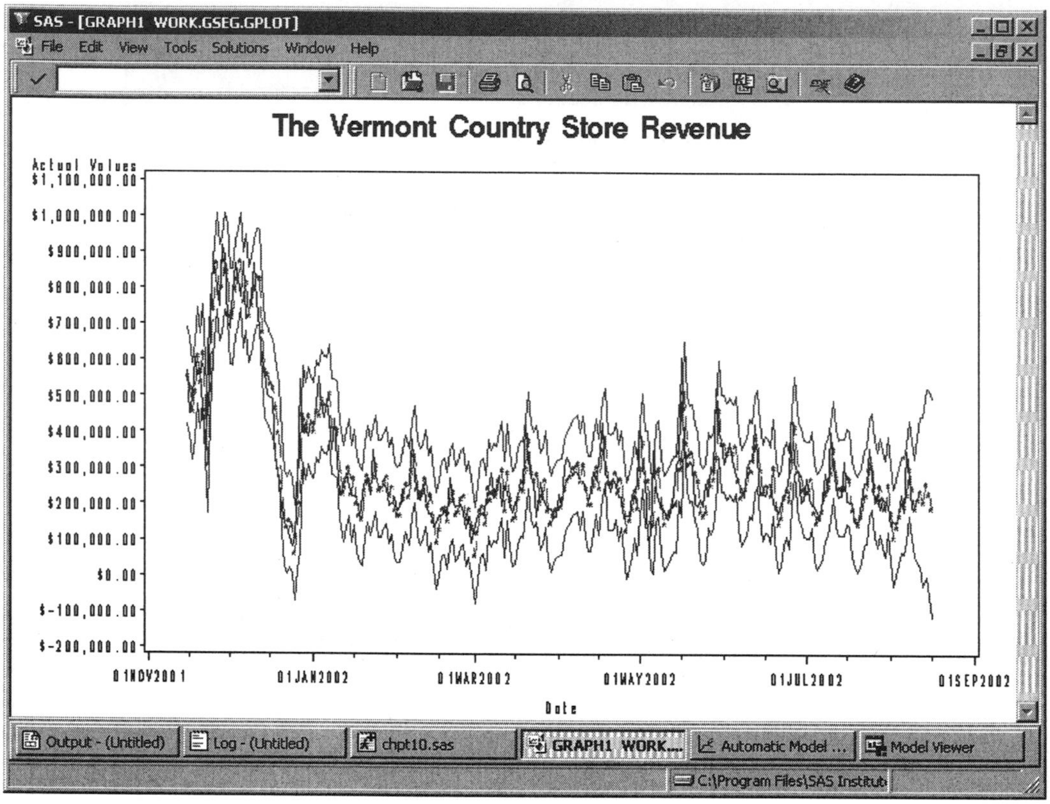

8.5 Scorecard Development

Each day the Vermont Country Store is provided with a report called a "scorecard" that examines its key metrics (variables of interest). The revenue is denoted Revenue (TARGET). The actual value for the day is removed and then forecasted using the HPF procedure. Since the current day's value is removed (9Aug02 in this case), the standard error and forecast estimate are independent of today's observed value. Standardized differences denoted "Difference" $(Y_t - \hat{Y}_t)/s_{\hat{y}_t}$ are also displayed for each metric.

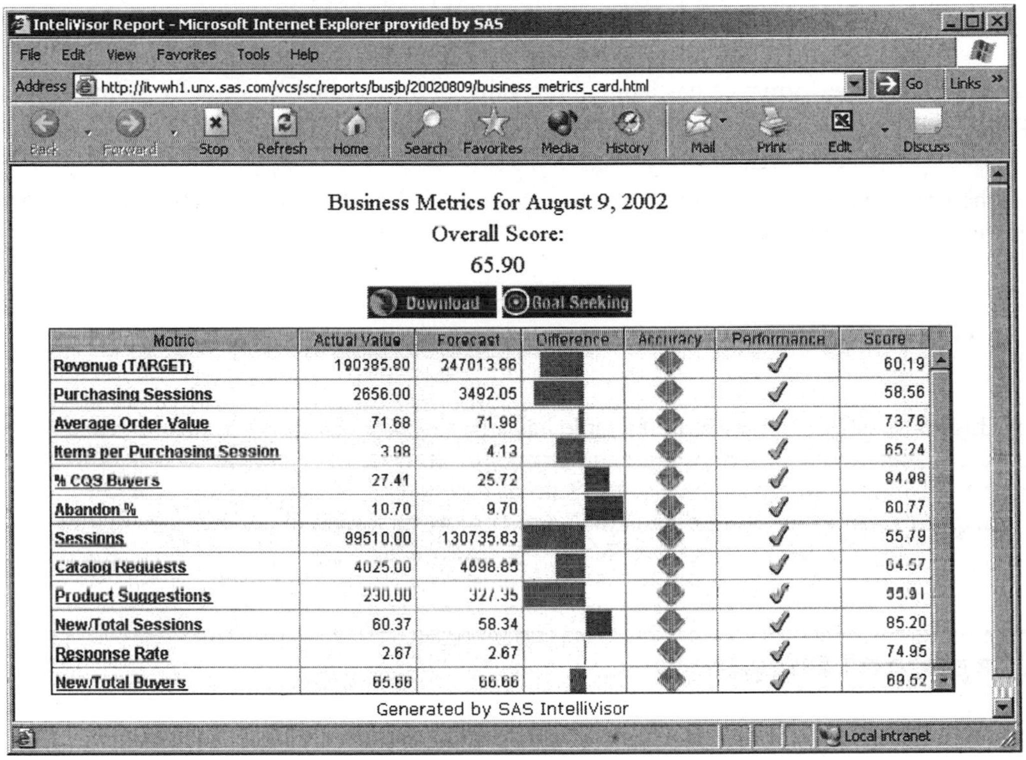

Metric	Actual Value	Forecast	Difference	Accuracy	Performance	Score
Revenue (TARGET)	190385.80	247013.86			✓	60.19
Purchasing Sessions	2656.00	3492.05			✓	58.56
Average Order Value	71.68	71.98			✓	73.76
Items per Purchasing Session	3.98	4.13			✓	65.24
% CQS Buyers	27.41	25.72			✓	84.98
Abandon %	10.70	9.70			✓	60.77
Sessions	99510.00	130735.83			✓	55.79
Catalog Requests	4025.00	4698.85			✓	64.57
Product Suggestions	230.00	327.35			✓	55.91
New/Total Sessions	60.37	58.34			✓	85.20
Response Rate	2.67	2.67			✓	74.95
New/Total Buyers	65.66	66.66			✓	69.52

8.6 Business Goal Performance Metrics

From a retailing business perspective, often you would like the actual values of a metric like Buyer Percent to be larger than the Forecast value so that you are doing better than expected. For a metric like Error Page Percent, smaller values are preferred. For each metric a directional business performance measure is computed for the day. If the preferred direction is greater than the forecast, the calculation is

$$(1/\sqrt{2\pi}\int_{-\infty}^{x} e^{-t^2/2}\,dt/2 + .5\,)*100 \quad \text{where } x = ((\text{Y} - \hat{\text{Y}})/s_{\hat{\text{Y}}})$$

Thus the Business performance has a minimum value of 50% (when Y is small). When Y matches the prediction, the Business performance statistic has a value of 75%; it increases toward 100% as Y gets larger than the prediction.

When the preferred direction of the business movement is less than the prediction, the Business performance measure is calculated as

$$(1-(1/\sqrt{2\pi}\int_{-\infty}^{x} e^{-t^2/2}\,dt/2))*100 \quad \text{where } x = ((\text{Y} - \hat{\text{Y}})/s_{\hat{\text{Y}}})$$

Using this approach, each metric in the table has a Business performance measure. The AUTOREG procedure is then applied by regressing the target (Revenue in this example) on the other metrics and treating 1-pvalues as weight statistics. The sum of products of weight statistics and Business Performance measures gives an overall daily mean score as shown in the previous display.

8.7 Graphical Displays

You can go to the scorecard table and select each metric to display the predictions and limits in a graphical format. In the following display, the scattered black dots represent the observed data, and the dots connected by lines represent the predictions. On the target day (9Aug02) the observed value is removed, so we designate the forecasts and upper and lower 95% prediction intervals with plus signs. Throughout the other historical data, the forecasts and forecast bounds are based on a model developed from the full data set that includes 9Aug02. The same is true of the forecasts and bounds beyond 9Aug02.

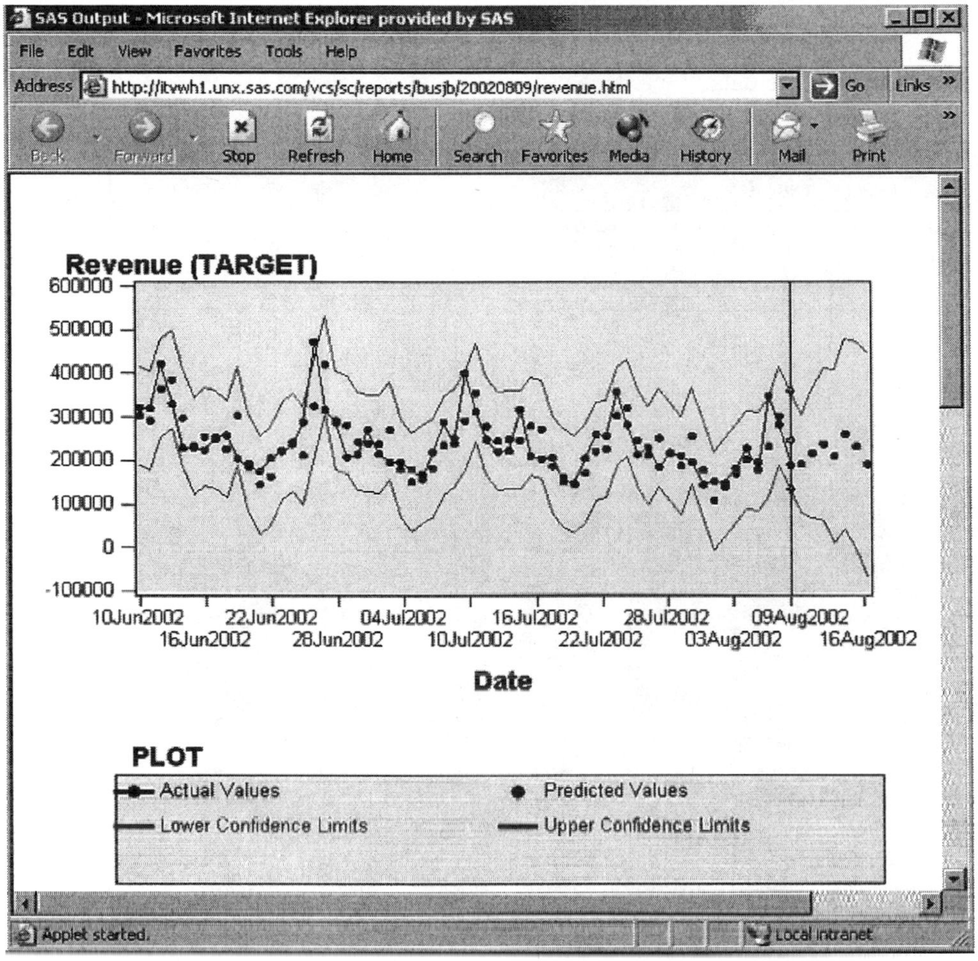
SAS Output - Microsoft Internet Explorer provided by SAS
File Edit View Favorites Tools Help
Address http://itvwh1.unx.sas.com/vcs/sc/reports/busjb/20020809/revenue.html
Go Links
Back Forward Stop Refresh Home Search Favorites Media History Mail Print
Revenue (TARGET)
600000
500000
400000
300000
200000
100000
0
-100000
10Jun2002 22Jun2002 04Jul2002 16Jul2002 28Jul2002 09Aug2002
16Jun2002 28Jun2002 10Jul2002 22Jul2002 03Aug2002 16Aug2002
Date
PLOT
Actual Values
Predicted Values
Lower Confidence Limits
Upper Confidence Limits
Applet started.
Local intranet

If the HPF procedure selects a seasonal model, you will see a display of the daily averages, as shown below. By clicking on a given day of the week, you can also see the associated history for that day over the past history of the data.

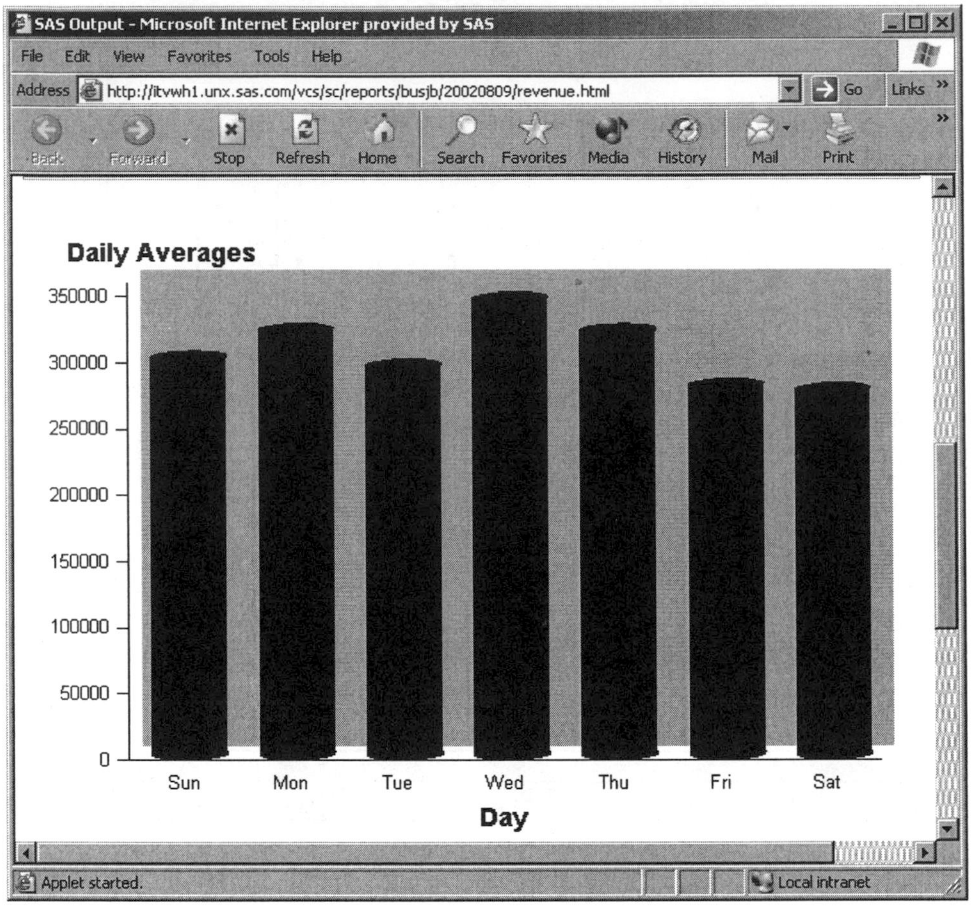

The drop in revenue is also displayed in the chart of past Sunday revenues.

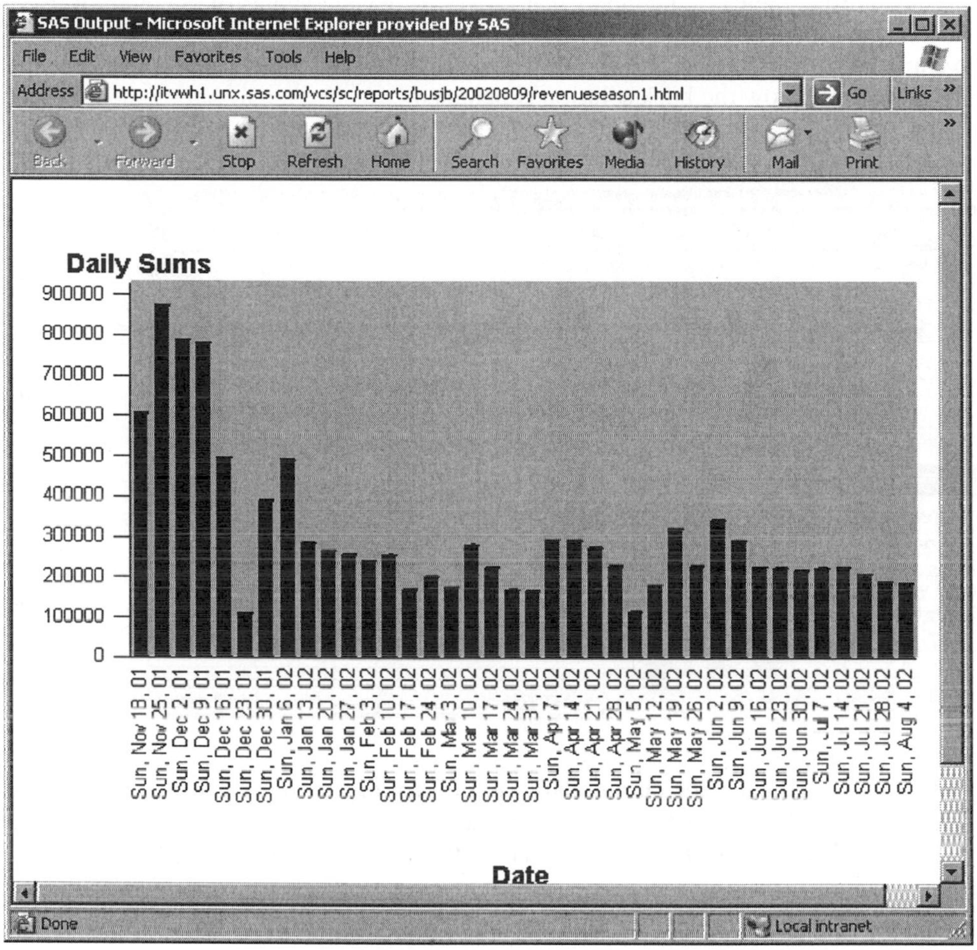

The scorecard also supports the output from a regression with autocorrelation and the ability to solve for inputs one at a time when seeking input values that deliver a specified level of a target. This is done using the SOLVE statement in PROC MODEL. By simply selecting the Goal Seeking Scorecard, you can find values of the inputs that satisfy increasing values of the target Revenue. An example of fitting a model and using it to later solve for values of the inputs is illustrated below. We restrict the explanatory variables to Purchasing Sessions, Average Order Value, and Product Suggestions to illustrate how the back solution is obtained.

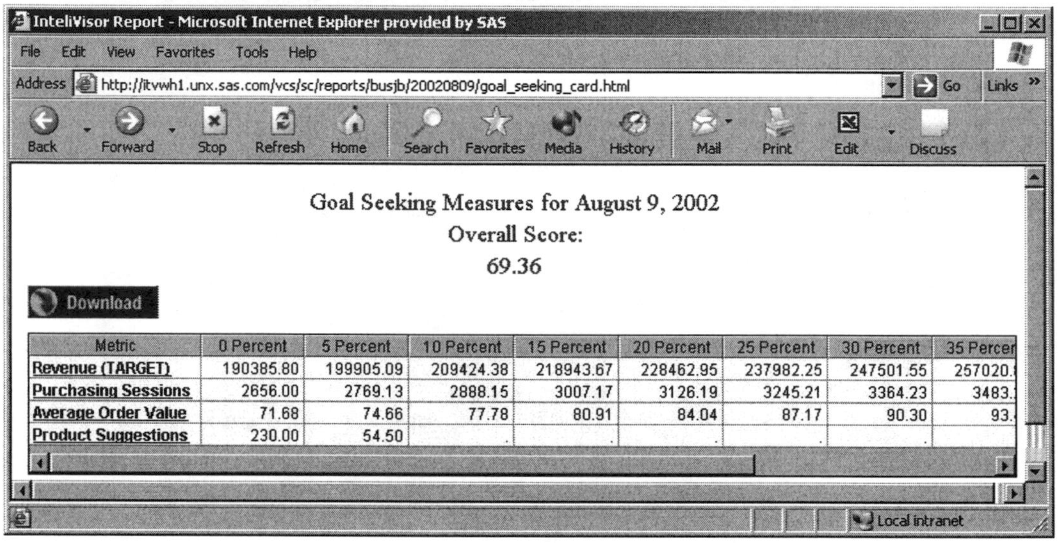

Goal Seeking Measures for August 9, 2002

Overall Score:

69.36

Metric	0 Percent	5 Percent	10 Percent	15 Percent	20 Percent	25 Percent	30 Percent	35 Percen
Revenue (TARGET)	190385.80	199905.09	209424.38	218943.67	228462.95	237982.25	247501.55	257020.
Purchasing Sessions	2656.00	2769.13	2888.15	3007.17	3126.19	3245.21	3364.23	3483.
Average Order Value	71.68	74.66	77.78	80.91	84.04	87.17	90.30	93.
Product Suggestions	230.00	54.50	.	.	.	.	.	

The 0 Percent column indicates the current daily settings for the metrics on 09Aug02. Increasing the target by 5% would set revenue at $199,905.09. To achieve this goal would require 2769 purchasing sessions, assuming all the other inputs remain at their 0 percent level (i.e., the 9Aug01 value).

It is interesting to note that the number of product suggestions would need to drop to 54.5 to achieve this 5% increase. In other words, fewer visitors would be suggesting alternative products to the site and would be more apt to purchase the observed products. Based on the regression results, the number of product suggestions becomes negative (unreasonable) as revenue increases beyond 5%. The display uses metadata (data that characterize positive and negative business directions and acceptable ranges, etc.) that describe reasonable values and set the corresponding negative values to missing. The increasing values for purchasing sessions and average order size provide reasonable results.

8.8 Goal-Seeking Model Development

The MODEL procedure analyzes models in which the relationships among the variables comprise a system of one or more nonlinear equations. The %AR macro can be used to specify models with autoregressive error processes similar to the AUTOREG procedure. In this case we are regressing revenue on buyer, dollar_per_purch_session, and productsuggestion_pages.

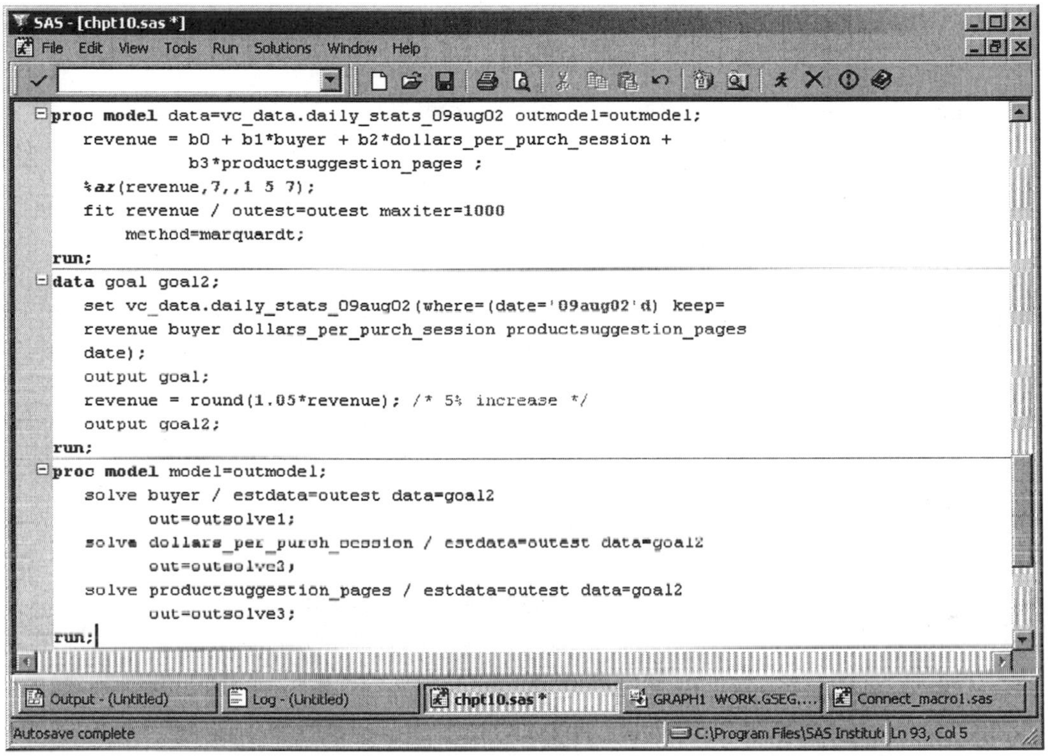


```
proc model data=vc_data.daily_stats_09aug02 outmodel=outmodel;
   revenue = b0 + b1*buyer + b2*dollars_per_purch_session +
             b3*productsuggestion_pages ;
   %ar(revenue,7,,1 5 7);
   fit revenue / outest=outest maxiter=1000
       method=marquardt;
run;
data goal goal2;
   set vc_data.daily_stats_09aug02(where=(date='09aug02'd) keep=
   revenue buyer dollars_per_purch_session productsuggestion_pages
   date);
   output goal;
   revenue = round(1.05*revenue); /* 5% increase */
   output goal2;
run;
proc model model=outmodel;
   solve buyer / estdata=outest data=goal2
         out=outsolve1;
   solve dollars_per_purch_session / estdata=outest data=goal2
         out=outsolve2;
   solve productsuggestion_pages / estdata=outest data=goal2
         out=outsolve3;
run;
```

The SOLVE data set is created to view values of the input variables that satisfy the 5% increase for the target variable Revenue.

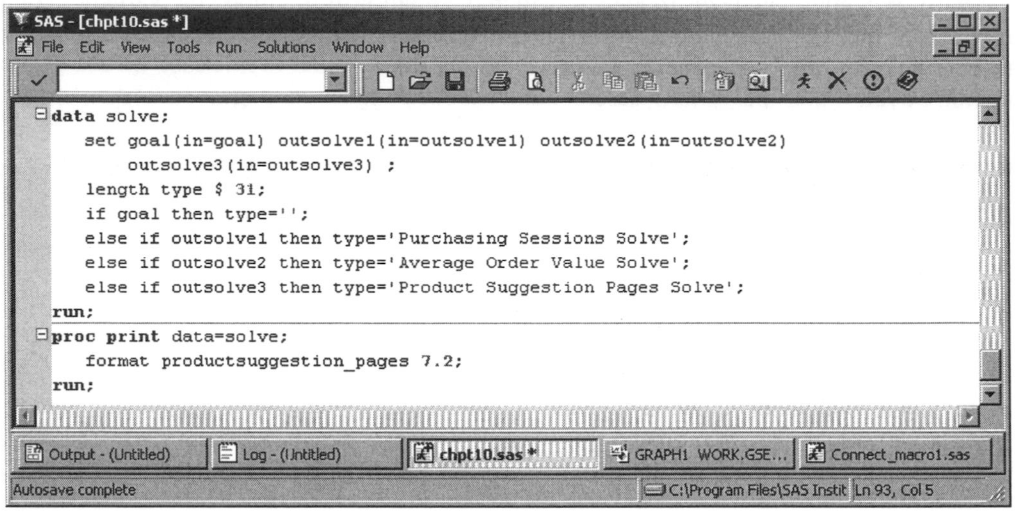

```
data solve;
   set goal(in=goal) outsolve1(in=outsolve1) outsolve2(in=outsolve2)
       outsolve3(in=outsolve3) ;
   length type $ 31;
   if goal then type='';
   else if outsolve1 then type='Purchasing Sessions Solve';
   else if outsolve2 then type='Average Order Value Solve';
   else if outsolve3 then type='Product Suggestion Pages Solve';
run;
proc print data=solve;
   format productsuggestion_pages 7.2;
run;
```

The output below examines the parameter estimates and test statistics. Lags 1, 5, and 7 for the autoregressive errors are statistically different from 0. The signs of the coefficient for purchasing sessions and average order value are positive and negative for product suggestions. The R square and significant parameters and AR terms suggest a reasonable model.

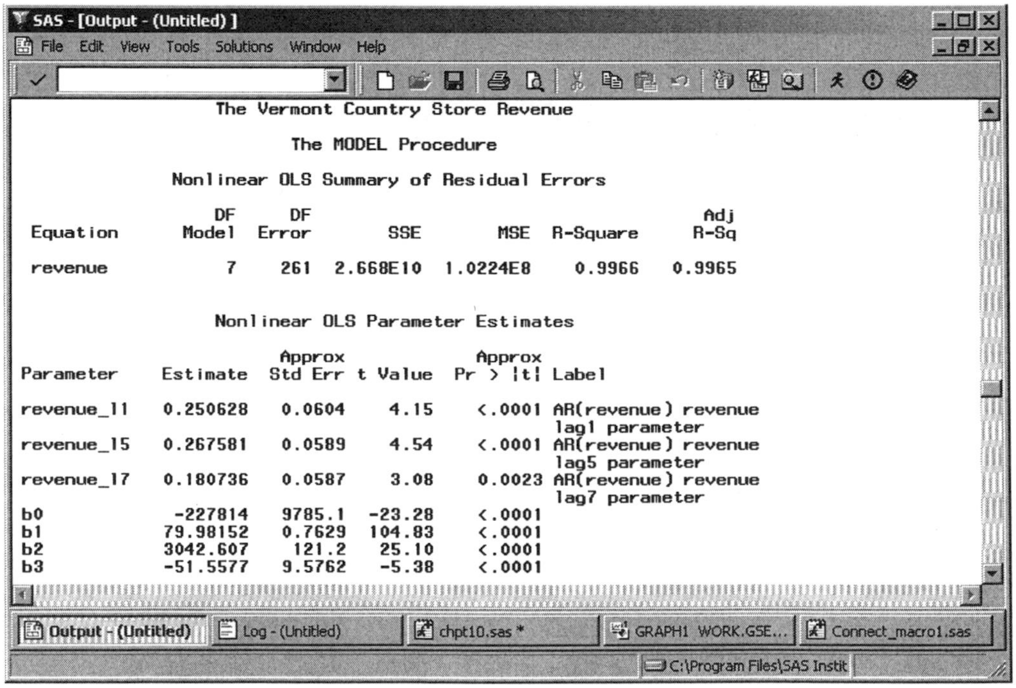

The Vermont Country Store Revenue

The MODEL Procedure

Nonlinear OLS Summary of Residual Errors

Equation	DF Model	DF Error	SSE	MSE	R-Square	Adj R-Sq
revenue	7	261	2.668E10	1.0224E8	0.9966	0.9965

Nonlinear OLS Parameter Estimates

Parameter	Estimate	Approx Std Err	t Value	Approx Pr > |t|	Label
revenue_l1	0.250628	0.0604	4.15	<.0001	AR(revenue) revenue lag1 parameter
revenue_l5	0.267581	0.0589	4.54	<.0001	AR(revenue) revenue lag5 parameter
revenue_l7	0.180736	0.0587	3.08	0.0023	AR(revenue) revenue lag7 parameter
b0	-227814	9785.1	-23.28	<.0001	
b1	79.98152	0.7629	104.83	<.0001	
b2	3042.607	121.2	25.10	<.0001	
b3	-51.5577	9.5762	-5.38	<.0001	

Observation 1 in the SOLVE data set shows the values for current values for the four variables for 09AUG2002. Using the fitted model with autoregressive errors, observations 2 through 4 demonstrate the changes in each individual input required to achieve a 5% increase in revenue, assuming the other inputs are at their current levels. These match the Goal Seeking Scorecard results.

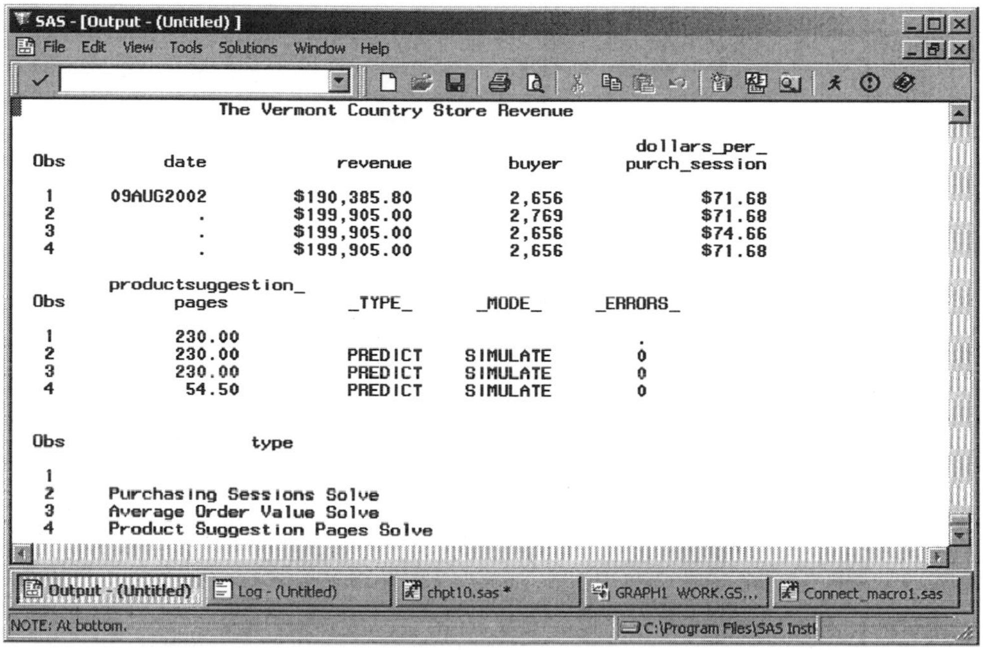

The Vermont Country Store Revenue

Obs	date	revenue	buyer	dollars_per_purch_session
1	09AUG2002	$190,385.80	2,656	$71.68
2	.	$199,905.00	2,769	$71.68
3	.	$199,905.00	2,656	$74.66
4	.	$199,905.00	2,656	$71.68

Obs	productsuggestion_pages	_TYPE_	_MODE_	_ERRORS_
1	230.00			.
2	230.00	PREDICT	SIMULATE	0
3	230.00	PREDICT	SIMULATE	0
4	54.50	PREDICT	SIMULATE	0

Obs	type
1	
2	Purchasing Sessions Solve
3	Average Order Value Solve
4	Product Suggestion Pages Solve

8.9 Summary

This example illustrates how you can apply automated forecasting techniques in a data mining environment. SAS IntelliVisor for Retail through the ASP delivery channel requires the ability to construct analytic results quickly in a batch environment without user intervention. The use of a daily scorecard allows the consumer to focus on what's important and how things are changing over time. By focusing on a goal-seeking report, you can set goals and determine the changes required to produce increasing returns on investment.

References

Akaike, H. 1974. "Markovian Representation of Stochastic Processes and Its Application to the Analysis of Autoregressive Moving Average Processes." *Annals of the Institute of Statistical Mathematics* 26:363–386.

Akaike, H. 1976. "Canonical Correlations of Time Series and the Use of an Information Criterion." In *Advances and Case Studies in System Identification*, ed. R. Mehra and D. G. Lainiotis. New York: Academic Press.

Akdi, Y., and D. A. Dickey. 1997. "Periodograms of Unit Root Time Series: Distributions and Tests." *Communications in Statistics* 27:69–87.

Anderson, T. W. 1971. *The Statistical Analysis of Time Series*. New York: Wiley.

Bailey, C. T. 1984. "Forecasting Industrial Production 1981–1984." *Proceedings of the Ninth Annual SAS Users Group International Conference,* Hollywood Beach, FL, 50–57.

Bartlett, M. S. 1947. "Multivariate Analysis." *Supplement to the Journal of the Royal Statistical Society*, Series B, IX:176–197.

Bartlett, M. S. 1966. *An Introduction to Stochastic Processes*. 2d ed. Cambridge: Cambridge University Press.

Bolerslev, Tim. 1986. "Generalized Autoregressive Conditional Heteroskedasticity." *Journal of Econometrics* 31:307–327.

Box, G. E. P., and D. R. Cox. 1964. "An Analysis of Transformations." *Journal of the Royal Statistical Society* B26:211.

Box, G. E. P., and G. M. Jenkins. 1976. *Time Series Analysis: Forecasting and Control.* Rev. ed. Oakland: Holden-Day.

Box, G. E. P., G. M. Jenkins, and G. C. Reinsel. 1994. *Time Series Analysis: Forecasting and Control.* 3d ed. Englewood Cliffs, NJ: Prentice Hall.

Brillinger, D. R. 1975. *Time Series: Data Analysis and Theory.* New York: Holt, Rinehart & Winston.

Brocklebank, J., and D. A. Dickey. 1984. *SAS Views: SAS Applied Time Series Analysis and Forecasting.* Cary, NC: SAS Institute Inc.

Chang, M. C., and D. A. Dickey. 1993. "Recognizing Overdifferenced Time Series." *Journal of Time Series Analysis* 15:1–8.

Chavern, J. 1984. "On the Limitations of Akaike's Information Criterion and Its Use in PROC STATESPACE." *Proceedings of the Ninth Annual SAS Users Group International Conference,* Hollywood Beach, FL, 106–111.

Cohen, H., ed. 1981. *Metal Statistics.* New York: Fairchild Publications.

Croston, J. D. 1977. "Forecasting and Stock Control for Intermittent Demands." *Operations Research Quarterly* 23, no. 3.

Davis, H. T. 1941. *The Analysis of Economic Time Series.* Chicago: Principia Press.

Dickey, D. A., W. R. Bell, and R. B. Miller. 1986. "Unit Roots in Time Series Models: Tests and Implications." *American Statistician* 40:12–26.

Dickey, D. A., and W. A. Fuller. 1979. "Distribution of the Estimators for Autoregressive Time Series with a Unit Root." *Journal of the American Statistical Association*, 427–431.

Dickey, D. A., and W. A. Fuller. 1981. "Likelihood Ratio Statistics for Autoregressive Time Series with a Unit Root." *Econometrica* 49:1057–1072.

Dickey, D. A., D. P. Hasza, and W. A. Fuller. 1984. "Testing for Unit Roots in Seasonal Time Series." *Journal of the American Statistical Association* 79:355–367.

Dickey, D. A., D. W. Janssen, and D. L. Thornton. 1991. "A Primer on Cointegration with an Application to Money and Income." *Review of the Federal Reserve Bank of St. Louis* 73:58–78.

Draper, N., and H. Smith. 1998. *Applied Regression Analysis.* 3d ed. New York: Wiley.

Durbin, J. 1960. "The Fitting of Time Series Models." *International Statistical Review* 28:233–244.

Engle, Robert. 1982. "Autoregressive Conditional Heteroskedasticity with Estimates of the Variance of United Kingdom Inflation." *Econometrica* 50:987–1007.

Engle, R. F., and C. W. J. Granger. 1987. "Cointegration and Error Correction: Representation, Estimation, and Testing." *Econometrica* 55:251–276.

Evans, B. 1998. "Estimation and Hypothesis Testing in Nonstationary Time Series Using Frequency Domain Methods." Ph.D. diss., North Carolina State University.

Fountis, N. G., and D. A. Dickey. 1989. "Testing for a Unit Root Nonstationarity in Multivariate Autoregressive Time Series." *Annals of Statistics* 17:419–428.

Fuller, W. A. 1976. *Introduction to Statistical Time Series.* New York: Wiley.

Fuller, W. A. 1986. "Using PROC NLIN for Time Series Prediction." *Proceedings of the Eleventh Annual SAS Users Group International Conference,* Atlanta, GA, 63–68.

Fuller, W.A. 1996. *Introduction to Statistical Time Series*. 2d ed. New York: Wiley.

Hall, A. 1992. "Testing for a Unit Root in Time Series with Data-Based Model Selection." *Journal of Business and Economic Statistics* 12:461–470.

Hamilton, J. D. 1994. *Time Series Analysis*. Princeton, NJ: Princeton University Press.

Hannan, E. J., and J. Rissanen. 1982. "Recursive Estimation of Mixed Autoregressive-Moving Average Order." *Biometrika* 69, no. 1 (April): 81–94.

Harvey, A. C. 1981. *Time Series Models.* Oxford: Philip Allan Publishers.

Jarque, C. M., and A. K. Bera. 1980. "Efficient Tests for Normality, Homoskedasticity and Serial Independence of Regression Residuals." *Economics Letters* 6:255–259.

Jenkins, G. M., and D. G. Watts. 1968. *Spectral Analysis and Its Applications*. Oakland: Holden-Day.

Johansen, S. 1988. "Statistical Analysis of Cointegrating Vectors." *Journal of Economic Dynamics and Control* 12:312–254.

Johansen, S. 1991. "Estimation and Hypothesis Testing of Cointegrating Vectors in Gaussian Vector Autoregressive Models." *Econometrica* 59:1551–1580.

Johansen, S. 1994. "The Role of the Constant and Linear Terms in Cointegration Analysis of Non-Stationary Variables." *Econometric Reviews* 13:205–230.

Johnston, J. 1972. *Econometric Methods*. 2d ed. New York: McGraw-Hill.

Jones, R. H. 1974. "Identification and Autoregressive Spectrum Estimation." *IEEE Transactions on Automatic Control*, AC-19:894–897.

Liu, Shiping, Ju-Chin Huang, and Gregory L. Brown. 1988. "Information and Risk Perception: A Dynamic Adjustment Process." *Risk Analysis* 18:689–699.

Ljung, G. M., and G. E. P. Box. 1978. "On a Measure of Lack of Fit in Time Series Models." *Biometrika* 65:297–303.

McSweeny, A. J. 1978. "Effects of Response Cost on the Behavior of a Million Persons: Charging for Directory Assistance in Cincinnati." *Journal of Applied Behavior Analysis* 11:47–51.

Nelson, D. B. 1991. "Conditional Heteroskedasticity in Asset Returns: A New Approach." *Econometrica* 59:347–370.

Nelson, D. B., and C. Q. Cao. 1992. "Inequality Constraints in the Univariate GARCH Model." *Journal of Business and Economic Statistics* 10:229–235.

Pham, D. T. 1978. "On the Fitting of Multivariate Process of the Autoregressive-Moving Average Type." *Biometrika* 65:99–107.

Priestley, M. B. 1980. "System Identification, Kalman Filtering, and Stochastic Control." In *Directions in Time Series*, ed. D. R. Brillinger and G. C. Tiao. Hayward CA: Institute of Mathematical Statistics.

Priestley, M. B. 1981. *Spectra Analysis and Time Series.* Volume 1: *Univariate Series.* New York: Academic Press.

Robinson, P. M. 1973. "Generalized Canonical Analysis for Time Series." *Journal of Multivariate Analysis* 3:141–160.

Said, S. E., and D. A. Dickey. 1984. "Testing for Unit Roots in Autoregressive Moving Average Models of Unknown Order." *Biometrika* 71, no. 3: 599–607.

SAS Institute Inc. 1984. *SAS/ETS User's Guide, Version 5 Edition.* Cary, NC: SAS Institute Inc.

SAS Institute Inc. 1985. *SAS/GRAPH User's Guide, Version 5 Edition*. Cary, NC: SAS Institute Inc.

SAS Institute Inc. 1985. *SAS Introductory Guide.* 3d ed. Cary, NC: SAS Institute Inc.

SAS Institute Inc. 1985. *SAS User's Guide: Basics, Version 5 Edition.* Cary, NC: SAS Institute Inc.

SAS Institute Inc. 1985. *SAS User's Guide: Statistics, Version 5 Edition.* Cary, NC: SAS Institute Inc.

Singleton, R. C. 1969. "An Algorithm for Computing the Mixed Radix Fast Fourier Transform." *IEEE Transactions of Audio and Electroacoustics*, AU-17:93–103.

Stock, J. H., and W. W. Watson. 1988. "Testing for Common Trends." *Journal of the American Statistical Association* 83:1097–1107.

Tsay, Ruey S., and George C. Tiao. 1984. "Consistent Estimates of Autoregressive Parameters and Extended Sample Autocorrelation Function for Stationary and Nonstationary ARMA Models." *Journal of the American Statistical Association* 79, no. 385 (March): 84–96.

Tsay, Ruey S., and George C. Tiao. 1985. "Use of Canonical Analysis in Time Series Model Identification." *Biometrika* 72, no. 2 (August): 299–315.

U.S. Bureau of Census. 1982. "Construction Workers in Thousands." *Construction Review.*

U.S. Department of Labor. 1977. "Publishing and Printing Nonproduction Workers 1944–1977." *Handbook of Labor Statistics.*

Whittle, P. 1963. "On the Fitting of Multivariate Autoregressions and the Approximate Canonical Factorization of a Spectral Density Matrix." *Biometrika* 50:129–134.

Index

A

D

E

F

J

K

L

M

Output Delivery System: The Basics
by **Lauren E. Haworth**

Painless Windows: A Handbook for SAS® Users
by **Jodie Gilmore**
(for Windows NT and Windows 95)

Painless Windows: A Handbook for SAS® Users, Second Edition
by **Jodie Gilmore**
(updated to include Version 7 features)

PROC TABULATE by Example
by **Lauren E. Haworth**

Professional SAS® Programmer's Pocket Reference, Fourth Edition
by **Rick Aster**

Professional SAS® Programmer's Pocket Reference, Second Edition
by **Rick Aster**

Professional SAS® Programming Shortcuts
by **Rick Aster**

Programming Techniques for Object-Based Statistical Analysis with SAS® Software
by **Tanya Kolosova**
and **Samuel Berestizhevsky**

Quick Results with SAS/GRAPH® Software
by **Arthur L. Carpenter**
and **Charles E. Shipp**

Quick Results with the Output Delivery System
by **Sunil K. Gupta**

Quick Start to Data Analysis with SAS®
by **Frank C. Dilorio**
and **Kenneth A. Hardy**

Reading External Data Files Using SAS®: Examples Handbook
by **Michele M. Burlew**

Regression and ANOVA: An Integrated Approach Using SAS® Software
by **Keith E. Muller**
and **Bethel A. Fetterman**

Reporting from the Field: SAS® Software Experts Present Real-World Report-Writing Applications

SAS® Applications Programming: A Gentle Introduction
by **Frank C. Dilorio**

SAS® for Forecasting Time Series, Second Edition
by **John C. Brocklebank,**
and **David A. Dickey**

SAS® for Linear Models, Fourth Edition
by **Ramon C. Littell, Walter W. Stroup,**
and **Rudolf J. Freund**

SAS® for Monte Carlo Studies: A Guide for Quantitative Researchers
by **Xitao Fan, Ákos Felsővályi, Stephen A. Sivo,**
and **Sean C. Keenan**

SAS® Macro Programming Made Easy
by **Michele M. Burlew**

SAS® Programming by Example
by **Ron Cody**
and **Ray Pass**

SAS® Programming for Researchers and Social Scientists, Second Edition
by **Paul E. Spector**

SAS® Software Roadmaps: Your Guide to Discovering the SAS® System
by **Laurie Burch**
and **SherriJoyce King**

SAS® Software Solutions: Basic Data Processing
by **Thomas Miron**

SAS® Survival Analysis Techniques for Medical Research, Second Edition
by **Alan B. Cantor**

SAS® System for Elementary Statistical Analysis, Second Edition
by **Sandra D. Schlotzhauer**
and **Ramon C. Littell**

SAS® System for Forecasting Time Series, 1986 Edition
by **John C. Brocklebank**
and **David A. Dickey**

SAS® System for Mixed Models
by **Ramon C. Littell, George A. Milliken, Walter W. Stroup,**
and **Russell D. Wolfinger**

SAS® System for Regression, Third Edition
by **Rudolf J. Freund**
and **Ramon C. Littell**

SAS® System for Statistical Graphics, First Edition
by **Michael Friendly**

The SAS® Workbook and *Solutions* Set
(books in this set also sold separately)
by **Ron Cody**

Selecting Statistical Techniques for Social Science Data: A Guide for SAS® Users
by **Frank M. Andrews, Laura Klem, Patrick M. O'Malley, Willard L. Rodgers, Kathleen B. Welch,**
and **Terrence N. Davidson**

Solutions for Your GUI Applications Development Using SAS/AF® FRAME Technology
by **Don Stanley**

Statistical Quality Control Using the SAS® System
by **Dennis W. King**

A Step-by-Step Approach to Using the SAS® System for Factor Analysis and Structural Equation Modeling
by **Larry Hatcher**

A Step-by-Step Approach to Using the SAS® System for Univariate and Multivariate Statistics
by **Larry Hatcher**
and **Edward Stepanski**

Step-by-Step Basic Statistics Using SAS®: Student Guide and *Exercises*
(books in this set also sold separately)
by **Larry Hatcher**

Strategic Data Warehousing Principles Using SAS® Software
by **Peter R. Welbrock**

Survival Analysis Using the SAS® System: A Practical Guide
by **Paul D. Allison**

Table-Driven Strategies for Rapid SAS® Applications Development
by **Tanya Kolosova**
and **Samuel Berestizhevsky**

Tuning SAS® Applications in the MVS Environment
by **Michael A. Raithel**

Univariate and Multivariate General Linear Models: Theory and Applications Using SAS® Software
by **Neil H. Timm**
and **Tammy A. Mieczkowski**

Using SAS® in Financial Research
by **Ekkehart Boehmer, John Paul Broussard,**
and **Juha-Pekka Kallunki**

Using the SAS® Windowing Environment: A Quick Tutorial
by **Larry Hatcher**

Visualizing Categorical Data
by **Michael Friendly**

Working with the SAS® System
by **Erik W. Tilanus**

Your Guide to Survey Research Using the SAS® System
by **Archer Gravely**

JMP® Books

Basic Business Statistics: A Casebook
by **Dean P. Foster, Robert A. Stine,**
and **Richard P. Waterman**

Business Analysis Using Regression: A Casebook
by **Dean P. Foster, Robert A. Stine,**
and **Richard P. Waterman**

JMP® Start Statistics, Second Edition
by **John Sall, Ann Lehman,**
and **Lee Creighton**

Regression Using JMP®
by **Rudolf J. Freund, Ramon C. Littell,**
and **Lee Creighton**

WILEY SERIES IN PROBABILITY AND STATISTICS

Established by WALTER A. SHEWHART and SAMUEL S. WILKS

A complete list of the titles in this series appears at the end of this volume.

WILEY SERIES IN PROBABILITY AND STATISTICS
ESTABLISHED BY WALTER A. SHEWHART AND SAMUEL S. WILKS

The ***Wiley Series in Probability and Statistics*** is well established and authoritative. It covers many topics of current research interest in both pure and applied statistics and probability theory. Written by leading statisticians and institutions, the titles span both state-of-the-art developments in the field and classical methods.

Reflecting the wide range of current research in statistics, the series encompasses applied, methodological and theoretical statistics, ranging from applications and new techniques made possible by advances in computerized practice to rigorous treatment of theoretical approaches.

This series provides essential and invaluable reading for all statisticians, whether in academia, industry, government, or research.

ABRAHAM and LEDOLTER · Statistical Methods for Forecasting
AGRESTI · Analysis of Ordinal Categorical Data
AGRESTI · An Introduction to Categorical Data Analysis
AGRESTI · Categorical Data Analysis, *Second Edition*
ANDĚL · Mathematics of Chance
ANDERSON · An Introduction to Multivariate Statistical Analysis, *Second Edition*
*ANDERSON · The Statistical Analysis of Time Series
ANDERSON, AUQUIER, HAUCK, OAKES, VANDAELE, and WEISBERG · Statistical Methods for Comparative Studies
ANDERSON and LOYNES · The Teaching of Practical Statistics
ARMITAGE and DAVID (editors) · Advances in Biometry
ARNOLD, BALAKRISHNAN, and NAGARAJA · Records
*ARTHANARI and DODGE · Mathematical Programming in Statistics
*BAILEY · The Elements of Stochastic Processes with Applications to the Natural Sciences
BALAKRISHNAN and KOUTRAS · Runs and Scans with Applications
BARNETT · Comparative Statistical Inference, *Third Edition*
BARNETT and LEWIS · Outliers in Statistical Data, *Third Edition*
BARTOSZYNSKI and NIEWIADOMSKA-BUGAJ · Probability and Statistical Inference
BASILEVSKY · Statistical Factor Analysis and Related Methods: Theory and Applications
BASU and RIGDON · Statistical Methods for the Reliability of Repairable Systems
BATES and WATTS · Nonlinear Regression Analysis and Its Applications
BECHHOFER, SANTNER, and GOLDSMAN · Design and Analysis of Experiments for Statistical Selection, Screening, and Multiple Comparisons
BELSLEY · Conditioning Diagnostics: Collinearity and Weak Data in Regression
BELSLEY, KUH, and WELSCH · Regression Diagnostics: Identifying Influential Data and Sources of Collinearity
BENDAT and PIERSOL · Random Data: Analysis and Measurement Procedures, *Third Edition*
BERRY, CHALONER, and GEWEKE · Bayesian Analysis in Statistics and Econometrics: Essays in Honor of Arnold Zellner
BERNARDO and SMITH · Bayesian Theory
BHAT and MILLER · Elements of Applied Stochastic Processes, *Third Edition*
BHATTACHARYA and JOHNSON · Statistical Concepts and Methods
BHATTACHARYA and WAYMIRE · Stochastic Processes with Applications
BILLINGSLEY · Convergence of Probability Measures, *Second Edition*
BILLINGSLEY · Probability and Measure, *Third Edition*
BIRKES and DODGE · Alternative Methods of Regression
BLISCHKE AND MURTHY (editors) · Case Studies in Reliability and Maintenance
BLISCHKE AND MURTHY · Reliability: Modeling, Prediction, and Optimization
BLOOMFIELD · Fourier Analysis of Time Series: An Introduction, *Second Edition*
BOLLEN · Structural Equations with Latent Variables
BOROVKOV · Ergodicity and Stability of Stochastic Processes
BOULEAU · Numerical Methods for Stochastic Processes
BOX · Bayesian Inference in Statistical Analysis
BOX · R. A. Fisher, the Life of a Scientist
BOX and DRAPER · Empirical Model-Building and Response Surfaces
*BOX and DRAPER · Evolutionary Operation: A Statistical Method for Process Improvement
BOX, HUNTER, and HUNTER · Statistics for Experimenters: An Introduction to Design, Data Analysis, and Model Building
BOX and LUCEÑO · Statistical Control by Monitoring and Feedback Adjustment
BRANDIMARTE · Numerical Methods in Finance: A MATLAB-Based Introduction
BROWN and HOLLANDER · Statistics: A Biomedical Introduction

*Now available in a lower priced paperback edition in the Wiley Classics Library.

BRUNNER, DOMHOF, and LANGER · Nonparametric Analysis of Longitudinal Data in Factorial Experiments
BUCKLEW · Large Deviation Techniques in Decision, Simulation, and Estimation
CAIROLI and DALANG · Sequential Stochastic Optimization
CHAN · Time Series: Applications to Finance
CHATTERJEE and HADI · Sensitivity Analysis in Linear Regression
CHATTERJEE and PRICE · Regression Analysis by Example, *Third Edition*
CHERNICK · Bootstrap Methods: A Practitioner's Guide
CHERNICK and FRIIS · Introductory Biostatistics for the Health Sciences
CHILÈS and DELFINER · Geostatistics: Modeling Spatial Uncertainty
CHOW and LIU · Design and Analysis of Clinical Trials: Concepts and Methodologies
CLARKE and DISNEY · Probability and Random Processes: A First Course with Applications, *Second Edition*
*COCHRAN and COX · Experimental Designs, *Second Edition*
CONGDON · Bayesian Statistical Modelling
CONOVER · Practical Nonparametric Statistics, *Second Edition*
COOK · Regression Graphics
COOK and WEISBERG · Applied Regression Including Computing and Graphics
COOK and WEISBERG · An Introduction to Regression Graphics
CORNELL · Experiments with Mixtures, Designs, Models, and the Analysis of Mixture Data, *Third Edition*
COVER and THOMAS · Elements of Information Theory
COX · A Handbook of Introductory Statistical Methods
*COX · Planning of Experiments
CRESSIE · Statistics for Spatial Data, *Revised Edition*
CSÖRGŐ and HORVÁTH · Limit Theorems in Change Point Analysis
DANIEL · Applications of Statistics to Industrial Experimentation
DANIEL · Biostatistics: A Foundation for Analysis in the Health Sciences, *Sixth Edition*
*DANIEL · Fitting Equations to Data: Computer Analysis of Multifactor Data, *Second Edition*
DASU and JOHNSON · Exploratory Data Mining and Data Cleaning
DAVID · Order Statistics, *Second Edition*
*DEGROOT, FIENBERG, and KADANE · Statistics and the Law
DEL CASTILLO · Statistical Process Adjustment for Quality Control
DETTE and STUDDEN · The Theory of Canonical Moments with Applications in Statistics, Probability, and Analysis
DEY and MUKERJEE · Fractional Factorial Plans
DILLON and GOLDSTEIN · Multivariate Analysis: Methods and Applications
DODGE · Alternative Methods of Regression
*DODGE and ROMIG · Sampling Inspection Tables, *Second Edition*
*DOOB · Stochastic Processes
DOWDY and WEARDEN · Statistics for Research, *Second Edition*
DRAPER and SMITH · Applied Regression Analysis, *Third Edition*
DRYDEN and MARDIA · Statistical Shape Analysis
DUDEWICZ and MISHRA · Modern Mathematical Statistics
DUNN and CLARK · Applied Statistics: Analysis of Variance and Regression, *Second Edition*
DUNN and CLARK · Basic Statistics: A Primer for the Biomedical Sciences, *Third Edition*
DUPUIS and ELLIS · A Weak Convergence Approach to the Theory of Large Deviations
*ELANDT-JOHNSON and JOHNSON · Survival Models and Data Analysis
ENDERS · Applied Econometric Time Series
ETHIER and KURTZ · Markov Processes: Characterization and Convergence
EVANS, HASTINGS, and PEACOCK · Statistical Distributions, *Third Edition*
FELLER · An Introduction to Probability Theory and Its Applications, Volume I, *Third Edition,* Revised; Volume II, *Second Edition*
FISHER and VAN BELLE · Biostatistics: A Methodology for the Health Sciences
*FLEISS · The Design and Analysis of Clinical Experiments
FLEISS · Statistical Methods for Rates and Proportions, *Second Edition*
FLEMING and HARRINGTON · Counting Processes and Survival Analysis
FULLER · Introduction to Statistical Time Series, *Second Edition*
FULLER · Measurement Error Models
GALLANT · Nonlinear Statistical Models
GHOSH, MUKHOPADHYAY, and SEN · Sequential Estimation
GIFI · Nonlinear Multivariate Analysis
GLASSERMAN and YAO · Monotone Structure in Discrete-Event Systems
GNANADESIKAN · Methods for Statistical Data Analysis of Multivariate Observations, *Second Edition*
GOLDSTEIN and LEWIS · Assessment: Problems, Development, and Statistical Issues
GREENWOOD and NIKULIN · A Guide to Chi-Squared Testing
GROSS and HARRIS · Fundamentals of Queueing Theory, *Third Edition*
*HAHN and SHAPIRO · Statistical Models in Engineering
HAHN and MEEKER · Statistical Intervals: A Guide for Practitioners
HALD · A History of Probability and Statistics and their Applications Before 1750

*Now available in a lower priced paperback edition in the Wiley Classics Library.

HALD · A History of Mathematical Statistics from 1750 to 1930
HAMPEL · Robust Statistics: The Approach Based on Influence Functions
HANNAN and DEISTLER · The Statistical Theory of Linear Systems
HEIBERGER · Computation for the Analysis of Designed Experiments
HEDAYAT and SINHA · Design and Inference in Finite Population Sampling
HELLER · MACSYMA for Statisticians
HINKELMAN and KEMPTHORNE: · Design and Analysis of Experiments, Volume 1: Introduction to Experimental Design
HOAGLIN, MOSTELLER, and TUKEY · Exploratory Approach to Analysis of Variance
HOAGLIN, MOSTELLER, and TUKEY · Exploring Data Tables, Trends and Shapes
*HOAGLIN, MOSTELLER, and TUKEY · Understanding Robust and Exploratory Data Analysis
HOCHBERG and TAMHANE · Multiple Comparison Procedures
HOCKING · Methods and Applications of Linear Models: Regression and the Analysis of Variance, *Second Edition*
HOEL · Introduction to Mathematical Statistics, *Fifth Edition*
HOGG and KLUGMAN · Loss Distributions
HOLLANDER and WOLFE · Nonparametric Statistical Methods, *Second Edition*
HOSMER and LEMESHOW · Applied Logistic Regression, *Second Edition*
HOSMER and LEMESHOW · Applied Survival Analysis: Regression Modeling of Time to Event Data
HØYLAND and RAUSAND · System Reliability Theory: Models and Statistical Methods
HUBER · Robust Statistics
HUBERTY · Applied Discriminant Analysis
HUNT and KENNEDY · Financial Derivatives in Theory and Practice
HUSKOVA, BERAN, and DUPAC · Collected Works of Jaroslav Hajek—with Commentary
IMAN and CONOVER · A Modern Approach to Statistics
JACKSON · A User's Guide to Principle Components
JOHN · Statistical Methods in Engineering and Quality Assurance
JOHNSON · Multivariate Statistical Simulation
JOHNSON and BALAKRISHNAN · Advances in the Theory and Practice of Statistics: A Volume in Honor of Samuel Kotz
JUDGE, GRIFFITHS, HILL, LÜTKEPOHL, and LEE · The Theory and Practice of Econometrics, *Second Edition*
JOHNSON and KOTZ · Distributions in Statistics
JOHNSON and KOTZ (editors) · Leading Personalities in Statistical Sciences: From the Seventeenth Century to the Present
JOHNSON, KOTZ, and BALAKRISHNAN · Continuous Univariate Distributions, Volume 1, *Second Edition*
JOHNSON, KOTZ, and BALAKRISHNAN · Continuous Univariate Distributions, Volume 2, *Second Edition*
JOHNSON, KOTZ, and BALAKRISHNAN · Discrete Multivariate Distributions
JOHNSON, KOTZ, and KEMP · Univariate Discrete Distributions, *Second Edition*
JUREČKOVÁ and SEN · Robust Statistical Procedures: Aymptotics and Interrelations
JUREK and MASON · Operator-Limit Distributions in Probability Theory
KADANE · Bayesian Methods and Ethics in a Clinical Trial Design
KADANE AND SCHUM · A Probabilistic Analysis of the Sacco and Vanzetti Evidence
KALBFLEISCH and PRENTICE · The Statistical Analysis of Failure Time Data, *Second Edition*
KASS and VOS · Geometrical Foundations of Asymptotic Inference
KAUFMAN and ROUSSEEUW · Finding Groups in Data: An Introduction to Cluster Analysis
KEDEM and FOKIANOS · Regression Models for Time Series Analysis
KENDALL, BARDEN, CARNE, and LE · Shape and Shape Theory
KHURI · Advanced Calculus with Applications in Statistics, *Second Edition*
KHURI, MATHEW, and SINHA · Statistical Tests for Mixed Linear Models
KLUGMAN, PANJER, and WILLMOT · Loss Models: From Data to Decisions
KLUGMAN, PANJER, and WILLMOT · Solutions Manual to Accompany Loss Models: From Data to Decisions
KOTZ, BALAKRISHNAN, and JOHNSON · Continuous Multivariate Distributions, Volume 1, *Second Edition*
KOTZ and JOHNSON (editors) · Encyclopedia of Statistical Sciences: Volumes 1 to 9 with Index
KOTZ and JOHNSON (editors) · Encyclopedia of Statistical Sciences: Supplement Volume
KOTZ, READ, and BANKS (editors) · Encyclopedia of Statistical Sciences: Update Volume 1
KOTZ, READ, and BANKS (editors) · Encyclopedia of Statistical Sciences: Update Volume 2
KOVALENKO, KUZNETZOV, and PEGG · Mathematical Theory of Reliability of Time-Dependent Systems with Practical Applications
LACHIN · Biostatistical Methods: The Assessment of Relative Risks
LAD · Operational Subjective Statistical Methods: A Mathematical, Philosophical, and Historical Introduction
LAMPERTI · Probability: A Survey of the Mathematical Theory, *Second Edition*
LANGE, RYAN, BILLARD, BRILLINGER, CONQUEST, and GREENHOUSE · Case Studies in Biometry
LARSON · Introduction to Probability Theory and Statistical Inference, *Third Edition*
LAWLESS · Statistical Models and Methods for Lifetime Data, *Second Edition*
LAWSON · Statistical Methods in Spatial Epidemiology
LE · Applied Categorical Data Analysis
LE · Applied Survival Analysis
LEE and WANG · Statistical Methods for Survival Data Analysis, *Third Edition*
LEPAGE and BILLARD · Exploring the Limits of Bootstrap
LEYLAND and GOLDSTEIN (editors) · Multilevel Modelling of Health Statistics
LIAO · Statistical Group Comparison

*Now available in a lower priced paperback edition in the Wiley Classics Library.

LINDVALL · Lectures on the Coupling Method
LINHART and ZUCCHINI · Model Selection
LITTLE and RUBIN · Statistical Analysis with Missing Data, *Second Edition*
LLOYD · The Statistical Analysis of Categorical Data
MAGNUS and NEUDECKER · Matrix Differential Calculus with Applications in Statistics and Econometrics, *Revised Edition*
MALLER and ZHOU · Survival Analysis with Long Term Survivors
MALLOWS · Design, Data, and Analysis by Some Friends of Cuthbert Daniel
MANN, SCHAFER, and SINGPURWALLA · Methods for Statistical Analysis of Reliability and Life Data
MANTON, WOODBURY, and TOLLEY · Statistical Applications Using Fuzzy Sets
MARDIA and JUPP · Directional Statistics
MASON, GUNST, and HESS · Statistical Design and Analysis of Experiments with Applications to Engineering and Science, *Second Edition*
McCULLOCH and SEARLE · Generalized, Linear, and Mixed Models
McFADDEN · Management of Data in Clinical Trials
McLACHLAN · Discriminant Analysis and Statistical Pattern Recognition
McLACHLAN and KRISHNAN · The EM Algorithm and Extensions
McLACHLAN and PEEL · Finite Mixture Models
McNEIL · Epidemiological Research Methods
MEEKER and ESCOBAR · Statistical Methods for Reliability Data
MEERSCHAERT and SCHEFFLER · Limit Distributions for Sums of Independent Random Vectors: Heavy Tails in Theory and Practice
*MILLER · Survival Analysis, *Second Edition*
MONTGOMERY, PECK, and VINING · Introduction to Linear Regression Analysis, *Third Edition*
MORGENTHALER and TUKEY · Configural Polysampling: A Route to Practical Robustness
MUIRHEAD · Aspects of Multivariate Statistical Theory
MURRAY · X-STAT 2.0 Statistical Experimentation, Design Data Analysis, and Nonlinear Optimization
MYERS and MONTGOMERY · Response Surface Methodology: Process and Product Optimization Using Designed Experiments, *Second Edition*
MYERS, MONTGOMERY, and VINING · Generalized Linear Models. With Applications in Engineering and the Sciences
NELSON · Accelerated Testing, Statistical Models, Test Plans, and Data Analyses
NELSON · Applied Life Data Analysis
NEWMAN · Biostatistical Methods in Epidemiology
OCHI · Applied Probability and Stochastic Processes in Engineering and Physical Sciences
OKABE, BOOTS, SUGIHARA, and CHIU · Spatial Tesselations: Concepts and Applications of Voronoi Diagrams, *Second Edition*
OLIVER and SMITH · Influence Diagrams, Belief Nets and Decision Analysis
PANKRATZ · Forecasting with Dynamic Regression Models
PANKRATZ · Forecasting with Univariate Box-Jenkins Models: Concepts and Cases
*PARZEN · Modern Probability Theory and Its Applications
PEÑA, TIAO, and TSAY · A Course in Time Series Analysis
PIANTADOSI · Clinical Trials: A Methodologic Perspective
PORT · Theoretical Probability for Applications
POURAHMADI · Foundations of Time Series Analysis and Prediction Theory
PRESS · Bayesian Statistics: Principles, Models, and Applications
PRESS · Subjective and Objective Bayesian Statistics, *Second Edition*
PRESS and TANUR · The Subjectivity of Scientists and the Bayesian Approach
PUKELSHEIM · Optimal Experimental Design
PURI, VILAPLANA, and WERTZ · New Perspectives in Theoretical and Applied Statistics
PUTERMAN · Markov Decision Processes: Discrete Stochastic Dynamic Programming
*RAO · Linear Statistical Inference and Its Applications, *Second Edition*
RENCHER · Linear Models in Statistics
RENCHER · Methods of Multivariate Analysis, *Second Edition*
RENCHER · Multivariate Statistical Inference with Applications
RIPLEY · Spatial Statistics
RIPLEY · Stochastic Simulation
ROBINSON · Practical Strategies for Experimenting
ROHATGI and SALEH · An Introduction to Probability and Statistics, *Second Edition*
ROLSKI, SCHMIDLI, SCHMIDT, and TEUGELS · Stochastic Processes for Insurance and Finance
ROSENBERGER and LACHIN · Randomization in Clinical Trials: Theory and Practice
ROSS · Introduction to Probability and Statistics for Engineers and Scientists
ROUSSEEUW and LEROY · Robust Regression and Outlier Detection
RUBIN · Multiple Imputation for Nonresponse in Surveys
RUBINSTEIN · Simulation and the Monte Carlo Method
RUBINSTEIN and MELAMED · Modern Simulation and Modeling
RYAN · Modern Regression Methods
RYAN · Statistical Methods for Quality Improvement, *Second Edition*
SALTELLI, CHAN, and SCOTT (editors) · Sensitivity Analysis
*SCHEFFE · The Analysis of Variance

*Now available in a lower priced paperback edition in the Wiley Classics Library.

SCHIMEK · Smoothing and Regression: Approaches, Computation, and Application
SCHOTT · Matrix Analysis for Statistics
SCHUSS · Theory and Applications of Stochastic Differential Equations
SCOTT · Multivariate Density Estimation: Theory, Practice, and Visualization
*SEARLE · Linear Models
SEARLE · Linear Models for Unbalanced Data
SEARLE · Matrix Algebra Useful for Statistics
SEARLE, CASELLA, and McCULLOCH · Variance Components
SEARLE and WILLETT · Matrix Algebra for Applied Economics
SEBER and LEE · Linear Regression Analysis, *Second Edition*
SEBER · Multivariate Observations
SEBER and WILD · Nonlinear Regression
SENNOTT · Stochastic Dynamic Programming and the Control of Queueing Systems
*SERFLING · Approximation Theorems of Mathematical Statistics
SHAFER and VOVK · Probability and Finance: It's Only a Game!
SMALL and McLEISH · Hilbert Space Methods in Probability and Statistical Inference
SRIVASTAVA · Methods of Multivariate Statistics
STAPLETON · Linear Statistical Models
STAUDTE and SHEATHER · Robust Estimation and Testing
STOYAN, KENDALL, and MECKE · Stochastic Geometry and Its Applications, *Second Edition*
STOYAN and STOYAN · Fractals, Random Shapes and Point Fields: Methods of Geometrical Statistics
STYAN · The Collected Papers of T. W. Anderson: 1943–1985
SUTTON, ABRAMS, JONES, SHELDON, and SONG · Methods for Meta-Analysis in Medical Research
TANAKA · Time Series Analysis: Nonstationary and Noninvertible Distribution Theory
THOMPSON · Empirical Model Building
THOMPSON · Sampling, *Second Edition*
THOMPSON · Simulation: A Modeler's Approach
THOMPSON and SEBER · Adaptive Sampling
THOMPSON, WILLIAMS, and FINDLAY · Models for Investors in Real World Markets
TIAO, BISGAARD, HILL, PEÑA, and STIGLER (editors) · Box on Quality and Discovery: with Design, Control, and Robustness
TIERNEY · LISP-STAT: An Object-Oriented Environment for Statistical Computing and Dynamic Graphics
TSAY · Analysis of Financial Time Series
UPTON and FINGLETON · Spatial Data Analysis by Example, Volume II: Categorical and Directional Data
VAN BELLE · Statistical Rules of Thumb
VIDAKOVIC · Statistical Modeling by Wavelets
WEISBERG · Applied Linear Regression, *Second Edition*
WELSH · Aspects of Statistical Inference
WESTFALL and YOUNG · Resampling-Based Multiple Testing: Examples and Methods for *p*-Value Adjustment
WHITTAKER · Graphical Models in Applied Multivariate Statistics
WINKER · Optimization Heuristics in Economics: Applications of Threshold Accepting
WONNACOTT and WONNACOTT · Econometrics, *Second Edition*
WOODING · Planning Pharmaceutical Clinical Trials: Basic Statistical Principles
WOOLSON and CLARKE · Statistical Methods for the Analysis of Biomedical Data, *Second Edition*
WU and HAMADA · Experiments: Planning, Analysis, and Parameter Design Optimization
YANG · The Construction Theory of Denumerable Markov Processes
*ZELLNER · An Introduction to Bayesian Inference in Econometrics
ZHOU, OBUCHOWSKI, and McCLISH · Statistical Methods in Diagnostic Medicine

*Now available in a lower priced paperback edition in the Wiley Classics Library.